International Nuclear Science and Technology Conference (INST 2016)

AA006097

Journal of Physics: Conference Series Volume 860

Bangkok, Thailand
4 – 6 August 2016

ISBN: 978-1-5108-4247-2
ISSN: 1742-6588

Printed from e-media with permission by:

Curran Associates, Inc.
57 Morehouse Lane
Red Hook, NY 12571

CURRAN ASSOCIATES INC. proceedings.com

Some format issues inherent in the e-media version may also appear in this print version.

Printed by Curran Associates, Inc. (2017)

Institute of Physics
Dirac House, Temple Back
Bristol BS1 6BE UK

Phone: 44 1 17 929 7481
Fax: 44 1 17 920 0979

techtracking@iop.org

Additional copies of this publication are available from:

Curran Associates, Inc.
57 Morehouse Lane
Red Hook, NY 12571 USA
Phone: 845-758-0400
Fax: 845-758-2633
Email: curran@proceedings.com
Web: www.proceedings.com

Table of contents

Volume 860

International Nuclear Science and Technology Conference 2016
4–6 August 2016, Bangkok, Thailand

Accepted papers received: 30 May 2017
Published online: 8 June 2017

Preface

Agriculture and food applications

Environmental applications

Radiation processing and industrial applications

Instrumentation and other related topics

Medical and nutritional applications

Nuclear physics and engineering

Nuclear and radiation safety

IOP Conf. Series: Journal of Physics: Conf. Series **860** (2017) 011001 doi:10.1088/1742-6596/860/1/011001

About Thailand Institute of Nuclear Technology (Public Organization)

Established in 21 April 2006, Thailand Institute of Nuclear Technology (Public Organization) or TINT is an independent organization separated from Office of Atoms for Peace to carry out research and development and to promote the utilization of nuclear technology. TINT also provides nuclear-related services and technology transfer to the public to help improve the quality of life and national development.

Vision

To be a leading nuclear solution-based research institute for Nation

Mission

1. Carry out the research and development on nuclear science and technology for sustainable development of the country
2. Transfer technology and provide consultancy services regarding the utilization of nuclear technology for socio-economic and environmental development
3. Administer and operate the research reactor and other nuclear facilities, and provide nuclear technology and nuclear safety services to the public
4. Promote a nuclear network and cooperate with organizations and research institutes, both domestic and international
5. Disseminate and build up public acceptance on the utilization of nuclear science and technology for national development

Strategy

1. Focus on research and development projects which meet the socio-economic and environmental benefits
2. Promote cooperation networks with domestic and international organizations
3. Develop the efficiency and quality of services of the nuclear science and technology, and the nuclear safety
4. Promote the cooperation network for communication, public relations, and knowledge dissemination to build up understanding and acceptance from all stakeholders and the public
5. Make use of advanced and efficient information technology for administration of databases and knowledge, and of technology transfer to the public
6. Develop and organizational management system with flexibility for efficient and effective cooperation

IOP Conf. Series: Journal of Physics: Conf. Series **860** (2017) 011001 doi:10.1088/1742-6596/860/1/011001

Research and Development

Research on nuclear science and technology

- Medical and public health research
- Biotechnology and Agriculture research
- Materials science and industry research
- Environment and safety research
- Nuclear physics and engineering research

Nuclear Technology Services

- Nuclear analysis, test and measurement services
- Radioisotope products services
- Irradiation services (foods, agricultural produces and medical devices, etc.)
- Gems irradiation services
- Waste management services
- Production, calibration, and maintenance of survey meters
- Nuclear technology transfer

Thailand Institute of Nuclear Technology (Public Organization)
9/9 Moo7, Sai moon, Ongkharak, Nakorn Nayok, 26120
Tel. +66 3739 2901 (to 6) Fax. +66 3739 2913 www.tint.or.th

IOP Conf. Series: Journal of Physics: Conf. Series **860** (2017) 011001 doi:10.1088/1742-6596/860/1/011001

Welcome address

Nikom Prasertchiewchan
Chairperson, INST 2016

Dear colleagues and friends,

On behalf of the organizing committee and Thailand Institute of Nuclear Technology, I would like to welcome you to International Nuclear Science and Technology Conference 2016 (INST2016) in Bangkok, Thailand, from 4-6 August 2016.

INST2016 is the second international conference in its series that covers topics in nuclear science, technology and applications. It serves a venue where colleagues from all breadths of nuclear and radiation research gather together, present and share their work at the conference.

This year, the organizing committee has made a goal to promote the conference theme "Nuclear for Better Health and Environment". Our comprehensive program has brought together over 400 participants from 14 countries and 2 international organizations. The program includes 12 technical sessions with 39 oral presentations and 94 poster presentations. It is complemented by extensive exhibitions. In addition, the panel discussion on August 6th will address how accelerators can contribute to sustainable development of many countries.

Through the conference, the organizing committee hopes that we are successful in providing a stage where new ideas are generated, innovation created, knowledge discussed and exchanged, friends made and strengthened, and collaboration initiated.

The organizing committee thanks Ministry of Science and Technology of Thailand, Office of Atoms for Peace, Synchrotron Light Research Institute, Nuclear Society of Thailand, Chulalongkorn University, Kasetsart University, Khon Kaen University and Thammasat University for co-hosting this event. We also thank Electricity Generating of Thailand, Pondpol Analytical Co., Ltd., Thai Sterilization Service, Co., Ltd., and Synergy Health (Thailand) Ltd., for sponsoring this conference.

We look forward to your participation and welcoming you to Bangkok.
With warmest regards,

Nikom Prasertchiewchan
Chair, INST2016
Manager, Radioactive Waste Management Center
Thailand Institute of Nuclear Technology

IOP Conf. Series: Journal of Physics: Conf. Series **860** (2017) 011001 doi:10.1088/1742-6596/860/1/011001

Welcome address

Somporn Chongkum
Editor-in-Chief, JPCS- INST 2016 Proceedings

It has been my great privilege to serve as the Co-Editor-in-Chief of the Journal of Physics Conference Series (JPCS), proceedings publication of International Nuclear Science and Technology 2016 (INST2016). INST2016 was the second of the INST conference series evolving from a notable national conference series on nuclear science and technology. INST2016 which was held in August 2016, Bangkok, Thailand had achieved fruitful outcomes. We would like to thank Thailand Institute of Nuclear Technology (TINT) for hosting this meeting, especially to Mr. Nikom Prasertchiewcharn, the organizing committee chairperson, and Dr. Phiriyatorn Suwanmala, the academic committee chairperson. We are also grateful to the members of the editorial board and the reviewers for their effort and dedication to bring the review process to success and ensure the quality of the research papers.

An increase of the number of articles accepted for publication from 31 (INST2014) to 45 (INST2016) indicates a growing interest among researchers to publish their work in the INST proceedings. The INST2016 proceedings publication has covered a mixed variety of articles in nuclear science and technology by authors from ASEAN and other countries including Japan, Korea and Sweden.

We hope to be able to initiate gradual changes in the near future for a more successful meeting and more importantly, for a higher level of proceedings publication.

We hope the members shall meet again and that the next meeting of INST series (INST2018) which will be held in Bangkok, Thailand in 2018 will also be fruitful and successful.

Thank you.

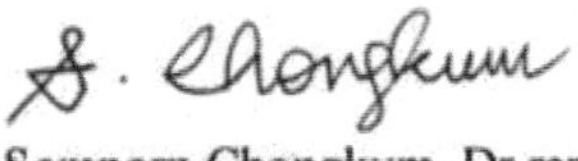

Somporn Chongkum, Dr.rer.nat.
Editor-in-Chief, JPCS – INST2016 Proceedings

International Nuclear Science and Technology Conference 2016 IOP Publishing
IOP Conf. Series: Journal of Physics: Conf. Series **860** (2017) 011001 doi:10.1088/1742-6596/860/1/011001

Keynote Speaker:

IAEA
International Atomic Energy Agency

'Roles of Nuclear Technology and IAEA Contribution to Health and Environmental Issues in Developing Countries'

Yukiya Amano
Director General, the International Atomic Energy Agency (IAEA)

Yukiya Amano is Director General of the International Atomic Energy Agency. The IAEA, an intergovernmental organization based in Vienna, is the global centre for cooperation in nuclear applications, energy, science and technology. Established in 1957, the Agency works with its Member States and partners to promote safe, secure and peaceful nuclear technologies and prevent the proliferation of nuclear weapons.

Career Summary

August 2005 - August 2009: Permanent Representative and Ambassador Extraordinary and Plenipotentiary of Japan to the International Organizations in Vienna and Governor to the IAEA

April 2007: Chairman of the First Session of the Preparatory Committee for the 2010 NPT Review Conference

September 2005 - September 2006: Chairman of the Board of Governors of the IAEA

August 2004: Ambassador, Director-General for Disarmament, Non-Proliferation and Science Department, Japanese Ministry of Foreign Affairs

Publications

- A Japanese View on Nuclear Disarmament (The Non-Proliferation Review, 2002)
- The Significance of the NPT Extension (Future Restraints on Arms Proliferation, 1996)

IOP Conf. Series: Journal of Physics: Conf. Series **860** (2017) 011001 doi:10.1088/1742-6596/860/1/011001

- La Non Proliferation Nucleaire en Exteme-Orient (Proliferation et Non-Proliferation Nucleaire, 1995)
- Sea Dumping of Liquid Radio Active Waste by Russia (Gaiko Jiho, 1994)

Education and Academic Experience

2001-2002:	Visiting Scholar, Monterey Institute of International Studies, USA
2001:	Fellow, Weatherhead Center for International Affairs, Harvard University, USA
2000-2001:	Lecturer, International Politics, Sophia University, Japan
1991-1992:	Lecturer, International Politics, Yamanashi University, Japan
1988 - 1990:	Director for Research Coordination and Senior Research Fellow, Japan Institute of International Affairs, Japan
1974-1975:	Studies at the University of Nice, France
1973-1974:	Studies at University of Besancon, France
1972:	Graduated from Faculty of Law, University of Tokyo, Japan

IOP Conf. Series: Journal of Physics: Conf. Series **860** (2017) 011001 doi:10.1088/1742-6596/860/1/011001

Conference

Nuclear technology has played an important role in many aspects of our lives, including agriculture, medicine and healthcare, materials, environment, forensics, energy, and frontier advancement. The International Nuclear Science and Technology Conference (INST) aims to bring together scientists, engineers, academics and students to share knowledge and experiences about all aspects of nuclear sciences.

INST2016 was the second of the INST conference series organized by Thailand Institute of Nuclear Technology. INST has evolved from a national conference series on nuclear science and technology that was held every two years in Bangkok for over a twenty-year period.

INST2016 was held from 4 - 6 August 2016 in Bangkok, Thailand, under the central theme "Nuclear for Better Life". The conference working language was English. The oral and poster research presentations covered seven major topics:

- Nuclear physics and engineering (PHY)
- Nuclear and radiation safety (SAF)
- Medical and nutritional applications (MED)
- Environmental applications (ENV)
- Radiation processing and industrial applications (IND)
- Agriculture and food applications (AGR)
- Instrumentation and other related topics (INS)

Keynote Speaker:

Yukiya Amano, Director General, International Atomic Energy Agency (IAEA)

Invited Speakers:

- Emanuele Tsesmelis, European Organization for Nuclear Research (CERN)
- Nguyen Tuan Khai, Vietnam Atomic Energy Institute (VINATOM), Vietnam
- Monique Lacroix, Institut National de La Recherche Scientifique, Canada
- Noriaki Seko, Japan Atomic Energy Agency, Japan
- Dheerawan Boonyawan, Chiang Mai University, Thailand
- Teerachat Sae-heng, Thammasat University, Thailand
- Waranont Anukool, Chiang Mai University, Thailand
- Dhanaj Saengchantr, Thailand Institute of Nuclear Technology, Thailand
- Edy Giri Rachman Putra, National Nuclear Energy Agency of Indonesia (BATAN), Indonesia
- Naruhiro Matsufuji, National Institute of Radiological Sciences, Japan
- Jong-Seo Chai, Sungkyunkwan University, Republic of Korea

IOP Conf. Series: Journal of Physics: Conf. Series **860** (2017) 011001 doi:10.1088/1742-6596/860/1/011001

Panel Discussion:

- Jong-Seo Chai, Sungkyunkwan University, Republic of Korea
- Emanuele Tsesmelis, European Organization for Nuclear Research (CERN)
- Naruhiro Matsufuji, National Institute of Radiological Sciences, Japan
- Ananya Ruangma, Wattanosoth Cancer Hospital, Thailand
- Pornthep Nisamaneepong, Thailand Institute of Nuclear Technology, Thailand
- Thiansin Liamsuwan, Thailand Institute of Nuclear Technology, Thailand

IOP Conf. Series: Journal of Physics: Conf. Series **860** (2017) 011001 doi:10.1088/1742-6596/860/1/011001

Committee

INST2016 was organized by Thailand Institute of Nuclear Technology (TINT).

Honorable Chairs

- Pornthep Nisamaneephong (Executive Director)
- Hannarong Shamsab (Deputy Director)
- Nipavan Poramatikul (Deputy Director)

Organizing Committee

- Nikom Prasertchiewcharn (Chairperson)
- Phiriyatorn Suwanmala (Vice Chair)
- Kanokporn Boonsirichai (Secretary)
- Dhanaj Saengchantr
- Panita Ruenbanthoeng
- Kanchalika Dechates
- Watin Chinangkoonphiwat
- Nongnuch Jangsawang
- Wannipa Phianphak
- Roppon Picha
- Tippanan Ngamprayad
- Voraluck Lertanuntrakul
- Wanwisa Prasartthammaporn
- Jaruratana Eamsiri
- Dussadee Rattanapra
- Kanokrat Tiyapun
- Saowaluck Phuengyat
- Natthamon Nunchukan

Academic Committee

- Phiriyatorn Suwanmala (Chairperson)
- Sasiphan Khaweerat (Secretary)
- Moleephan Dangprasert
- Boonsom Porntepkasemsan
- Kanokrat Tiyapun
- Kasinee Hemvichian
- Dussadee Rattanapra
- Roppon Picha
- Saensuk Wetchagarun
- Sarinrat Wonglee

IOP Conf. Series: Journal of Physics: Conf. Series **860** (2017) 011001 doi:10.1088/1742-6596/860/1/011001

IOP Conf. Series: Journal of Physics: Conf. Series **860** (2017) 011001 doi:10.1088/1742-6596/860/1/011001

Program

International Nuclear Science and Technology Conference (INST 2016)
4-6 Aug 2016, Bangkok
Tentative Program

Day 1: Thursday, 4th August 2016

Time	Topic
7:15-8:00	Registration and coffee break (participants ready in the hall by 8:00)
9:00-10:00	Opening ceremony • Arrival of HRH Princess Maha Chakri Sirindhorn • Report by Minister of Science and Technology • Presentation of TINT Invention Awards (College Level) and Young Nuclear Scientist Award • Royal Opening Address of HRH Princess Maha Chakri Sirindhorn • Photo session
10:00-10:40	Keynote speaker: Yukiya Amano (Director General of the International Atomic Energy Agency (IAEA)) *Roles of Nuclear Technology and IAEA Contribution to Health and Environmental Issues in Developing Countries*
10:40-12:00	Exhibition session
12:00-13:30	Lunch break
Session I: PHY (Chair: Rattachat Mongkolnavin, Physics CU, Thailand)	
13:30-14:00	Invited speaker PHY15: Emmanuele Tsesmelis (CERN, Switzerland) *CERN - A Gateway to Science and Technology*
14:00-14:30	Oral presentation PHY06: Noritaka Yusa (Department of Quantum Science and Energy Engineering, Tohoku University, Japan) *Probability of Detection Model for the Non-Destructive Inspection of Steam Generator Tubes of PWRs*
14:30-15:50	Poster session I and coffee break (1 h 20 min)
Session II: SAF (Chair: Pantip Ampornrat, OAP, Thailand)	
15:50-16:20	Invited speaker SAF15: Nguyen Tuan Khai (VINATOM, Vietnam) *Assessment for Released Source Term and Consequently Exclusive Area Boundary (EAB) and Precautionary Action Zone (PAZ) in Scenario of Severe Accidents at Ninh Thuan 1 nuclear power plant*
16:20-16:40	Oral presentation SAF14: Kampanart Silva (TINT, Thailand) *Establishment of ASEAN Network for Nuclear Power Plant Safety Research*
16:40-17:00	Oral presentation SAF08: Pawitra Aim-O (Department of Nuclear Engineering, Chulalongkorn University, Thailand) *Monte Carlo Simulation of Innovative Neutron and Photon Shielding Material Composing of High Density Concrete, Waste Rubber, Lead and Boron Carbide*

IOP Conf. Series: Journal of Physics: Conf. Series **860** (2017) 011001 doi:10.1088/1742-6596/860/1/011001

Day 2: Friday, 5th August 2016 (2 parallel sessions)

	Session III: AGR (Chair: Panatsada Akikunprasert, KKU, Thailand)	Session IV: ENV (Chair: Pannee Pakkong, KU, Thailand)
9:00-9:30	Invited speaker AGR15: Monique Lacroix (INRS, Canada) *The Use of Suitable Packaging and Irradiation Process to Assure Food Safety*	Oral presentation ENV09: Ryuta Hazama (Faculty of Human Environment, Osaka Sangyo University, Japan) *Recent Tritium Levels in Tap Waters Collected at the Eastern part of China*
9:30-9:50	Oral presentation AGR06: Syukria Kurniawati (BATAN, Indonesia) *Heavy Metals Cr, Co, Hg and Zn in Foods from the Traditional Market in Jakarta*	Oral presentation ENV08: Prasong Kessaratikoon (Thaksin University, Songkhla, Thailand), *Specific Activities and Radiological Hazard Assessment in Beach Sand Samples in Songkhla Province, Thailand after Fukushima Dai-Ichi Nuclear Power Plant Accident in Japan*
9:50-10:10	Oral presentation AGR02: Kanchana Chahorm (TINT, Thailand) *Improvement of Microbiological Qualities of Namphrik by Gamma Irradiation*	Oral presentation ENV06: Natalia Adventini (Center for Applied Nuclear Science and Technology, Jawa Barat, Indonesia) *Pb Identification in Soil surrounding a Used Lead Acid Battery Smelter in West Java*
10:10-10:30		Oral presentation ENV07: Piyawan Krisanangkura (OAP, Thailand) *Fast and efficient uranium(VI) removal from simulated wastewater with and without water hardness by easily collected graphene oxide sponge*
10:30-10:40	Coffee break	
	Session V: Biodosimetry (MED/SAF) (Chair: Sirirat Biramontri, OAP, Thailand)	Session VI: IND (Chair: Sunchai Nilsuwankosit, Nuclear Eng. CU, Thailand)
10:40-11:10	Oral presentation SAF02: Issariya Chairam (OAP, Thailand) *Thailand Biological Dosimetry Network for Emergency Preparedness and Response*	Invited speaker IND17: Noriaki Seko (JAEA, Japan) *Radiation Processing of Polymer and Its Environmental Applications*
11:10-11:30	Oral presentation SAF01: Benchawan Rungsimaporn (Ramathibodi Hospital, Thailand) *Preparedness for Radiological Emergency in Thailand: Establishment of Dose-Response Curves for Dicentrics and PCC Rings*	Oral presentation IND09: Watcharee Katinonkul (Department of Science Service, Ministry of Science and Technology, Thailand) *Effect of Gamma Irradiation Combined with Ionic Liquid Pretreatment of Corn Shell on Enzymatic Digestibility*
11:30-11:50	Oral presentation MED15: Rujira Wanotayan (Mahidol University, Thailand) *Evaluation of XRCC4 extremely C-terminal (XECT) domain in DNA double-strand break repair of radiation-induced DNA damage*	Oral presentation IND04: Myo Chit Aung (Mandalay Technological University, Myanmar) *Laboratory Inspection of Single-Pass Tray Column by Gamma Column Scanning Technique*

IOP Conf. Series: Journal of Physics: Conf. Series **860** (2017) 011001 doi:10.1088/1742-6596/860/1/011001

11:50-12:10	Oral presentation MED08: Pimpon Uttayarat (TINT, Thailand) *Development of γ-H2AX Assay as a Biomarker in Response to Ionizing Radiation*	
12:10-13:40	Lunch break	
	Session VII: PHY/INS/IND (Chair: Edy Giri Rachman Putra (BATAN, Indonesia)	Session VIII: MED (Chair: Thiansin Liamsuwan, TINT, Thailand)
13:40-14:40	Invited speaker PHY17: Dheerawan Boonyawan (Plasma & Beam Physics Research, Chiang Mai University, Thailand) *Innovative Research of Plasma Physics for Life Sciences*	Invited speaker MED11: Teerachat Sae-heng (Thammasat University, Thailand) *The Mathematical Model of P53 Regulatory Circuit Manipulation in Cancer Therapy*
	Invited speaker PHY08: Waranont Anukool (Department of Physics and Materials Science, Chiang Mai University, Thailand) *CMU Cold-atom Facility and Application on Various Scientific Disciplines*	Oral presentation MED01: Irena Gudowska (Department of Physics, Stockholm University, Sweden) *Out-of- field dose in proton spot scanning beams*
		Oral presentation MED20: Siritorn Buranurak (Khon Kaen University, Thailand) *Development and commissioning of a new technique of total body irradiation treatment using real-time in vivo dosimetry with fiber-coupled Al2O3:C luminescence detectors*
14:40-15:00	Oral presentation IND20: Dhanaj Saengchantr (Thailand Institute of Nuclear Technology, Thailand) *Role of Radioactive Material Applications for Industries in Thailand*	Oral presentation MED05: Narongchai Autsavapromporn (Division of Therapeutic Radiology and Oncology, Department of Radiology, Faculty of Medicine, Chiang Mai University, Thailand) *Late Effects in the Progeny of Bystander Human Cells after Carbon Ions are Dependent on Radiation Quality: The Relevance to Cancer Risk*
15:00-16:20	Poster session II and coffee break (1 h 20 min)	
	Session IX: INS (Chair: Manit Sonsuk, Nuclear Society of Thailand)	Session X: MED (Chair: Ananya Ruangma, Wattanosoth Cancer Hospital, Thailand)
16:20-16:50	Invited speaker INS18: Edy Giri Rachman Putra (BATAN, Indonesia) *Neutron Scattering Facilities at G. A. Siwabessy Reactor Indonesia for Materials Science and Biology Research*	Invited speaker X: MED22: Pleumchitt Rojanapanthu (Thammasat University, Thailand) *Drug Discovery and Development Center*
16:50-17:10	Oral presentation INS12: Kiadtisak Saenboonruang (Kasetsart University, Thailand) *The Current Status of a Gas Electron*	Oral presentation MED13: Thititip Tippayamontri (Université de Sherbrooke, Canada) *Targeting the DNA of cancer cells in*

IOP Conf. Series: Journal of Physics: Conf. Series **860** (2017) 011001 doi:10.1088/1742-6596/860/1/011001

	Multiplier (GEM) Research at Kasetsart University	*chemoradiation therapy: Intratumoral administration of 5FU chemotherapeutic drug and 18 F-FLT positron source in a colorectal cancer mouse model*
17:10-17:30	Oral presentation INS08: Mohd M. Sabri (Malaysian Nuclear Agency, Selangor, Malaysia) *Adaptive and Self-Tuning Control of the TRIGA Mark II Reactor*	
18:00	Reception + Dinner talk (Kasidete Teeranitayatarn)	

Day 3: Saturday, 6th August 2016

Session XI: PHY/SAF/IND (Chair: Thawatchai Onjun, SIIT, Thailand)	
9:00-9:20	Oral presentation IND21: Thein Zaw Win (Mandalay Technological University, Myanmar) *Isolation and Identification of Chitin from Prawn Shell Wastes*
9:20-9:40	Oral presentation SAF03: Kasama Durongsak (Department of Nuclear Engineering, Chulalongkorn University, Thailand) *Development of remaining wall thickness measurement system for boiler wall tube using gamma scattering technique*
9:40-10:00	Oral presentation SAF06: Nanthavan Yaanant (TINT, Thailand) *Study on Developing Safety Infrastructure for Mineral Processing Waste (NORM Waste) and Contamination Monitoring at the TINT Rare Earth Research & Development Center, Klong 5, Pathumthani*
10:00-10:20	Oral presentation PHY10: Sasiphan Khaweerat (TINT, Thailand) *Development of a Neutron Tomography System in Thailand*
10:20-10:30	Coffee break
Session XII (Chair: Prapong Klysuban, SLRI, Thailand)	
10:30-11:00	Oral presentation of Young Nuclear Scientist Award Winner Wanvimol Pasanphan (Kasetsart University, Thailand) *Light stabilizer-conjugated-stearylate chitosan nanoparticles: A bio-based additive for free radical stabilization of healthcare plastics under irradiation*
11:00-11:30	Invited speaker MED18: Naruhiro Matsufuji (NIRS, Japan) (MED) *Overview summary of clinical heavier-ion progress in Japan*
11:30-12:00	Invited speaker INS19: Jong-Seo Chai (SKKU, South Korea) (INS) *Cyclotron Development in Korea*
12:00-13:30	Lunch break
13:30-15:00	Panel discussion: How can accelerators contribute to sustainable development of countries? 1. Jong-Seo Chai (SKKU, Korea) 2. Emmanuel Tsesmelis, (CERN, Switzerland) 3. Naruhiro Matsufuji (NIRS, Japan) 4. Ananya Ruangma (Wattanosoth Cancer Hospital, Thailand) 5. Pornthep Nisamaneepong (TINT, Thailand) 6. Thiansin Liamsuwan (TINT, Thailand), Moderator
15:00	Announcement of Oral and Poster Presentation Awards
15:10	Summary and closing ceremony

IOP Conf. Series: Journal of Physics: Conf. Series **860** (2017) 011001 doi:10.1088/1742-6596/860/1/011001

Presentations

Section 1: Invited and oral presentations

Code	Speaker	Organization	Title
AGR02	Kanchana Chahorm	Thailand Institute of Nuclear Technology (Public Organization)	Improvement of microbiological qualities of namphrik by gamma irradiation.
AGR06	Syukria Kurniawati	National Nuclear Energy Agency (BATAN)	Heavy metals Cr, Co, Hg and Zn in foods from the traditional market in Jakarta
AGR15	Monique Lacroix	Institut National de La Recherche Scientifique	The use of suitable packaging and irradiation process to assure food safety
ENV06	Natalia Adventini	National Nuclear Energy Agency (BATAN)	Pb identification in soil surrounding a used lead acid battery smelter in west Java, Indonesia
ENV07	Piyawan Krisanangkura	Office of Atoms for Peace	Fast and efficient uranium (VI) removal from simulated wastewater with and without water hardness by easily collected graphene oxide sponge
ENV08	Prasong Kessaratikoon	Thaksin University	Specific activities and radiological hazard assessment in beach sand samples in Songkhla province, Thailand after Fukushima Dai-Ichi nuclear power plant accident in Japan
ENV09	Ryuta Hazama	Osaka Sangyo University	Recent tritium levels in tap waters collected at the eastern part of China
IND02	Dherendra Gihwala	Cape Peninsula University of Technology	The use of PIXE and SEM to determine the correlation of the agglomeration of metal oxides with the mechanical properties of nano-composites containing TiO2 particles
IND04	Myo Chit Aung	Mandalay Technological University	Laboratory inspection of single-pass tray experimental model column by gamma column scanning technique
IND09	Watcharee Katinonkul	Department of Science Service, Ministry of Science and Technology, Thailand	Effect of gamma irradiation combined with ionic liquid pretreatment of corn shell on enzymatic digestibility
IND17	Noriaki Seko	JAEA	Radiation processing for environmental applications
IND20	Dhanaj Saengchantr	Thailand Institute of Nuclear Technology (Public Organization)	Role of radioactive material applications for industries in Thailand
IND21	Thein Zaw Win	Mandalay Technological University	Isolation and identification of chitin from prawn shell wastes

IOP Conf. Series: Journal of Physics: Conf. Series **860** (2017) 011001 doi:10.1088/1742-6596/860/1/011001

Code	Speaker	Organization	Title
INS08	Mohd Sabri Bin Minhat	Malaysian Nuclear Agency	Adaptive and self-tuning control of the TRIGA Mark II reactor
INS12	Kiadtisak Saenboonruang	Kasetsart University	The current status of a gas electron multiplier (GEM) research at Kasetsart university, Thailand
INS18	Edy Giri Rachman Putra	National Nuclear Energy Agency (BATAN)	Neutron scattering facilities at G. A. Siwabessy reactor Indonesia for materials science and biology research
INS19	Jong-Seo Chai	Sungkyunkwan University	Cyclotron development in Korea
MED01	Irena Gudowska	Stockholm University	Out-of-field dose in proton spot scanning beams
MED05	Narongchai Autsavapromporn	Chiang Mai University	Late effects in the progeny of bystander human cells after carbon ions are dependent on radiation quality: the relevance to cancer risk
MED08	Pimpon Uttayarat	Thailand Institute of Nuclear Technology (Public Organization)	Development of g-H2AX assay as a biomarker in response to ionizing radiation
MED11	Teerachat Saeheng	Nagasaki University	The mathematical model of P53 regulatory circuit manipulation in cancer therapy
MED13	Thititip Tippayamontri	University of Sherbrooke	Targeting the DNA of cancer cells in chemoradiation therapy: intratumoral administration of 5FU chemotherapeutic drug and 18F-FLT positron source in a colorectal cancer mouse model
MED15	Rujira Wanotayan	Mahidol University	Evaluation of XRCC4 extremely C-terminal (XECT) domain in DNA double-strand break repair of radiation-induced DNA damage
MED18	Naruhiro Matsufuji	National Institute of Radiological Sciences, Japan	Overview summary of clinical heavier-ion progress in Japan
MED20	Siritorn Buranurak	Khon Kaen University	Development and commissioning of a new technique of total body irradiation treatment using real-time in vivo dosimetry with fiber-coupled Al2O3:C luminescence detectors
MED22	Pleumchitt Rojanapanthu	Thammasat University	Drug discovery and development center
PHY06	Noritaka Yusa	Umaru Musa Yar'adua University	Probability of detection model for the non-destructive inspection of steam generator tubes of PWRs
PHY08	Woranont Anukool	Chiang Mai University	CMU cold-atom facility and application on various scientific disciplines
PHY10	Sasiphan Khaweerat	Thailand Institute of Nuclear Technology (Public Organization)	Development of a neutron tomography system in Thailand

IOP Conf. Series: Journal of Physics: Conf. Series **860** (2017) 011001 doi:10.1088/1742-6596/860/1/011001

Code	Speaker	Organization	Title
PHY15	Emmanuel Tsesmelis	CERN	CERN - A gateway to science and technology
PHY17	Dheerawan Boonyawan	Chiang Mai University	Innovative research of plasma physics for life sciences
SAF01	Benchawan Rungsimaphorn	KasetsartUniversity Mahidol University	Preparedness for radiological emergency in Thailand: establishment of dose-response curves for dicentrics and PCC rings
SAF02	Issariya Chairam	Office of Atoms for Peace	Thailand biological dosimetry network for emergency preparedness and response
SAF03	Kasama Durongsak	Chulalongkorn University	Development of remaining wall thickness measurement system for boiler wall tube using gamma scattering technique
SAF06	Nanthavan Yaanant	Thailand Institute of Nuclear Technology (Public Organization)	Study on developing safety infrastructure for mineral processing waste (NORM Waste) and contamination monitoring at the TINT rare earth research & development center, Klong 5, Pathumthani.
SAF08	Pawitra Aim-O	Chulalongkorn University Synchrotron Light Research Institute	Monte Carlo simulation of innovative neutron and photon shielding material composing of high density concrete, waste rubber, lead and boron carbide
SAF14	Kampanart Silva	Thailand Institute of Nuclear Technology (Public Organization)	Establishment of ASEAN network for nuclear power plant safety research
SAF15	Nguyen Tuan Khai	Institute for Nuclear Science and Technology, VINATOM	Assessment for released source term and consequently exclusive area boundary (EAB) and precautionary action zone (PAZ) in scenario of severe accidents at Ninh Thuan 1 nuclear power plant

Section 2: Poster presentations

Code	Presenter	Organization	Title
AGR01	Chutima Kranrod	Chulalongkorn University	Preliminary survey of radioactivity level in Thai medicinal herb plants
AGR03	Panchalee Prakhongsil	Thailand Institute of Nuclear Technology (Public Organization)	Effects of gamma and electron beam irradiation on reduction of microbial load and antioxidant properties of Cassia alata (L.) Roxb.
AGR04	Supalak Kongsri	Thailand Institute of Nuclear Technology (Public Organization)	Instrumental neutron activation analysis for elemental composition in Thai rice samples
AGR05	Saovapong Charoen	Thailand Institute of Nuclear Technology (Public Organization)	In vitro antagonism test of irradiated Bacillus subtilis culture for suppression of rice blast fungi
AGR07	Wanitch Limohpasmanee	Thailand Institute of Nuclear Technology (Public Organization)	The control of brown planthopper (Nilaparvata lugens Stål) by the sterile insect technique
AGR08	Surasak Sajjabut	Thailand Institute of Nuclear Technology (Public Organization)	Effect of gamma irradiation on beta glucan and total water soluble polysaccharide in split gill mushroom (Shizophyllum commune)
AGR09	Sirilak Chookaew	Thailand Institute of Nuclear Technology (Public Organization)	Application of electron beam to improve the microbiological safety of Andrographis paniculata
AGR11	Huaili Qin	Nuctech Company Limited, China	The largest phytosanitary irradiation dedicated EB facility in the world
AGR12	Boonsom Porntepkasemsa n	Thailand Institute of Nuclear Technology (Public Organization)	Soil to rice transfer factors and annual effective doses assessment for 238U and 232Th
AGR13	Jaruratana Eamsiri	Thailand Institute of Nuclear Technology (Public Organization)	Efficacy of electron beam and gamma irradiation on microbial reduction and antioxidant activities of Butea superba. Roxb
AGR14	Wachiraporn Pewlong	Thailand Institute of Nuclear Technology (Public Organization)	Electron beam radiation effects on antioxidant properties and microbiological of Glycyrrhiza glaba (licorice)
ENV01	Muhayatun Santoso	National Nuclear Energy Agency of Indonesia	Comparison of XRF and PIXE techniques for elemental characterization of airborne particulate matter
ENV02	Dyah K. Sari	National Nuclear Energy Agency of Indonesia	Quality assurance assessment of EDXRF for airborne particulate matter analysis
ENV03	Indah Kusmartini	National Nuclear Energy Agency of Indonesia	Elemental characterization of Indonesian volcanic ash Sinabung mountain by neutron activation analysis

Code	Presenter	Organization	Title
ENV05	Muhayatun Santoso	National Nuclear Energy Agency of Indonesia	Characterization of airborne particulate matter samples from urban areas in Indonesia using nuclear analytical techniques
ENV10	Santi Raksawong	Prince of Songkla University	The 7Be profile in the undisturbed soil used for reference site to estimate the soil erosion
ENV11	Wannee Srinuttrakul	Thailand Institute of Nuclear Technology (Public Organization)	Determination of 2H/1H and 18O/16O in water used for rice cultivation by cavity ring-down spectroscopy
ENV12	Suprawee Siriboonprapob	Office of Atoms for Peace	A preliminary study of radon concentration in ground water from Songkhla by using a portable LSC
ENV13	Tarika Thumvijit	Chiang Mai University	Survey of indoor radon concentration at hot spring area, Doi Saket district, Chiang Mai
ENV14	Suputtra Visetpotjanakit	Office of Atoms for Peace	Participation in proficiency test for tritium strontium and cesium isotopes in seawater 2015 (IAEA-RML- 2015-02)
ENV15	Suputtra Visetpotjanakit	Office of Atoms for Peace	Prompt determination of 137Cs in large volume seawater using Cu-hexacyanoferrates cartridges
ENV16	Uthaiwan Injarean	Thailand Institute of Nuclear Technology (Public Organization)	Comparison of NAA, ICP-MS, ICP-OES and WD-XRF techniques for determination of rare earths, uranium, and other elements in geological samples
ENV17	Phachirarat Sola	Thailand Institute of Nuclear Technology (Public Organization)	Investigation of radon level in air and water of workplaces at Thailand Institute of Nuclear Technology, Thailand
ENV18	Rawiwan Kritsananuwat	Chulalongkorn University	Transfer factor of 226Ra, 232Th and 40K from soil to Alpinia Galangal plant grown in northern Thailand
ENV19	Wannee Srinuttrakul	Thailand Institute of Nuclear Technology (Public Organization)	Determination of stable cesium and strontium in rice samples by inductively coupled plasma mass spectrometry
ENV20	Wutthikrai Kulsawat	Thailand Institute of Nuclear Technology (Public Organization)	d2H and d18O in precipitation, irrigation and pond water in Suphanburi rice paddy field
ENV21	Nitipon Noipow	Thailand Institute of Nuclear Technology (Public Organization)	Potential of using tritium and stable isotope characteristics for illegal groundwater pumping investigation in Bangkok metropolitan area

IOP Conf. Series: Journal of Physics: Conf. Series **860** (2017) 011001 doi:10.1088/1742-6596/860/1/011001

Code	Presenter	Organization	Title
ENV22	Kiattipong Kamdee	Thailand Institute of Nuclear Technology (Public Organization)	Exploration and evaluation of groundwater recharge mechanism of the upper Chao Phraya Basin (Bangrakam district) by nuclear techniques
ENV23	Junghyup Lee	Korea Institute of Nuclear Safety (KINS)	Real-time underwater radiation monitoring system operation for environmental surveillance program in Korea
IND01	Apichate Maneewong	Thailand Institute of Nuclear Technology (Public Organization)	Validation of 182Ta, 54Mn and 46Sc measurements in blue topaz using gamma spectrometric analysis
IND03	Akara Akaranate	Thailand Institute of Nuclear Technology (Public Organization)	Determination of liquid level adsorbed in the thermal insulation using neutron back scattering technique
IND05	Pakorn Pakaiphuek	Thailand Institute of Nuclear Technology (Public Organization)	The switching development on the MOSFET by using 8 MeV electron irradiation
IND06	Prartana Kewsuwan	Thailand Institute of Nuclear Technology (Public Organization)	Effect of concentrations of different molecular weight irradiated chitosan on radiation synthesis of silver nanoparticles
IND08	Thamaborn Ploykrathok	Chulalongkorn University	Determining the bio-based content of bio-plastics used in Thailand by radiocarbon analysis
IND11	Phiriyatorn Suwanmala	Thailand Institute of Nuclear Technology (Public Organization)	Dye adsorbent prepared by radiation-induced graft polymerization of acrylic acid onto carboxymethyl cellulose
IND12	Anantachai Pechrak	Thailand Institute of Nuclear Technology (Public Organization)	A study of residence time distribution using radiotracer technique in the large scale plant facility
IND13	Kokiat Sukrod	Thailand Institute of Nuclear Technology (Public Organization)	Conceptual study of fan-beam computed tomography technique for on-field industrial applications
IND14	Joo Yeon Kim	Korean Association for Radiation Application	Development of an advanced system for supporting the radiation technology information in Korea
IND18	Chalermphong Polee	Chulalongkorn University	Investigation of a hallow in tree trunk using the developed gamma-ray transmission technique
IND19	Nongnut Jangsawang	Thailand Institute of Nuclear Technology (Public Organization)	Study on optical and radiological properties of aquamarine irradiated by electrons from linear electron accelerator
INS01	Anith Ghazali	Malaysian Nuclear Agency	Adaptive PI controller for a nuclear process control using MRAC approach
INS02	Apisara Charoensri	Office of Atoms for Peace	Measurement of fast neutron spectrum by using CR-39 detectors

IOP Conf. Series: Journal of Physics: Conf. Series **860** (2017) 011001 doi:10.1088/1742-6596/860/1/011001

Code	Presenter	Organization	Title
INS03	Anawat Rittirong	Kasetsart University	A calculation of efficiencies of a GEM-based neutron detector using different solid neutron converters
INS04	Endah Damastuti	National Nuclear Energy Agency (BATAN)	Assessment of NAA preparedness for elemental characterization toward certification of coal fly ash reference material candidate
INS05	Kalaya Changkrueng	Office of Atoms for Peace	Proficiency test for trace elements and uranium isotopes in water (Proficiency test IAEA-TEL-2015-01)
INS06	Kalaya Changkrueng	Office of Atoms for Peace	Column-separation for uranium-thorium in nitric acid matrix using ICP-MS determination techniques
INS07	Ladapa Srijittawa	Office of Atoms for Peace	U-235 and U-238 analysis by using gamma spectrometry technique and mathematical model for safeguards
INS09	Sahakan Monthonwattana	Thailand Institute of Nuclear Technology (Public Organization)	Gamma response characterization of optically stimulated luminescence (OSL) affects personal dosimetry
INS10	Thiansin Liamsuwan	Thailand Institute of Nuclear Technology (Public Organization)	Investigation of the response characteristics of OSL albedo neutron dosimeters in a 241 AmBe reference neutron field
INS11	Rattakarn Yensano	Khon Kaen University	The design and construction of a high volume air and fallout sampler for a radiation monitoring station
INS13	Somyot Srisatit	Chulalongkorn University	Threat liquid using X-Ray imaging technique
INS14	Woro Yatu Niken Syahfitri	National Nuclear Energy Agency (BATAN)	Macro elemental analysis food samples by nuclear analytical technique
INS15	Joo Yeon Kim	Korean Association for Radiation Application	Inverse tracking of unknown model parameters in atmospheric dispersion using a probabilistic inference
INS16	Woo Ho Shin	Korean Association for Radiation Application	The development of practice manual for LSC based on job analysis in radiation measurement and evaluation
INS17	Thiti Rungseesumran	Thailand Institute of Nuclear Technology (Public Organization)	Effect of the scattering radiation in air and two type of slap phantom between PMMA and the ISO water phantom for personal dosimeters calibration
INS20	Djoko Prakoso Dwi Atmodjo	National Nuclear Energy Agency (BATAN)	INAA application for trace element determination in biological reference material
MED06	Panatsada Awikunprasert	Khon Kaen University	Survey of radiation dose from CT examinations in Srinakarind hospital, Thailand

IOP Conf. Series: Journal of Physics: Conf. Series **860** (2017) 011001 doi:10.1088/1742-6596/860/1/011001

Code	Presenter	Organization	Title
MED07	Petcharakorn Hanpanich	Khon Kaen University	The use of Fourier transform infrared (FTIR) spectroscopy to differentiate malignant and non-malignant bile samples
MED09	Srisawan Somkit	Office of Atom for Peace	Effect to health from quantity of Polonium-210 coming from smoking cigarettes of commercialized brands in Thailand
MED10	Wiranee Sriwiang	Thailand Institute of Nuclear Technology (Public Organization)	Radiolabeling of 90Y-DOTA-[Pro1, Tyr4]-Bombesin: Optimization condition to diminish metabolite peptide
MED12	Thanete Duangta	Thailand Institute of Nuclear Technology (Public Organization)	Preparation of lyophilized Hynic-TOC kit to extend the shelf life and more convenient for the transport
MED14	Thititip Tippayamontri	University of Sherbrooke	Absolute electron scattering cross sections from DNA constituents for targeted radionuclide therapy: Condensed tetrahydrofuran (THF)
MED16	Sarinya Wongsanit	Thailand Institute of Nuclear Technology (Public Organization)	Evaluation of potential stabilizers for suppression of side-products formed during the labeling of 68Ga-DOTA-Bombesin
MED21	Wiranee Sriwiang	Thailand Institute of Nuclear Technology (Public Organization)	Synergistic of active targeting DOTANOC with passive targeting AuNPs to specific uptake somatostatin receptors: Preliminary results of conjugation and characterization process
PHY01	Boonyarith Chatthong	Prince of Songkla University	Investigations of impurity effects on toroidal flow and internal transport barrier in ITER and DEMO
PHY02	Jintana Pakdeewanich	Prince of Songkla University	Investigations of ETB and ITB formations based on bifurcation concept
PHY03	Kanokrat Tiyapun	Thailand Institute of Nuclear Technology (Public Organization)	Neutronics and thermal hydraulic analysis of TRR-1/M1 reactor using MCNPX and COOLOD-N2 computer code
PHY04	Wichian Ratanatongchai	Thailand Institute of Nuclear Technology (Public Organization)	Measurements of TRR-1/M1 prompt gamma neutron activation beam
PHY05	Vithit Pungkun	Office of Atoms for Peace	The effect of angle and size on the efficiency and energy resolution of cadmium telluride detectors
PHY07	Siriyaporn Sangaroon	Mahasarakham University	The MCNP simulation of PGNAA system at TRR-1/M1

IOP Conf. Series: Journal of Physics: Conf. Series **860** (2017) 011001 doi:10.1088/1742-6596/860/1/011001

Code	Presenter	Organization	Title
PHY09	Roppon Picha	Thailand Institute of Nuclear Technology (Public Organization)	Renovation status of Thailand Reactor neutron radiography
PHY11	Jatechan Channuie	Thailand Institute of Nuclear Technology (Public Organization)	Comparison of photon attenuation coefficients of various barite concrete formulas, ultra-high density concrete and lead by experimental data
PHY12	Nguyen Quyen	Sirindhorn International Institute of Technology	Effect of water quality to acidification caused by dielectric barrier discharge plasma
PHY13	Ahmad Usman	University of Malaya Umaru Musa Yar'adua University	Cyclotron production of 48V via nat Ti(d,x) 48V nuclear reactions; a promising radionuclide
PHY16	Wittawat Kanjanaput	Sirindhorn International Institute of Technology	Saturated neoclassical tearing modes in Tokamak plasma simulated by using modified island module
PHY20	Suksit Sangkaew	Office of Atoms for Peace	Fuel burnup analysis for Thai research reactor by using MCNPX computer code
SAF04	Kotchaphan Kanjana	Thailand Institute of Nuclear Technology (Public Organization)	Gamma-radiation-induced corrosion of aluminum alloy: Low dose effect
SAF05	Kotchaphan Kanjana	Thailand Institute of Nuclear Technology (Public Organization)	High-temperature setup for investigation of fission product behavior
SAF07	Pantip Ampornrat	Office of Atoms for Peace	Determination of degradation of core-structural materials of Thailand research reactor TRR-1/M1
SAF09	Darunee Peekhunthod	Office of Atoms for Peace	Baseline quantity of 131I, 137Cs, 134Cs and 40K in Thai urines and internal exposure dose
SAF10	Vithit Pungkun	Office of Atoms for Peace	xPU/BaSO4 and xPU/Bi2O3 composite sheet for X-ray attenuation
SAF11	Jeelawat Esor	Thailand Institute of Nuclear Technology (Public Organization)	Comparing conversion coefficients from air kerma to personal dose equivalent Hp(3) for eye lens dosimetry
SAF13	Kampanart Silva	Thailand Institute of Nuclear Technology (Public Organization)	Preliminary results of consequence assessment of a hypothetical severe accident using Thai meteorological data
SAF16	Yutthana Tumnoi	Office of Atoms for Peace	5 years after the Fukushima-Daiichi nuclear accident; has the coastal & marine environment of Thailand been radioactively effected?

IOP Conf. Series: Journal of Physics: Conf. Series **860** (2017) 011001 doi:10.1088/1742-6596/860/1/011001

Code	Presenter	Organization	Title
SAF17	Anan O-Manee	Thailand Institute of Nuclear Technology (Public Organization)	Determination of distribution coefficient of non-radioactive cesium (Cs) between an aqueous Solution of very low Cs concentration and pure materials, such as SiO2 and bentonite (Kunipia F)
SAF18	Jittima Bangvirunrak	Office of Atoms for Peace	Internal radiation exposure due to intake I-131 for nuclear medicine workers in Thailand

Section 3: Selected articles

IOP Conf. Series: Journal of Physics: Conf. Series **860** (2017) 011001 doi:10.1088/1742-6596/860/1/011001

IOP Conf. Series: Journal of Physics: Conf. Series **860** (2017) 011001 doi:10.1088/1742-6596/860/1/011001

IOP Conf. Series: Journal of Physics: Conf. Series **860** (2017) 011002 doi:10.1088/1742-6596/860/1/011002

Peer review statement

All papers published in this volume of *Journal of Physics: Conference Series* have been peer reviewed through processes administered by the proceedings Editors. Reviews were conducted by expert referees to the professional and scientific standards expected of a proceedings journal published by IOP Publishing.

IOP Conf. Series: Journal of Physics: Conf. Series **860** (2017) 012001 doi:10.1088/1742-6596/860/1/012001

Preliminary survey of radioactivity level in Thai medicinal herb plants

C Kranrod[1], S Chanyotha[1,4], R Kritsananuwat[1], T Ploykrathok[1], P Pengvanich[1], Y Tumnoi[2], T Thumvijit[3] and S Sriburee[3]

[1] Department of Nuclear Engineering, Faculty of Engineering, Chulalongkorn University, Bangkok, Thailand
[2] Bureau of Technical Support for Safety Regulation, Office of Atomic for Peace, Chatuchak, Bangkok, Thailand
[3] Department of Radiological Technology, Faculty of Medicine Technology, Chiang Mai University, Chiang Mai, Thailand

[4] Corresponding author e-mail: supitcha.c@chula.ac.th

Abstract. In this research, the natural radioactivity concentrations and their respective annual effective dose of the naturally occurring radionuclides ^{226}Ra, ^{228}Ra and ^{40}K in selected medicinal herb plants were investigated. Seven kinds of popular Thai medicinal herb plants had been studied: turmeric, ginger, safflower, moringa, gotu kola, garlic and alexandria senna. The radiological risk associated with the use of these medicinal plants was assessed. The activity concentrations of ^{226}Ra, ^{228}Ra and ^{40}K were determined using the gamma-ray spectrometry technique. The radioactivity concentrations were found to range from less than 0.20 to 6.67 Bqkg^{-1} for ^{226}Ra, less than 0.10 to 9.69 Bqkg^{-1} for ^{228}Ra, and from 159.42 to 1216.25 Bqkg^{-1} for ^{40}K. Gotu kola showed the highest activity concentrations of ^{226}Ra and ^{228}Ra, while ginger showed the highest activity concentration of ^{40}K. The total annual effective dose due to ingestion of these herb plants were found to range from 0.0028 to 0.0097 mSvy^{-1} with an average value of 0.0060±0.0001 mSvy^{-1}. The results conclude that the Thai medicinal herb plants samples from this research are considered safe in terms of the radiological hazard.

1. Introduction

Thailand has practiced traditional medicine for more than a thousand years with the known record that date back to the Sukhothai period (1238-1377). The knowledge of traditional medicine based on the usage of Thai medicinal herbs has been gradually developed, systematized, revised, recorded, and passed on from generation to generation throughout the country's history.[1] Thai medicinal herbs, known as Sa-mun-phrai in the local language, have been used to treat various symptoms and ailments. After the World Health Organization began to promote national traditional heritage in 1977, the Foundation for the Promotion of Thai Traditional Medicine was established, and over the years the interest and importance of herbal healing has continued to grow. [2]

Raw parts of plants and their extracts are used in medicinal products all over the world. It is estimated that 25% of modern medicines are derived from medicinal plants, of which most are flowering plants. About 80% of the world population (especially in developing countries) use herbal medicine plants as their primary source of healthcare [3]. There are over 250,000 flowering plant species that serve as resources for the production of new drugs. Plant research has increased globally and the large body of collected evidence shows the immense potential of medicinal plants used in various traditional systems [4].

Various environmental conditions can affect the medical properties and effectiveness of medicinal plants, and may also have a direct influence on their growth in the place of their occurrence and cultivation [5]. An emerging problem in many developing countries is the industrial pollution that threatens the health of the local environment and can lead to the contamination of vegetation by heavy metals, pesticides, and radioactivity [6]. The contamination level in the medicinal plants should therefore be monitored and analysed.

According to the International Food Safety Authorities Network [7], plants used as food commonly have ^{40}K, ^{232}Th and ^{238}U and their progenies. It is expected that similarities would be found in plants used for medicinal purposes since plants are the primary pathway of natural radionuclides to enter the human body via the food chain. In a variety of concentrations, the Naturally Occurring Radioactive Materials (NORMs) have always been present in every part of the earth and in the tissue of all living beings. Natural radionuclides such as ^{238}U, ^{232}Th and ^{40}K are found in every constituent of the environment (air, water, soil, and food) and in humans, and thereby subjecting human beings to certain degree of radiation exposure. [8] The estimation of risk to humans from medicinal plants through ingestion requires a quantitative understanding of the interrelated pathways by which the radionuclides are eventually ingested by humans [9]. The assessment of radioactive contamination of medicinal plants not only contributes to determination of the quality of the plant material, but also provides useful information on the safety of their consumption by humans.

2. Methods and Materials

Samples of medical herb plants were collected from various local markets and drugstores in Thailand. All samples had been certified and licensed by the Thai Food and Drug Administration under the Ministry of Public Health. They included 7 kinds of popular Thai medicinal herb plants (24 samples) used extensively for treating various diseases or complementary medicine. Table 1 lists the medicinal plants selected for the present study, as well as their common names, scientific names, medical uses, and the particular part of the plant used. Each sample had been dried, blended and packaged inside a capsule. All samples have patented license and been certified by the Food and Drug Administration, Ministry of Public Health, Thailand.

2.1. Sample preparation

The Thai medicinal herb plant samples, which were already in powder form, were removed from inside of their plastic capsules. They were dried at a temperature of 80 °C in an oven for 6 hours. The dry mass value was used to determine the radionuclide concentration. Each dried sample, about 100 g, was put in a cylindrical plastic container (2.5 inch in height and 3 inch in diameter). All samples were sealed and kept for at least 30 days in order to ensure the radioactive secular equilibrium before taken for measurement by gamma spectrometric method [10].

2.2. Measurement of the samples

The gamma-ray spectroscopy system consisted of a high purity germanium (HPGe) detector with a relative efficiency of 60% and 1.85 keV energy resolution (FWHM) at the 1332 keV energy peak (of ^{60}Co isotope); a multichannel analyzer (MCA) with 4096 channel (ORTEC's system); and a detector shield with adequate size of cavity to accommodate large sample measurement. The shield was made of lead with 10 cm thickness, and lined inside with graded absorber made of Cd (~1.6 mm.) and Cu (~0.4 mm.).

2.3. Quality assurance and Quality control

Quality assured standard materials procured from IAEA were used for the calibration of the detector. In order to have the density of the reference material similar to that of the dried sample powder, the reference material power, IAEA-RGU-1 and IAEA-RGTh-1, were diluted and mixed with the laboratory grade boric acid powder with the ratio of 1:3. The weight of diluted reference standard powder used for the detector calibration is about 100 g similar to the weight of the herb sample. All of standards and samples were filled in the cylindrical plastic container, which has an identical size of 2.5 inch in height and 3 inch in diameter, and the container cap was sealed with tape and silicone. Thus, the geometrical factor of the standard will be the same as the sample when they were measured.

Based on the measured gamma-ray peaks emitted by the daughter radionuclides in the ^{232}Th and ^{238}U decay series, and the peak of ^{40}K, the concentrations of ^{226}Ra, ^{228}Ra, and ^{40}K were determined. The calculations relied on the establishment of secular equilibrium in the samples, due to the much smaller lifetime of the daughter radionuclides in the decay series of ^{232}Th and ^{238}U. The gamma-rays of ^{228}Ac (338.4 and 911 keV) were used to determine the ^{228}Ra concentration. The gamma-rays of ^{214}Pb (351.3 keV) and ^{214}Bi (609.3 keV) were used to determine the ^{226}Ra. The 1461 keV was used to determine the concentration of ^{40}K in different samples. Each sample was counted for 172,800 sec. (48 hours) in the same geometry setting.

QA/QC included efficiency checks performed weekly and monthly background counts. In addition, comparison measurements were carried out to find the reliability of the measurements. Some of the samples were sent to the Radiation Measurement Group, Bureau of Technical Support for Safety Regulations, Office of Atoms for Peace, Thailand for the analysis of radioactivity by gamma spectrometry. The results of these comparison measurements showed good agreement. The deviation in the results of two laboratories was within 10%. Our laboratory has also participated in the inter-comparison exercises and proficiency testing for gamma spectrometry organized by the IAEA since 2012. The results from these inter-comparisons demonstrated satisfactory performance with respect to the quality of the analytical results generated by the laboratory.

2.4. Calculation of radionuclides and annual committed effective dose

Following the spectrum analysis, counting rates for each detected photo peak and activity per mass unit for each of the detected nuclides are calculated. The specific activity (in Bq/kg) is given by the expression (1).

$$A_{sp} = \left(\frac{N}{t} - \frac{N_0}{t_0}\right) \Big/ \left(I_\gamma \bullet \varepsilon \bullet m\right) \tag{1}$$

Where, N is the net counts of a given peak, t = 48 hour is the counting time for the sample, N is the background of the given peak, t_0 = 48 hours is the counting time for the background, ε is the detection efficiency, I_γ is the number of gamma photons per disintegration and m is the mass in kg of the measured sample.

If there is more than one peak in the energy analysis range for a nuclide, an average of the peak activities is made, and the result is then the weighted average nuclide activity. The total uncertainty value is composed of the random and systematic errors in all the factors involved in producing the final nuclide concentration result listed in Table 2.

Having obtained the values for the specific activity concentrations of the individual naturally occurring radionuclides in the medicinal plants, the average annual committed effective dose, E_{ave}, for ingestion of naturally occurring radioactive materials (NORMs) in the Thai medicinal herb plants were calculated using the expression (2) given by [11].

$$E_{ave} = I_p . DCF_{ing} . A_{sp} \tag{2}$$

Where DCF_{ing} is the dose conversion factor for ingestion, for each radionuclide (i.e., 2.8×10^{-4} mSvBq^{-1}, 6.7×10^{-4} mSvBq^{-1}, and 6.2×10^{-6} mSvBq^{-1}for ^{226}Ra, ^{228}Ra, and ^{40}K, respectively, for an adult) [12,13], I_p is the consumption rate from intake of NORMs in medicinal plants, A_{sp} is the activity concentration in the plant sample.

IOP Conf. Series: Journal of Physics: Conf. Series **860** (2017) 012001 doi:10.1088/1742-6596/860/1/012001

Table 1. Scientific name and uses of selected traditional medicinal plants.

Scientific name	Common name	Medicinal uses	Parts used
1. Curcuma longa	Turmeric	Laxative, tonic, alterative, antipyretic, antibacterial, detergent (an agent that cleanses boils, ulcers, stops bleeding, wounds, etc.), ophthalmicum (remedy for diseases of the eye), antiperiodic (prevents the periodic recurrence of attacks of a disease; as in malaria), aperient (mild or gentle laxative), diuretic, antiseptic, and deobstruent (removes obstructions by opening the natural passages or pores of the body).	Roots
2. Zingiber officinale Roscoe	Ginger	Antispasmodic, antiemetic, analgesic, antiseptic, appetizer, aromatic, carminative, condiment, diaphoretic, expectorant, febrifuge, pungent, sialagogue, stimulant Topically: increases blood flow to an area	Roots and rhizomes
3. Carthamus tinctorius L.	Safflower	Diaphoretic, diuretic, emmenagogue, analgesic, carminative	Flowers
4. Moringa oleifera Lam.	Moringa	anti-diabetic effects, cardio tonic	Seed
5. Centella asiatica (L.) Urb.	Gotu kola	Alterative, antipyretic, diuretic, febrifuge, antispasmodic, nervine, sedative, tonic	Whole plant
6. Allium sativum L.	Garlic	Alterative, anthelmintic, antispasmodic, carminative, cholagogue, digestive, expectorant, febrifuge, antibiotic, antiseptic, stimulant	Bulb
7. Senna alexandrina Mill.	Alexandria Senna	A laxative substance, anthelmintic	Leaves

3. Results and Discussions

Table 2 shows the radioactivity concentrations (Bqkg^{-1}) of the natural radionuclides ^{226}Ra, ^{228}Ra and ^{40}K in the medicinal plants selected for this study. All natural radionuclides were determined in 7 kinds of Thai medicinal herb plants in a total of 24 samples. The data showed that all Thai medicinal herb plants had activity concentrations in the ranged of less than 0.20 to 6.67 Bqkg^{-1} for ^{226}Ra, less than 0.10 to 9.69 Bqkg^{-1} for ^{228}Ra, and from 159.42 to 1216.25 Bqkg^{-1} for ^{40}K. Gotu kola had the highest activity concentration of ^{226}Ra and ^{228}Ra, while ginger had the highest activity concentration of ^{40}K.

The variations in the activity concentrations could be due to differences in the geological location of the plants and the radiochemical composition of the soils in which these medicinal plants were grown or cultivated. The levels of activity concentration of natural radionuclides are not normalized across the globe, and each plant has ability to absorb particular elements more than the others [11].

In this paper, the assumption that a unit consumption rate (I_p) of 1 kg per annum was used. The number is derived from the medical labels which received the certificate and license from The Thai Food and Drug Administration of the Ministry of Public Health. The average total annual committed effective doses due to ingestion of ^{226}Ra, ^{228}Ra, and ^{40}K in the medicinal herb plants were calculated using the corresponding values of activity concentrations for each of the medicinal plants. The results are presented in Table 2 and Figure 1. The results varied from 0.0028 to 0.0097 mSvy^{-1} with an average value of 0.0060±0.0001 mSvy^{-1}. The highest average was recorded for Turmeric, while Safflower and Alexandria Senna showed the lowest average.

From the viewpoint of the annual effective dose and the activity concentration of ^{226}Ra, ^{228}Ra, and ^{40}K in the medicinal herb plants compared with this study (the conservative comparison), Table 2 showed the results from published work with the annual committed effective dose and the activity concentration of NORMs in the different medicinal herb plants conducted in India and Nigeria [14,15]. The average activity concentrations of both ^{226}Ra and ^{228}Ra in our study appear to be lower than the values obtained in South India [14] and Nigeria [15]. However, our study showed lower average activity concentration of ^{40}K than the South India study, but higher average concentrations than the Nigeria study. Thus, the average annual committed effective dose due to the ingestion of ^{226}Ra, ^{228}Ra, and ^{40}K in Thai medicinal herb plants in this study was slightly higher than the results reported in Nigeria [15] but much lower than that reported in South India [14].

Table 2. Radioactivity concentration of ^{226}Ra, ^{228}Ra, and ^{40}K (in Bqkg^{-1}) and annual committed effective dose (mSvy^{-1}) in the Thai medicinal herb plant samples (dry weight)

Medicinal plants (number of Sample)	Average of Activity concentration (in Bq kg^{-1})			Average Annual effective dose (mSvy^{-1})
	^{226}Ra	^{228}Ra	^{40}K	
Present study				
1.Turmeric (8)	1.94±0.62	3.90±0.36	749.87±8.79	0.0097±0.0004
2.Ginger (3)	2.98±0.56	4.31±0.50	669.35±12.03	0.0092±0.0003
3.Safflower (2)	2.44±1.10	2.74±0.43	115.74±3.55	0.0028±0.0004
4.Moringa (3)	1.01±0.45	3.13±0.33	368.40±5.82	0.0046±0.0003
5.Gotu kola (3)	3.86±0.21	8.38±0.76	396.53±5.37	0.0086±0.0004
6.Garlic (2)	0.51±0.24	0.13±0.08	370.43±6.31	0.0042±0.0005
7.Alexandria Senna (3)	1.03±0.11	1.09±0.27	271.95±7.46	0.0028±0.0003
Average	**1.97±0.34**	**3.38±1.19**	**420.33±3.77**	**0.0060±0.0001**
South India study [14]				
1.Justica adhatoda	11.27±1.59	5.07±2.15	6831.40±490.28	0.1067±0.0136
2.Ficus racemosa	4.51±0.23	8.72±0.44	832.95±59.69	0.0084±0.0005
3.Eupatorium odoratum	11.15±0.66	4.87±0.79	1305.00±94.74	0.0436±0.0041
4.Ziziphus rugosa	5.98±2.56	5.10±0.31	143.99±10.83	0.0091±0.0016
5.Cittrus limon	6.08±0.73	5.67±1.00	1910.00±139.09	0.0148±0.0012
6.Bauhinia acuminate	2.66±0.21	2.42±0.36	3724.00±27.44	0.0580±0.0032
7.Mimosa pudica	5.35±0.34	4.14±0.44	320.86±2.40	0.0235±0.0019
8.Mesua ferrea	3.73±0.17	4.43±0.26	93.79±7.19	0.0075±0.0007
Average	**6.34±0.85**	**5.05±0.63**	**1895.25±163.33**	**0.0340±0.0043**
Nigeria study [15]				
1.Cashew nut tree	40.08±4.12	38.69±0.71	74.59±2.19	0.0069±0.0004
2. Neem tree	19.22±2.12	31.69±0.66	324.18±8.69	0.0049±0.0002
3. African copaiba	30.69±2.56	30.47±0.24	123.34±6.22	0.0055±0.0003
4. Moringa	13.71±1.96	39.48±0.90	184.59±3.43	0.0047±0.0003
5. Guava	10.79±4.24	36.53±1.10	219.98±9.28	0.0043±0.0005
6. Umbrella tree	18.18±3.52	41.05±0.92	145.59±7.19	0.0051±0.0003
7. Shea butter tree	42.47±3.76	27.76±1.02	129.78±5.63	0.0064±0.0004
Average	**25.02±0.95**	**35.10±0.29**	**171.72±2.60**	**0.0054±0.0001**

4. Conclusion

The average annual effective dose determined in this study due to the ingestion of natural radionuclides in Thai medicinal herb plants is far below the average annual effective dose of 0.3 mSv/y received per person worldwide due to the internal exposures other than radon and thoron that is reported UNSCEAR 2000 report [13]. The study has found insignificant annual committed effective dose due to the use of these Thai medicinal herb plants into Plant Medicine, therefore the radiological hazard associated with intake of the natural radionuclides in the medicinal plants is trivial.

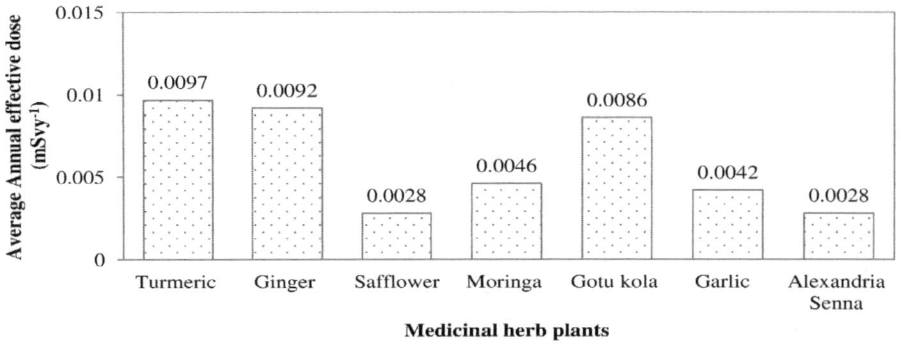

Figure 1. The variation of the annual committed effective dose from natural radionuclides (^{226}Ra, ^{228}Ra, and ^{40}K) in the Thai medicinal herb plant samples.

Acknowledgments
The authors wish to gratefully acknowledge the generous funding from Agricultural Research Development Agency (Public Organization), and to especially thank our colleagues for their interest and support of this research.

References
[1] Pennapa S and Anchalee C, 2014 Traditional medicine in Kingdom of Thailand. *Thai Massage and Medicine* **1** 97-120.
[2] Robin M, Sebastian A, Brad M, John O, Chris G and Morley SM 2007 Medicinal Plants of Thailand *Asia-Pacific Biotech News* **11** 508. DOI: http://dx.doi.org/10.1142/S0219030307000584
[3] Chan K 2003 Some aspects of toxic contaminants in herbal medicines. *Chemosphere* **52** 1361-71.
[4] Nomita DK, Nandakumar SH and Sanjiv K 2008 Estimation of essential and trace elements in some medicinal plants by PIXE and PIGE techniques *Nucl. Instrum. Methods Phys. Res., Sect. B* **266** 1605-10.
[5] Salamon I and Haban M 2004 The radioactivity evaluation of selected species of medicinal plants. *Acta Fytotecnica and Zootecnica* **4** 103-6.
[6] Duffy S, Simon SL and Whicker FW 1999 ^{137}Cs contamination of plants used for traditional medicine and implication for human exposure. *J Environ Radioact.* **46** 27-44.
[7] World Health Organization (WHO) 2011. Information on nuclear accidents and radioactive contamination of foods *International Food Safety Authorities Network*. Geneva.
[8] Varier MK 2009 *Nuclear radiation detection: Measurements and analysis* New Delhi India: Narosa Publishing House.
[9] Eisenbud M 1987 *Environmental radioactivity from natural, industrial, and military sources* (3rd ed.) INC: Academic Press.
[10] Kranrod C, Chanyotha S, Pornnumpa C, Kritsananuwat R and Sriploy P 2015 *Radiat. Prot. Dosimetry* **167(1-3)** 270-5.
[11] Tettey-Larbi L, Darko EO, Schandorf C and Appiah AA. 2013 Natural radioactivity levels of some medicinal plants commonly used in Ghana. *Springerplus* **157(2)** 1-9.
[12] World Health Organization (WHO) 2007 *WHO guidelines for assessing quality of herbal medicines with reference to contaminants and residues* WHO Press Spain 118.
[13] United Nations Scientific Committee on the Effects of Atomic Radiation 2000 *Sources and effects of ionizing radiation* New York United Nations.

IOP Conf. Series: Journal of Physics: Conf. Series **860** (2017) 012001 doi:10.1088/1742-6596/860/1/012001

[14] Chandrashekara K and Somashekarappa HM 2016 Estimation of radionuclides concentration and average annual committed effective dose due to ingestion for some selected medicinal plants of South India. *J. Radiat. Res. Appl. Sci.* **9** 68-77.

[15] Njinga RL, Jonah SA and Gomina M 2015 Preliminary investigation of naturally occurring radionuclides in some traditional medicinal plants used in Nigeria. *J. Radiat. Res. Appl. Sci.* **8** 208-215.

IOP Conf. Series: Journal of Physics: Conf. Series **860** (2017) 012002 doi:10.1088/1742-6596/860/1/012002

Improvement of microbiological qualities of namphrik by gamma irradiation

K Chahorm[1], N Neramitmansook, N Kongsang and J Ko

Thai Irradiation Center, Thailand Institute of Nuclear Technology (Public Organization), Pathumthani, Thailand

[1] Corresponding author e-mail: kanchanac@tint.or.th

Abstract. Twenty samples of Namphrik from commercial markets were evaluated the microbiological qualities. It was found that 15 samples did not meet Thai Community Product Standard. The total plate count (TPC) in 15 samples were higher than the maximum limits ($1.60x10^4$ – $4.4x10^5$ CFU/g). In addition, the other pathogens were higher than the maximum limits such as *B.cereus* in 11 samples ($2.10x10^3$ – $6.10x10^4$ CFU/g) *S.aureus* in 2 samples (15 – 40 CFU/g) *Clostridium perfringens* in 4 samples ($1.00x10^2$ – $8.8x10^3$ CFU/g) and yeast&mold in 9 samples (3.00 $x10^2$ – $9.00x10^3$ CFU/g). To reduce TPC and pathogenic bacteria, the gamma irradiation were applied at 3.28- 4.43 kGy. The results indicated that the irradiation can reduce the TPC around 1.2 – 3.9 log cycles and eliminate pathogens bacteria in the product to make all of 15 samples qualified to the standard. The sensory evaluation was conducted in Namphrik Narok by using difference from control test to determine whether the consumers can differentiate between the non-irradiated and irradiated. The result showed that the consumers can significantly differentiate the color, odor and flavor ($p<0.05$). However, the preference test showed that there was no significant preferences at $p>0.05$. Both non-irradiated and irradiated were scored at 6.4 (slightly to moderately preference). Thus the gamma irradiation can be used as a tool to improve the microbiological qualities of the Namphrik Narok product without effecting the consumer preference.

1. Introduction

Namphrik is traditional Thai foods consisting of several dried ingredients such as chili, garlic, scallion, herbs, blended fish or shrimp etc. Spices and herbs may be contaminated to a wide range of microorganism during pre- and post-harvest. Such contamination may occur during processing, storage, distribution, sale and/or use [1]. Spices have been used in many industries, with the food industry and catering. Dried materials from plant origin and spices are commonly heavily contaminated with moulds and bacteria [2]. About 138 samples of 15 different spices obtained from common markets were evaluated fungi. The result show that *Aspergillus, Penicillium,*and *Rhizopus* were detected in range of 5325–6800 cfu/g [3]. Thai people preferred to eat Namphrik with hot cooked rice, fresh vegetables and other meals. Due to consisting of dried ingredients, Namphrik trends to be high microbial contamination. In addition, there was not any heat processing before eating that may be cause diarrhea in human. Several methods have been using for decontamination in herbs and spices such as carbon dioxide under pressure, microwaves, high pressure and ionizing radiation [4] etc. Irradiation was suitable to reduce the contaminated microorganism. In pararell, we were studied the effect of irradiation on sensory acceptance.

2. Materials and methods

2.1. Market survey

Twenty samples of Namphrik were collected from commercial market and evaluated the microbiological qualities according to Thai community product standard (No.130/2556) as shown in table 1.

Table 1. Thai community product standard (No.130/2556)

Types of Microorganisms	**Thai community product standard (No.130/2556)**
Total Plate Count	< 10^4 CFU/g
Total Yeast & Mold	< 100 CFU/g
Escherichia coli	< 3 MPN/g
Staphylococcus aureus	< 10 CFU/g
Bacillus cereus	< 10^3 CFU/g
Clostridium perfringens	< 100 CFU/g
Salmonella spp.	Not detected in 25 g

2.2. Gamma irradiation process

Gamma irradiation process was conducted at Thai Irradiation Center, Thailand Institute of Nuclear Technology (Public Organization) using cobalt-60 gamma source (The multipurpose Irradiator JS 8900). Namphrik was irradiated with gamma rays at 3.28- 4.43 kGy Irradiated and non-irradiated samples of Namphrik were stored at room temperature. The evaluation of microbiological properties and sensory were evaluated after irradiated 5 and 8 days, respectively.

2.3. Determination of microbiological

Total plate count (TPC), Total Yeast & Mold (TYM), *Bacillus cereus*, *Clostridium perfringens* and *Salmonella* spp. were determined according to the method of Bacteriological Analytical Manual Chapter 3 [5], Chapter 18 [6], Chapter 14 [7], Chapter 16 [8] and Chapter 5 [9] respectively. *Staphylococcus aureus* and *Escherichia coli* were determined according to the method of AOAC (2003.11) [10] and AOAC (2005.03) [11].

2.4. Sensory evaluation

Sensory evaluation of irradiated and non-irradiated samples was conducted by 18 panelists after irradiated for 8 days. Difference from control test and preference tests were conducted in Namphrik to determine whether there is a statistically significant difference and preference between the non-irradiated and irradiated groups. Difference from control test was evaluated color, odor and flavor. Preference test was evaluated using a 9 point hedonic scale (1 = Extremely Dislike to 9 = Extremely like). The data were analyzed by MINITAB software. Mean values and standard error of mean are reported and significant difference was defined at $P < 0.05$.

3. Results and discussion

3.1. Microbiological qualities of namphrik in the market

Twenty samples of Namphrik from commercial market were evaluated for microbiological qualities. It was found that 15 samples were not complied with the Thai community product standard No. 130/2556 because TPC was higher than the maximum limits (15 samples) and also the pathogenic microorganisms was exceeded the limits such as *Bacillus cereus* (11 samples), TYM (9 samples), *Clostridium perfringens* (4 samples), *Staphylococcus aureus* (2 samples) as shown in table 2 and 3. The market surveys in qualities of Namphrik have not widely studied. In agreement with Thai Health Promotion Foundation reported that 15 of 35 samples were nonqualified to the standard due to exceeding of TPC and pathogenic microorganisms exceeding the limits such as *Bacillus cereus* (13 samples), *Clostridium perfringens* (1 samples), *Escherichia coli* (14 samples) and Coliform bacteria (7 samples) [12]. It seems that the results of this study supported the report of Thai Health Promotion Foundation. However, *Escherichia coli* and *Salmonella* spp. were not detected in this study. The report of Thai Health Promotion Foundation also reported that 0.8% of consumers having diarrhea every times and 13.7% of consumers sometimes having diarrhea after eating Namphrik.

Table 2 .The pathogenic microorganisms exceeding the limits in non- irradiated Namphrik

Microorganisms	Nonqualified samples	CFU/g
Total plate count	15	$1.60x10^{4}$ – $4.4x10^{5}$
Bacillus cereus	11	$2.10x10^{3}$ – $6.10x10^{4}$
Total Yeast & Mold	9	$3.00\ x10^{2}$ – $9.00x10^{3}$
Clostridium perfringens	4	$1.00x10^{2}$ – $8.8x10^{3}$
Staphylococcus aureus	2	15 – 40
Escherichia coli	0	-
Salmonella spp.	0	-

3.2. Effect of irradiation on Microbiological qualities

The irradiation technology is one of method to use to decontamination in many foods. It has been suggested as the nonthermal method for destroying pathogenic and spoilage microorganisms in the final product. The nutritional properties of irradiated food is the same as non-irradiated product [13, 14]. The sensory characteristics of foods may/may not alter. Nonqualified samples from market survey were irradiated at 3.28 to 4.43 kGy. The microbiological tests were conducted after 5 day of irradiation. TPC of nonqualified samples were reduced to the acceptable level of the standard. The applied doses can lower TPC from 1.2 to 3.9 log cycle and the average reduction was 2.4 log cycle reduced from the original level. The effect of radiation on pathogenic decontamination was shown in table 3. All of TYM, *Staphylococcus aureus*, 10 of 13 *Bacillus cereus* and 3 of 4 *Clostridium perfringens* were reduced lower than the detection limits which were <10 CFU/g or <100 CFU/g depending on methods. Irradiation at the dose of 3.28-4.43 kGy was sufficient to improve the microbiological qualities of unqualified samples.

Table 3. *Microbial contamination in Namphrik (CFU/g)*

Standard	Sample	No.1	No.2	No.3	No.4	No.5	No.6	No.7	No.8	No.9	No.10	No.11	No.12	No.13	No.14	No.15	No.16	No.17	No.18	No.19	No.20
Total Plate Count (<10,000 CFU/g)	Pre	2.50×10^{5}	5.00×10^{4}	1.40×10^{5}	4.30×0^{3}	3.60×10^{4}	6.90×10^{4}	1.60×10^{4}	6.70×10^{3}	5.70×10^{4}	8.00×10^{2}	6.30×10^{4}	9.60×10^{4}	7.00×10^{4}	1.80×10^{5}	1.60×10^{5}	4.80×10^{2}	1.30×10^{5}	1.00×10^{5}	4.40×10^{5}	6.70×10^{3}
	Post	38	9.10×10^{2}	1.30×10^{3}	2.60×10^{2}	1.90×10^{2}	80	20	13	4.00×10^{2}	53	7.90×10^{2}	1.40×10^{2}	2.40×10^{2}	1.10×10^{3}	1.60×10^{3}	33	1.40×10^{2}	2.10×10^{2}	60	43
Salmonella spp. (not detected in 25 g)	Pre	ND	ND	ND	ND	ND	ND	ND	ND	ND	ND	ND	ND	ND	ND	ND	ND	ND	ND	ND	ND
	Post	-	-	-	-	-	-	-	-	-	-	-	-	-	-	-	-	-	-	-	-
S.aureus (<10 CFU/g)	Pre	<10	<10	<10	<10	<10	<10	<10	<10	<10	<10	<10	<10	<10	<10	<10	<10	<10	<10	40	15
	Post	-	-	-	-	-	-	-	-	-	-	-	-	-	-	-	-	-	-	<10	<10
B.cereus (<1,000 CFU/g)	Pre	2.00×10^{2}	1.40×10^{4}	5.10×10^{3}	<100	1.40×10^{4}	1.80×10^{4}	<100	<100	3.90×10^{3}	<100	1.40×10^{4}	1.30×10^{4}	8.60×10^{3}	2.10×10^{3}	3.60×10^{3}	4.00×10^{2}	6.10×10^{4}	<100	<100	<100
	Post	-	1.50×10^{2}	1.00×10^{2}	-	<100	<100	-	-	<100	-	<100	<100	<100	3.50×10^{2}	<100	-	<100	-	-	-
Clostridium perfringens (< 100 CFU/g)	Pre	45	3.90×10^{2}	60	1.00×10^{2}	75	1.10×10^{2}	<10	<10	<10	<10	<10	<10	8.80×10^{3}	60	35	<10	<10	<10	<10	<10
	Post	-	<10	-	<10	-	<10	-	-	-	-	-	-	65	-	-	-	-	-	-	-
E.coli (<3 CFU/g)	Pre	<10	<10	<10	<10	<10	<10	<10	<10	<10	<10	<10	<10	<10	<10	<10	<10	<10	<10	<10	<10
	Post	-	-	-	-	-	-	-	-	-	-	-	-	-	-	-	-	-	-	-	-
Yeast&Mold (<100 CFU/g)	Pre	<100	5.50×10^{2}	1.70×10^{3}	<100	<100	<100	<100	<100	<100	<100	3.00×10^{2}	3.00×10^{2}	9.00×10^{3}	1.50×10^{3}	1.00×10^{3}	4.50×10^{2}	3.50×10^{2}	<100	<100	<100
	Post	-	<100	<100	-	-	-	-	-	-	-	<100	<100	<100	<100	<100	<100	<100	-	-	-

3.3. Sensory evaluation

3.3.1. Difference from control test. The difference test was evaluated the changes in color, odor and flavour. The panels compared the aspects to give scores how much the difference between the control (non irradiated) sample and the test samples. The result showed that the panelists can significantly differentiate the color, ordor and flavor ($p<0.05$). But the level was at slight (score=1) to very slight (score=2) differences in irradiated samples as shown in table 4.

Table 4. The result of Difference from control test of irradiated and non-irradiated samples.

	Non-Irradiated	Irradiated	P-value
Color	0.611 ± 0.916	1.111 ± 1.023	0.035
Order	0.500 ± 0.786	0.944 ± 1.056	0.007
Flavor	0.611 ± 0.916	1.000 ± 1.188	0.004

3.3.2. Preference test of irradiated and non-irradiated samples. The test was designed to evaluate the preferences of panellists to the control and irradiated products. The preference test showed that there was no significant changes in preference at ($p>0.05$). Non-irradiated and irradiated samples were scored at 6.4 (slightly to moderately like) as shown in table 5. Although the results of the difference from control test showed that the panel can differentiate sensory aspects such as color, odor and flavor. The irradiation doesn't affect the preferences of the product.

Table 5. The result of Preference test in of irradiated and non-irradiated samples.

	Non-Irradiated	Irradiated	P-value
Preference	6.444 ± 2.502	6.444 ± 2.036	1.000

4. Conclusion

From this study, it was found that 15 from 20 samples did not meet Thai community product standard. The irradiation at doses 3.28-4.43 kGy can reduce TPC (1-3 log cycles) to be less than standard and eliminate pathogenic bacteria in all of the samples to make all of non qualifies samples to comply with the standard. Thus the dose of 3-4.5 kGy was appropriate to use as the generic dose to reduce microorganisms and pathogens in Namphrik. Although irradiation alters slightly sensory aspects, it doesn't affect the preferences of the irradiated product. The gamma irradiation was suitable to use as a tool to improve the microbiological qualities of Namphrik product without altering the preference of customers.

5. Future study

Investigation of microbiological and physical quality of Namphrik Narok (non-irradiated and irradiated) after 3 and 6 months will be studied. We plan to study the effect of difference packages among glass bottle, plastic bag and aluminum bag on microbiological and physical quality of Namphrik Narok (non-irradiated and irradiated) to find the most suitable package for the product.

References

[1] McKee L H 1995 Microbial contamination of spices and herbs: a review **28** 1-11.
[2] Romagnoli B, Menna V, Gruppioni N and Bergamini C 2007 *Aflatoxins in spices, aromatic herbs, herbs–teas and medicinalplants marketed in Italy* (Food Control vol 18) pp 697-701.
[3] Mohamed H, Saad A 2010 *Egypt. J. Biological Sciences*. **17** 167-175
[4] Agnieszka B and Krzysztof S 2014 *Comparision of methods of herbs and spices decontamination* (CHEMIK) pp 100-102.

[5] U.S. Food and Drug Administration (USFDA). Bacteriological Analytical Manual Online (BAM online): Aerobic Plate Count . [Online]. 2001. [cited 2016 July 15] Available from: http://www.fda.gov/Food/FoodScienceResearch/LaboratoryMethods/ucm063346.htm.

[6] U.S. Food and Drug Administration (USFDA). Bacteriological Analytical Manual Online (BAM online): Yeasts, Molds and Mycotoxins. [Online]. 2001. [cited 2016 July 15] Available from: http://www.fda.gov/Food/FoodScienceResearch/LaboratoryMethods/ucm071435.htm.

[7] U.S. Food and Drug Administration (USFDA). Bacteriological Analytical Manual Online (BAM online): *Bacillus cereus*. [Online]. 2012. [cited 2016 July 15] Available from: http://www.fda.gov/Food/FoodScienceResearch/LaboratoryMethods/ucm070875.htm.

[8] U.S. Food and Drug Administration (USFDA). Bacteriological Analytical Manual Online (BAM online): *Clostridium perfringens*. [Online]. 2001. [cited 2016 July 15] Available from: http://www.fda.gov/Food/FoodScienceResearch/LaboratoryMethods/ucm070878.htm.

[9] U.S. Food and Drug Administration (USFDA). Bacteriological Analytical Manual Online (BAM online): *Salmonella*. [Online]. 2015. [cited 2016 July 15] Available from: http://www.fda.gov/Food/FoodScienceResearch/LaboratoryMethods/ucm070149.htm.

[10] AOAC Official Method 2003.11 Petrifilm™ Staph Express Count Plate.

[11] AOAC Official Method 2005.03 Detection and Confirmed Quantitation of Coliforms and *E.coli* in Foods SimPlate Coliform and *E.coli* Color Indicator.

[12] Thai Health Promotion Foundation's [Online]. [cited 2016 July 15] Available from: http://resource.thaihealth.or.th/hilight/12329.

[13] Ahn H-J, Kim J-H, Lee J-W, Jo C and Byun M-W 2004 Application of gamma irradiation for the reduction of residual nitrite and nitrosamine in meat products *Korean J. Food Sci.* **24** 367-372.

[14] WHO (World Health Organization) 1999 *High dose irradiation* (WHO Technical Report Series 890) pp 9-37.

IOP Conf. Series: Journal of Physics: Conf. Series **860** (2017) 012003 doi:10.1088/1742-6596/860/1/012003

Effects of gamma ray and electron beam irradiation on reduction of microbial load and antioxidant properties of Chum-Hed-Thet (*Cassia alata* (L.) Roxb.)

P Prakhongsil[1], W Pewlong, S Sajjabut and S Chookaew

Thailand Institute of Nuclear Technology (Public Organization), Ongkhalak, Nakhon nayok 26120, Thailand

[1] Corresponding author e-mail: punpanchalee@hotmail.com

Abstract. Considering the growing demands of herbal medicines, *Cassia alata* (L.) Roxb. has been reported to have various phytochemical activities. It has also been called in Thai as Chum-Hed-Thet. In this study, *C. alata* (L.) Roxb. powder were exposed to gamma and electron beam irradiation at doses of 0, 5, 10, 15 and 20 kGy. At the dose of 10 kGy, both of gamma and electron beam irradiation were sufficient in reducing microbial load of irradiated samples as specified in Thai pharmacopoeia (2005). These include the total aerobic microbial count of $< 5.0x10^5$ CFU/g, total fungi count of $< 5.0x10^4$ CFU/g, bile tolerant gram negative bacteria of $< 10^4$ (per g). In addition, pathogenic *Clostridium* spp. (per 10 g), *Salmonella* spp. (per 10 g), *S. aureus* (per 1g) and *E.coli* (per 1g) were absence. In terms of the bioactive molecules, the total phenolic content, DPPH free radical scavenging activity and ferric reducing antioxidant potential of unirradiated and irradiated samples were 19.32-22.44 mg gallic acid equivalent/g, 5.20-7.82 mg ascorbic acid equivalent/g and 69.46-82.06 μmol $FeSO_4$/g, respectively. However, there were no significant differences between unirradiated and irradiated samples (p>0.05). Therefore, both of radiation by gamma ray or electron beam at 10 kGy was sufficient in elimination of microbial flora and did not significantly affected the total phenolic content and antioxidant activities of *C. alata* (L.) Roxb.

1. Introduction

Cassia alata (L.) Roxb. is commonly known in Thai as Chum-Hed-Thet. The common names are candle bush, ringworm bush and Christmas candle[1]. This plant is widely found in tropical regions, sometimes cultivated for medicinal purposes. The main parts used as herbal medicine are the leaves, the barks and the roots. They are very high medicinal values like antimicrobial property and antioxidant activity and traditionally being used in the treatment of skin infections in man[2]. Leaf extract is a good antioxidant[3] and the compound obtained from has been identified as a flavonol compound and named as Kaempferol[4]. Due to its origin it is generally highly contaminated by microorganisms. While the decontamination technologies for herbs and spices, like ethylene oxide, thermal inactivation or the use of steam have disadvantages. The fumigation by ethylene dioxide is prohibited in Europe due to carcinogenic potential in human. However, herbal substances are heat sensitive products, thus the application of steam can cause alteration of aroma and odor, as well as the reduction of volatile compounds[5]. The application of gamma irradiation for the decontamination is a quite efficient process. Recently, electron beam irradiation has been introduced as an alternative method, as though that it being environmental friendly and rapid method for controlling spoilage of foods and improve the hygienic qualities of pharmaceutical and agricultural products.

The aim of this work was to study the potential application of gamma ray and electron beam irradiation in order to investigate the effects on the microbial contamination, the total phenolic content and antioxidant activity of *C. alata* (L.) Roxb. powder.

2. Materials and methods

2.1. Materials

Powder of *C. alata* (L.) Roxb. was purchased from retailer for this present study, which was procured from Sub-Central Thailand region, in March to April, 2014. One hundred grams of dry samples were packed and sealed in aluminium foil packets. Ten packets of dry samples was placed in polyethylene bags and stored at ambient temperature.

Figure 1. Flowers and leaves of Chum-Hed-Thet (*C. alata* (L.) Roxb.).

2.2. Gamma ray and electron beam irradiation

Gamma ray and electron beam irradiation were performed at Gems Irradiation Centre, Thailand Institute of Nuclear Technology (Public Organization), Ongkharak district, Nakhon nayok province, Thailand. Samples (1,000 g) were exposed to radiation at various doses of 5, 10, 15 and 20 kGy. Gamma irradiation were carried out by gamma irradiator from source of Co-60 model Type II (ANS N43.10.), Power Plus System Ltd., UK. Electron beam irradiation were carried out by electron beam accelerator at energy of 5.3 MeV, model 20/16, Mevex, Canada. The unirradiated samples served as controls.

2.3. Microbial analysis

The microbial study was done to enumerate the total aerobic microbial count (CFU/g), total fungi count (CFU/g). Specific microorganisms such as bile tolerant gram negative bacteria (per 1 g), pathogenic *Clostridium* spp.(per 10 g), *Salmonella* spp. (per 10 g), *Staphylococcus aureus* (per 1 g) and *Escherichia coli* (per 1 g) were determined. Control and the irradiated samples were kept at ambient temperature until the analysis were carried out. The samples were taken after 5 days of irradiation. All test were run in triplicate samples and mean values were used. The level of microbial contaminants of unirradiated and irradiated samples were analyzed by method of microbial limit test indicated in Thai pharmacopoeia (2005)[6].

2.4. Total phenolic content and antioxidant analysis

2.4.1. Sample preparation. The samples were taken after 7 days of irradiation. 0.1 g of samples were extracted for 1 h at room temperature in ultrasonic bath with 10 ml of methanol. The sample suspensions were centrifuged, and the supernatant was filtered and the filtrate stores at -20 ^{0}C until analysis.

2.4.2. Total phenolic content. The content of reducing components was estimated using the Folin-Ciocateau assay according to the method developed by Velioglu *et al*, 1998[7]. The total phenolic content was expressed as gallic acid equivalent (GAE) mg/g of dry sample. The estimation of the phenolic compounds was carried out in triplicate.

2.4.3. Free radical scavenging assay with DPPH. This assay was modified from that described in a previous report by Khattak *et al*, 2008[8] with slight modifications. All test were run in triplicate. The absorbance of each solution was determined by spectrophotometer at 517 nm. The free radical scavenging activity was expressed as ascorbic acid equivalent.

2.4.4. Ferric reducing potential (FRAP). Ferric reducing potential (FRAP) assay was performed according to the method described by Benzei and Strain, 1996 [9]. The absorbance was read at 596 nm. using a spectrophotometer. The antioxidant potential of the samples was analyzed based on a calibration curve plotted using $FeSO_4.7H_2O$ at concentrations ranging from 400 to 2,000 μM. All test were run in triplicate and mean values were used to calculate concentration of an antioxidant. A ferric reducing ability was expressed as μmol $FeSO_4$/g.

2.4.5. Statistical analysis. Data (2.4.2, 2.4.3 and 2.4.4) were analyzed by one-way ANOVA and differences among treatment means were determined by Duncan's new multiple-range test.

3. Results and Discussion

3.1. Microbial analysis

As show in table 1, the initial levels of total aerobic microbial count (TAMC) and the total fungi count of *C. alata* (L.) Roxb. in this present study were quite high at the levels of 7.60 x 10^6 and 1.10 x 10^5 CFU/g, respectively. The results showed that the initial microorganisms levels were exceed the acceptable limits for microbial contamination as indicated in Thai pharmacopoeia (2005) (TAMC < 5.0x10^5 CFU/g, total fungi count < 5.0x10^4 CFU/g, not more than 10^4 bile tolerant gram negative bacteria (per 1 g) and specific microorganisms were absence). Gamma ray and electron beam irradiation at dose of 10, 15 and 20 kGy was effective in reduction of the initial of total aerobic microbial count by decreased 2-5 log cycle and eliminated total fungi count to less than 10 CFU/g in irradiated samples. The results indicated that the gamma ray and electron beam irradiation significantly reduced microbial load in samples in dose dependent manner. For specific microorganisms, gamma ray and electron beam irradiation at dose of 5 kGy completely eliminated bile tolerant gram negative bacteria to less than 10 (per 1g), whereas *Salmonella* spp. (per 10g), pathogenic *Clostridium* spp. (per 10g), *S. aureus* (per 1 g) and *E. coli* (per 1 g) were not found in unirradiated and irradiated samples.

These results indicated that gamma ray and electron beam irradiation at dose of 10 kGy was sufficient in reduction of microbial contamination of *C. alata* (L.) Roxb. as indicated in Thai pharmacopoeia(2005). However, in this present study, dosage of greater than 20 kGy was needed to completely decontaminated of *C. alata* (L.) Roxb.

The effects of gamma ray and electron beam irradiation on the microbial contamination of herbs are dependent on dose and the type of microorganism present. The effects on bacterial reduction are relatively dose dependent manner, so the end result depends on the initial contamination level[10] [11].

Phianphak et al. 2007 reported seventeen species of herbs established in Thai traditional remedies were microbially decontaminated by gamma radiation doses of 7.7 and 8.8 kGy. All year round post-harvested and powdered herbs in the study were contaminated with microorganisms, as shown by total bacteria count in the range of 10^5-10^{11} CFU/g and mold counts in the range of 10^3-10^5 CFU/g, *Staphylococcus* spp.10^3-10^7 CFU/g, *Salmonella* spp.10^4-10^7 CFU/g and coliform bacteria 10^3-10^7 CFU/g. No microorganisms were found after gamma treatment [12]. Eamsiri et al. 2016 reported the initial microorganisms of unirradiated sample of *Butea Superba*. Roxb. (Kwao Khrua Dang) root extract were exceed the permissible limits of world health organization (WHO). The results of total viable bacterial counts indicated that the non-irradiated samples of Kwao Khrua Dang powder were highly contaminated with bacteria at the level of 3.9×10^6 CFU/g. Electron beam irradiation at doses of

Table 1. Microbial analysis of *C. alata* (L.) Roxb. powder irradiated at varied doses of gamma ray and electron beam irradiation.

Microbial Flora	Type	Dose (kGy)				
		0	5	10	15	20
TAMC (CFU/g)	Gamma	7.60×10^6	5.30×10^5	4.20×10^4	1.0×10^3	10
	E-beam		5.60×10^5	2.40×10^4	2.0×10^2	10
Total fungi count (CFU/g)	Gamma	1.10×10^5	1.4 x10	< 10	< 10	< 10
	E-beam		10	< 10	< 10	< 10
Bile tolerant gram negative bacteria (per1g)	Gamma	Less than 10^3 more than 10^2	< 10	< 10	< 10	< 10
	E-beam		< 10	< 10	< 10	< 10
Pathogenic *Clostridium* spp.(per 10 g)	Gamma	NF	NF	NF	NF	NF
	E-beam		NF	NF	NF	NF
Salmonella spp. (per10 g)	Gamma	NF	NF	NF	NF	NF
	E-beam		NF	NF	NF	NF
S. aureus (per 1 g)	Gamma	NF	NF	NF	NF	NF
	E-beam		NF	NF	NF	NF
E. coli (per 1 g)	Gamma	NF	NF	NF	NF	NF
	E-beam		NF	NF	NF	NF

All test were run in triplicate samples and mean values were used. NF: Not Found.

5 kGy reduced the microbial contamination to the standard level, complete microbial decontaminations were obtained at 10 kGy irradiation. Irradiated doses of 5, 10, 15 and 20 kGy did not affect total phenolic content and antioxidant activity [13]. Prajna *et al*, 2012 reported electron beam irradiation at dosage of 8 kGy was sufficient to destroy all fungal contaminants in six different herbal raw materials whereas dosage of 12 kGy was needed to completely hygienise. Doses selected in the study were not affected the nutritional quality like carbohydrates, proteins, phenolics and tannins of herbs to a larger extent [10].

3.2. Total phenolic content and antioxidant activity

As show in table 2, the total phenolic content, the DPPH free radical scavenging activity and the ferric reducing antioxidant potential (FRAP) of unirradiated and irradiated samples exhibited ranges from 19.32-22.44 mg gallic acid equivalent/g, 5.20-7.82 mg ascorbic acid equivalent/g and 69.46 to 82.06 μmol $FeSO_4$/g, respectively. The results indicated that no significant differences in the phenolic content, antioxidant activities by DPPH and the ferric reducing antioxidant potential (FRAP) between the un irradiated and irradiated samples of gamma ray and electron beam irradiation at doses of 5, 10, 15 and 20 kGy ($p > 0.05$).

This results according to the other studies, gamma irradiated at 1, 3, 5 and 10 kGy did not show significant differences in antioxidant activity compared with the unirradiated of seven dessert spices (anise, cinnamon, ginger, licorice, mint, nutmeg and vanilla) [14] and tea (*Camillia sinensis*) [15]. Gamma irradiation of sage and oregano at 30 kGy did not have any significant effect on DPPH radical scavenging activity, reducing power and total phenolic content ($p > 0.05$)[16]. Gamma radiation processing at doses of 5 kGy complete decontamination of microorganisms in Triplala (a mixture of *Emblica officinalis*, *Terminalia chebula* and *Terminalia bellirica*) and irradiated samples showed linearly increasing concentration of gallic acid (3.3 to 4.5 times), total phenolic contents (2.16 to 2.87 times) and antioxidant properties with increasing radiation dose up to 25 kGy[17]. Radiation treatment with 10 kGy and 20 kGy of gamma radiation and electron beam did not affect significantly the yield of extraction of chamomile (*Matricaria chamomilla* L.) dried powder and sensory qualities of all extracts were not significantly different depending on irradiation types and doses[18].

Table 2. Analysis of total phenolic content, DPPH and FRAP of *C. alata* (L.) Roxb. irradiated at varied doses of gamma ray and electron beam irradiation.

Ionizing radiation	Dose (kGy)	Total phenolic content (mg gallic acid equivalent/g)	DPPH (mg ascorbic acid equivalent/g)	FRAP (µmol $FeSO_4$/g)
	0	22.08 ± 0.98 [ab]	7.21 ± 0.67 [ab]	82.06 ±13.30 [a]
Gamma	5	21.14 ± 0.84 [ab]	6.69 ± 1.37 [ab]	75.83 ± 1.74 [a]
	10	22.13 ± 1.02 [ab]	7.82 ± 0.54 [b]	76.21 ± 1.31 [a]
	15	19.32 ± 0.44 [a]	6.11 ± 0.73 [ab]	69.46 ± 0.85 [a]
	20	20.37 ± 1.05 [ab]	6.81 ± 0.45 [ab]	70.74 ± 1.86 [a]
Electron	5	22.44 ± 2.42 [b]	6.81 ± 1.28 [ab]	81.53 ±14.27 [a]
	10	21.22 ± 2.91 [ab]	5.64 ± 1.45 [a]	79.22 ±13.39 [a]
	15	21.14 ± 1.35 [ab]	5.20 ± 1.22 [a]	79.99 ±12.14 [a]
	20	21.33 ± 1.47 [ab]	5.62 ± 1.82 [a]	78.31 ±10.35 [a]

Each value is expressed as mean ± standard deviation (n=3).
Mean with difference small letters within column attribute are significantly different ($p < 0.05$)

4. Conclusions

The results of this work showed that gamma ray and electron beam irradiation could be considered as effective method to improve the microbial qualities of *C. alata* (L.) Roxb. powder. Both of gamma ray and electron beam irradiation with doses of 10 kGy led to desired microbial reduction with acceptable variation of antioxidant activity, whereas dosage of greater than 20 kGy was needed to completely hygienise. Dose selected in this study at 5, 10, 15 and 20 kGy were no significant differences on total phenolic content and antioxidant activity of *C. alata* (L.) Roxb.

Acknowledgements

The authors would like to offer particular thanks to staffs at Gems Irradiation Centre, Thailand Institute of Nuclear Technology (Public Organization), responsible for aiding irradiation process. We also thank Dr. Pimpon Uttayarat for help in editing abstracts and special thanks to the academic committee of The International Nuclear Science and Technology Conference 2016, Thailand for editing the manuscripts.

References

[1] Chatterjee S, Chatterjee S and Dutta S 2012 *Inter. J. Res. Bot.* **2** 13-9
[2] Chatterjee S Chatterjee S and Dutta S 2013 *Med. Aromat. Plants.* **2** 1000e143
[3] Sarkar B, Khodre S, Patel, P and Mandaniya M 2014 *Asian J. Pharm. Sci. & Tech.* **4** 4-7
[4] Moriyama H, Moriyama H, Iizuka T, Nagai M, Miyataka H and Satoh T 2003 *Yakugakuzasshi* **123** 607-11
[5] Schweiggert U, Carle R and Schieber A 2007 *Trends Food Sci. Tech.* **18** 260-8
[6] Ministry of Public Health 2005 *Thai pharmacopoeia Supplement* Vol I and II (Bangkok : Office of National Buddishm Press) pp 11-27
[7] Velioglu S, Mazza G, Gao L and Oomah B 1998 *J. Agri. Food Chem.* **46** 4113-7
[8] Khattak K F, Simpson T J and Ihasnullah 2008 *Food Chem.* **11** 967-72
[9] Benzie I F F and Strain J J 1996 *Anal. Biochem.* **239** 70-6
[10] Prajna P S, Rama B P, Ganesh S, Subhramanya P and Dharmendra K M 2012 *Adv. Biores.* **3** 99-106
[11] Zaman S, Khorshed Alam Md, Mortuza F and Bari Md. L, 2015 *J. Food Nutri. Sci.* **3** 165- 170
[12] Phianphak W, Rengpipat S and Cherdshewasart W 2007 *Songklanakarin J. Sci. Technol.* **29** 157-66
[13] Eamsiri J, Sajjabut S, Pewlong W and Chookaew S 2016 *Proc. Int. Conf. on Nat. Prod. for Health and Beauty (NATPRO6). January 21-23 (Khonkaen)* 354-7

IOP Conf. Series: Journal of Physics: Conf. Series **860** (2017) 012003 doi:10.1088/1742-6596/860/1/012003

[14] Murcia M A, Egea I, Romojaro J, Parras P, Jimenez, A M and Martinez M 2004 *J. Agri. Food Chem.* **52** 1872-81
[15] Mishra B B, Gautam S and Sharma A 2006 *J. Food Sci.* **71** M151-6
[16] Perez M B, Banek S A and Croci C B 2011 *Food Chem.* **126** 121-6
[17] Kumari N A, Kumari P, Mitra D, Prasad B, Tiwary B N and Varshney L 2009 *J. Food Sci.* **74** M109-13
[18] Bachir M A 2014 *Inno. Romanian Food Biotech.* **15** 31-9

IOP Conf. Series: Journal of Physics: Conf. Series **860** (2017) 012004 doi:10.1088/1742-6596/860/1/012004

The effect of gamma radiation on sterility and mating ability of brown planthopper, *Nilaparvata lugens*(Stål) in field cage

W Limohpasmanee[1], T Kongratarpon and T Tannarin
Thailand Institute of Nuclear Technology, Ongkharak, Nakhon Nayok, 26120 Thailand

[1] Corresponding author e-mail: wanitch1@yahoo.co.th

Abstract. The brown planthopper, *Nilaparvata lugens*(Stål) is the major rice pest in Thailand. Adults and nymphs suck the sap from the rice plant causing it to wilt and transmitting the grassy stunt and the ragged stunt diseases. The population suppression by the sterile insect technique is overwhelmingly a function of mating between sterile males and wild females. The objectives of these experiments were to determine the suitable dose which induces partially sterile in *N. lugens* and their effect on wild population in the field cages. One-day-old 4th and 5thinstar nymphs and adults were irradiated in a ^{60}Co irradiator at the doses of 30, 60, 90 and 120 Gy. It was found that irradiation at the dose of 90 Gy induced complete sterility in female and 78.47 % sterility in males. The inherited sterility were transferred to their progenies and induced 51.46 and 77.00 % sterility in F-1 males and females. The irradiation as the mention dose did not affect mating ability. The competitiveness index was increased when the ratio of irradiated males per normal male was increased. The releasing irradiated males at 10 fold of normal males in field cages could suppress F-1 population 80.11 % and suppress F-2 population 80.32 % when compare with the control. This technique may be applied to delay and/or reduce seasonal increase of brown planthopper.

1. Introduction

Brown planthopper (BPH, *Nilaparvata lugens* (Stål)) is the major rice pest. Adults and nymphs suck the sap from the rice plant causing it to wilt and transmitting the grassy stunt and the ragged stunt diseases. In Thailand, *N. lugens* was found in the central, the lower north and some parts of north eastern regions. In 2012, the outbreak areas were covered 970,620 rai, which was approximated to 10.07 % of the total cultivated area. The usual chemical method of control has not always prevented crop damage but to some extent may have even magnified the problem, through the development of insect resistance to the chemicals and the elimination of the hopper's natural enemies. The sterile insect technique (SIT) was first developed in 1927 and could be applied as a component of an integrated program to control several major insect pests. The advantages of this technology are species-targeted, environmental-friendly, highly efficient when the population density of the target pest is partly reduced, and replacement application of the harmful insecticide. The technique was first applied to screwworms (*Cochliomyia hominivorax*) in south-eastern USA and Mexico. SIT is also being used to suppress codling moths, (*Cydia pomonella* (L.)) in Canada, fruit flies (*Bactrocera* spp.) in Argentina, Chile, USA, Mexico, Japan, Israel and Jordan, and tsetse flies (*Glossina* spp.) in many African countries. In Thailand, Thailand Institute of Nuclear Technology (Public Organization) has conducted an integrated control of fruit flies by using the sterile insect technique together with other control methods in several areas including Doi Ang Khang in Chiang Mai province, Troknong Subdistrict in Chantaburi province and Long district in Phrae province. Successful application of SIT requires the ability to rear, sterilize and distribute sufficient number of insects to achieve a sufficiently high "overflooding" (sterile: wild insect) ratio in the field for the sterile males to successfully compete and mate with their wild counterparts. However, because Lepidoptera and Hemiptera are radioresistant, irradiation at a full-sterility dose will cause adverse effects such as morphological and behavioral abnormality as well as a decrease in mating ability (Lance and McInnis 2005). But irradiation at sub-sterility dose could induce partially sterile in parent and transfer inherited sterility to F1 offspring. The lower dose of radiation used to induce F1 sterility increases the quality and competitiveness of the released insects (North 1975).

In general, males irradiated at the completely sterilizing dose often cannot compete with untreated males in mating although irradiation did not significant reduce longevity of the males (Economopoulos, 1972). In addition, mass rearing of insect at constant temperature and using artificial diets for many generations affects physiological and behavioral characteristics. Kakinohana (1980) reported mass rearing with artificial diet in laboratory caused the qualitative changes in oviposition, sexual maturation, diurnal rhythm of mating, mating competitiveness and dispersal ability in the field of the melon fly. Hibino and Iwahashi (1992) showed differences in mating behavior between mass-reared melon flies and wild flies. Iwahashi (1993) showed a non-synchronized daily activity pattern with wild melon flies of the released flies. Sexual compatibility is the degree to which two sympatric groups of insect tend to mate randomly without regard to their group of origin rather than mating selectively with members of their own groups. Due to sterile insects did not reproduce, and therefore their fitness was largely a matter of survival, dispersion, adequate behavioral response, habitat finding, and successful mating, i.e. fitness ends with mating or insemination (LaChance 1979). Therefore the general sexual competitiveness of sterile insect was largely defined and influenced by components such as survival, mating propensity, mating compatibility, post-mating factors, etc. (FAO/IAEA/USDA 2003; Lance and McInnis, 2005).

Field releases of partially sterile insects have demonstrated the usefulness of inherited sterility to control many lepidopteran pests, including the cabbage looper *Trichoplusia ni* (Hubner) (North and Holt 1969), corn earworm *Helicoverpa zea* (Boddie) (Carpenter and Gross 1993), gypsy moth *Lymantria dispar* (L.) (Mastro 1993), codling moth (Bloem et al. 2001). Season-long releases of irradiated *H. zea* in mountain valleys of North Carolina, USA, delayed and/or reduced seasonal increases of wild *H. zea* males.

The objectives of these experiments were to determine the suitable dose of gamma radiation which induces partially sterile in *N. lugens* and their effect on wild population in the field cages.

2. Methods

The brown planthoppers used in these experiments were obtained from Pathumthani Rice Research Center and mass-reared at Thailand Institute of Nuclear Technology (Public Organization) Ongkarak, Nakhon Nayok, Thailand. One-day-old 4th and 5thinstar nymphs and adults were irradiated in a ^{60}Co irradiator (GC-5000, BRIT, India) at the doses of 30, 60, 90 and 120 Gy; the dose rate was 4 kGy/h. At the 5th nymph stage, the insects were sexed and reared separately. To assess the sterility of parental males and females, irradiated males were mated with normal females and normal males were mated with irradiated females. While normal males were mated with normal females in the control cage, the number of the offspring was counted every 3 days and the progenies obtained from each cages were reared on the rice seedlings until they emerge to adult. To assess the sterility of F-1 males and females, F-1 males were mated with normal females and F-1 females were mated with normal males. The number of the offspring was counted every 3 days. For the remaining calculations, data were angle (arcsine) transformed and replications were polled.

To assess sexual competitiveness, one-day-old adults were irradiated at the doses of 90 and 120 Gy and placed as ratio 1:1 and 3:1 of normal males in mating cages which 10 virgin females were provided. While normal males were mated with normal females and irradiated males were mated with normal females in the control cages. The number of F-1 progenies was counted every 3 days and reared with rice seedling until all released insects died. The number of offspring of normal and irradiated males which mated with normal females in each treatment was given, as well as the resulting competitiveness values, estimated by modified the determination of sterile insect competitiveness (Fried, 1971).

$$\mathrm{CI} = \frac{(Ha - Hb)}{(Hb - Hc)} \times \frac{N}{S}$$

Where CI = Competitiveness Index;
H_a = number of offspring from a mating with normal males;
H_b= number of offspring from a mating with irradiated and normal males for an N/S ratio;
H_c= number of offspring a mating with irradiated males;
N = Number of normal males
S = Number of irradiated males

Value of 1 indicates the equal competitiveness between the sterile and normal males, and< 1 indicates normal females mated with sterile males less than normal males.

To assess effect of partially-sterile insect releasing on population of brown planthopper in the100x100x150 cm. field cages. The 14 days rice seedlings were provided in the cages as their food. The one-day-old adults were irradiated at the doses of 90 Gy and released 0, 1, 5 and 10 fold of normal males in the cages, which 100 normal males and females were released before. The number of F-1 and F-2 progenies were counted every 7 days and reared with rice seedling for 60 days. The data were analyzed with ANOVA.

3. Results

When the 4th instar nymphs of *N. lugens* were irradiated at the doses of 30 and 60 Gy, the resulting females only laid a few eggs after mating with normal males, and all eggs did not hatch. On the other

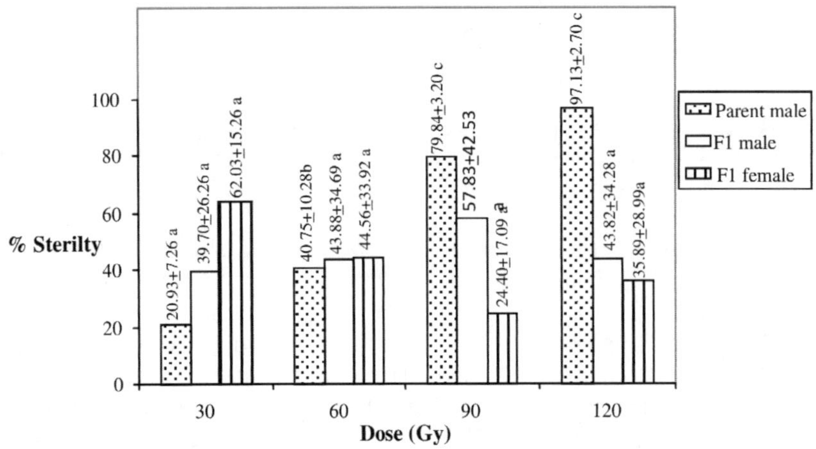

Means in the same pattern columns followed by the same letter are not significantly different ($p < 0.05$).
Figure1. Effects of gamma radiation on the sterility of *N. lugens* when irradiated the 5th nymph.

hand, males irradiated at the same doses exhibited only 66.99 and 90.96 % sterility, respectively. At the doses of 90 and 120 Gy, all irradiated nymphs died before adult emergence.

The results involving the irradiation of *N. Lugens* at the 5th instar nymph stage are shown in figure 1. The females irradiated at the doses of 30, 60 and 90 Gy laid only a few eggs, and all of the eggs did not hatch. Females irradiated at the dose of 120 Gy did not laid any eggs. The average sterility of males irradiated at the doses of 30, 60, 90 and 120 Gy was 20.93, 40.75, 79.84 and 97.13 % respectively. Radiation-induced sterility increased with increasing doses; the relationship between the radiation dose and sterility could be expressed as y = 0.0181x+3.244 with $R^2 = 0.9527$. The sterility of the corresponding F1 males whose male parents had been irradiated at the doses of 30, 60, 90 and 120 Gy were 39.70, 43.88, 57.83 and 43.82 %, respectively. On the other hand, the sterility of F1 females whose male parents had been irradiated at the doses of 30, 60 90 and 120 Gy were 62.03, 44.56, 24.40 and 35.89 %, respectively.

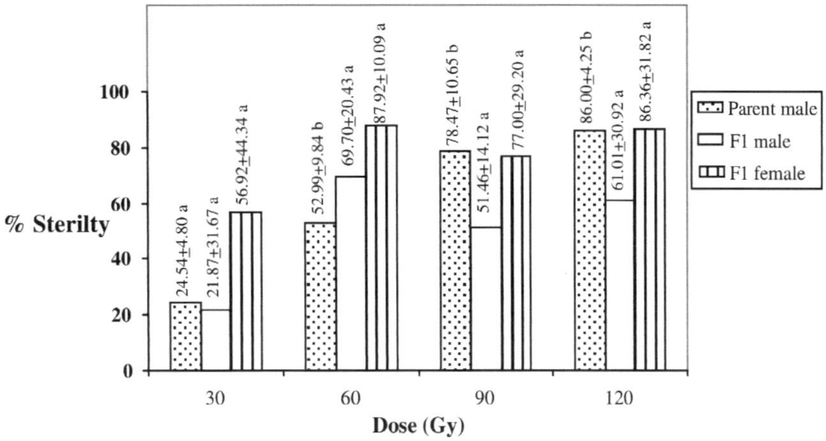

Means in the pattern color columns followed by the same letter are not significantly different ($p < 0.05$).
Figure2. Effects of gamma radiation on the sterility of *N. lugens* when irradiated the adults.

The result of adult irradiation was shown in the figure 2. The females irradiated at the doses of 30, 60, 90 and 120 Gy laid a few eggs and all eggs did not hatch. The average sterility of males irradiated at the doses of 30, 60, 90 and 120 Gy was 24.54, 52.99, 78.47 and 86.00 % respectively. The sterility increased with increasing doses, the relationship between doses and sterility were y = 46.008ln(x) – 132.54, $R^2 = 0.9883$. The sterility of the corresponding F1 males whose male parents had been irradiated at the doses of 30, 60, 90 and 120 Gy were 21.87, 69.70, 51.46 and 61.01 %, respectively. On the other hand, the sterility of F1 females whose male parents had been irradiated at the doses of 30, 60 90 and 120 Gy were 56.92, 87.92, 77.00 and 86.36 %, respectively.

The results of the sexual competitiveness assay were shown in table 1. The competitiveness indices of the 90 Gy irradiated males were 1.03 and 3.43 when released at the ratios 1:1 and 3:1 overnormal males. The competitiveness index increased as the ratio of irradiated males per normal male increased. These values were truncated to 1 and equal competitiveness between sterile and normal males was assumed. The competitiveness indices of the 120 Gy irradiated males were 0.99 and 0.95 when released at the ratios 1:1 and 3:1 over normal males. The competitiveness index decreased slightly when the ratio of irradiated males per normal males increased. The results also indicated that gamma irradiation at the dose of 90 Gy had no effects on mating ability of *N. lugens* males, while irradiation at the dose of 120 Gy slightly decreased the competitiveness.

The results of partially-sterile insect releasing on population of brown plant hopper in the field cages were shown in table 2. The average number of F-1 nymph at the peak were decreased every treatments but only the treatment which released irradiated males 10 fold of normal males was highly significant (F= 7.19, P<0.01).The average number of F-1 females and F-2 nymph at the peak of all treatments which released irradiated males were decreased high significantly (F= 11.52, P<0.01 ; F=40.58, P<0.01) when compare with control, but were non-significant among the treatments. The 90 Gy irradiated males releasing at 1, 5 and 10 fold of normal males suppressed F-1 population 3.77, 33.86 and 80.11 %respectively and suppressed F-2 population 75.44, 76.58, 80.32 % respectively.

Table 1. The average number of offspring and the competitiveness index of irradiated males at different ratios of normal and irradiated males.

Treatments	Average number of offspring (nymphs)	Sterility (%)	Competitiveness index
10 NM x 10 NF	58.40	-	-
10 $IR_{90}M$ x 10 NF	10.60	81.88 ± 3.37b	-
10 NM x 10 NF x 30 $IR_{90}M$	15.00	74.37 ± 2.82b	3.43 ± 0.84b
10 $IR_{120}M$ x 10 NF	5.80	89.98 ± 4.32b	
10 NM x 10 NF x 10 $IR_{120}M$	34.00	41.35 ± 16.79a	0.99 ± 0.61a
10 NM x 10 NF x 30 $IR_{120}M$	21.60	62.75 ± 12.26b	0.95 ± 0.60a

Means in the same column followed by the same letter are not significantly different ($p < 0.05$)
NM : normal male NF : normal female $IR_{90}M$: 90 Gy irradiated male $IR_{120}M$: 120 Gy irradiated male

Table 2. Effect of sterile insects releasing on brown planthopper population in the field cages.

Treatment	Maximum number of the brown planthopper					
	Nymphs (F-1)		Females (F-1)		Nymphs (F-2)	
	Mean	SD	Mean	SD	Mean	SD
100 NMx100 NF	3509.75a	517.45	1104.75a	130.08	6035.50a	1304.76
100 NMx100 NFx100 IRM	3377.50a	1138.90	355.50b	118.21	1482.50b	306.00
100 NMx100 NFx500 IRM	2321.25ab	1299.26	398.75b	387.28	1413.25b	591.26
100 NMx100 NFx1000 IRM	698.00b	89.35	355.50b	85.25	1188.00b	501.18

Means in the same column followed by the same letter are not significantly different ($p < 0.05$)
NM : normal male NF : normal female IRM : 90 Gy irradiated male

4. Discussion and conclusion

Adult of *N. lugens* was more resistant to radiation than 4th and 5th nymph, due to somatic division is at a minimum (Robinson 2005). Furthermore, radiosensitivity appears to be influenced by cell repopulation, tissue and organ regeneration ability, and biological repair. And males were more resistant to radiation sterilization than females, like the most species. Due to the basic difference between male and female germ cells (sperm are haploid post-meiotic, whereas mature eggs are pre-meiotic). Females were completely sterile when irradiated at the dose of 30 Gy. While males were sterile only 24.54 %, when irradiated at the same dose. Mochida (1973b) found that egg formation was interrupted when fifth-instar female nymphs were irradiated at 15 to 20 krad.

The Lepidoptera and Hompotera which have the diffuse centromeres, some dominant lethal mutation may be expressed just prior to hatch, but most of the broken chromosome pass on to the F-1 generation. It is mainly in the F1 generation of Lepidoptera that induced chromosome abnormalities express themselves as dominant lethal mutations (Carpenter 2005). Insect in this group was radiation resistant, the irradiation at the fully sterility dose caused adverse effects on insect such as abnormality, mating ability decrease. To maximize genetic damage to mature germ cells and minimize somatic damage, insect should be irradiated at 1-day adult as the dose of 90 Gy although irradiation at the dose of 120 Gy gave higher sterility both in parent and F-1. Irradiation adult at the dose of 90 Gy, females were completely sterile while male were sterile 78.47 %. and F-1 males and females were sterile 51.46 and 77.00 % respectively.

The sexual competitiveness assay indicated that gamma irradiation at the dose of 90 Gy had no effects on mating ability of *N. lugens* males, while irradiation at the dose of 120 Gy slightly decreased the competitiveness. And the competitiveness index increased as the ratio of irradiated males per normal male increased when released the 90 Gy irradiated males. When 90 Gy irradiated males were released at the ratio of 3:1 over normal males, 74.37% suppression of F1 fertility was achieved. When the dose of 120 Gy was used, the suppression level was reduced to 62.75%, indicating that 90 Gy irradiation showed a higher efficiency of controlling *N. Lugens* populations. At lower doses, irradiated insects were partially sterile and would carry minimal somatic and morphological damage. The partially sterile males appeared more active, dispersed over a greater distance, and generally were more competitive than males with a higher level of sterility (Bloem et. al. 2001). The releasing 90 Gy males at 1, 5 and 10 fold of normal males in field cages showed 3.77, 33.86 and 80.11 %suppression in F-1 population and 75.44, 76.58, 80.32 % suppression in F-1 population. Their efficiency was decreased when compare results with mating competitiveness assay, because of the difference of space and population density. However, this technique could reduce the increasing rate of brown planthopper only. But the number of insects in filed cages still increases because the fraction of males that remained fertile was higher than the fraction of increasing rate. This technique was applied to delay and/or reduce seasonal increases of wild corn earworm *Helicoverpa zea* (Boddie) in mountain valleys of North Carolina, USA. Release ratios average less than 5:1 overall, but reduce the wild population of *H. zae* by more than 70 %. In another case, season-long field studies of the codling moth were conducted in apple orchards in Washington State, USA, both release of partially sterile codling moth and combination of mating disruption plus the release of partially sterile codling moth gave the significantly lower fruit damage when compared with control plots located outside the treatment areas.

The suitable dose of gamma radiation which induces partially sterile in *N. lugens* adults was 90 Gy, irradiated females were completely sterile and irradiated males were sterile 78.47 %. The abnormal chromosome were transfer to their progeny and induced 51.46 and 77.00 % sterility in F-1 males and females. The irradiation as the mentioned dose did not affect mating ability and the competitiveness index was increased when the ratio of irradiated males per normal male was increased. The releasing irradiated males at 10 fold of normal males in field cages could suppress F-1 population 80.11 % and suppress F-2 population 80.32 %, when compare with the control.

References

[1] FAO 2005 *Glossary of phytosanitary terms* Secretariat of the International Plant Protection Convention (IPPC), FAO

[2] LaChance LE 1967 *Genetics of insect vectors of disease* (Elsevier, Amsterdam, The Netherlands) pp 617-50

[3] LaChance LE, CH Schmidt, and RC Bushland *Pest control Biological, physical, and selected chemical methods* (Academic Press, New York, USA) pp 147-196

[4] LanceDR and DOMcINNIS 2005*Sterile Insect Technique* (Springer, The Netherlands) pp 70-85

[5] FAO/IAEA/USDA 2003*Manual for Product Quality Control and Shipping Procedures for Sterile Mass-Rearing Tephritid Fruit Flies*, Version **5.0**. IAEA. 85 pp

[6] North DT 1975 *Annual Review of Entomology***20** 167-82

[7] BloemS, K ABloem, JE Carpenter and CO Calkins 2001*Environmental Entomology* **30** 763-69

[8] Carpenter JE, ANSaprks and HLCromroy 1987a*J. of Econ.Ent.* 801233-37

[9] Carpenter JE., and HR Gross 1993*Environmental Entomology***22** 1084-91

[10] Economopoulus AP 1972*Environmental Entomology***1**:490-97

[11] KakinohanaH 1980 Qualitative change in the mass-reared melon fly, *Dacus cucurbitae* Coq. *Proc. of Symp.on Fruit Fly Problems*27-36

[12] Hibino Y and OIwahashi 1992*Appl. Ent. Zool.***24**152-54

[13] Iwahashi O 1993*Follow-up Mission Report* for THA/5/038/02.IAEA. 33p

[14] La ChanceLE1979 *Genetics in relation to insect management* (The Rockefeller Foundation, New York, N.Y., USA) pp 8-19

[15] FAO/IAEA/USDA 2003 Product quality control and shipping procedures for sterile mass-reared tephritd fruit flies.Version 5.0 IAEA, Vienna, Austria Http://www.iaea.org/programmes/nafa/d4/index.html

[16] Fried M 1971 *J. of Econ.Ent.***64**869–72

[17] Bauer H 1967*Chromosoma***22**102-25

[18] Klassen W and J F Creech 1971*Miscellaneous Publication USDA/ARS* **Number 1182**

[19] North DT and G G Holt 1969 *Canada Entomologist***101**513-20

[20] Mastro VC1993 *Final Research Co-ordination Meeting IAEA* STI/PUB/929 pp 125-129

[21] Muller HJ 1954 *Radiation Biology***1**351-473

[22] LaChance L E and J G Riemann 1964*Mutation Research***I** 318-33

[23] Robinson AS 2005*Sterile Insect Technique* (Springer, The Netherlands) pp 70-85

[24] Mochida O 1973 Effect of gamma radiation on the development and reproduction of the brown planthopper, *Nilaparvata lugens* (Stal) (Homoptera: Delphacidae) 42A4DC1A93E823038C

[25] Carpenter JE, S Bloem and F Marec 2005*Sterile Insect Technique* (Springer, The Netherlands) pp 115-46

[26] Carpenter JE, Hidryani N Nelly, and BG Mullinix 1997 *J. of Econ.Ent.***90**: 444

IOP Conf. Series: Journal of Physics: Conf. Series **860** (2017) 012005 doi:10.1088/1742-6596/860/1/012005

Elemental characterization of Mt. Sinabung volcanic ash, Indonesia by Neutron Activation Analysis

I Kusmartini[1], W Y N Syahfitri, S Kurniawati, D D Lestiani, M Santoso

Center for Applied Nuclear Science and Technology
Tamansari 71, 40132, Bandung, Indonesia

[1] Corresponding author e-mail: indahkus@batan.go.id

Abstract. Mount Sinabung is a volcano located in North Sumatera, Indonesia which has been recorded not erupted since 1600. However in 2013 it has been erupted and cause of black thick smog, rain sand and volcanic ash. Volcanic ash containing trace elements material that can be utilized in various applications but still has potential danger of heavy metals. In order to obtain an elemental composition data of volcanic ash, the characterization of volcanic ash were carried out using Neutron Activation Analysis. The volcanic ash was taken from Mt. Sinabung eruption. Samples were irradiated at the rabbit system in the reactor G.A Siwabessy facilities with neutron flux ~ 1013 n.cm-2.s-1 and then counted using HPGe detector. Method validation was carried out by SRM NIST Coal Fly Ash 1633b and NIST 2711a Montana II Soil with recovery values were in the range of 96–108% and 95–106% respectively. The results showed that major elements; Al, Na, Ca and Fe, concentrations were 8.7, 1.05, 2.98 and 7.44 %, respectively, minor elements K, Mg, Mn, Ti, V and Zn were 0.87%, 0.78%, 0.18%, 0.62%, 197.13 ppm and 109.35 ppm, respectively, heavy metals; As, Cr, Co and Sb, contents were 4.48, 11.75, 17.13 and 0.35 ppm, respectively while rare earth elements such as Ce, Eu, La, Nd, Sm, Yb were 45.33, 1.22, 19.63, 20.34, 3.86, and 2.57 ppm respectively. The results of the elemental contents of volcanic ash that has been obtained can be used as the scientific based data for volcanic material utilization by considering the economic potential of elements contained and also the danger of the heavy metals content.

1. Introduction

Mount Sinabung is a stratovolcano, located in Karo Regency, North Sumatra Province of Indonesia. It is a solitary volcano that has a single peak, and classified as B -Type, because since the 1600's there was no record of eruptive activity [1]. But on November 24, 2013, Mount Sinabung were erupted and reach the highest alert level (level 4) since it spat out black and thick smog, followed by rain sand and volcanic ash which covered thousands of hectares of farmer crops under the radius of six kilometers [2]. Volcanic ash is fragments of magma and consists of mineral, volcanic glass and also a material containing high silica and aluminum [2,3]. The composition of particles from volcanic eruptions should reflect the matrix composition of the magma. SiO_2 contents of 40–55 wt% and CaO plus K_2O between 4 and 6 wt% match the composition of basaltic to basalt andesitic magma [4].

Related of that there are several studies of the characterization of volcanic ash., F. Ruggieri, *et al* (2012) had been researched mineral compositions of volcanic ash samples of Chaiten volcano, Chile Southern, such as SiO_2, Al_2O_3, Fe_2O_3, MnO, MgO, CaO, Na_2O, K_2O, TiO_2 were 72.06, 14.66, 1.88, 0.07, 0.6, 1.99, 4.18, 2.89, 0.2 % respectively [5]. These materials in the ash can provide important information on the nature of magma because chemical compositions of magma usually show distinct features of each volcano and because the assemblages and compositions of minerals reflect their host [6]. The other study had been done by Bonnie Canion, *et al* (2012) that found of trace elements in volcanic ash from Mount Merapi, such as As, Co, Cr, Sb, V, Zn [7]. General features of the studied volcanic ashes from the Puna in Northwestern Argentina by F Ruggieri *et al* (2010) also obtained rare earth elements such as La, Ce, Nd, Sm, Eu, Yb [8]. The rare earth elements (REEs) or rare earth metals (REMs) as defined by IUPAC, represent a group of 17 transient metals. They comprise scandium, yttrium and further 15 elements called lanthanides, ranging from the most abundant cerium and lanthanum to the less frequent lutetium. The natural abundances of REEs greater than their name would suggest, with several elements being present at an average content of 30 mg/kg on the Earth's crusth [9]. Therefore explosive eruptions are capable of causing significant environmental repercussions, affecting simultaneously many cycles in the Earth's system, impacting on water, soil and surface sediments. An important reflection of volcanic activity is the natural contribution of chemical elements [10].

Neutron Activation Analysis (NAA) is a sensitive analytical technique which useful for performing both qualitative and quantitative multi-elemental analysis of major, minor and trace elements in samples from almost every field of scientific or technical interest. Particularly advantage is taken of the fact that the samples do not have to undergo any chemical treatment, non destructive which in many materials belong to the major matrix components. It enables the observation of trace elements, often at amounts in the order of microgram to nanogram. The high selectivity of gamma-ray spectrometry allows for simultaneous determination of many radionuclides [22]. NAA is an established technique which can be used for the measurement of trace element levels in various matrices. It involves neutron activation of the different trace elements in the sample and measurement of the radioactivity of each radioisotope formed in order to measure the individual concentrations of elements. NAA facilitates of elements in samples of different matrices especially of environmental origin. It is one of the most mature analytical methods currently used and yet remains highly compete with others in term of accuracy, detection limits and multi elemental capabilities [11,12].

2. Methodology

2.1. Sample Collection and Preparation
The samples of ash were collected in yard of residential dwelling as far as 6 km of the volcano. These sites were located in mountain side of Sinabung Karo Regency, North Sumatra Province. Volcanic ash samples were taken about 1 kg of the 2013 eruption. The volcanic ash samples were dried at temperature 45–60 ^{0}C using infrared lamps. After it dried reduction of particles size were done using electrical pulverisette and 200 mesh sieve to get $\leq 74\mu m$ particles size of volcanic ash.

2.2. Standard Preparation
About 25–40 mg of volcanic ash were weighted and put into 0.3 mL polyethylene vials. Certified reference material from National Institute of Standards and Technology (NIST), SRM NIST 1633b and SRM NIST 2711a Montana II Soil were used to validate the method. Mix standards for relative method were prepared and placed in the same size vials. Preparation of NAA standards was carried out by pipetting 100 μL of ICP multi element standard (E-Merck) then dripped in the polyethylene vial. The vial dried under the infrared lamp then sealed by heating. Weights of elements in the prepared standards were shown in Table 1.

2.3. Irradiation and Gamma Measurement
The samples, certified reference materials and standards were irradiated in the rabbit system of The Multi-Purpose Reactor G.A Siwabessy in Serpong, Banten Province of Indonesia with neutron flux ~ 10^{13} n.cm^{-2}.s^{-1}. The condition of irradiation were shown in Table 2. After cooled the samples dan standard measured using a spectrometer gamma ray, HPGe detector manufactured by Canberra. The spectrum analysis was carried out using GENIE 2000.

Table 1. The weight of elements for preparation NAA standards

IOP Conf. Series: Journal of Physics: Conf. Series **860** (2017) 012005 doi:10.1088/1742-6596/860/1/012005

Elements	Weight (ug)
Al	21
Na	20.98
Ca	101.2
Fe	49.9
K	21
Mg	21
Mn	1
Ti	10
V	1.01
Zn	10
As	9.8
Cr	1.01
Co	1
Sb	1
La	0.1
Sm	0.1
Eu	1
Yb	1
Ce	1
Nd	1

Table 2. Irradiation conditions

Element	Half Time	Irradiation Time	Delay/Cooling	Counting Time
Al, Ca, Ti, Mn, V, Mg	Short-lived radionuclide	1 minute	1 minute	200 seconds
Na, K, As, Sm, La	Medium-lived radionuclide	10 minutes	1 day	2000 seconds
Fe, Co, Cr, Sb, Eu, Yb, Ce, Nd, Zn	Long-lived radionuclide	1 hour	1 week	4000 seconds

Reliability of volcanic ash analysis results depends on instruments performance as well as personal capability and was proven by its accuracy. Accuracy can be defined as a measure of the closeness of the results with its true value while precision can be considered as a measure of the repeated and reproducibility of a result. Accuracy can be stated as % recovery. Recovery is the ratio between measurement result with certificate value [13], expressed by the formula (1):

$$\%recovery = \frac{value_{analyt}}{value_{certificate}} \times 100\% \qquad (1)$$

The accuracy of method were determined by comparing results of the SRM 1633b and SRM NIST 2711a Montana II Soil with its certificate value respectively. Acceptable criteria for accuracy were based on AOAC *guidelines for single laboratory validation* shown in Table 3 [13].

IOP Conf. Series: Journal of Physics: Conf. Series **860** (2017) 012005 doi:10.1088/1742-6596/860/1/012005

Table 3. Acceptable criteria for accuracy [13]

Concentration	% Recovery
100%	98-101
10%	95-102
1%	92-105
0,1%	90-108
0,01%	85-110
10 ppm	80-115
1 ppm	75-120
10 ppb	70-125

3. Result and Discussion

Validation of NAA was carried out using NIST Coal Fly Ash 1633b and NIST 2711a Montana II Soil Standards Reference Material (SRM). Table 4 showed the results of measurement of SRM NIST Coal Fly Ash 1633b with %recovery were in the range of 96–108%. These values were in the %recovery of acceptance criteria for 0.1% concentration 90–108% [13]. While in the Table 5 showed the results of measurement of 2711a Montana II Soil by NAA method with %recovery were in the range of 95–106%. These values were also in the %recovery range of acceptance criteria for 0.1% concentration that is 90–108% [13]. It might be said that the results for Coal Fly Ash 1633b and 2711a Montana II Soil are in good agreements with the recommended certified value and therefore the NAA method was quite accurate for determining the elemental contents of similar matrices of samples that being analyzed.

In this studies NAA method was applied to determine the concentrations of 20 elements (Al, Na, Ca, Fe, K, Mg, Mn, Ti, V, Zn, As, Cr, Co, Sb, La, Sm, Eu, Yb, Ce and Nd) shown in Table 6 and the figure of gamma rays spectrum were shown in Figure 1. The results of this sudy were compared to those found by other published studies ilustrating the elemental mean concentration levels in volcanic ash of other regions of some prominent states of Europe, Africa, South America and Asia regions [12]. The mean value of Al concentration in Mt. Sinabung volcanic ash, Indonesia were found to be similar with permian Trassic's China [15], while the element of Ca 3 times lower than found in Mt. Shihan, Jordan [16]. Sodium contents of Mt. Sinabung volcanic ash were three times lower than in volcanic

Table 4. Elemental concentration (ppm) in NIST Coal Fly Ash 1633b Standard

Element	Unit	Measured (ppm)	Certificate (ppm)	% Recovery
Al	%	14.4 ± 0.2	15.05 ± 0.27	96%
Ca	%	1.5 ± 0.2	1.51 ± 0.06	100%
Fe	%	7.8 ± 0.2	7.78 ± 0.04	100%
K	%	1.9 ± 0.03	1.95± 0.03	100%
Na	%	0.2 ± 0.0008	0.201± 0.003	100%
Mg	%	0.46 ± 0.11	0.48 ± 0.008	96%
Ti	%	0.7 ± 0.06	0.79 ± 0.014	98%
V	%	0.03 ± 0.001	0.03 ± 0.003	100%
Ce	ppm	190 ± 1.24	190	100%
Cr	ppm	200 ± 6.35	198 ± 4.7	101%
Co	ppm	48 ± 0.5	50	97%
La	ppm	94 ± 1.47	94	100%
Sc	ppm	40 ± 0.1	41	99%
Sm	ppm	20 ± 0.2	20	100%
Zn	ppm	227 ± 63	210	108%

Table 5. Elemental concentration (ppm) in NIST 2711a Montana II Soil Standard

Element	Unit	Measured (ppm)	Certificate (ppm)	%Recovery
Al	%	6.64 ± 0.15	6.72 ± 0.06	99%
Ca	%	2.35 ± 0.13	2.42 ± 0.06	97%
Fe	%	2.83 ± 0.04	2.82 ± 0.04	100%
K	%	2.68 ± 0.06	2.53 ± 0.1	106%
Mg	%	1.07 ± 0.16	1.07 ± 0.06	100%
Mn	%	0.06 ± 0.0016	0.06 ± 0.0018	100%
Na	%	1.21 ± 0.003	1.2 ± 0.01	102%
Ti	%	0.32 ± 0.04	0.31 ± 0.008	99%
As	ppm	107 ± 4	107 ± 5	100%
Ce	ppm	70 ± 5.7	70	100%
Co	ppm	9.89 ± 0.2	9.89 ± 0.18	100%
Cr	ppm	52.7 ± 1.21	52.3 ± 2.9	100%
Eu	ppm	1.1 ± 0.7	1.1 ± 0.2	100%
La	ppm	37 ± 1	38 ± 1	98%
Sc	ppm	8.4 ± 0.1	8.5 ± 0.1	99%
Sm	ppm	5.8 ± 0.22	5.9 ± 0.28	98%
V	ppm	76.6 ± 5.7	80.7 ± 5.7	95%
Zn	ppm	424 ± 10	414 ± 11	102%

ash of Mt. Puna, Argentina [14], while concentration of Fe in Mt. Sinabung volcanic ash, Indonesia and Mt. Grímsvötn Denmark were almost have similar value. Further, Mg in Mt. Sinabung's volcanic ash, Indonesia were has concentration 4 times lower than Mg concentration of Mt. Shihan's, Jordan. However, concentration of Mn in Mt. Sinabung's volcanic ash was the highest among others.

The value of trace elements; V, Zn, As, Cr, Co and Sb concentrations and its comparisons with other results were also shown in Table 6. The content of V and Zn in Mt. Sinabung's volcanic ash was the highest compared with other results, but if it compared with Cordón Caulle Volcanic Complex, Argentina both Zn concentrations tend to have similar value. The mean value of As was 9 times lower than in Chaitén Volcano Chile and 4 times higher than in Alphine Anthering volcano Austria. For Cr the mean value in Indonesia was 23 times lower than Mt. Shihan's, Jordan and 2 times higher than Puna Volcano, Argentina. Concentration of Co was three times lower than in Mt. Shihan's Jordan however Sb contains in Mt. Sinabung Indonesia has the smallest concentration among other results. The particular attention should be given to the concentration of potential pollutants semi-volatile elements; As and Sb, which can be assosiated considering their affinity of fluid or gas phases or forming sulphate or halide volatile compounds at magmatic temperatures, being adsorbed onto ash particles. These elements especially As are potential toxic trace elements for the environtmental [14]. Determination results of rare earth elements (La, Sm, Eu, Yb, Ce and Nd) in volcanic ash were also shown in Table 6. The concentration of La in Mt. Sinabung, Indonesia was three times lower than in Alphine Anthering Volcano Austria also for the element of Sm was three times lower than in Eyjafjallajökull Volcano, Iceland. The mean value of Eu in Mt. Sinabung, Indonesia and Mt. Shihan, Jordan were found to be similar. For Ce in Mt. Sinabung, Indonesia was three times lower than in Alphine Anthering Volcano Austria however with Chaitén Volcano, Chili and Mt. Shihan, Jordan the concentrations were in the same range. Concentration of Nd in Mt. Sinabung, volcanic ash was two times lower than found in Alphine Anthering Volcano Austria, Cordón Caulle Volcanic Complex, Argentina and Eyjafjallajökull Volcano Iceland but four times higher than Permian-Triassic, China. Rare earth elements and their compounds, due to the unique magnetic, electric and optical properties are used in many kinds of industrial products such as, the photo luminescence materials, magnetic alloys, catalysts for petroleum refining, carrier materials of automobile exhaust catalysts and high-refractive index low-dispersion optical glasses. Moreover, $LaCl_3$ is added to the chemical fertilizers to increase the yields of crop [9].

IOP Conf. Series: Journal of Physics: Conf. Series **860** (2017) 012005 doi:10.1088/1742-6596/860/1/012005

The differences concentration of elements for country to each volcanic ash depend on the physical fractination of magma during the explosive events and the subaerial transport by wind currents could have changed the original chemical composition of volcanic ash [19]. These results support earlier statement that there is a difference in the concentration of trace elements. This could be due to the geological conditions and geochemical factors such as hydrothermal fluids have been reported to be responsible for the enrichment of trace elements. Beside the trace elements available these ashes are also provided with macro and micronutrients (Mn, Mg and Fe) evidencing the fertilising potential of the ashes. Although the total contributions of such elements are relatively low in ashes, their deposition could provide an external critical nutrient source in low nutrient areas [6].

Many other elements studied are potentially toxic trace elements (PTTEs) and some of them are included in the drinking water guidelines due to their potential toxicity even at very low concentrations and must be especially monitored in the environmental assessment of these ash fall deposits. Human and animal health effects by expose of ash, linked to respiratory illness, is closely associated with particle size, which are especially anticipated for diameters $\leq$ 10 μm because aerodynamically fine particles successfully negotiate the curves of the throat and are carried to the lungs. An aircrafts operations (especially commercial jets) is threatened by volcanic cloud encounters, which a variety of hazards can exist such as airborne very fine ash that enters and melts in jet turbines and causing failure [20]. Volcanic ash emissions are episodic events that contribute approximately an order of magnitude less silicate particles to the atmosphere on average that contributed by the mineral dust cycle. To place this global context, the total mass of global emissions of fine ash may be estimated from historic records of eruption frequency and assumptions on the proportion of fine ash generated by different styles of explosive volcanism [21].

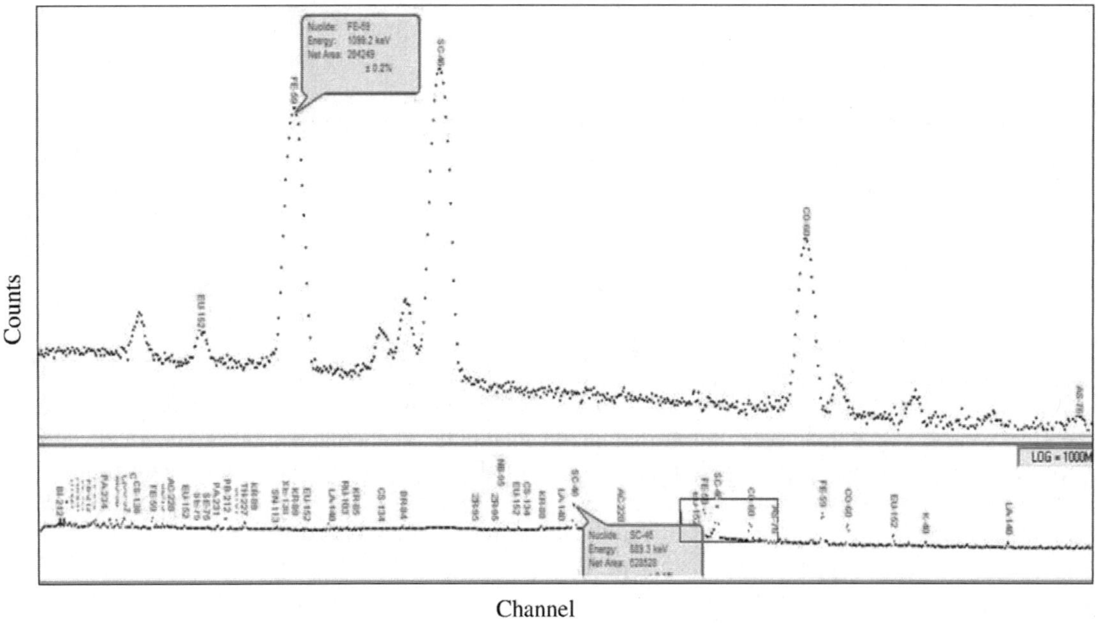

Figure 1: Gamma-rays spectrum from sofware GENIE 2000

Table 6. Elemental concentrations in volcanic ash in this study and other regions

Element	Unit	Present study, Sinabung Volcanic Indonesia 2013	LOD	*n*	Cordón Caulle Volcanic Complex Argentina 2011[a]	Chaitén Volcano Chili 2008[b]	Shihan Volcano Jordania 2007[c]	Grímsvötn Volcano, Denmark 2011[d]	Eyjafjallajö kull Volcano Iceland 2010[e]	Permian-Triassic Boundary China 2015[f]	Alphine Anthering Volcano, Austria 2003[g]	Puna Volcano Argentina 2010[h]
Al	%	8.70 ± 2.26	1.E-03	3	7.56	3.88	4.08	4.68		7.89	5	3.29
Na	%	1.05 ± 0.12	2.E-06	5	3.88	1.55	1.40	0.81		0.05	0.15	1.45
Ca	%	2.98 ± 0.04	1.E-02	3	1.95	1.42	7.02	0.5		0.61	1.52	0.63
Fe	%	7.44 ± 0.5	2.E-06	5	3.37	0.66	3.97	7.7		0.89	1.5	0.28
K	%	0.87 ± 0.13	2.E-04	5	2.73	1.19	0.50	1.19		2.61	0.27	1.86
Mg	%	0.78 ± 0.11	3.E-03	3	0.55	0.36	4.17	1.8		1.34	1.2	0.09
Mn	%	0.18 ± 0.02	1.E-05	3	0.09	0.05	0.13				0.03	0.06
Ti	%	0.62 ± 0.09	1.E-03	3	0.38	0.12	1.04			0.34	0.46	0.06
V	ppm	197 ± 5.12	1.E-01	3	33	13.8	176				27	6.6
Zn	ppm	109 ± 20.5	2.E-03	5	104	51.2	94				50	39.8
As	ppm	4.48 ± 0.79	2.E-03	3	15	38.4					< 1	5.75
Cr	ppm	11.8 ± 2	2.E-04	5	9.1	19	281				10	4.49
Co	ppm	17.1 ± 1.07	5.E-02	5	6.25	2.57	42				2.38	0.82
Sb	ppm	0.35 ± 0.004	2.E-06	2	1.02	1.51					0.44	1.13
La	ppm	19.6 ± 2.2	5.E-03	5	37	24.4	25		40.9	3.77	63.6	14.5
Sm	ppm	3.86 ± 0.43	7.E-04	5	9.43	2.83	4.8		11.4	2.25	9.84	5.26
Eu	ppm	1.22 ± 0.06	9.E-06	5	1.82	0.66	1.2		3.27	0.83	2.42	0.79
Yb	ppm	2.57 ± 0.79	1.E-02	5	6.86	1.44	1.1		4.78	4.36	3.09	1.77
Ce	ppm	45.3 ± 5.27	2.E-04	5	88	46.5	48		93	9.81	135	33.7
Nd	ppm	20.3 ± 1.84	6.E-04	5	44.8	16.5	17		48.6	4.7	53.1	12.7

[a] Daga R et al., 2014 [14].
[b] Ruggieri F et al., 2012 [6].
[c] Ibrahim KM et al., 2014 [16].
[d] Lieke KI et al., 2013 [4].
[e] Tepe N and Bau M. Et al., 2014 [18].
[f] Liao Z et al., 2016 [15].
[g] Huber H et al., 2003 [17].
[h] Ruggieri F et al., 2010 [8].

4. Conclusions

The characterization results showed that Mt. Sinabung volcanic ash contains major elements; such as Al, Na, Ca and Fe with their concentrations were 8.7, 1.05, 2.98 and 7.44 %, respectively, minor elements; K, Mg, Mn, Ti, V and Zn with their concentration were 0.87, 0.78, 0.18, 0.62 %, 197.13 and 109.35 ppm, respectively while heavy metals As, Cr, Co and Sb were 4.48, 11.75, 17.13 and 0.35 ppm, respectively and rare earth elements Ce, Eu, La, Nd, Sm, Yb were respectively 45.33, 1.22, 19.63, 20.34, 3.86, and 2.57 ppm. The differences concentration of volcanic ash of country to each other depend on the physical fractination of magma during the explosive events and the subaerial transport by wind currents that could have changed the original chemical composition of volcanic ash. The results support earlier statement that there is a difference in the concentration of trace elements that could be due to the geological conditions and geochemical factors which have been reported to be responsible for the enrichment of trace elements. Related to that, the results of elemental contents of volcanic ash obtained can be used as the scientific based data for volcanic material utilization by considering the economic potential of rich elements and its danger of the heavy metal contents.

References

[1] Sutawidjaja IS, Prambada O, Siregar DA 2013 The August 2010 Phreatic Eruption of Mount Sinabung, North Sumatra. *Indonesian Journal of Geology*,Vol **8** p 55-61.

[2] Karolina R, Syahrizal, Putra MA, Prasetyo TA 2015 Optimization of the use of volcanic ash of Mount Sinabung eruption. *Procedia Engineering* **125** p 669 – 674.

[3] Hardianita S, Bosas RM, Nuraini Y, 2015 Application of organic matter and biofertilizer to improve growth and yield of maize on soil damaged by volcanic ash of Mount kelud in East Java. *Journal of Degreaded and Mining Lands Management* Vol **2**, number 4 : p 403 – 408.

[4] Lieke KI, Kristensen TB, Korsholm US, SØrensen JH, Kandler K, weinbruch S, Ceburnis D, Ovadnevaite J, O'Dowd CD, Bilde M 2013 Characterization of volcanic ash from the 2011 Grimsvotn eruption by means of single-particle analysis. *Atmospheric Environment* **79** p 411 – 420.

[5] Nakagawa M, Ohba T 2002 Mineral in Volcanic Ash 1: Primary Minerals and Volcanic Glass. Division of Earth and Planetary Sciences. Hokkaido.

[6] Ruggieri F, Fernandes-Turiel JL, Saavedra J, Gimeno D, Polanco E, Amigo A, Galindo G, Caselli Al 2012 Contribution of volcanic ashes to the regional geochemical balance: The 2008 eruption of Chaiten volcano, Southern Chile. *Scince of Total Environtment* **425**. P 57 – 88.

[7] Canion B, Jacques C, Landsberger S, Taftazani A 2012 Trace analysis of Indonesian volcanic ash using thermal and epithermal neutron activation analysis. *Nukleotika* **57**(4): p 585-589.

[8] Ruggieri F, Saavedra J, Fernandes-Turiel JL, Gimeno D, Garcia-Valles M, 2010 Environmental geochemistry of ancient volcanic ashes. *Journal of Hazardous Materials* **183** p 353–365.

[9] Pyrzynska K, Kubiak A, Wysocka I 2016 Application of solid phase extraction procedure for rare earth elements determination in environmental samples. *Talanta* **154** p 15 – 22.

[10] Ghermandi L, Gonzalez S, Franzese J, Oddi F, 2015 Effects of volcanic ash deposition on the early recovery of gap vegetation in Northwestern Patagonian steppes. *Journal of Arid Environments* **122** p 154-160.

[11] Glascock MD 2004 An Overview of Neutron Activation Analysis university of Missouri Research Reactor.

[12] Bouhila Z, Mouzai M, Azli T, Nedjar A, Mazouzi C, Zergoug Z, Boukhadra D, Chegrouche S, Lounici H 2015 Investigation of aerosol trace element concentration nearby Algiers for environtmental monitoring using instrumental neutron activation analysis. *Atmospheric Research* **166** p 49–59.

[13] Gonza´lez AG and Herrador MA. 2007 A practical guide to analytical method validation, including measurement uncertainty and accuracy profiles, Trends in Analytical Chemistry Vol. 26, No. 3 p 227-238

[14] Daga R, Guevara SR, Poire DG, Arribére M 2014 Characterization of tephras dispersed by the recent eruptions of volcanoes Calbuno (1961), Chaiten (2008) and Cordon Caulle Complex

(1961 and 2011), in Northern Patagonia. *Journal of South American Earth Sciences* **49** p 1 – 14.

[15] Liao Z, Hu W, Cao J, Wang X, Yao S, Wu H, Ye W 2016 Heterogeneous volcanism across the Permian-Triassic Boundary in South China and implications for the Lates Permiant Mass Extintion: New evidence from volcanic ash layers in the Lower Yangtze Region. *Journal of Asian Earth Sciences.*

[16] Ibrahim KM, Moh'd BK, Masri AI, Al-Taj MM, Musleh SM, Alzughoul KA 2014 Volcanotectonic evolution of central jordan: Evidence from the Shihan Volcano. *Journal of African Earth Sciences* **100** p 541 – 553.

[17] Huber H. Koeberl c, Egger H 2003 Geochemical study of lower Eocene volcanic ash layers from the Alpine Anthering Formation, Austria. *Geochemical Journal*, vol **37** p 123 – 134.

[18] Tepe N, Bau M, 2014 Importance of nanoparticles and colloids from volcanic ash for riverine transport of trace elements to the ocean: Evidence from glacial-fed rivers after the 2010 erupton of Eyjafjallajokull Volcano, Iceland. *Science of the Total Environtment* **488-489** p 243-251.

[19] Tapia FV, Martinez-Paco M, et al 2016 Altered volcanic ash layers of the Late Cretaceous San Felipe Formation, Sierra Madre Oriental (Northeastern Mexcico): U – Pb geochronology, provenance and tectonic setting. *Journal os South American Earth Sciences* **70** p 18 – 35.

[20] Rose WI and Durant AJ. 2009. Fine ash content of explosive eruptions. *Journal of Volcanology and Geothermal Research.*

[21] Durant AJ Villarosa G, Rose WI, Delmelle P, Prata AJ, Viramonte JG 2012 Long-range volcanic ash transport and fallout during the 2008 eruption of Chaiten volcano, Chile. *Phisics and Chemistry of the Earth* **45-46** p 50-16.

[22] Greenberg RR, Bode P, Elisabete A, Fernandes DN 2011 Neutron activation analysis: A primary method of measurement. *Spectrochimica Acta Part* **B 66** 193-241.

IOP Conf. Series: Journal of Physics: Conf. Series **860** (2017) 012006 doi:10.1088/1742-6596/860/1/012006

Lead identification in soil surrounding a used lead acid battery smelter area in Banten, Indonesia

N Adventini[1,3], M Santoso[1], D D Lestiani[1], W Y N Syahfitri[1] and L Rixson[2]

[1] Center for Applied Nuclear Science and Technology, National Nuclear Energy Agency, Jln. Tamansari No.71, Bandung 40132, West Java, Indonesia
[2] Center for Isotop and Radiation, National Nuclear Energy Agency, Jl. Lebak Bulus Raya No. 49, Jakarta 12440, West Java, Indonesia

[3] Corresponding author e-mail: nadventini@yahoo.com

Abstract. A used lead acid battery smelter generates particulates containing lead that can contaminate the surrounding environment area. Lead is a heavy metal which is harmful to health if it enters the human body through soil, air, or water. An identification of lead in soil samples surrounding formal and informal used lead acid battery smelters area in Banten, Indonesia using EDXRF has been carried out. The EDXRF accuracy and precision evaluated from marine sediment IAEA 457 gave a good agreement to the certified value. A number of 16 soil samples from formal and informal areas and 2 soil samples from control area were taken from surface and subsurface soils. The highest lead concentrations from both lead smelter were approximately 9 folds and 11 folds higher than the reference and control samples. The assessment of lead contamination in soils described in C_f index was in category: moderately and strongly polluted by lead for formal and informal lead smelter. Daily lead intake of children in this study from all sites had exceeded the recommended dietary allowance. The HI values for adults and children living near both lead smelter areas were greater than the value of safety threshold 1. This study finding confirmed that there is a potential health risk for inhabitants surrounding the used lead acid battery smelter areas in Banten, Indonesia.

1. Introduction

Lead acid battery is one of the supporting components motor vehicle acts as a power supply. Nowadays, automotive industry growth in Indonesia is increasing significantly influences the development of lead acid battery industry. Lead is the main component of the lead acid battery. The high cost production as well as restrictions import of lead by the government had caused recycling of lead acid batteries carried out in many countries including in Indonesia [1]. Several common recycling of lead acid battery in order to meet the factory needs are reduction-oxidation and electrolysis methods [2]. In Indonesia, besides the government licensed lead acid battery smelter plants, there are also other activities such as used lead acid batteries burning in the open air by local people. The latter activities do not use any technologies and facilities smelting which environmental friendly. Both activities process produce the formation of sulfides, named lead slag, gas and particulate matter emissions. The particulate matters contain lead derived especially from anthropogenic activities spread out through the air potentially contaminating soil, water, food and the atmosphere. Soil is one part of the heavy metal cycle. Lead will accumulate in human body tissues through the soil ingestion causing toxicity to human if exceeds the tolerance limit [3]. The amount of lead in soil can indicate present of lead contamination originating from recycling activities of lead acid batteries. Zeinab Safar *et al* [4] in 2014 examined lead in soil from Awadulla secondary lead smelter in Egypt. About 50 percent of soil samples showed a high degree of lead contamination in the range of 302.275-782.625 mg/kg. The results of this study strongly indicated that there was a potential contamination derived from the smelter that should be further examined [4].

Soil analysis with lead pollutant generated by such activities has rarely been studied by people in Indonesia. Therefore, in this study, an identification lead in soil surrounding both formal and informal lead acid battery smelter areas in Banten, Indonesia was carried out. An examination to investigate the degree of pollution, the daily intake amount of lead as a toxic element and to present health risk assessment for lead through soil exposure for inhabitants living surrounding this used lead acid battery smelter area were also studied. These activities were based on a study of ambient air level concentrations of lead in this area which found were quite high [5] so it became a necessity to examine the environmental contamination caused by this smelter which could be the major source of lead in this surrounding area. The method used in this study is energy dispersive X-ray fluorescence (EDXRF). This method is simple in sample preparation, could analyze elements simultaneously as well as has good capability in measuring samples in low concentration in the range of ppm. Lead has high atomic number can be well measured using this method. The results obtained from this study hopefully can be used as a science based information for the authorities to assess the impact of lead contamination to human in surrounding a used lead acid battery smelter area in Indonesia.

2. Materials and Methods

2.1. Site location

Soil samples were collected not only from a used lead acid battery smelter which has license from Ministry of Environment and Forestry Republic of Indonesia but also from an informal lead smelter. Both of them are located in Tangerang, Banten, Indonesia. The climate in this area is categorized as a tropical zone, with an annual average humidity, temperature and rainfall commonly of 78.7%, 27.7°C and 1922 mm, respectively. Two control samples were collected at 40.5 km from Bogor considered as background area. Figure 1 showed the location of sampling site. A soil sampling standard developed in accordance with United States Environmental Protection Agency (US EPA) was adopted [7]. All experimental steps meet the quality assurance of ISO/IEC 17025:2005 which is implemented in our laboratory.

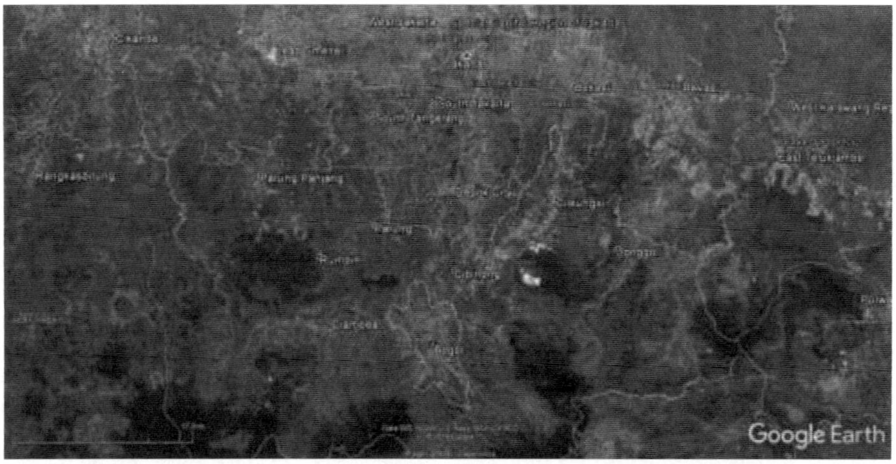

Figure 1. The location of sampling site in Banten, Indonesia

2.2. Sample collection

Total of 16 soil samples and 2 control samples were taken using a lead free shovel at approximate depth of 0-10 cm (surface) and 10-15 cm (subsurface). The particular samples were collected about 250 grams. Soil samples from formal smelter were taken at a distance of 320 m and 577 m in Southwest direction. Meanwhile the samples from the informal smelter areas were taken at a distance less than 100 m of the smelter in 3 directions: West, North and South. Two samples from South were taken at 30 km from the smelter. All soil samples were then placed in plastic bags and were given a laboratory number. The same procedure was applied to the control samples.

2.3. Sample preparation

Soil samples were placed on plastic tray and were sun-dried for 4-5 days. After drying, clods and large aggregates were crushed and mixed using a glass pestle and mortar. A coning and quartering method was used to obtain about 25 grams of homogenous samples. The samples were pulverized and sieved to reduce particle size of soil to less than 0.074 mm.

2.4. Sample measurement

About 1 g of non-pressed standard reference material (SRM) or soil samples with 3 replicates for each sample were placed in plastic sample holders and loaded into the EDXRF Spectrometer MiniPal 4 (PANalytical, Netherland). A soil sediment application method was developed to analyze the elements in SRM and soil samples. The optimum measurement condition of Lead was measured with parameter measurements as follows: voltage 30 kV, current 150 μA, filter material Al and air as medium. Lead was measured with parameter measurements as follows: voltage 30 kV, current 150 μA, filter material Al and air as medium. The measurement time was 300 seconds. A calibration curve used for the purpose of quantitative analysis was obtained from the measurement of the following reference material; soil 7-IAEA (International Atomic Energy Agency), estuarine sediment-NIST 1646 (National Institute of Standards and Technology), estuarine sediment-NIST 1646a and montana soil-NIST 2711a.

2.5. Quality control

Quality control of the method was checked by measuring RM marine sediment-IAEA 457, then evaluated using verification parameters accuracy and precision which are implemented in the laboratory. Acceptable range of the recovery limit is 85-110% for 0.01% concentration [8]. Laboratory precision parameter is described as %RSD. RSD Horwitz (RSDr) was calculated using this formula:

$$RSD_r,\% = 2^{(1-0.5\log C)} \qquad (1)$$

where C is the concentration of analyte expressed as dimensionless mass fraction. Accepted values of %RSD are supposed to be in 0.5 up to 2 RSDr [9].

2.6. Assessment of lead contaminations in soil

The potential ecological risk of heavy metal pollutants in the surface soils of this study was evaluated using the ecological risk index as a contamination factor introduced by Hakanson, 1980 [10] using this formula:

$$C_f = \frac{C_m}{B_m} \qquad (2)$$

where C_f= contamination factor of the interested element; B_m=background concentration from this study; C_m= concentration of the element in the sample. Category of C_f values: 0-1=unpolluted to moderately polluted; 1-2=moderately polluted; 2-3=moderately to strongly polluted; 3-4=strongly polluted.

2.7. Human health risk assessment

In this study, human health risk assessment of lead mainly focused only on the soil exposure assessment and risk characterization. During the exposure assessment stage, an average daily intake dose (ADD in mg/(kg.d)) is used to quantify the oral exposure dose of deleterious substances [11]. The ADD through soil ingestion can be calculated using the following formula:

$$ADD = \frac{C.IR.ED.EF}{BM.AT} \qquad (3)$$

The variables C, IR, ED, BM, EF and AT represent heavy metals content (mg/kg), ingestion rate (mg/(kg.d)), exposure duration (day), reference body mass (kg), exposure frequency (age) and average time (day), respectively; the values of these parameters are listed in Table 1 [12, 13]. Ingestion rate (IR) soil values in this activity for adult and children are 150 mg/d and 200 mg/d, respectively [14].

Table 1. Parameters used in exposure assessment

Parameter	Value	
	Adult	Children
Reference body mass (BM)	60 kg	25 kg*
Exposure duration (ED)	365 d	365 d
Exposure frequency (EF)	70 a	70 a*
Average time (AT)	25550 d	25550 d

Note: *Daping *et al*, 2015[14]

Hazard quotient (HQ) or hazard index (HI) is a ratio of ADD to reference dose (RfD) characterizes the health risk of non-carcinogenic adverse effects due to exposure to toxicants calculated by the following formulas:

$$HQ = ADD/RfD \qquad (4)$$

$$HI = \Sigma HQ \qquad (5)$$

The RfD represents a toxicity index of a daily exposure to the population in comparison to a safe level of exposure orally over a lifetime. An index value < 1 is assumed there is no risk for adverse health effects from the ingestion of soil exposed to heavy metal for local residents in an anthropogenic activity area [14]. The health risks for inhabitants in surrounding smelter area resulting from the lead contaminated soil ingestion were assessed based on the index. The oral reference dose (RfD for lead in food is 3.50×10^{-3} (mg/(kg·d)) [15].

3. Results and discussion

3.1. Calibration curve

A deconvolution for fitting the spectral of soil 7-IAEA, estuarine sediment-NIST 1646, estuarine sediment-NIST 1646a and montana soil-NIST 2711a was carried out after running the soil sediment application program to construct a standard curve. Lead, the element of interest, was identified using two line energies, L_α and L_β at 10.5 keV and 12.6 keV, respectively. In EDXRF, the determination of lead in soil when arsenic is present in relatively high concentrations is difficult because the energy of arsenic K_α completely overlaps the lead L_α line. To overcome this obstacle, the accurate lead identification was performed using the ratio of lead L_α and L_β lines by applying line overlap (LOV) program.

The determination of lead in soil using calibration curve of standard reference materials is acceptable elsewhere [13]. To construct the calibration curve, the XRF signal intensity of L_α and L_β lines was plotted against the certified value of lead in each SRM. The Pearson correlation coefficient (r^2 value) was 0.9990.

3.2. Quality control result

To check the accuracy and precision of XRF, RM marine sediment IAEA 457 were examined. The accuracy and precision evaluated gave accuracy and precision were 100% and 0.1% respectively. The recovery values for the concentration of 0.01% were in accordance with the accuracy acceptance criteria [8]. Precision of the reference material fulfilled requirement implemented in the laboratory. It means the performance of EDXRF Spectrometer gave a good agreement with the value of standard material.

3.3. Concentration and contamination factor of lead

Concentrations, depth and distance ratio as well as contamination factor of lead in soil investigated in this activity are shown in Table 2.

Table 2. Summary of concentration, depth and distance ratio and C_f of lead in soil samples

Location	Distance	Direction	Concentration (mg/kg)		SD	Concentration ratio between two depths	C_f	Category
			Surface (max)	Subsurface (min)	(mg/kg)	ratio	max	
Formal lead smelter	324 m	Southeast	1784	451	943	4	8	Strongly polluted
	577 m	Southeast	385	238	104	2	2	Moderately polluted
Informal lead smelter	65 m	North	868	159	501	5	4	Strongly polluted
	75 m	North	447	89	253	5	2	Moderately polluted
	50 m	West	1350	195	817	7	6	Strongly polluted
	100 m	West	154	135	14	1	1	Unpolluted to moderately polluted
	75 m	South	2125	171	1382	12	10	Strongly polluted
	30 km	South	192	184	6	1	1	Unpolluted to moderately polluted

From Table 2, the ratios of soil lead concentration samples from surface and subsurface in sampling site near the formal smelter area ranged from 2 to 4. It indicated that the lead contamination in surface was higher than subsurface area which means that the lead contaminations were decreased as the depths increased. It also can be found in the soil from informal lead smelters. Similar results of the lead contamination decreased as the function of depth was also studied by Li, *et al* [17]. Some studies reported, lead does not move to any great extent in soils and unless mixing occurs. It generally concentrated near the surface. Therefore, the lead concentrations in the subsurface soil were not significantly different and almost samples showed similar level. The experimental result indicated that concentrations of lead decreased significantly to the distance from the metallurgical works. It can be seen in the formal lead smelters that the lead concentrations in the surface soil collected near the smelter (324 m) was five times higher than the soil at the distance away from the smelter (577m). The lead concentrations of soil collected from informal lead smelter also showed the similar trends toward the distances. The lead concentrations from the West direction decreased almost nine times from 1350 mg/kg to 154 mg/kg. From the South direction, the lead concentrations decreased more than eleven times from 2125 mg/kg (75m from the smelter area) to 192 mg/kg (30 km from the smelter area. The results showed that the lead concentrations in surface soil from both formal and informal lead smelters were inversely proportional to the distance from the smelter. Therefore, it can be confirmed that the high lead contamination was originated from the lead smelter [5, 6].

Lead concentration from the surface soil in the formal smelter and informal smelter were 1784 mg/kg and 2125 mg/kg, respectively, exceeding the reference value which in the range 2-200 mg/kg of leads in urban soils [18]. They were also higher than the levels detected in the control soils collected from a background area situated far from industrial activity. Control samples taken from Bogor gave concentration of 217 mg/kg which was within reference range. Lead concentration from the formal smelter was found in a significant proportion approximately 9 times higher than the reference and control samples. In general, lead concentration from all direction of informal smelter at a distance < 100 m were also higher 4-11 folds than the reference and control samples. Compared to the previous study by Phing *et al* [19], maximum lead concentration from both formal and informal smelters were higher than that in garden soils in mining area, South China (297 mg/kg).

The lead contamination factor (C_f) in soil from formal and informal lead smelters presented in the Table 2. In addition, $C_{f\ max}$ was 8 for the soil collected at 324 m Southeast of formal smelter. Meanwhile the samples from informal lead smelter at 75m South, 50m West, 65m and 75m North resulted $C_{f\ max}$ values of 10, 6, 4 and 2, respectively. These C_f values indicated that the assessment of lead contamination in soils from formal and informal lead smelters was category: moderately and strongly polluted with lead.

3.4. Human health assessment

Daily intake lead of local population and soil exposure risk characterization for adults and children from the investigated area were presented in Table 3 and Table 4.

Table 3. Daily intake lead of local population via soil exposure in this activity

Location	Distance	Direction	ADD µg /(kg d)			
			Adults	Children	Adults	Children
			Surface soil		Subsurface soil	
Control site	40.5 km	South	0.39	1.57	0.22	0.71
Formal lead smelter	324 m	Southeast	4.46	14.27	1.13	3.61
	577 m	Southeast	0.96	3.08	0.40	1.59
Informal lead smelter	65 m	North	5.31	17.00	0.43	1.37
	75 m	North	1.12	3.58	0.15	0.59
	50 m	West	3.38	10.80	0.49	1.56
	100 m	West	0.39	1.23	0.34	1.08
	75 m	South	2.17	6.94	0.40	1.27
	30 km	South	0.48	1.54	0.46	1.47

The results showed ADD of lead for adults living in 324 m Southeast and 65 m North were 4.46 µg/(kg·d) and 5.31 µg/(kg·d), respectively. The investigated areas showed higher ADD of lead than the report of smelting area in Suxian County, South China which was only 0.776 µg/(kg·d) [14]. The ADD of lead for children living in the same area were 14.27 µg/(kg·d) and 17.00 µg/(kg·d). Other direction of informal lead smelter from the 50 m West and 75 m South gave ADD of lead for children were 10.80 µg/(kg·d) and 6.94 µg/(kg·d), respectively. The ADD of lead for adults and children in both areas indicated the exceeding recommended dietary allowance value which concentration is 3.5 µg/(kg·d). In addition, the ADD of lead for children was 3 folds higher than those of adults due to the greater amount of soil ingestion in children and the children's weights were lighter. Children are prone to be at the highest risk from intake lead in contaminated soils.

The HI risk characterization was obtained from the sum of Total HQ's at two depths of each distance. The hazard index demonstrated in Table 4 indicated the different HI values for adults and children. The highest value was found in adults and children living in Southeast and North direction at the nearest distance from both lead smelters which were also significant higher than the value of safety threshold in no smelting activity area.

Table 4. Lead in soil risk characterization in this study

Location	Distance	Direction	HQ				HI	
			Adults	Children	Adults	Children	Adults	Children
			Surface soil		Subsurface soil			
Control site	40.5 km	South	0.11	0.45	0.06	0.20	0.18	0.65
Formal lead smelter	324 m	Southeast	1.27	4.08	0.32	1.03	1.60	5.11
	577 m	Southeast	0.28	0.88	0.11	0.45	0.39	1.33
Informal lead smelter	65 m	North	1.52	4.86	0.12	0.39	1.64	5.25
	75 m	North	0.32	1.02	0.04	0.17	0.36	1.19
	50 m	West	0.96	3.09	0.14	0.45	1.10	3.53
	100 m	West	0.11	0.35	0.10	0.31	0.21	0.66
	75 m	South	0.62	1.98	0.11	0.36	0.73	2.35
	30 km	South	0.14	0.44	0.13	0.42	0.27	0.86

This study indicated that the potential health risks for inhabitants in the smelter area are higher than control area. Risk evaluation via incidental direct soil ingestion exposed to heavy metals was considered to be an important part of hazard assessment, especially for young children with regular hand-mouth activity. Based on HI value in this study, children exposed to heavy metals to a greater extent than adults, which may harm brain and nervous system development. This study showed, human exposure to heavy metal in soils might occur directly through the ingestion of soil. Long term lead exposure via soil ingestion poses potential health problems to the inhabitants living in the surrounding of these used lead acid battery smelter areas in Tangerang, Banten, Indonesia.

4. Conclusion

The highest lead concentrations were 1784 mg/kg and 2125 mg/kg resulted from the formal lead smelter and informal lead smelter, respectively. Assessment lead contamination in soils in this study from formal and informal lead smelter were category: moderately and strongly polluted by lead stated in C_f index. The average daily intake value of children in this study from Southeast, North, West and South had exceeded the recommended dietary allowance. The highest ADD value for adults and children were 5.31 μg/kg.d and 17.00 μg/kg.d respectively. The HI values for adults and children living from Southeast, North and West direction at the nearest distance from both lead smelter area were considerably higher than the value of safety threshold. These results could be a fundamental data for local government and could be applied to evaluate the surrounding smelter area throughout the country for reducing the lead pollution in the region.

5. References

[1] Esquivel-Hernández G, Bolaños-Ulloa P, Navarro-Monge M, Alfaro-Solís R, Sibaja-Brenes J P, Mora-Barrantes J C and Valdés-González J 2012 *J.Waste Management and the Environment* **VI**251-258

[2] J Pan, Zhang, Y Sun, Z Wang and Y Yang 2012 *J. Electrochem. Commun***19** 70–72

[3] Mackay C A K, Taylor M P, Munksgaard N C, Hudson-Edwards K A, and Burn-Nunes L 2013

*J.Elsevier Environmental Pollution***180** 304-311

[4] Zeinab S, Mounir W L, Wegdan L and Mohammed H K 2014 *The Journal of field actions***7** 1-8

[5] Muhayatun S., Diah DL, Rita M, Esrom H, Halimah S, Andreas M and Philip KH 2011 *J.Atmospheric Pollution Research* **2** 190-196

[6] L Rixson1, E Riani and Muhayatun S2015 *A Scientific Journal for The Applications of Isotopes and Radiation* **11**51-64

[7] US EPA SESDPROC-300-R3 2014 *Operating Procedur Soil Sampling* Effective Date: Aug 21

[8] Anonymous 2002 *Guidelines for Single Laboratory Validation of Chemical Methods for Dietary Supplements and Botanicals* (AOAC)

[9] Anonymous2004*Definitions and Calculations of Horrat Values From Intralaboratory Data* [serial online] [cited january10th, 2011];Available from: URL :http://www.aoac.org/dietsupp6/Dietary-Supplement-website/HORRAT_SLV.pdf

[10] Hakanson, L1980 *Water Resource Journal***14** 975–1001.

[11] World Health Organization (WHO)*Evaluation of Certain Food Additives and Contaminants 41st Report of the Joint FAO/WHO Expert Committee on Food Additives)*(WHO Technical Report)

[12] United States Environmental Protection Agency (US EPA) 2011 *Risk-Based Concentration TableWashington Agency*(Washington, DC, USA:United States Environmental Protection Agency)

[13] United States Environmental Protection Agency (US EPA) 1989 *Risk Assessment Guidance for Superfund* Vol. I (Washington DC, USA: Human Health Evaluation Manual Office of Emergency and Remedial Response)

[14] Daping S, Dafang Z, Dong J, Jingying F and Qiao W 2015*Int. J. Environ. Res. Public Health* **12** 7100-7117 doi:10.3390/ijerph120707100

[15] United States Environmental Protection Agency (US EPA) 2002 *Supplemental Guidance for Developing Soil Screening Levels for Superfund Sites* (Washington, DC, USA:Office of Emergency and Remedial response)

[16] Dana J D, Alison R K and Neil F 2014 Calibration and Evaluation of an X-Ray Fluorescence Method for the Determination of Lead and Arsenic in Soils *J Environ Anal Chem*.Jreac(2014)1000103

[17] Li, P., Lin, C., Cheng, H., Duan, X., Lei, K.2015. Contamination and health risks of soil heavy metals around a lead/zinc smelter in southwestern China. *Ecotoxicology and Environmental Safety*113, pp. 391-399

[18] United States Environmental Protection Agency (US EPA) 2011 Risk Hazard Standards for Lead in Paint, Dust and Soil (TSCA Section 403)

[19] Phing Z, Huanping L, Zhian L, Bi Zhou and Muuray B M 2014*Multiple Exposure and Effects Assessment of Heavy Metals in the Population near Mining Area in South China* Published online 2014 Apr 11. doi: 10.1371/journal.pone.0094484

Specific activities and radiological hazard assessment in beach sand samples in Songkhla province,Thailand after Fukushima Dai-Ichi nuclear power plant accident in Japan

P Kessaratikoon[1,4], R Boonkrongcheep[1], N Choosiri[1], M Daoh[2], S Udomsomporn[3]

[1] Nuclear and Material Physics Research Unit (NuMPRU) Department of Physics Faculty of Science Thaksin University, Muang, Songkhla, 90000, Thailand
[2] Program of Physics and General Science Faculty of Science and Technology Songkhla Rajabhat University, Muang, Songkhla, 90000, Thailand
[3] Office of Atoms for Peace (OAP) 16 Vibhavadi Rangsit Rd., Chatuchak, Bangkok, 10900, Thailand

[4] Corresponding author e-mail: prasong_mi@hotmail.com

Abstract. According to the Fukushima Dai-Ichi Nuclear Power Plant Accident in Japan on March 11, 2011, specific activities of natural (^{40}K, ^{226}Ra and ^{232}Th) and anthropogenic (^{137}Cs) radionuclides in 210 beach sand samples which were collected from Maharat, Sai Kaew, Samila, Chalatat, Na Thab, Sakom and Soi Sawan beaches in Songkhla Province along the eastern coast of Thailand, have been studied and evaluated. The specific activity levels of ^{40}K, ^{226}Ra, ^{232}Th and ^{137}Cs in the area were found in the range 110.13 – 4574.12, 6.99 – 451.87, 3.75 – 665.76 and 0.58– 9.91 Bq/Kg, respectively. Furthermore, the median values of specific activities of ^{40}K, ^{226}Ra, ^{232}Th and ^{137}Cs were calculated and compared with other research data in Thailand as well as global measurements and evaluations. Moreover, four radiological hazard indices for the studied area were also assessed by using the median values of specific activities of ^{40}K, ^{226}Ra and ^{232}Th.

1. Introduction

The Great Tohoku earthquake that occurred off the Japanese northeast coast on March 11, 2011 and the resulting tsunami that damaged the Fukushima Dai-ichi nuclear power plant (FDNPP) resulted in several explosions at the reactors between March 12 and 15. These explosions led to significant releases of radionuclides into the atmosphere [1]. The famous nuclear power plant accident was obviously occurred with a broader disaster. This accident resulted in a release of radioactive materials into the atmosphere and marine environment and evacuation of residents living near the nuclear power plants. This accident has raised concerns around the world about the safety of nuclear power generation [2]. Hence, several studies around the world have been performed to evaluate the specific activities of some anthropogenic radionuclides after the FDNPP accident [3–8]. The radioactivity monitoring result suggests that the Fukushima-derived radionuclides spread in the north part of the Northern Hemisphere atmosphere [9, 10]. In ASEAN counties, Vietnam is the first country which interested in studying the effect from the FDNPP disaster. Radionuclides from the reactor accident at the FDNPP were observed in the surface air at stations in Hanoi, Dalat, and Ho Chi Minh City (HCMC) in Vietnam, about 4500 km southwest of Japan, during the period from March 27 to April 22, 2011 [4]. No studies to date in Thailand were performed and measured the radioactivity in some environment samples from dry deposition of radionuclides from the FDNPP accident.

Naturally Occurring Radioactive Materials (NORM) and Technologically Enhanced Naturally Occurring Radioactive Materials (TENORM) consist of materials, usually industrial wastes or by-products enriched with radioactive elements found in the environment, such as uranium, thorium and potassium and any of their decay products. Since 1970s, NORM and TENORM have been attracted to scientists and researchers around the world to monitor and measure the level of radioactivity in both physical and biological environment samples for evaluating the radiation dose for people who use or live in the study area. Besides, there are many scientists and researchers in several countries have studied the levels of natural background radiation by in situ measurements or by analysis of specific activities of natural (^{40}K, ^{226}Ra and ^{232}Th) and anthropogenic (^{137}Cs) radionuclides in surface soil or beach sand samples [11–15]. Moreover, oil and gas industry areas should be monitored and evaluated to NORM and TENORM concentration.

Songkhla province is known for its commercial, agro industrial and socio-economic activities in the southern region of Thailand. Majority of the social, economic, and industrial activities in the south of Thailand takes place in this province. Located about 950 kilometers from Bangkok, Songkhla province, with a population of about 1.4 million, is a major seaside province, connecting with Kedah State of Malaysia. Hat Yai district is the best-known business center in the southern region of Thailand. It plays an important role in the southern economy and serves as the junction of communication and a gateway welcoming Malaysian and Singaporean tourists. Consequently, Songkhla is regarded as the capital province of the south of Thailand. The province is strategically located along the eastern coast in the lower southern region of Thailand and also connected to the Gulf of Thailand. There are many activities of oil and gas industry for Thailand kingdom in the Gulf of Thailand. According to the connection with the Pacific Ocean, the Gulf of Thailand might have accumulated some of the radioactive contaminations from the FDNPP accident in Japan. As a result, Songkhla province, especially in some popular beaches for Thai and foreign tourists, should be the best location for NORM/ TENORM and fissionable materials evaluation in the southern region of Thailand.

In response to the FDNPP accident, the 2015 research project to study the increasing of natural and anthropogenic radionuclides in some attractive beaches along the eastern coast of Thailand in Songkhla province have been initiated and performed by collaboration of Thaksin University (TSU), Songkhla Rajabhat University (SRU) and Office of Atoms for Peace (OAP).

This present paper presents the specific activities values of natural (^{40}K, ^{226}Ra and ^{232}Th) and anthropogenic (^{137}Cs) radionuclides in 210 beach sands samples collected from seven beaches (Maharat, Sai Kaew, Samila, Chalatat, Na Thab, Sakom and Soi Sawan beaches) which are famous and beautiful beaches in Songkhla province along the eastern coast and in the southern region of Thailand. Furthermore, the distributions and the appropriate medium values of the specific activities and radiological hazard indices for the investigated area were also studied and evaluated. Moreover, the results were also compared with the Office of Atoms for Peace (OAP) annual report data, Thailand and global radioactivity measurement and evaluations.

2. Experimental

2.1. Sample collection and preparation

All of surface beach sand samples (210) were collected from Maharat beach (30) in Sathing Phra distict, Sai Kaew beach (30) in Singhanakhon district, Samila (30) and Chalatat Beaches (30) in Muang Songkhla district, Na Thab beach (30) in Chana distric, Sakom (30) and Soi Sawan beaches (30) in Thepha district along the eastern coast of Songkhla province in the south of Thailand. Each sample was dried up at room temperature after collection and sieved through a 2 mm mesh-sized sieve to remove stone, pebbles and other macro-impurities. Before the measurement and analysis, all beach samples were oven dried at a temperature of 100°C for 3 hours for removing moisture. The homogenized sample was placed in a PVC cylindrical container. The container was sealed hermitically and externally using a cellophane tape and kept aside for about a month to ensure equilibrium between ^{226}Ra and its daughters and ^{228}Ra and its daughters before being taken for measurement and analysis by using gamma spectrometry technique.

2.2. Measurement and analysis

The specific activities of natural (^{40}K, ^{226}Ra and ^{232}Th) and anthropogenic (^{137}Cs) radionuclides in all 210 beach sand samples were determined by employing a high-purity germanium detector (HPGe, CANBERRA Model GC 2018) and gamma spectrometry analysis system at Nuclear and Material Physics Research Unit (NuMPRU), Department of Physics, Faculty of Science, Thaksin University, Songkhla Campus. The detector was enclosed in a massive 10 cm thick lead shielding. Geometric efficiency for beach sand matrices in the container was determined by the KCL, IAEA/RGU-1 and IAEA/RGTh-1 reference materials (International Atomic Energy Agency IAEA, Vienna, Austria). The spectra were analyzed using the program GENIE 2000. The specific activity of ^{40}K was determined from its 1460 keV γ-line. The specific activities of ^{226}Ra and ^{232}Th were determined by their decay products ^{214}Pb (351.9 keV) and ^{208}Tl (583.2 keV), respectively. The specific activity of ^{137}Cs was also determined from its 661.7 keV γ-line. The statistic computer program was employed to analyze the frequency distribution of specific activities of natural (^{40}K, ^{226}Ra and ^{232}Th) and anthropogenic (^{137}Cs) radionuclides in all surface beach sand samples for the investigated area. Furthermore, the results were also compared with some research data in Thailand as well as global measurements and evaluations. Moreover, four radiological hazard indices in the study area were also evaluated by using the appropriate medium values of the frequency distribution and the equations as shown in the following section.

2.3. Equations and formulas

The absorbed dose rate (D) in outdoor air at 1 m above the ground was calculated by using the conversion factors published in [16] with all medium values of specific activity of ^{40}K, ^{226}Ra, and ^{232}Th from this study and is given below

$$D = 0.461C_{Ra} + 0.623C_{Th} + 0.0414C_{K}, \tag{1}$$

where C_{Ra}, C_{Th} and C_K are the medium values of specific activity of ^{226}Ra, ^{232}Th and ^{40}K in Bq/kg, respectively. Furthermore, the radium equivalent activity (Ra_{eq}) was calculated through the following relationship [17, 18]:

$$Ra_{eq} = C_{Ra} + 1.43C_{Th} + 0.077C_{K} \tag{2}$$

Moreover, the external hazard index (H_{ex}) was also evaluated by using the equation which was defined as [5]:

$$H_{ex} = C_{Ra}/370 + C_{Th}/259 + C_{K}/4810 \leq 1 \tag{3}$$

The annual effective dose rate (AED_{out}) were calculated using the absorbed dose rates in air (D) obtained from the medium values of specific activity of natural (^{40}K, ^{226}Ra and ^{232}Th) radionuclides in beach sand samples, adopting the conversion factor of 0.7 Sv/Gy to convert from the absorbed dose in air to the effective dose received by adults and considering that people in Thailand, on average, spend approximately 20% of their time outdoors. The annual effective dose rate (AED_{out}) were calculated using the following equation [12, 16]:

$$AED_{out}\ (mSv/y) = D\ (nGy/h) \times 8760\ h \times 0.2 \times 0.7\ (Sv/Gy) \times 10^{-6} \tag{4}$$

All four radiological hazard indices and their average values were calculated and compared with the recommended values reported by UNSCEAR [19-21].

3. Results and Discussion

3.1. Specific activity values and ranges of natural and anthropogenic radionuclides

Specific Activity Values and Ranges of natural (^{40}K, ^{226}Ra, ^{232}Th) and anthropogenic (^{137}Cs) radionuclides in 210 beach sand samples collected from Maharat (30), Sai Kaew (30), Samila (30), Chalatat (30), Na Thab (30), Sakom (30) and Soi Sawan (30) beaches in Songkhla Province was measured and calculated by using HPGe detector and gamma spectrometry technique. All results were presented in table 1.

Table 1. Specific Activity Values and Ranges of ^{40}K, ^{226}Ra, ^{232}Th and ^{137}Cs in Bq/kg in 210 beach sand samples for Maharat, Sai Kaew, Samila, Chalatat, Na Thab, Sakom and Soi Sawan beaches in Songkhla Province.

Study Areas	Specific Activities (Bq/kg)			
	^{40}K	^{226}Ra	^{232}Th	^{137}Cs
Maharat beach (30)	648.99 – 1508.70	18.85 – 111.51	14.60 – 122.94	N.A.
Sai Kaew beach (30)	145.51 – 1154.30	6.99^{a} – 74.53	9.68 – 83.99	0.79 - 2.31
Samila beach (30)	384.56 – 4574.12^{b}	14.09 – 433.53	9.13 – 448.78	N.A.
Chalatat beach (30)	167.85 – 1904.85	7.85 – 451.87^{b}	13.45 – 665.76^{b}	0.58^{a} - 4.52
Na Thub beach (30)	110.13^{a} – 2307.65	9.25 – 258.87	11.68 – 235.12	1.28 - 4.91
Sakom beach (30)	122.06 – 1925.11	12.97 – 120.74	6.47 – 179.63	1.18 - 9.91^{b}
Soi Sawan beach (30)	512.00 – 3955.88	20.11 – 83.18	3.75^{a} – 94.68	1.77 - 9.30
Ranges (210)	110.13 - 4574.12	6.99 - 451.87	3.75 - 665.76	0.58 - 9.91

^{a}Minimum values ^{b}Maximum values

3.2. Frequency distributions of specific activities

The frequency distribution of specific activities of ^{40}K, ^{226}Ra, ^{232}Th and ^{137}Cs in 210 beach sand samples collected from all seven beaches in Songkhla province (Thailand), were studied and analyzed by using a statistical computer program. The results of Chalatat, Sakom and Soi Sawan beaches were chosen and presented in the following figure 1– figure 12. It was found that the frequency distribution of specific activities of ^{40}K, ^{226}Ra, ^{232}Th and ^{137}Cs in all beach sand samples collected from Songkhla province, were asymmetrical distribution.

3.3. Median value of specific activities

According to the asymmetry of frequency distributions of ^{40}K, ^{226}Ra, ^{232}Th and ^{137}Cs in all (210) beach sand samples as shown in figure 1– figure 12, the median values should be selected and used for the corresponding radiological hazard evaluation in the investigated area. All of these results were also compared with the Office of Atoms for Peace (OAP) annual report data, Thailand and global radioactivity measurements and evaluations as presented in table 2.

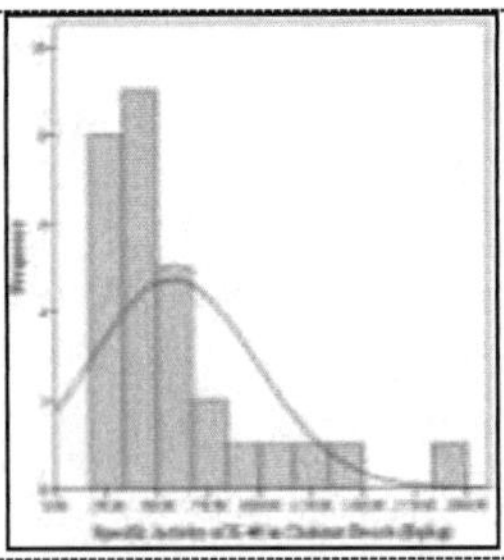

Figure 1. Frequency distribution of specific activity of ^{40}K in Chalatat beach sand samples.

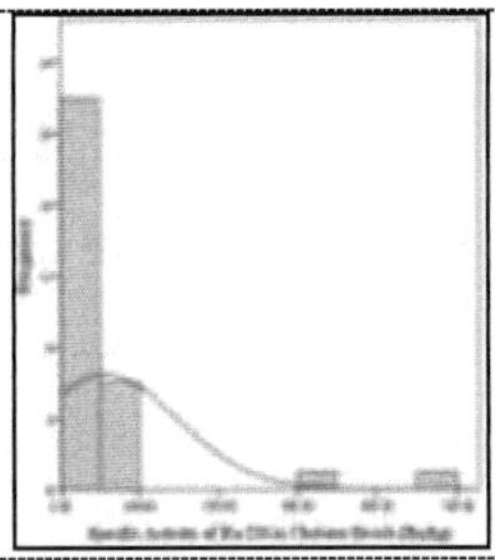

Figure 2. Frequency distribution of specific activity of ^{226}Ra in Chalatat beach sand samples.

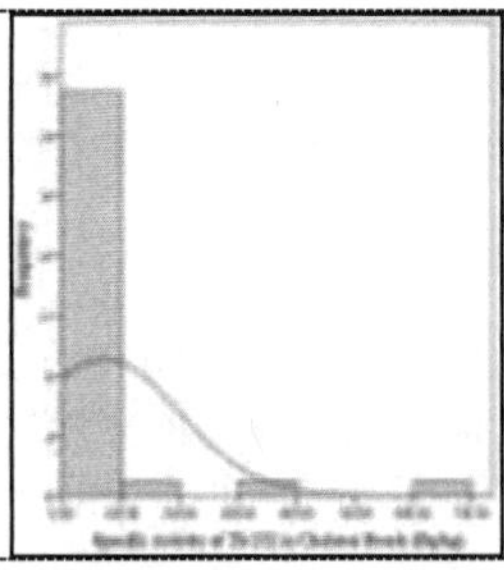

Figure 3. Frequency distribution of specific activity of ^{232}Th in Chalatat beach sand samples.

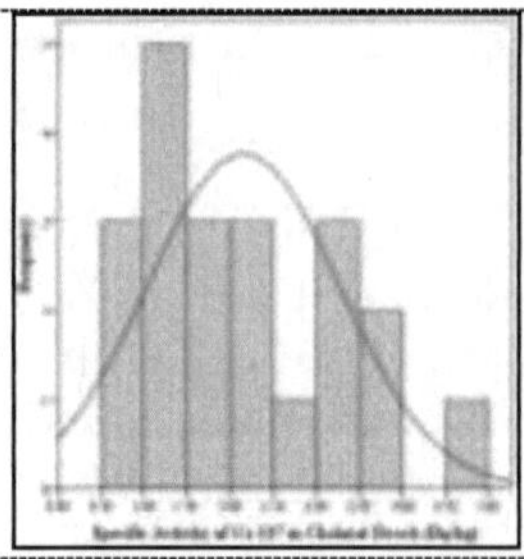

Figure 4. Frequency distribution of specific activity of ^{137}Cs in Chalatat beach sand samples.

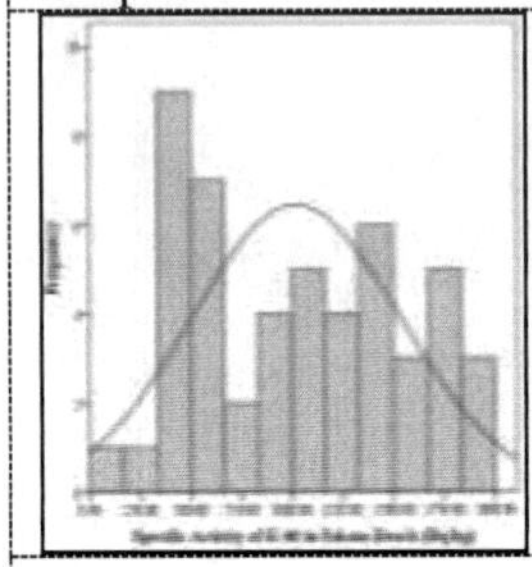

Figure 5. Frequency distribution of specific activity of ^{40}K in Sakom beach sand samples.

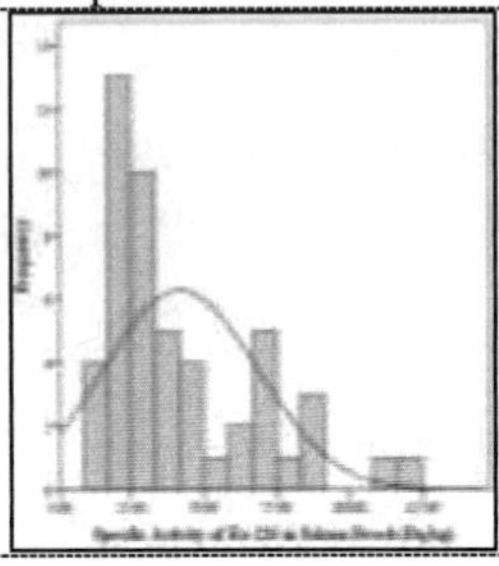

Figure 6. Frequency distribution of specific activity of ^{226}Ra in Sakom beach sand samples.

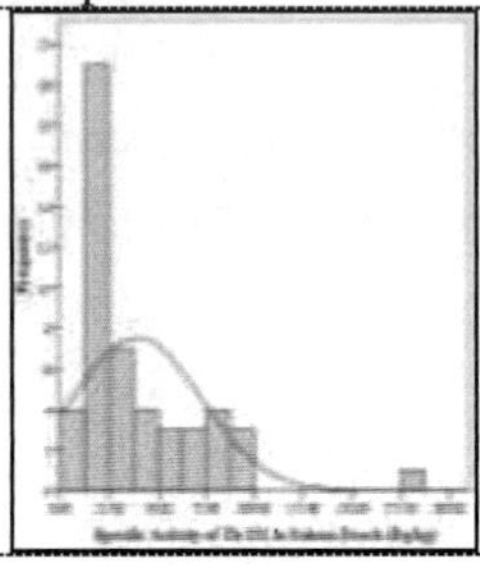

Figure 7. Frequency distribution of specific activity of ^{232}Th in Sakom beach sand samples.

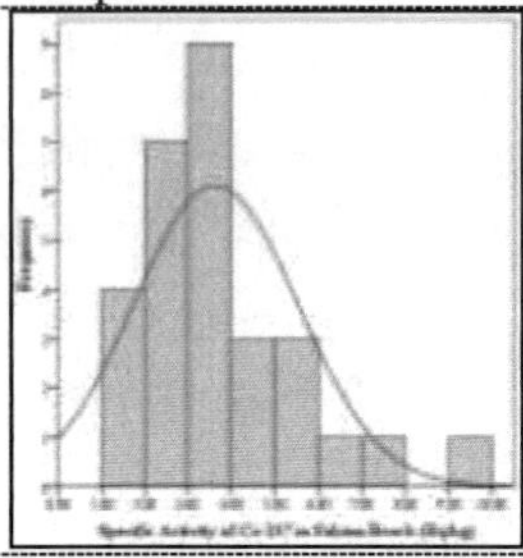

Figure 8. Frequency distribution of specific activity of ^{137}Cs in Sakom beach sand samples.

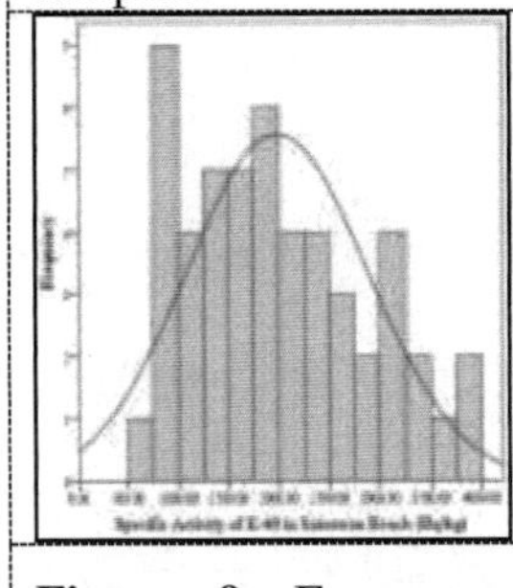

Figure 9. Frequency distribution of specific activity of ^{226}Ra in Soi Sawan beach sand samples.

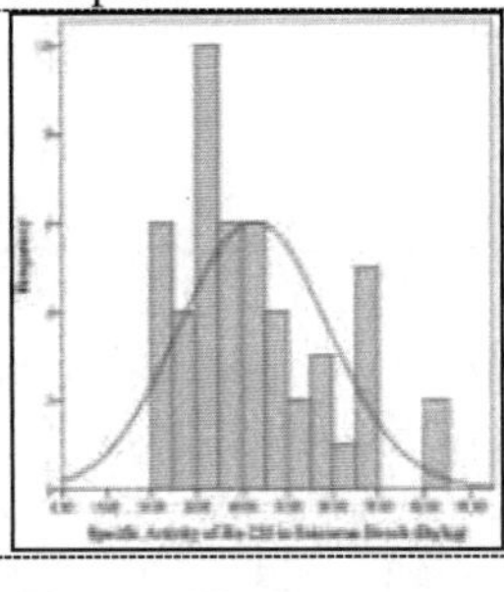

Figure 10. Frequency distribution of specific activity of ^{226}Ra in Soi Sawan beach sand samples.

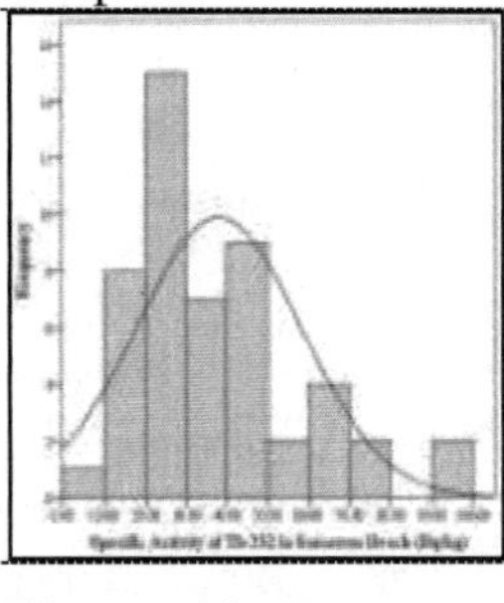

Figure 11. Frequency distribution of specific activity of ^{232}Th in Soi Sawan beach sand samples.

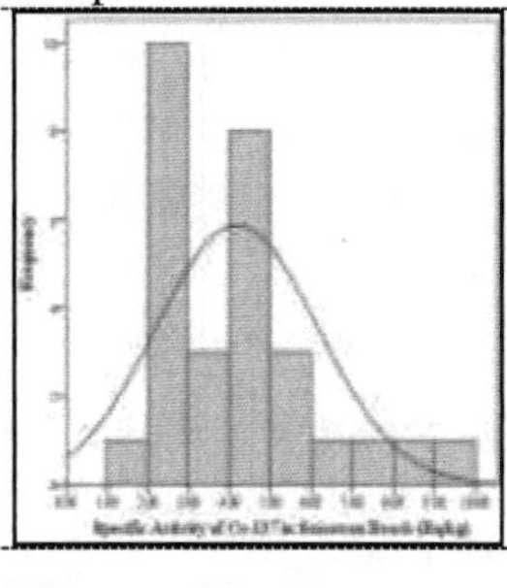

Figure 12. Frequency distribution of specific activity of ^{137}Cs in Soi Sawan beach sand samples.

3.4. Radiological hazard assessment

Furthermore, four radiological hazard indices which are gamma absorbed dose rate (D), radium equivalent activity (Ra_{eq})**,** external hazard index (H_{ex}) and annual external effective dose rate (AED_{out}), were also evaluated and presented in table 2 for Maharat, Sai Kaew, Samila, Chalatat, Na Thab, Sakom and Soi Sawan beaches in Songkhla Province by using the median values of specific activities of ^{40}K, ^{226}Ra and ^{232}Th. Moreover, the results were also compared with some research data in Thailand and UNSCEAR as shown in the same table.

Table 2. Comparison of median values of specific activity of ^{40}K, ^{226}Ra, ^{232}Th and ^{137}Cs in Bq/kg and radiological hazard indices in 7 beach sand samples in Songkhla province with OAP annual report data, Thailand and global radioactivity measurement and evaluations.

Locations	Specific Activity (Bq/kg)				Radiological Hazard Indices			
	^{40}K	^{226}Ra	^{232}Th	^{137}Cs	D (nGy/h)	Ra_{eq} (Bq/kg)	H_{ex}	AED_{out} (mSv/y)
Patong beach (Phuket province)	3538.09 ± 959.56	29.72 ± 11.48	32.71 ± 7.51	3.28 ± 1.03	180.56 ± 49.70	348.93 ± 96.11	0.94 ± 0.26	0.22 ± 0.06
Naiyang beach (Phuket province)	1648.27 ± 702.98	14.62 ± 6.54	18.10 ± 4.22	1.19 ± 0.66	86.25 ± 34.75	167.42 ± 66.70	0.45 ± 0.18	0.11 ± 0.04
Takua Pa and Thai Muang beach (Phang Nga province)	1069.99 ± 119.43	40.39 ± 4.30	41.15 ± 3.65	N.A.	88.55 ± 9.20	181.62 ± 18.71	0.49 ± 0.05	0.11 ± 0.01
Ao Nang beach (Krabi province)	330.73	4.60	5.78	N.A.	19.41	38.33	0.10	0.02
Noppharat Thara beach (Krabi province)	307.60	5.31	5.28	N.A.	18.47	36.55	0.10	0.02
Pakmeng beach (Trang province)	57.3	23.1	6.9	N.A.	17.32	37.38	0.10	0.02
Chaweng beach (Surat Thani province)	373.30	18.85	23.53	N.A.	38.80	81.24	0.22	0.05
Maharat beach[a] (Songkhla province)	895.97 ± 122.52	39.28 ± 3.27	42.06 ± 2.34	N.A.	81.40 ± 8.04	168.42 ± 16.05	0.45 ± 0.04	0.10 ± 0.01
Sai Kaew beach[a] (Songkhla province)	726.91 ± 115.41	21.79 ± 2.50	15.69 ± 1.72	1.52 ± 0.47	49.91 ± 7.00	100.20 ± 13.85	0.27 ± 0.01	0.06 ± 0.01
Samila beach (Songkhla province)	868.51 ± 118.42	30.50 ± 4.22	23.31 ± 2.22	N.A.	64.54 ± 8.23	130.71 ± 16.51	0.35 ± 0.04	0.08 ± 0.01
Chalatat beach[a] (Songkhla province)	488.05 ± 105.21	28.58 ± 3.84	30.11 ± 2.06	1.90 ± 0.65	52.14 ± 7.41	109.21 ± 14.89	0.29 ± 0.04	0.06 ± 0.01
Na Thub beach[a] (Songkhla province)	1044.54 ±122.25	23.97 ± 3.02	23.76 ± 2.04	2.04 ± 1.17	69.10 ± 7.72	138.38 ± 15.34	0.37 ± 0.04	0.08 ± 0.01
Sakom beach[a] (Songkhla province)	1028.92 ± 130.68	31.53 ± 3.18	24.98 ± 2.00	3.17 ± 0.95	72.69 ± 8.12	146.47 ± 16.09	0.40 ± 0.04	0.09 ± 0.01
Soi Sawan beach[a] (Songkhla province)	1869.32 ± 183.33	36.73 ± 3.88	31.61 ± 2.68	4.00 ± 1.10	114.02 ± 11.05	225.87 ± 21.83	0.61 ± 0.06	0.14 ± 0.01
OAP (Southern region of Thailand)[22]	511.04 ± 7.04	171.55 ± 3.13	211.19 ±1.98	1.13 ± 0.49	N.A.	N.A.	N.A.	N.A.
Worldwide mean [21]	400	35	30	N.A.	N.A.	N.A.	N.A.	N.A.
UNSCEAR [19-21]	N.A.	N.A.	N.A.	N.A.	55	370	1	0.48

[a]Present study

The ranges of specific activities of natural (^{40}K, ^{226}Ra, ^{232}Th) and anthropogenic (^{137}Cs) radionuclides in 210 beach sand samples collected from Maharat, Sai Kaew, Samila, Chalatat, Na Thab, Sakom and Soi Sawan beaches in Songkhla Province were 110.13 – 4574.12 Bq/kg for ^{40}K, 6.99 – 451.87 Bq/kg for ^{226}Ra, 3.75 – 665.76 Bq/kg for ^{232}Th and 0.58 – 9.91 Bq/Kg for ^{137}Cs as shown in table 1. The frequency distribution of specific activities of ^{40}K, ^{226}Ra, ^{232}Th and ^{137}Cs in beach sand samples in every location were studied and found to be asymmetrical distribution as shown in figure 1 – figure 12. Hence, the median values analyzed from all frequency distributions, were chosen to evaluate the radiological hazard indices for the investigated area. Furthermore, specific activities of ^{40}K in all beaches were mostly higher than some research in Thailand, OAP research data and worldwide mean but lower than in Phuket and Phang Nga province. The specific activities of ^{226}Ra and ^{232}Th were in the same order of research data in Thailand and worldwide mean but lower than OAP research data. Moreover, specific activities of ^{137}Cs in the investigated area were in the same range of all research data in Thailand and OAP research data as presented in table 2. According to Songkhla province is one of the province which is located in the eastern coast of Thailand and connected to the Pacific Ocean. The study area, Maharat, Sai Kaew, Samila, Chalatat, Na Thab, Sakom and Soi Sawan beaches which are located on the eastern coast of Songkhla province, certainly have some experiences of NORM and TENORM activities e.g. oil and gas industry wastes and FDNPP accident. For this reason, the values of gamma absorbed dose rate (D) in all beaches were higher than about 2.9 – 6.6 times to some research data in Thailand and UNSCEAR but mainly lower than in Patong and Naiyang beaches in Phuket province. The values of radium equivalent activity in all beaches were higher than about 2.7 – 6.2 times to some research data in Thailand but mostly lower than in Patong and Naiyang beaches in Phuket province which is less than 370 Bq/kg, which are the acceptable value for safe use. The values of external hazard index obtained in this study were ranged from 0.27 ± 0.01 to 0.61 ± 0.06

which were less than unity. The calculated annual effective dose rate for all beaches were also ranged from 0.06 ± 0.01 to 0.14 ± 0.01 mSv/y which lower than the worldwide average value of 0.48 mSv/y as presented by UNSCEAR.

4. Conclusions

The median values of specific activity of natural radionuclides (^{40}K, ^{226}Ra, ^{232}Th) in 210 beach sand samples collected from the monitored area (Maharat, Sai Kaew, Samila, Chalatat, Na Thab, Sakom and Soi Sawan beaches in Songkhla province) are at the typical level of radioactivity from natural background radiation. Furthermore, the median values of specific activity of ^{137}Cs in the study area were in the same level of research data in Thailand and OAP after the Fukushima Dai-ichi nuclear power plant accident. Moreover, there are three (Ra_{eq}, H_{ex} and AED_{out}) from four values of radiological hazard indices were lower than the recommend values which was reported by UNSCEAR. This means that the FDNPP accident was insignificant for the population living and spending sometimes in the investigated area. The results agree with the discussion "the effect of this radioactive cloud from FDNPP accident on low latitude areas seems to be rather minor [23].

Acknowledgements

The authors are grateful to Assoc. Prof. Dr. Tripop Bhongsuwan, Department of Physics, Faculty of Science, Prince of Songkhla University (PSU), Hat Yai Campus for his generous permission to us to use all of standard reference materials for calculation and analysis in this research. The authors are thankful to some of undergraduate and graduate students for their works and patience in samples collection, preparation, measurement and analysis some parts of these research data. The investigation is a research project which was fully supported by Office of the National Research Council of Thailand (NRCT) and Research and Development Institute Thaksin University (RDITSU).

References

[1] Ott A de V, Gurriaran A, Cagnat X and Masson O 2013 *J. Environ. Radioact.* 125 6–16
[2] Kim C K *et al.* 2012 *J. Environ. Radioact.* **111** 70–82
[3] Barsanti M *et al.* 2012 *J. Environ. Radioact.* **114** 126–30
[4] Ioannides K, Stamoulis K and Papachristodoulou C 2012 *J. Radioanal. Nucl. Chem.* **298(2)** 1207–13
[5] MacMullin S, Giovanetti G K, Green M P, Henning R, Holmes R, Vorren K and Wilkerson J 2012 *J. Environ. Radioact.* **112** 165–70
[6] Ozman S F, Cesar A, Boztosan I and Yavuz M 2014 *Radiat. Phys. Chem.* **103** 37–44
[7] Onda Y, Kato H, Hoshi M, Takahashi Y and Nguyen M L 2015 *J. Environ. Radioact.* **139** 300–307
[8] Satou Y, Suek K, Sasa K, Kitagawa J, Ikarashi S and Kinoshita N 2015 *J. Radioanal. Nucl. Chem.* **303** 1197–200
[9] Hirose K 2011 *J. Environ. Radioact.* **111** 13–7
[10] Long N Q, Truong Y, Hien P D, Binh, N T, Sieu L N, Giap T V and Phan N T 2012 *J. Environ. Radioact.* **111** 53–8
[11] Freitas A C and Alencar A S 2004 *J. Environ. Radioac.* **75** 211–23
[12] Mohanty A K, Sengupta D, Das S K, Vijayan V and Saha S K 2004 *Radiat. Meas.* **38** 153–65
[13] Yang Y X, Wu X M, Jiang Z Y, Wang W X, Lu J G, Lin J, Wang L M and Hsia Y F 2005 *Appl. Radiat. Isot.* **63** 255–9
[14] Korobova E M, Brown J B, Ukraintseva N G and Surkov V V 2007 *J. Environ. Radioac.* **96** 145–56
[15] Kessaratikoon P, Boonkrongcheep R, Benjakul S and Youngchauy U 2013 *J. Radioanal. Nucl. Chem.* **297(2)** 247–55
[16] Singh S, Rani A and Mahajan R K 2005 *Radiat. Meas.* **39** 431–9
[17] Veiga R *et al.* 2006 *Radiat. Meas.* **41** 189–96
[18] Beretka J and Mathew P J 1985 *Health. Phys.* **48** 87–95

[19] UNSCEAR 1988 *United Nations* New York
[20] UNSCEAR 1993 *United Nations* New York
[21] UNSCEAR 2000 *United Nations* New York
[22] OAP 1994 – 2002 *Office of Atoms for Peace* Ministry of Science and Technology Bangkok Thailand
[23] Takemura T, Nakamura H, Takigawa M, Kondo H, Satomura T, Miyasaka T and Nakajima T 2011 *SOLA* **7** 101–4

IOP Conf. Series: Journal of Physics: Conf. Series **860** (2017) 012008 doi:10.1088/1742-6596/860/1/012008

Transfer factor of ^{226}Ra, ^{232}Th and ^{40}K from soil to *Alpinia Galangal* plant grown in northern Thailand

R Kritsananuwat, S Chanyotha[1] , C Kranrod and P Pengvanich

Department of Nuclear Engineering, Chulalongkorn University, 254 Phayathai Road, Pathumwan, Bangkok, 10330, Thailand

[1] Corresponding author e-mail: supitcha.c@chula.ac.th

Abstract. This paper reports the activity concentration of three natural radionuclides, ^{226}Ra, ^{232}Th and ^{40}K, found in *Alpinia Galangal* plants which are widely used in various Asian cuisines and traditional medicine. The galangal plants and their relevant soils were collected from four provinces in the north of Thailand under natural field conditions. The activity concentration of radionuclides was determined using gamma-ray spectrometry. Soil-to-plant transfer factors (TFs) for ^{226}Ra, ^{232}Th and ^{40}K were investigated in rhizome and aerial parts of the galangal plants. The activity concentration in the soils ranged from 22 to 88 Bq kg^{-1} for ^{226}Ra, 27 to 157 Bq kg^{-1} for ^{232}Th and 58 to 1157 Bq kg^{-1} for ^{40}K. In *Alpinia Galangal*, the concentration ranged from < 0.2 to 2.0 Bq kg^{-1} for ^{226}Ra, < 0.1 to 2.9 Bq kg^{-1} for ^{232}Th and 205 to 2247 Bq kg^{-1} for ^{40}K. The TF ranged from < 0.002 to 0.073 for ^{226}Ra, < 0.001 to 0.061 for ^{232}Th and 0.26 to 7.9 for ^{40}K. The TFs in the aerial parts were higher than those for the rhizomes. The obtained values can be considered as a baseline data for activity concentrations of natural radionuclides and their TFs in Thailand for future environmental radiation monitoring. The Annual effective ingestion dose due to ingestion of ^{226}Ra, ^{232}Th and ^{40}K in galangals is significantly below the worldwide value reported by UNSCEAR 2000.

1. Introduction

The consumption of plants containing radionuclides represents the main way leading to human ingestion of radionuclides and consequently contributing to a long-term internal radiation exposure [1, 2, 3]. Radionuclides are incorporated into plants as a result of direct deposition from the atmosphere onto external plant surfaces and indirectly from re-suspension of soil on aerial parts as well as direct uptake of radionuclides through the root system [3, 4, 5]. Radionuclides in soil are then transferred to plants along with essential nutrients during mineral uptake and are trans-located to various parts of plants including the edible portions [1, 6, 7]. These plants when consumed by man would lead to continuous radiation dose. Thus, Soil-to-plant is recognized as one of the major pathways for transferring radionuclides to humans. Concentration of radionuclides in soil is a significant factor influencing the concentration in plants. Therefore, it is important to investigate the distribution of radionuclides in plants and associated soils together with the radionuclide long-term behaviour, such as mobility, transfers and translocation [5].

The uptake of radionuclides from soil to plants is difficult to quantify and complex [2]. Soil-to-plants transfer factor (TF) is a convenient approach used to describe the amount of radionuclides expected to enter a certain plant from the soil [8, 9]. It has been widely used in the estimation of the transport of radionuclides and other elements of interest through the food chain and also in assessment of the contribution to internal radiation dose through an ingestion pathway. TF is regarded as one of the most important parameters used in mathematical models for the radiological risk and environmental safety assessment needed for nuclear facilities [10]. The transfer factor is defined as the ratio of the concentration of a radionuclide in a plant to the activity concentration of that same radionuclide in the soil [11]. Since the TF can vary from location to location due to different climates, soil types, and vegetation, local transfer factors should also be observed. Several studies have been carried out in different regions of the world about the distribution and transfer of radionuclides from soils to plants [1-15]. However, there are only few data sets for tropical countries in South East Asia. Since there is no record of radioactive contamination in the environment of Thailand, it is a great deal to establish a baseline radioactivity data in the environmental samples in Thailand.

The terrestrial radionuclides, ^{238}U, ^{232}Th and their decay products as well as ^{40}K, occur naturally and exist in sufficient quantities to contribute a significant natural radiation dose to humans. They are considered to be the main source of internal and external exposure of terrestrial radiation except for radon [8]. Thus, determining the distribution of these radionuclides in environmental samples and their possible pathways to human exposure is necessary for assessing the effects of radiation exposure.

Alpinia galangal, commonly known as greater galangal, is a perennial, robust and tilling rhizomatous herb belonging to the ginger family (Zingiberaceae). The plant grows up to 3.5 m in height with a subterranean and branching aromatic rhizome. The rhizomes are 2.5–10.0 cm thick and reddish brown externally with branch out in several directions. The aerial leafy stem, pseudo-stem, is erect, formed by the rolled leaf sheaths. Leaves are long, narrow and green blade-shaped with 23–45 cm long and 3.8–11.5 cm board. Flowers develop on top of the flowering stem. They are white with red lip in terminal panicles and emit sweet aroma which attracts insects. Fruit is an ovuled capsule and 1-1.5 cm in diameter.

Galangal occurs wild, semi-wild and under commercial cultivation. It is also grown as a foliage garden plant along the fence lines. The optimum growing environment for galangal is tropical and sub-tropical conditions. It can be cultivated easily in sunny to moderately shady areas with deep, fertile and moist but not swampy soils. Sandy or clayey soils rich in organic matter and with good drainage are preferred. Galangal is cultivated in many parts of Thailand and throughout South and Southeast Asia [16]. The galangals cultivated in Thailand are not only distributed across the country but also exported to foreign countries such as Japan, America and European countries. It has been widely used as a culinary (e.g., Tom Yum and green curry) and as traditional medicinal spice in Asian and medieval European countries for a long time. Its well-documented human medicinal effects include antiallergic activity, antimicrobial activity, anti-inflammatory activity, carcinogenesis inhibition, gastroprotective effect, and cardiovascular effect [16]. From Thailand and other Asian countries, the uses of galangal spread throughout some European countries, America and Russia.

Therefore, the present study was aimed to establish a baseline data on the activity concentrations of ^{226}Ra, ^{232}Th and ^{40}K in soils and *Alpinia galangal* (Greater galangal) as well as soil-to-plant TF collected in four provinces of Northern Thailand under natural field conditions. Furthermore, to estimate the risk from internal radiation exposure through the ingestion of *Alpinia galangal*, an annual effective ingestion dose due to individuals consuming galangal in the study areas was determined.

2. Material and Methods

2.1. Sample collection

Twenty-five soil samples and ten *Alpinia galangal* plant samples were collected on March 3-13, 2015 from four selected provinces in the northern part of Thailand as shown in figure 1. The selected provinces include Chiang Mai, Chiang Rai, Lampang and Phayao. These provinces are the major plantation zones of Northern Thailand. There are a lot of vegetables, grain and fruits cultivated in these areas as well as the galangals. Galangal is an annual plant which can be found in almost all commercial farming and home gardening not only in the study areas but also across the country. All the parts of galangal both fresh and dry form can be used for medicinal and culinary purposes. There are dried sliced and powdered galangal produced from the northern part of Thailand are exported to the overseas markets. Therefore galangal plants and their cultivated soil in the study areas were proposed to collect. All part of galangal, leaves, stem and rhizome, as well as its relevant soil from 0 to 20 cm depth of top soil were taken.

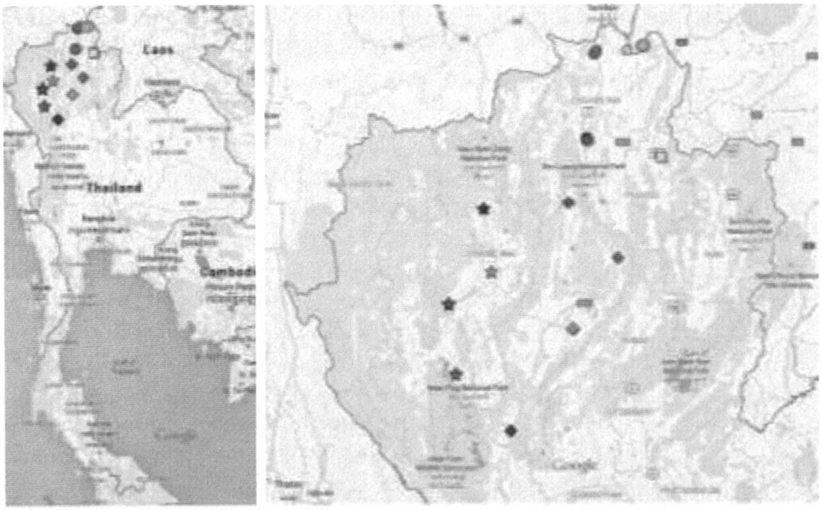

Figure 1. Sampling location in four study provinces in Thailand.

2.2. Sample preparation and analysis

Twenty-five soil samples were oven dried at a temperature of 110 °C until constant weight. The dried samples were pulverized into a fine powder and sieved through a 250 μm mesh size sediment sieve. Ten *Alpinia galangal* plant samples were washed with pure water for removing dust and soil and then were separated into rhizome, stem and leaves. The plant samples also were dried in an oven at 110°C until a constant dry weight was obtained and were ground into a fine powder. All homogenized samples were analyzed for activity concentration of ^{226}Ra, ^{232}Th and ^{40}K using gamma spectroscopy. The samples were packed into airtight plastic containers and sealed to prevent the escape of radon (^{222}Rn) and thoron (^{220}Rn) gases. Prior to measurement, the sample was stored for at least four weeks in order to establish secular equilibrium between ^{226}Ra and ^{232}Th and their radioactive progenies. Activity concentrations in the samples were measured using one of two high-purity germanium (HPGe) detectors with a relative efficiency of 25 and 30 % in a low background configuration. Energy and efficiency calibrations of the detector were carried out using three different IAEA standard reference materials including IAEA-RGU-1, IAEA-RGTh-1 and IAEA-RGK-1. Counting time for each sample was set at 12 hours for soil and 48-60 hours for plants. By assuming secular equilibrium with their progenies in the ^{238}U and ^{232}Th decay chain, the activity concentration of ^{226}Ra and ^{232}Th was calculated using gamma-rays associated with decays of ^{214}Bi (609.3 keV) and ^{228}Ac (911.2 keV), respectively. The activity concentration of ^{40}K was derived directly from the measured intensity of its gamma peak at energy 1,460.8 keV.

Statistical data, range, arithmetic mean and standard deviation, of activity concentrations in soil and plant samples were calculated to present the distribution of radionuclides in the samples and compared with the worldwide average values reported by UNSCEAR. In order to consider the significance differences in the mean, mean comparisons of the activity concentrations in samples among four provinces as well as between the obtained overall mean and worldwide values were performed by T-test using SPSS 19.0 (IBM Corp., Armonk, NY, USA).

Soil-to-plant transfer factor (TF), representing the transfer mechanism of radionuclides, is widely used to describe plant uptake from soil. The activity concentrations of radionuclides in galangals and their relevant soil are assumed to be linear and they were used for calculating TFs according to the following equation [1, 9, 14]:

$$TF = \frac{\text{Activity concentration of radionuclides in plant (Bg kg}^{-1}\text{ dry weight)}}{\text{Activity concentration of radionuclides in soil (Bg kg}^{-1}\text{ dry weight)}}$$

The annual effective ingestion dose to man due to consumption of galangal as medicine and food is calculated using the following equation [17]:

$$Dose\ (\mu Sv\ y^{-1}) = Activity\ concentration\ (Bq\ kg^{-1}) \times Annual\ intake\ (kg\ y^{-1}) \times DCF\ (\mu Sv\ Bq^{-1})$$

where DCF is the dose convection factor for ingestion, 0.28 μSv Bq^{-1} for ^{226}Ra, 0.23 μSv Bq^{-1} for ^{232}Th and 0.0014 μSv Bq^{-1} for ^{40}K [9].

The appropriate dose of galangal depends on several factors. At this time there is not enough scientific information to provide a basis of dosage recommendations for medicinal plants. Generally, for medicinal purpose, the recommended intake of dry galangal rhizome is 1–6 g daily. It is meant that the annual intake is approximately 0.4–2.2 kg y^{-1}. A survey of food ingredient labels available in domestic markets was found that galangal is used approximately 2.5-9 g in one dish. If man eats food contained galangal every day, in one year a consumption rate of galangal should be 0.9-3.3 kg y^{-1}. A well-accepted consumption rate of galangal as both medical plants and food is not available, however, the annual intake rate of 3.3 kg y^{-1} was assumed for all purpose used in this study.

3. Results and discussion

3.1. Activity concentration in soils

Mean and standard deviation in mean as well as range values of activity concentrations of natural radionuclide ^{226}Ra, ^{232}Th and ^{40}K in twenty-five soil samples collected from the four study provinces are summarized in table 1. The lowest concentration of ^{226}Ra (22±1 Bq kg^{-1}), ^{232}Th (27±1 Bq kg^{-1}) and ^{40}K (58±3 Bq kg^{-1}) was found in samples collected from Chiang Rai province. The highest activity concentrations of ^{226}Ra (88±1 Bq kg^{-1}), ^{232}Th (157±3 Bq kg^{-1}) and ^{40}K (1157±22 Bq kg^{-1}) was found in samples collected from Chiang Mai province. The activity concentrations of radionuclides were non uniform distribution.

Table 1. Means and ranges of activity concentration of natural radionuclides in soils.

Sampling location (no. of sample)	Concentration (Bq kg^{-1})		
	^{226}Ra	^{232}Th	^{40}K
Chiang Mai (4)	78±14 (56–88)	130±27 (94–157)	958±210 (706–1157)
Chiang Rai (9)	38±18 (22–75)	47±15 (27–69)	493±396 (58–1015)
Lampang (9)	37±7 (25–44)	53±11 (39–74)	404±180 (98–611)
Phayao (3)	40±3 (36–42)	50±2 (48–52)	378±102 (308–495)
Overall (25)	44±19 (22–88)	63±33 (27–157)	521±320 (58–1157)
Thailand [UNSCEAR 2000]	48 (11–78)	51 (7–120)	230 (7–712)
Worldwide [UNSCEAR 2000]	35 (16–110)	30 (11–64)	400 (140–850)

Assuming normal distribution, Independent two-sample *t*-test was performed to examine the mean difference of activity concentrations between all pairs of four provinces. The results of statistical analyses indicated that the mean concentrations of ^{226}Ra (78±14 Bq kg^{-1}), ^{232}Th (130±27 Bq kg^{-1}) and ^{40}K (958±210 Bq kg^{-1}) of soil samples collected from Chiang Mai province were significantly higher ($p<0.05$) than those collected from three other provinces while the mean concentrations among these others, Chiang Rai, Lampang and Phayao, were comparable. One-sample *t*-test was performed to determine the difference between the mean concentrations obtained from this study and the worldwide average values reported by UNSCEAR 2000. The difference of mean ^{232}Th concentration was significantly different with the worldwide average values for all provinces. The mean concentration of ^{226}Ra and ^{40}K of Chiang Mai were significantly higher than the worldwide average but others were comparable. The overall mean concentration of all radionuclides, 44±19 Bq kg^{-1} for ^{226}Ra, 63±33 Bq kg^{-1} for ^{232}Th and 521±320 Bq kg^{-1} for ^{40}K, were significantly higher than those in the southern Thailand obtained from the previous study (29 Bq kg^{-1} for ^{226}Ra, 49 Bq kg^{-1} for ^{232}Th and 344 Bq kg^{-1} for ^{40}K) [18]. Comparing with worldwide average values, only the overall mean of ^{226}Ra and ^{232}Th obtained from this study were significantly higher than the worldwide average values (35 Bq kg^{-1} for ^{226}Ra and 30 Bq kg^{-1} for ^{232}Th), whereas the overall mean of ^{40}K was not significantly different with the worldwide average values (400 Bq kg^{-1}).

The activity concentrations of natural radionuclides in all soil samples were in the order ^{40}K > ^{232}Th > ^{226}Ra. ^{40}K dominates over others since it is a minor isotope of naturally occurring potassium which is the seventh most abundant element on the earth, making up 2.6 % of the weight of the Earth's crust. The higher activity concentration of ^{232}Th compared to ^{226}Ra may be related to difference in chemical speciation and solubility of their parents, ^{232}Th and ^{238}U respectively, in a natural environment. In the Earth's crust, thorium and uranium tend to occur together due to similar ionic size and bond character. However, they are commonly fractionated during surficial processes, i.e. weathering, transportation and deposition. In almost all natural environments, thorium has a very low solubility and preferentially accumulated on particular phases due to limited stability of Th(IV) complexes; whereas under an oxidizing environment, U(IV) is oxidized to U(VI). The latter is chemically more soluble and mobile; therefore, well redistributed and transported in surface and groundwater under certain environmental conditions [19].

3.2. Activity concentration in Alpinia galangal plant

The activity concentrations of radionuclides in *Alpinia galangal* plant varied from sample to sample and location to location and were generally several orders of magnitude lower than in the associated soils as shown in table 2. The activity concentrations of ^{226}Ra in most galangal samples and those of ^{232}Th in few samples were lower than the detection limit, 0.2 Bq kg^{-1} and 0.1 Bq kg^{-1} for ^{226}Ra and ^{232}Th respectively, of the analytical techniques utilized in the present study. The activity concentration of ^{226}Ra, ^{232}Th and ^{40}K in all galangal samples were very low compared to their relevant soils, since only trace amounts of these radionuclides can diffuse through Casparian strips (a waxy layer structural for the control of uptake of substances into roots) and then may be transported to the aerial organs of plants. The plant roots serves as a semipermeable filter controlling the transport of radionuclides to the aerial parts [1].

Range and mean of activity concentrations in each galangal part, rhizome, stem and leaf are shown in table 3. The overall concentrations obtained from the present study ranged from <0.2 to 2.0 Bq kg^{-1} with a mean of 0.9±0.7 Bq kg^{-1} for ^{226}Ra, <0.1 to 2.9 Bq kg^{-1} with a mean of 0.9±0.9 Bq kg^{-1} for ^{232}Th and 205 to 2247 Bq kg^{-1} with a mean of 785±576 Bq kg^{-1} for ^{40}K. The mean value of ^{226}Ra was not significantly different with that of ^{232}Th ($p<0.05$). The activity concentrations of ^{40}K in all samples and their mean value were significantly higher than those of ^{226}Ra and ^{232}Th which indicate the essential nutrient status of potassium for the plants. Although ^{226}Ra and ^{232}Th are not essential to plant growth, plants may absorb these nuclides, which are similar to calcium, and translocate them to the aerial parts [13].

3.3. Transfer factors (TFs)

Estimated transfer factors are presented in table 4. There was a wide variation of TFs for the same plant or same species. The overall TF values ranged from <0.002 to 0.073 with a mean of 0.023±0.027 for ^{226}Ra, <0.001 to 0.061 with a mean of 0.015±0.018 for ^{232}Th and 0.26 to 7.9 with a mean of 2.2±2.1 for ^{40}K. TFs of ^{226}Ra and ^{232}Th obtained from the present study were not significantly different ($p<0.05$) and they were in the same range of soil-to-galangal TF reported in a previous study (0.004-0.068 for ^{238}U, parent nuclide of ^{226}Ra, and 0.003-0.059 for ^{232}Th) [2]. TFs for ^{40}K were significantly higher than those of the other radionuclides in all samples and most of them were higher than 1 which suggests higher levels of ^{40}K uptake. This may be due to the continuous accumulation of ^{40}K through root uptake over a period of time. It is well known that potassium is an essential macronutrient for metabolism and taken up by plants from soil in varied amounts depending upon their metabolism and ^{40}K respectively. From TFs values of each sample, it was observed that the activity concentrations of ^{226}Ra, ^{232}Th and ^{40}K and their TFs in the aerial part (leaf and stem) were higher than those in the rhizome, whereas *t*-test determined that there is no difference between the mean values among those parts for all radionuclides.

Table 2. Activity concentration of natural radionuclides in *Alpinia galangal* and their corresponding soil.

Plant Sample	Concentration (Bq kg^{-1})		
	^{226}Ra	^{232}Th	^{40}K
CMI-1-Soil	84	142	1157
CMI-1-Rhizome	0.2	1.1	767
CMI-1-Leaf	1.2	1.1	1038
CMI-2-Soil	88	157	1103
CMI-2-Rhizome	<0.2	0.6	599
CMI-2-Stem	1.0	0.7	1051
CMI-2-Leaf	<0.2	<0.1	619
CRI-1-Soil	75	27	58
CRI-1- Rhizome	0.4	0.2	205
CRI-1- Stem	<0.2	0.4	459
CRI-2-Soil	27	42	955
CRI-2- Rhizome	<0.2	1.5	294
CRI-2- Stem	<0.2	2.6	252
CRI-2-Leaf	2.0	0.4	275
CRI-3-Soil	61	65	517
CRI-3-Stem	<0.2	<0.1	1130
LPG-1-Soil	25	39	529
LPG-1- Rhizome	<0.2	0.1	624
LPG-1- Stem	0.8	0.4	1605
LPG-1- Leaf	<0.2	0.4	2127
LPG-2-Soil	26	46	98
LPG-2- Rhizome	<0.2	0.2	606
LPG-3-Soil	44	74	611
LPG-3- Rhizome	<0.2	0.2	678
LPG-3- Stem	<0.2	2.5	2247
LPG-4-Soil	42	55	153
LPG-4- Rhizome	<0.2	0.3	550
PYO-1-Soil	41	52	308
PYO-1- Rhizome	<0.2	0.6	372
PYO-1- Stem	<0.2	0.3	442
PYO-1- Leaf	<0.2	2.9	541

CMI= Chiang Mai; CRI= Chiang Rai; LPG= Lampang; PYO= Phayao

Table 3. Activity concentration of natural radionuclides in each part of *Alpinia galangal*.

Galangal part	Concentration (Bq kg^{-1})		
	^{226}Ra	^{232}Th	^{40}K
Rhizome	0.3±0.1 (<0.2–0.4) *n*=2	0.5±0.5 (0.1–1.5) *n*=9	522±188 (205–767) *n*=9
Stem	0.9±0.1 (<0.2–1.0) *n*=2	1.1±1.1 (<0.1–2.6) *n*=6	1026±719 (252–2247) *n*=7
Leaf	1.6±0.6 (<0.2–2.0) *n*=2	1.2±1.2 (<0.1–2.9) *n*=4	920±728 (275–2127) *n*=5
Overall (Present study)	0.9±0.7 (<0.2–2.0) *n*=6	0.9±0.9 (<0.1–2.9) *n*=19	785±576 (205–2247) *n*=21

n=number of samples

Table 4. Soil-to-plant transfer factors (TFs).

Plant Sample	^{226}Ra	^{232}Th	^{40}K
CMI-1-Rhizome	0.002	0.008	0.66
CMI-1-Leaf	0.014	0.008	0.90
CMI-2-Rhizome	<0.002	0.004	0.54
CMI-2-Stem	0.011	0.004	0.95
CMI-2-Leaf	<0.002	<0.001	0.56
CRI-1- Rhizome	0.005	0.006	3.51
CRI-1- Stem	<0.002	0.016	7.85
CRI-2- Rhizome	<0.002	0.036	0.31
CRI-2- Stem	<0.002	0.061	0.26
CRI-2-Leaf	0.073	0.010	0.29
CRI-3-Stem	<0.002	<0.001	2.18
LPG-1- Rhizome	<0.002	0.002	1.18
LPG-1- Stem	0.030	0.009	3.03
LPG-1- Leaf	<0.002	0.010	4.02
LPG-2- Rhizome	<0.002	0.004	6.19
LPG-3- Rhizome	<0.002	0.002	1.11
LPG-3- Stem	<0.002	0.033	3.68
LPG-4- Rhizome	<0.002	0.005	3.60
PYO-1- Rhizome	<0.002	0.012	1.21
PYO-1- Stem	<0.002	0.005	1.44
PYO-1- Leaf	<0.002	0.056	1.76

CMI= Chiang Mai; CRI= Chiang Rai; LPG= Lampang; PYO= Phayao

3.4. Annual effective ingestion dose

The Annual effective ingestion dose due to ingestion of ^{226}Ra, ^{232}Th and ^{40}K in galangals obtained from the present study are shown in table 5. The ingestion dose was calculated by assuming the annual intake rate is 3.3 kg y^{-1}. The annual effective ingestion dose varied from <0.2 to 1.9 μSv y^{-1} with a mean of 0.8±0.6 μSv y^{-1} for ^{226}Ra, <0.1 to 2.2 μSv y^{-1} with a mean of 0.6±0.7 μSv y^{-1} for ^{232}Th and 0.9 to 10 μSv y^{-1} with a mean of 3.6±2.7 μSv y^{-1} for ^{40}K. These obtained values is significantly below the total worldwide annual effective ingestion dose for ^{40}K and both uranium and thorium series (280 μSv y^{-1}) reported by UNSCEAR 2000.

Table 5. Annual effective ingestion dose of natural radionuclides in each part of *Alpinia galangal*.

Galangal part	Annual effective ingestion dose ($\mu Sv\ y^{-1}$)		
	^{226}Ra	^{232}Th	^{40}K
Rhizome	0.3±0.1 (<0.2–0.3) *n*=2	0.4±0.4 (0.1–1.1) *n*=9	2.4±0.9 (0.9–3.5) *n*=9
Stem	0.8±0.1 (<0.2–0.7) *n*=2	0.8±0.8 (<0.1–1.9) *n*=6	4.7±3.3 (1.2–10) *n*=7
Leaf	1.0±0.5 (<0.2–1.9) *n*=2	0.9±0.9 (<0.1–2.2) *n*=4	4.1±3.4 (1.3–9.8) *n*=5
Overall (Present study)	0.8±0.6 (<0.2–1.9) *n*=6	0.6±0.7 (<0.1–2.2) *n*=19	3.6±2.7 (0.9–10) *n*=21

n=number of samples

4. Conclusions

The activity concentration of ^{226}Ra, ^{232}Th and ^{40}K were determined in soil and *Alpinia galangal* collected from four provinces in the north of Thailand. Our results showed a large variation and non-uniform distribution in soil and *Alpinia galangal* plant samples for each of the natural radionuclides investigated. The activity concentrations of ^{226}Ra and ^{232}Th in *Alpinia galangal* were much lower than in the associated soils, whereas ^{40}K concentration of not all samples were higher than those of their associated soils. It also found that the activity concentration and the uptake of ^{40}K was significantly higher that other radionuclides in all galangal samples. The annual effective ingestion dose values obtained from the present study were much below the worldwide dose of 280 μSv y^{-1}. This study indicates the uses of the *Alpinia galangal* in cuisines and medicines should not pose any potential health hazards to people who consume galangal cultivated in the study area. The results obtained from this study can be considered as baseline data for TFs of natural radionuclides in Thailand and may also serve as a guideline for future monitoring and assessment of naturally occurring radioactive material (NORM) discharges from any industrial activities. The distribution of natural radionuclides in soil and plant is valuable information therefore we will carry out this study with various types of plant. Since plant uptake from soil was probably influenced by various factors such as soil characteristics, amount and physico-chemical form of radionuclides in soil, plant species, temperature, rainfall, and agricultural management, these parameters should be further investigated in the future.

Acknowledgements

The authors are very thankful to "Rachadapisek Sompote Fund, Chulalongkorn University" for the award of Postdoctoral Fellowship at Chulalongkorn University. This study carried out under "Construction of Baseline Data of Natural and Man-made Radiation Sources for Supporting Thailand's Radiation Emergency Preparedness (Phase 1)" by financial support from Thailand's budget in brief fiscal year 2015. We are also very grateful to Prof.Dr. William C. Burnett for making valuable suggestions of an earlier draft of this manuscript.

References

[1] Mollah A S 2014 Radionuclide uptake from soil to plants: Influence of soil classification *Radionuclide contamination and remediation through plants* vol 1 ed D K Gupta and C Walther (Switzerland: Springer)

[2] Krisananuwat R, Sahoo S K and Arae H 2015 Distribution of ^{238}U and ^{232}Th in selected soil and plant samples as well as soil to plant transfer factors around Southern Thailand *J. Radioanal. Nucl. Chem.* **303** 2571–77

IOP Conf. Series: Journal of Physics: Conf. Series **860** (2017) 012008 doi:10.1088/1742-6596/860/1/012008

[3] Al-Masri M S, Al-Akel B, Nashawani A, Amin Y, Khalifa K H and Al-Ain F 2008 Transfer of ^{40}K, ^{238}U, ^{210}Pb, and ^{210}Po from soil to plant in various locations in south of Syria *J. Environ. Radioact.* **99(2)** 322–31

[4] Al-Masri M S, Amin Y, Ibrahim S and Nassri M 2015 Transfer of ^{210}Po, ^{210}Pb and ^{238}U from some medicinal plants to their essential oils *J. Environ. Radioact.* **141** 51–6

[5] Vera Tome F, Blanco Rodríguez M P and Lozano J C 2003 Soil-to-plant transfer factors for natural radionuclides and stable elements in a Mediterranean area *J. Environ. Radioact.* **65** 161–75

[6] Marčiulionienė D, Lukšienė B and Jefanova O 2015 Accumulation and translocation peculiarities of 137Cs and 40K in the soil – plant system *J. Environ. Radioact.* **150** 86–92

[7] Vandenhove H, Olyslaegers G, Sanzharova N, Shubina O, Reed E, Shang Z and Velasco H 2009 Proposal for new best estimates of the soil-to-plant transfer factor of U, Th, Ra, Pb and Po. *J. Environ. Radioact.* **100** 721–32

[8] United Nations Scientific Committee on the Effects of Atomic Radiation 2000 *Sources and effects of ionizing radiation* New York United Nations

[9] International Atomic Energy Agency 2009 *Quantification of radionuclide transfer in terrestrial and freshwater environments for radiological assessments (IAEA-TECDOC-1616)* Vienna International Atomic Energy Agency

[10] Gregory A O and Agbalagba E O 2014 Assessment of natural radioactivity, associated radiological health hazards indices and soil-to-crop transfer factors in cultivated area around a fertilizer factory in Onne, Nigeria *Environ. Earth. Sci.* **71** 1541–9

[11] Aswood M Sh, Jaafar M S and Bauk S 2013 Assessment of radionuclide transfer from soil to vegetables in farms Cemeron Highlands and Penang, (Malaysia) using neutron activation analysis *Appl. Phys. Res.* **5(5)** 85–92

[12] International Atomic Energy Agency 1994 *Handbook of parameter values for prediction of radionuclide transfer in temperate environments (IAEA Technical report series No. 364)* Vienna International Atomic Energy Agency

[13] Mitchell N1, Pérez-Sánchez D and Thorne M C 2013 A review of the behaviour of U-238 series radionuclides in soils and plants *J. Radiol. Prot.* **33** R17-R48

[14] James J P, Dileep B N, Ravi P M, Joshi R M, Ajith T L, Hegde A G and Sarkar P K 2011 Soil to leaf transfer factor for the radionuclides ^{226}Ra, ^{40}K, ^{137}Cs and ^{90}Sr at Kaiga region, India *J. Environ. Radioact.* **102(12)** 1070–7

[15] Chen S B, Zhu Y G and Hu Q H 2005 Soil to plant transfer of ^{238}U, ^{226}Ra and ^{232}Th on uranium mining-impacted soil from southeastern China *J. Environ. Radioact.* **82** 223–36

[16] Ravindran P N, Beverages T G, Pillai G S and Balachandram I 2012 Galangal *Handbook of Herbs and Spices* vol 2 ed K V Peter (Cambridge: Woodhead Publishing) chapter 15 pp 313–8

[17] Chandrashekara K and Somashekarappa H M 2016 Estimation of radionuclides concentration and average annual committed effective dose due to ingestion for some selected medicinal plants of South India *J. Rad. Res. Appl. Sci.* **9** 68–77

[18] Kritsananuwat R, Arae H, Fukushi M, Sahoo S K and Chanyotha S 2015 Natural radioactivity survey on soils originated from southern part of Thailand as potential sites for nuclear power plants from radiological viewpoint and risk assessment *J. Radioanal. Nucl. Chem.* **305** 487–99

[19] Kritsananuwat R, Arae H, Sahoo S K, Fukushi M, Pangza K and Chanyotha S 2015 Radiological risk assessment of ^{238}U, ^{232}Th and ^{40}K in Thailand coastal sediments at selected areas proposed for nuclear power plant sites *J. Radioanal. Nucl. Chem.* **303** 325–34

The ^{7}Be profiles in the undisturbed soil used for reference site to estimate the soil erosion

S Raksawong[1], M Krmar[2], T Bhongsuwan[1,3]

[1] Nuclear Physics Research Laboratory and Geophysics Research Center, Department of Physics, Faculty of Science, Prince of Songkla University, Thailand
[2] Department of Physics, Faculty of Science, University of Novi Sad, Trg D. Obradovica 4, Novi Sad, 21000, Serbia

[3] Corresponding author e-mail: tripop.b@psu.ac.th

Abstract. The cosmogenic radionuclide ^{7}Be is increasingly used to obtain information on event-related soil erosion rates within agricultural landscapes. In this study, we select two undisturbed and flat areas to calculate the reference inventory and relaxation mass depth by using ^{7}Be technique to document short-term erosion. Our results showed that the depth distribution of ^{7}Be in undisturbed soil profiles was 1.0 cm in sites S02 and S03; the initial activities were 31.6 and 38.8 Bq.kg^{-1}, respectively. The relaxation mass depths were 5.4 and 7.2 kg.m^{-2} and the measured reference ^{7}Be inventories were 71 and 110 Bq.m^{-2} for sites S02 and S03, respectively. The difference values of the relaxation mass depth and the reference inventory of both sites implied that for determining a short term soil erosion using ^{7}Be, the reference site was suggested to be selected as close as possible to the study site.

1. Introduction

Some agricultural practices such as burning of stubble and crop residues after harvest and subsequent ploughing and disc harrowing for seed bed preparation leave large areas of bare soil have increased the soil erosion problem that relating to the reduction in soil productivity and to the off-site effects of eroded sediments such as transportation of the sediment-associated nutrients and contaminants (e.g. agricultural pesticides) through terrestrial and aquatic ecosystems [1-3]. Therefore, the reliable information on rates of soil loss is important for an improved understanding of sediment transfer and storage in catchments and river basins to provide a basis for sustainable soil management. The environmental radionuclides ^{137}Cs ($T_{1/2}$ = 30 years) and unsupported ^{210}Pb ($T_{1/2}$ = 22 years) are widely used to document rates of soil redistribution on agricultural land, but both these radionuclides have provided the retrospective estimates in the medium-term (i.e. ca. 50 years for ^{137}Cs and up to 100 years for unsupported ^{210}Pb). The obtained values from both radionuclides are the average rates taking account of inter-annual variability in the magnitude and frequency of erosional events. If the magnitudes of soil loss associated with individual events or periods characterized by specific land use conditions are required, the short-lived radionuclide ^{7}Be is used to calculate the soil erosion rates. This paper reports a preliminary investigation of undisturbed and flat areas aimed at exploring the potential for using ^{7}Be measurements as the reference inventory and relaxation mass depth to document the short-term events of soil erosion.

2. Methodology

2.1 Background and initial depth distribution of the beryllium-7 in the soil

^{7}Be is produced by cosmic-rays spallation reactions in the atmosphere. These reactions occur primarily in the stratosphere and upper troposphere, where charged particles (alpha particles, electrons and protons) induce nuclear reactions with oxygen and nitrogen atoms. The ^{7}Be attaches to airborne particles and its deposition is continuously delivered to the earth's surface by wet and dry fallout [2, 4-6]. Although it is quickly sequestered to the Be^{2+} ion by slightly acid rainfall, its ion is rapidly and strongly fixed onto the clay minerals in the soil [2, 7]. Therefore, cosmogenic ^{7}Be has been widely used as a tracer to document both the magnitudes and the spatial patterns associated with short-term or event based soil redistribution on agricultural land [1-2, 8]. The ^{7}Be technique is, principally, based on the comparison of ^{7}Be inventory (Bq.m^{-2}) between sampling sites and reference site. The reference site is selected from undisturbed area located near sampling sites. The ^{7}Be inventory for sampling site is depleted relative to the reference inventory that can be referred to occurring erosive processes, whereas areas of deposition can be located by increased inventories [1-2, 8].

There are three key assumptions for using ^{7}Be measurements to estimate both the soil erosion and sedimentation rates [1-2, 8]. The deposition of ^{7}Be fallout from atmosphere to the ground surface is, first, spatially uniform. Secondly, preexisting ^{7}Be in the surface soil in study area is uniformly distributed across the area. Finally, the ^{7}Be deposited during event is rapidly and strongly absorbed in couple centimeters of surface soil and can be mobilized by erosion. From these assumptions, it can be assumed that the initial vertical distribution of the ^{7}Be activity $C(x)$, (Bq.kg^{-1}) within the soil will be characterized by an exponential decrease with mass depth x, kg.m^{-2}[1-2]. The sample model to estimate the soil erosion and sedimentation rates is described below,

$$C(x) = C(0)\exp(-x/h_o) \tag{1}$$

where $C(0)$ is the initial activity of the surface soil (at $x = 0$) and h_o, kg.m^{-2}, is the relaxation mass depth.

The reference inventory of ^{7}Be, A_{ref}, Bq.m^{-2}, is defined as the initial total areal activity at an uneroded stable site or reference site in the study area:

$$A_{ref} = A(0) = \int_0^\infty C(x)dx = h_o C(0) \tag{2}$$

The areal activity of ^{7}Be below depth x, $A(x)$, Bq.m^{-2}, for the initial distribution is therefore:

$$A(x) = \int_x^\infty C(x)dx = A_{ref}\exp(-x/h_o) \tag{3}$$

The relaxation mass depth describes the shape of the initial depth distribution of both the activity as shown in Eq. (1) and areal activity as shown in Eq. (3) in the soil.

By measuring the activity of ^{7}Be, C, in different depth increments of soil collected from the reference site and establishing the mass depth of each depth increment, the values of $A(x)$ for corresponding mass depths x down the reference profile can be calculated. Logarithmically transforming Eq. (3), h_o and the reference inventory A_{ref} can be deduced from a linear regression between $ln[A(x)]$ and x.

2.2 *Study site*

Atmospheric deposition fluxes of ^{7}Be in the study areas, the fallout samples during December, 2012 to December, 2013 for periods from 1 to 31 days depending on the frequency of rain were collected at the top roof of the Physics building called S01 site (7°00'24.6" N, 100°29'57.8" E) where is the control site to estimate the running inventory of ^{7}Be. Two sampling sites for rain water samples were at selected reference sites 7°00'30.9" N, 100°30'22.0" E (S02) and 6°59'32.1" N, 100°33'04.8" E (S03). Meteorological data were provided by the Thai Meteorological Department at the Songkhla Meteorological station (WMO Index: 568501/48568, located at 7°12'14" N, 100°36'11" E). Rain water samples were collected after rainfall event. The water sample was filtered through the MnO_2-fiber (5 g) which was in the cylinder of 15 cm long and 1.8 cm inner diameter.

The climate of the Songkhla province, Thailand is controlled by tropical monsoons that consist of the southwest (May – October) and northeast (October – February) monsoons. There are generally two seasons – the rainy season and dry season. The rainy season is influenced by the southwest monsoon (May – September) that brought the moist and warm air from the Indian Ocean whereas the northeast monsoon (October – January), brought the moist air from the Gulf of Thailand.

a) b)

Figure. 1 Photographs showing **a)** soil core sampling; **b)** cutting the cover grasses

Soil samples were collected from two selected reference sites S02 and S03 in the U-Tapao subcatchment, Songkhla province, Southern Thailand. Using a hand-operated corer equipped with the PVC pipe (8.5 cm internal diameter, 20 cm length), 8 soil cores were collected from two selected reference sites S02 and S03 on January 10, 2014. Figure. 1 shows the soil core sampling and cutting the cover grasses. All the soil cores were kept vertically and returned to the laboratory, and then sliced at 0.5 cm intervals through its depth using the hand-held extruder. All the cut samples were dried at 105 °C for 24 hours by electric oven, grinded by ceramic mortar, sieved to < 2 mm fraction, homogenized, weighed and put into a polyethylene bottle.

2.3 Measurement and analyze method

Beryllium-7 activity concentrations in fallout (wet and dry) samples, soil samples and grass were determined by measuring its gamma line at 477.6 keV. The gamma-ray measurements were performed using gamma spectrometer with an HPGe detector (GC7020, Canberra Industries, USA) in a low-background cylindrical shield (Model 747, Canberra, USA). Data were accumulated in a multichannel analyzer (DSA1000, Canberra, USA) based on a personal computer. The energy resolution was 0.88 keV (FWHM) at 122 keV (^{57}Co), and 1.77 keV (FWHM) at 1332 keV (^{60}Co). The minimum counting-time for each sample was set in 40,000s. For samples having an analytical error more than 1σ, the counting-time will be added until 100,000s to provide reasonable analytical error (less than 25% or 1σ). The spectrum analysis was performed by the commercial software Genie 2000 (Canberra, USA). The absolute efficiency curve of detector used in this work was fitted from activity concentrations of discrete gamma-ray lines in the IAEA TEL 2011-03 WWOPT soil-04 sample with certified radionuclide activities, and it was used to calculate the relative efficiency of ^{7}Be at 477.7 keV. The ^{7}Be activities were corrected for decay to the sampling time on January 10, 2014.

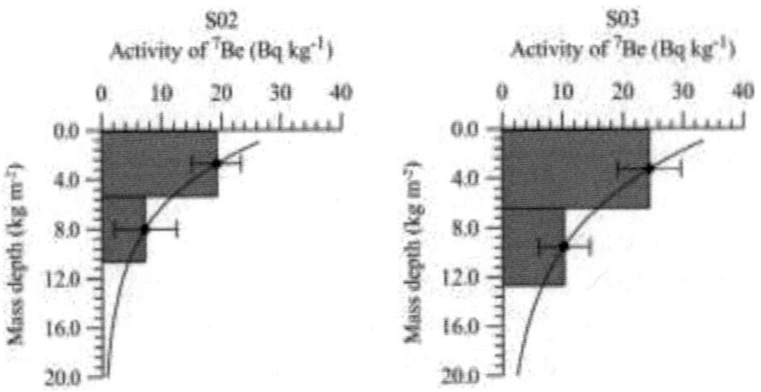

Figure 2. The depth distribution of ^{7}Be in the soil profiles within the selected reference sites.

3. Results and Discussions

The results reported in this paper were obtained from the selected reference sites (S02 and S03) where located in the undisturbed, non-tilled areas at the U-Tapao subcatchment, Songkhla province, Southern Thailand. Our results show that the depth distribution of ^{7}Be in undisturbed soil profiles is 1.0 cm in both sites (S02 and S03), the initial activities are 31.6 and 38.8 Bq.kg^{-1} for S02 and S03 sites, respectively. The ^{7}Be activities for both locations declined exponentially with mass depth as shown in Figure 2. Following from Eq.(1), the relaxation mass depths were, therefore, calculated from linear regression between $ln[C(x)]$ and x. These obtained results are 5.4 and 7.2 kg.m^{-2} for S02 and S03 sites, respectively.

The measured reference inventories of ^{7}Be in soil profiles are 71 and 110 Bq.m^{-2} for S02 and S03 sites, respectively. The calculated reference inventories from Eq. (2) are 170 and 279 Bq.m^{-2} for sites S02 and S03, respectively. Both important parameters, relaxation mass depth and reference inventory, can be used to estimate the magnitudes of soil loss or deposition. The soil mass eroded per unit area, R, kg.m^{-2}, or soil erosion rate as provided from Blake et al. [1] can be calculated by using both values as:

$$R = h_o \ln\left[A_{ref} / A\right] \tag{4}$$

Moreover, the magnitude of deposited sediment rate, R', kg.m^{-2}, can be calculated by:

$$R' = \left(A' - A_{ref}\right)/C_d \tag{5}$$

when A', Bq.m^{-2}, is the areal activity of ^{7}Be at a sampling point located in sediment deposition zone, C_d is the mean ^{7}Be activity of deposited sediment that can be estimate by

$$C_d = \int_S C_e R dS \Big/ \int_S R dS \tag{6}$$

when C_e, Bq.kg^{-1}, is the weighted mean ^{7}Be activity of mobilized sediment from eroding area S, m^{-2}, which can be:

$$C_e = (A_{ref} - A)/R = A_{ref}\left[1 - \exp(-R/h_o)\right]/R \tag{7}$$

From procedure outlined above, the h_o and A_{ref} are crucial important for calculating the magnitudes of soil erosion and deposition rates at sampling points for study area. The relaxation mass depths and ^{7}Be inventories in the soil obtained from selected reference sites are different because the atmospheric depositional fluxes of ^{7}Be are controlled by several factors depending on location, precipitation, the cover crops etc. Lohaiza et al. reported the monthly relaxation mass depths and inventory of ^{7}Be from one reference site changing with season, the maximum value in heavy rain period and low value in dry period [7]. Moreover, the relaxation mass depths in some locations where selected reference site are significantly higher than that in others. Therefore, the running atmospheric deposition flux of ^{7}Be and the ^{7}Be activity in the covered grasses were used to compare with the inventory in soil for the reference study site.

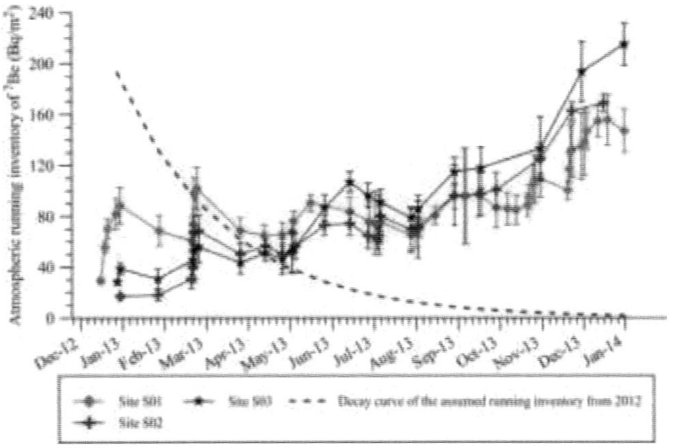

Figure 3. Atmospheric running inventory of ^{7}Be collected from S01, S02 and S03 sites.

In this study, the atmospheric depositional fluxes of ^{7}Be have been decay-corrected to the collecting time of rainfall sample, and the results are shown in Figure 3. In order to calculate the atmospheric deposition running inventory of ^{7}Be following Zhu and Olsen [9], the calculated running inventory values range 29 ± 1 to 156 ± 14 Bq.m^{-2}, with an average value of 92.2 Bq.m^{-2} for site S01, 2.0 ± 0.5 to 96 ± 12 Bq.m^{-2}, with an average value of 19.6 Bq.m^{-2} for site S02, and 2.9 ± 1 to 103 ± 15 Bq.m^{-2}, with an average value of 27.1 Bq.m^{-2} for S03. The estimated ^{7}Be running inventories at the time of sediment sampling are 156 ± 20 Bq.m^{-2}, 169 ± 7 Bq.m^{-2} and 215 ± 17 Bq.m^{-2} for S01, S02 and S03, respectively. Atmospheric running inventories for the selected reference sites S02 and S03 are shown in the Table 1 and Figure 4.

Table 1. ^{7}Be inventories in soil cores, areal activities in cover grasses and atmospheric running inventories which were corrected to the soil sampling time on January 10, 2014.

Type/Location	S02	S03
Measured sediment inventory (Bq.m^{-2})	71 ± 18	110 ± 26
Areal activity in grasses (Bq.m^{-2})	112 ± 9	108 ± 9
Atmospheric running inventory (Bq.m^{-2})	169 ± 7	215 ± 17

The measured ^{7}Be inventories in soil cores of 71 ± 18 and 110 ± 26 Bq.m^{-2} were stacked with the areal activities in cover grasses of 112 ± 9 and 108 ± 9 Bq.m^{-2} for the selected reference sites S02 and S03, respectively. The atmospheric running inventories are similar to the stacked activities between the areal activities in cover grasses and the measured ^{7}Be inventories in soil cores for both sites S02 and S03 as shown in Figure 4. The relaxation mass depths and the reference inventories of both selected reference sites showed the different values which had been influenced from the different soil types and the different atmospheric depositions of ^{7}Be. Therefore, if this method is used to determine the short term soil erosion, the reference site is suggested to be as close as possible to the study area.

IOP Conf. Series: Journal of Physics: Conf. Series **860** (2017) 012009 doi:10.1088/1742-6596/860/1/012009

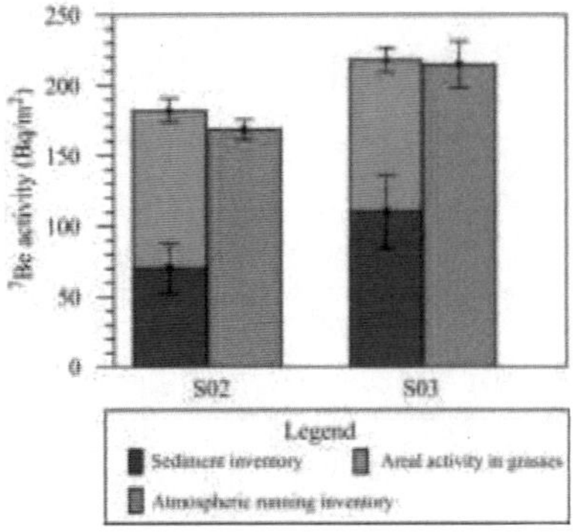

Figure 4. The bar charts comparing the atmospheric running inventory and the stacked columns between measured sediment inventory and areal activity in grasses.

Although, the calculated reference inventories, 170 and 279 Bq.m^{-2}, are larger than the measured inventories. The measured inventories in soil cores, 71 ± 18 and 110 ± 26 Bq.m^{-2} for both sites S02 and S03, may be used as the reference inventories, A_{ref}, in the model to calculate the soil erosion rates in study area. This is in fact the calculated reference inventories are not possible to be larger than the atmospheric running inventory. This problem may occur due to the thicker section of soil cores. In the future work, the thin sliced section of soil cores will be made, because there will be more than two points in the ^{7}Be profiles with mass depth to estimate the calculated reference inventory and relaxation mass depth.

4. Conclusions

This preliminary study is successful to detect ^{7}Be in the soil and atmospheric deposition in the South of Thailand located near equatorial zone where is known that very low ^{7}Be activity and the ^{7}Be measurement in soil can be used to estimate important parameters (relaxation mass depth and reference inventory) for approached model of the soil erosion. Further work, we will investigate the magnitudes of both soil erosion and deposition obtained from the ^{7}Be measurement technique in this study area.

Acknowledgments

This research was financially supported by the Prince of Songkla University (project code: SCI550111S), the National Research Council of Thailand (project code: SCI560125S), the Physics Department, Geophysics Research Center, Faculty of Science and the graduate fellowship from the Graduate School, Prince of Songkla University. In addition, the first author would like to thank Prince of Songkla University, Graduate Studies Grant.

References

[1] Blake W H, Walling D E and He Q 1999 Appl. Radiat. Isot. **51** 599–605

[2] Schuller P, Iroumé A, Walling D E, Mancilla H B, Castillo A and Trumper R E 2006 J. Environ. Qual. **35** 1756–1763

[3] Sepulveda A, Schuller P, Walling D E and Castillo A 2008 J. Environ. Radioact. **99** 35–49

[4] Ioannidou A and Papastefanou C 2006 J. Environ. Radioact. **85**, 121–136

[5] Ishikawa Y, Murakami H, Sekine T and Yoshihara K 1995 J. Environ. Radioact. **26** 19–36

[6] Papastefanou C and Ioannidou A 1995 J. Environ. Radioact. **26** 273–282

[7] Lohaiza F, Velasco H, Juri Ayub J, Rizzotto M, Di Gregorio D E, Huck H and Valladares D L 2014 J. Environ. Radioact. **130** 72–77.

[8] Zapata F, Garcia-Agudo E, Ritchie J C and Appleby P G 2002 Introduction *Handbook for the Assessment of Soil Erosion and Sedimentation Using Environmental Radionuclides.* Ed Zapata F (Dordrecht: Kluwer Academic Publishers) chapter 1 pp 1–13

[9] Zhu J and Olsen C R 2009 J. Environ. Radioact. **100** 192–197

IOP Conf. Series: Journal of Physics: Conf. Series **860** (2017) 012010 doi:10.1088/1742-6596/860/1/012010

Participation in proficiency test for tritium strontium and caesium isotopes in seawater 2015 (IAEA-RML-2015-02)

S Visetpotjanakit[1] and S Kaewpaluek

Bureau of Technical Support for Safety Regulation , Office of Atoms for Peace, 16 Vibhavadi rangsit road Ladyao Chatuchak, Bangkok 10900, Thailand

[1] Corresponding author e-mail: suputtra.v@oap.go.th

Abstract. A proficiency test (PT) exercise has proposed by the International Atomic Energy Agency (IAEA) in the frame of the IAEA Technical Cooperation project RAS/7/021 "Marine benchmark study on the possible impact of the Fukushima radioactive releases in the Asia-Pacific Region for Caesium Determination in Sea Water" since 2012. In 2015 the exercise was referred to Proficiency Test for Tritium, Strontium and Caesium Isotopes in Seawater 2015 (IAEA-RML-2015-02) to analyse ^{3}H, ^{134}Cs, ^{137}Cs and ^{90}Sr in a seawater sample. OAP was one of the 17 laboratories from 15 countries from Asia-Pacific Region who joined the PT exercise. The aim of our participation was to validate our analytical performance for the accurate determination of radionuclides in seawater by developed methods of radiochemical analysis. OAP submitted results determining the concentration for the three elements i.e. ^{134}Cs, ^{137}Cs and ^{90}Sr in seawater to the IAEA. A critical review was made to check suitability of our methodology and the criteria for the accuracy, precision and trueness of our data. The results of both ^{134}Cs and ^{137}Cs passed all criteria which were assigned "Accepted" statuses. Whereas ^{90}Sr analysis did not pass the accuracy test therefore it was considered as "Not accepted" Our results and all other participant results with critical comments were published in the IAEA proficiency test report.

1. Introduction

Accurate and precise radiation monitoring is critical to determine the radionuclide level in the environment for assessment of the radiological impact and risk to the public and environment. This is especially important during nuclear power plant accidents. For instance the recent nuclear power plant accident happened in Japan, Fukushima on March 2011. It released radionuclides to the environment i.e. to the atmosphere, seawater, river and land. In such cases, nearby countries must monitor the situation by estimating radioactivity levels in the environment and the impact of the contamination on humans, animals and the environment to protect their countries. Moreover it is important to be able to share the radiological impact within regions. Those radiochemical laboratories should have the same standard and capability to determine radionuclides in environment in order to share and compare the radiological impact for comprehensive protection and preparedness. The International Atomic Energy Agency (IAEA) is an inter phase organisation who is deeply concerned about radioactive data quality. It has regularly conducted interlaboratory comparisons and proficiency tests on radionuclides in various samples. Due to the Fukushima accident, it recently organised a new proficiency test in the frame of the IAEA Technical Cooperation project RAS/7/021 "Marine benchmark study on the possible impact of the Fukushima radioactive releases in the Asia-Pacific Region for Caesium Determination in Sea Water".

Office of Atoms for Peace (OAP) has a radiochemistry laboratory composed of gamma spectrometer, alpha spectrometer, liquid scintillation counter and gross alpha-beta counting system to carry out our radiation monitoring program by both radiochemical analysis and direct measurement. We also have a remote ambient gamma dose rate and radionuclides stations all around Thailand for establishment of radiation baseline information and assessment of the radiological impact on the environment in the case of any accidents. We however need to approve our radionuclide analysis method to assure data quality to be used as an accurate estimation for any nuclear and radiation circumstances. We have participated in a number of exercises supported by the IAEA. Lastly we were one of 17 laboratories from 15 countries from the Asia-Pacific Region who participated in the new exercise which was referred to Proficiency Test for Tritium, Strontium and Caesium Isotopes in Seawater 2015 (IAEA-RML-2015-02). We used our developed methods for the radionuclide analysis of ^{134}Cs, ^{137}Cs and ^{90}Sr. All participant results were published in the IAEA proficiency test report [1].

2. Experiment

2.1. PT sample description

The five litre proficiency test sample containing ^{134}Cs, ^{137}Cs, ^{90}Sr and ^{3}H was received from IAEA. The picture of PT sample can be seen from figure 1. The sample was of unknown activity and was determined in its ^{134}Cs, ^{137}Cs and ^{90}Sr activities. The results were submitted to the IAEA to evaluate accuracy and quality control.

2.2. ^{134}Cs and ^{137}Cs determination method

2.2.1. Chemicals and reference solution for ^{134}Cs and ^{137}Cs determination using AMP pre-concentration technique. Ammonium phosphomolybdate (AMP), CsCl, HCl and NaOH used were analytical grade. The ^{134}Cs and ^{137}Cs reference solution used to prepare calibration source was obtained from Eckert and Ziegler Isotope Product.

2.2.2. Sample preparation and analysis for ^{134}Cs and ^{137}Cs determination using AMP pre-concentration technique. The method for radiochemical analysis of caesium in seawater was developed from Hirose's technique [2-3]. The three repeated samples were prepared by taking five hundred grams of aliquots into separate beakers. The samples were acidified with HCl to pH 1.6. Then 0.26 g of CsCl (caesium carrier) and 4 g of AMP were added. The mixture was stirred for one hour and left overnight to allow the precipitate to settle. The precipitate samples were centrifuged at 3000 rpm, 5 min and washed with 1 M HCl twice. The precipitate samples were dissolved with 10 M NaOH. The solution was heated at 80°C, 15 min to remove ammonia. The precipitation step was repeated. The second precipitate samples were dissolved with 10 M NaOH. The solution was transferred to calibrated containers (the same shape and type as prepared calibration source i.e. 60 ml polypropylene bottle) and made up the same volume as the calibration source i.e. 20 ml. The pictures of AMP pre-concentration method are shown in figure 2.

2.2.3. Calibration source preparation for ^{134}Cs and ^{137}Cs determination using AMP pre-concentration technique. The reference solution which contained 1.077±0.010 Bq of ^{134}Cs and 1.006±0.005 Bq of ^{137}Cs was spiked into 20 ml of 1 M HCl in the calibrated container (60 ml polypropylene bottle).

2.2.4. Counting equipment and measurement for ^{134}Cs and ^{137}Cs determination using AMP pre-concentration technique. The gamma-ray spectrometry system, HPGe (Li) detector (CANBERRA) with MAESTRO software was used. The prepared calibration source was used to calibrate counting efficiency. The samples in calibrated containers were measured with the same gamma-ray spectrometry for 80000 s.

Figure 1. The five litre proficiency test sample from IAEA.

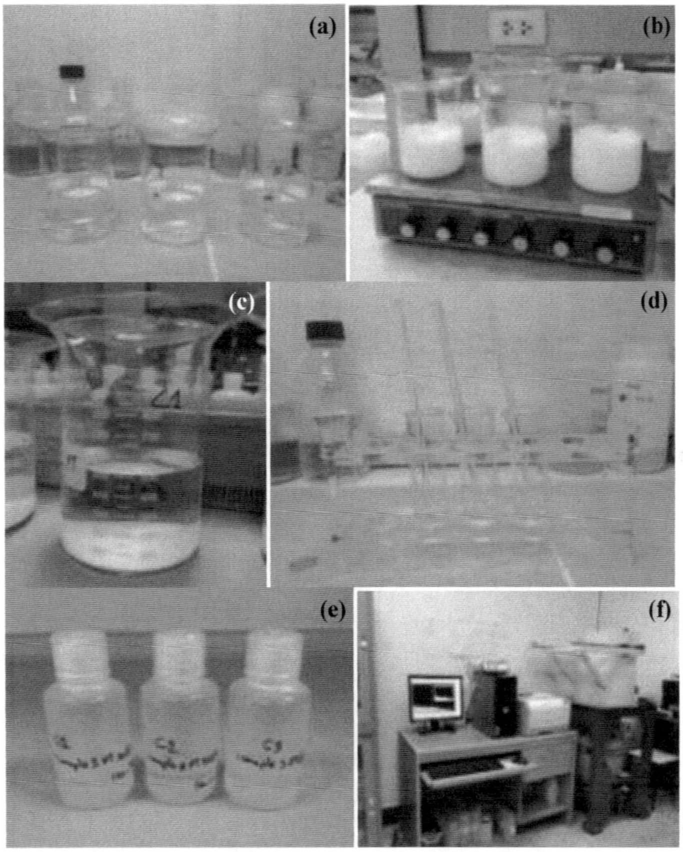

Figure 2. ^{134}Cs and ^{137}Cs determination using AMP pre-concentration technique (a) The 500 gram aliquots in beakers, (b) Precipitating the aliquots by adding CsCl and AMP, (c) The AMP precipitates being left to settle, (d) Dissolving the AMP precipitate with NaOH, (e) The sample solutions in calibrated containers and (f) Gamma-ray spectrometer used for measuring in ^{134}Cs and ^{137}Cs activity in the samples.

IOP Conf. Series: Journal of Physics: Conf. Series **860** (2017) 012010 doi:10.1088/1742-6596/860/1/012010

2.3. ^{90}Sr determination method

2.3.1. Chemicals and reference solution for ^{90}Sr determination using ion exchange chromatographic technique. $Sr(NO_3)_2$, $Ba(NO_3)_2$, $Y(NO_3)_3$, $Ca(NO_3)_2$, NaOH, HCl, EDTA, NaCl, Na_2CO_3, HNO_3, $(NH_4)_2SO_4$, sodium acetate buffer solution used were analytical grade. The cation resin used was Dowex 50WX8 hydrogen form. The ^{90}Sr certified reference solution used to prepare calibration source was purchased from Eckert and Ziegler Isotope Product.

2.3.2. Sample preparation and analysis for ^{90}Sr determination using ion exchange chromatographic technique. The method for radiochemical analysis of strontium in seawater was developed from EPA-600/4-80-032 [4] and EMSL-LV-0539-17 [5]. A one kilogram aliquot of PT sample was used. $Sr(NO_3)_2$, $Ba(NO_3)_2$, and $Ca(NO_3)_2$ were added into the sample as carriers. To pre-concentrate strontium the sample was alkalised to pH 10 with NaOH and added with 3 N Na_2CO_3 to form carbonate precipitate. Strontium was separated using cation exchange chromatography. The Dowex 50WX8 hydrogen form was used. The loading solution was prepared by dissolving the carbonate precipitate with HCl and adding EDTA and sodium acetate buffer solution at pH 3.8. The resin was conditioned with 4 N NaCl and 5% NaOH. The resin was applied with the loading solution and then washed with 2% EDTA. Strontium fraction was eluted with 1.5 N HCl. $SrCO_3$ was formed by precipitating the strontium fraction with Na_2CO_3. The $SrCO_3$ precipitate was kept for 2 weeks for yttrium-ingrowth.Y-90 in equilibrium with ^{90}Sr was separated in hydroxide form. The $SrCO_3$ precipitate was dissolved with HNO_3 and then added with $Y(NO_3)_3$ and NaOH in order to form hydroxide precipitate. Its filtrate contained strontium fraction was used to determine strontium gravimetric yield by precipitating $SrCO_3$. Whereas the hydroxide precipitate was reprecipitated into oxalate form by adding HCl and sat. $(NH_4)_2C_2O_4$. The yttrium oxalate sample was filtrated and placed into planchet for measuring with gas proportional counter. Yttrium gravimetric yield was determined by weighting the sample. The pictures of ion exchange chromatographic technique are shown in figure 3.

2.3.3. Calibration source preparation for ^{90}Sr determination using ion exchange chromatographic technique. The ^{90}Sr certified reference solution which had activity 1.042±0.008 Bq was used to prepare yttrium oxalate calibration source using a short procedure of sample separation.

2.3.4. Counting equipment and measurement for ^{90}Sr determination using ion exchange chromatographic technique. Low background proportional counter, Berthold LB 770 with UMS programme was used. The prepared calibration source was used to determine counting efficiency. The sample was measured with the gas proportional counter for 100 min.

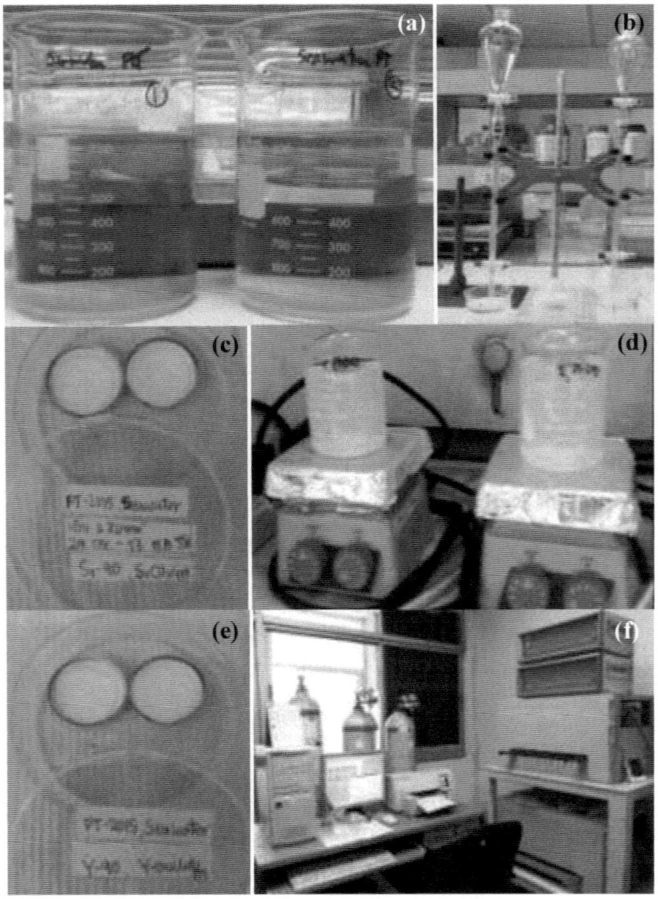

Figure 3. ^{90}Sr determination using ion exchange chromatographic technique (a) The one kilogram aliquot being precipitated as carbonate foam, (b) Ion exchange chromatography used to separate strontium, (c) $SrCO_3$ precipitates being kept for yttrium-ingrowth, (d) Precipitating yttrium in oxalate form, (e) Purified yttrium oxalate for activity measurement and (f) Gas proportional counter used for measuring in ^{90}Y activity in samples.

2.4. Data Evaluation

The PT results judged were based on IAEA criteria using different statistic evaluation criterion such as accuracy, precision and trueness as follow [1, 6-7];

2.4.1. Accuracy. The relative bias (RB) was the first step in producing a score for a result Value Analyst. RB between the Analyst's value and the IAEA target value was calculated and expressed as a percentage according to equation (1).

$$Bias_{relative} = \frac{Value_{Analyst} - Value_{IAEA}}{Value_{IAEA}} X\ 100\% \quad (1)$$

The absolute value of the relative bias was compared to the Maximal Accepted Relative Bias (MARB) which were 20% for both ^{134}Cs and ^{137}Cs and 25% for ^{90}Sr. Value analysts were scored as "Pass" for accuracy when:

$$|Bias_{relative}| \leq MARB \tag{2}$$

2.4.2. Precision and trueness. The precision, P, was calculated according to equation (3).

$$P = \left(\left(\frac{unc_{IAEA}}{Value_{IAEA}}\right)^2 + \left(\frac{unc_{Analyst}}{Value_{Analyst}}\right)^2\right)^{1/2} X\ 100\% \tag{3}$$

The precision P was compared to the Limit of Accepted Precision (LAP) which were 20% for both ^{134}Cs and ^{137}Cs and 25% for ^{90}Sr. Value analysts were scored as "Pass" for precision when:

$$P \leq \text{LAP} \tag{4}$$

Value analysts results for trueness were scored as "Pass" when:

$$|Bias_{relative}| \leq \frac{Value_{Analyst}}{Value_{IAEA}} 2.58\, P \tag{5}$$

2.4.3. Final evaluation. The final score can be summarised according to the detailed evaluation as follow:

"Accepted" when all three tests were passed.
"Not Accepted" when the accuracy test was failed.
"Warning" when accuracy test was passed but either precision or trueness test was failed.

3. Results and discussion

The IAEA assigned values were shown in table 1 and our reported results of three radionuclides can be seen in table 2 and 3.

Table 1. IAEA assigned values.

^{134}Cs activity concentration (Bq kg^{-1})	^{137}Cs activity concentration (Bq kg^{-1})	^{90}Sr activity concentration (Bq kg^{-1})
0.1544±0.0006	0.2044±0.0013	0.1009±0.0007

Table 2. Results of ^{134}Cs and ^{137}Cs analysis.

Radionuclide	Sample weight (kg)	Individual activity concentration (Bq kg^{-1})			Mean activity concentration (Bq kg^{-1})
		1	2	3	
^{134}Cs	0.5	0.174±0.008	0.144±0.007	0.161±0.008	0.160±0.008
^{137}Cs	0.5	0.214±0.010	0.181±0.009	0.212±0.009	0.202±0.009

Table 3. Results of ^{90}Sr analysis.

Radionuclide	Sample weight (kg)	Activity concentration (Bq kg^{-1})
^{90}Sr	1.0	0.840±0.110

For ^{134}Cs and ^{137}Cs results the three individual analysis looked close together. For ^{90}Sr result only one analysis was performed and its activity was quite high when compared with those of ^{134}Cs and

^{137}Cs activities. However the evaluation in those results shown in detail in statistical view is presented in table 4.

Table 4. Performance evaluation.

Radio-nuclide	Analyst Value (Bq kg^{-1})	Analyst Uncert. (Bq kg^{-1})	IAEA Value (Bq kg^{-1})	IAEA Uncert. (Bq kg^{-1})	RB (%)	P (%)	Trueness Limit (%)	Accuracy	Precision	Trueness	Final Score
^{134}Cs	0.160	0.008	0.1544	0.0006	3.4	4.9	13	Pass	Pass	Pass	Accepted
^{137}Cs	0.202	0.009	0.2044	0.0013	-1.0	4.5	12	Pass	Pass	Pass	Accepted
^{90}Sr	0.840	0.110	0.1009	0.0007	730	14	291	Fail	Pass	Fail	Not accepted

For determining caesium by the AMP precipitation technique, working in laboratory was quite simple. The procedure mainly involved precipitation steps to catch caesium in water media. Due to AMP being selective caesium scavenger and the simple separation steps the process loss was suitably negligible to obtain accurate and precise results. In addition for radioactivity measurement the use of prepared calibration source which had the same shape and type as sample containers potentially help to obtain correct counting efficiency, consequently obtaining the accurate final activities. Therefore the result of ^{134}Cs and ^{137}Cs passed the three tests which resulted in being assigned the status Accepted. However the result of ^{90}Sr did not pass accuracy test therefore it was considered as not accepted. Its relative bias was found to be very high i.e. 730 %. This may be due to two main possibilities. Firstly, the sample might contain contaminated ^{134}Cs and ^{137}Cs and this raised gross beta activity when using gas proportional counter. Secondly, the efficiency calculated from prepared calibration source was underestimated. Possibly the calibration source and sample had different precipitate thicknesses. The precipitate sample seemed to be thinner than those of calibration source. For measuring beta particle, the self-absorption difference needed to be corrected in order to determine proper efficiency, consequently obtaining an accurate final activity.

4. Conclusion

Our method for radiochemical analysis for caesium determination using AMP precipitation fully succeeded in ^{134}Cs and ^{137}Cs analysis since the results passed the three tests i.e. accuracy precision and trueness and then acquired the status Accepted. The AMP precipitation technique was proved to be suitable for determining ^{134}Cs and ^{137}Cs in seawater since AMP as scavenge was selective for caesium and the separation steps was not complicated. Also the other successful key was to use sample container having the same shape and type as prepared calibration source. This potentially helped to obtain corrected counting efficiency, hence obtaining the accurate final activities. In the case of ^{90}Sr the ion exchange chromatographic technique used resulted the inaccurate ^{90}Sr activity. To improve the analysis both aspects, radiochemical technique and radioactivity measurement should be considered. For the radiochemical technique, delicate separation steps must be used to remove other radionuclides from the final precipitate sample in order to avoid over beta particle counting using gas proportional counter. For measurement aspect, the different self- absorption between calibration source and sample has to be corrected in order to obtain accurate efficiency and therefore the calculated ^{90}Sr activity in the sample should be correct.

References

[1] Harms A V, Osvath I and Osborn D 2016 *IAEA-RML-2015-02 Proficiency Test for Determination of Radionuclides in Seawater* (Vienna: IAEA)

[2] Hirose K, Aoyama M, Igarashi Y and Komura K 2008 Improvement of ^{137}Cs analysis in small volume seawater samples using the Ogoya underground facility *J. Radioanal. Nucl. Ch.* **276**

pp 795-798

[3] Phaopeng N 2013 *The Measurement of Cs-137 Radioactivity in Seawater and the Oyster (Crassostrea belcheri) from the lower Gulf of Thailand for Radiation Dose Assessment in the General Public* (Bangkok: Kasetsart University)

[4] Herman L K and Earl L W 1980 *Prescribed Procedures for Measurement of Radioactivity in Drinking Water* (OHIO: EPA)

[5] EPA 1979 *Radiochemical Analytical Procedures for Analytical of Environmental Samples* (Nevada: EPA)

[6] Shakhashiro A, Sanaone U, Trinkl A, Makarewicz M, Yonezawa A C, Kim C K, Kis-Benedek G, Benesch T. and Schorn R 2007 *Report on the IAEA-CU-2006-04 ALMERA Proficiency Test on the Determination of Gamma Emitting radionuclides, IAEA/AL/170* (Vienna: IAEA)

[7] Pham M K, Osvath I and Sanchez-Cabeza J 2010 *Proficiency Test Determination of Radionuclides in Fish Flesh Samples IAEA/AQ/13* (Vienna: IAEA)

IOP Conf. Series: Journal of Physics: Conf. Series **860** (2017) 012011 doi:10.1088/1742-6596/860/1/012011

Determination of ^{137}Cs in large volume seawater using Cu-hexacyanoferrate cartridge filters

S Visetpotjanakit[1] and Y Tumnoi

Bureau of Technical Support for Safety Regulation , Office of Atoms for Peace, 16 Vibhavadi rangsit road Ladyao Chatuchak, Bangkok 10900, Thailand

[1] Corresponding author e-mail: suputtra.v@oap.go.th

Abstract. A simple method to determine ^{137}Cs in seawater has been developed based on the use of a Cu-hexacyanoferrate scavenger. The Cu-hexacyanoferrate supported on cotton wound cartridge filter was used to absorb ^{137}Cs from seawater by passing large volumes over the cartridge filters with flowrate of 240 L hr^{-1}. Results from the Cu-hexacyanoferrate method were proved acceptable for accuracy with bias below ± 20 % i.e. - 9.16 to + 18.55 % when compared with the traditional ammonium molybdophosphate pre-concentration method. This developed method is cost-effective and less time consuming. In addition it can be easily performed at sampling fields.

1. Introduction

Cs-137 is one of the fission products which has a long half life of 30 years. It can be released from a nuclear accident, nuclear waste discharge and atomic bomb test to the atmosphere and into the ocean. Therefore ^{137}Cs is one of our monitored radionuclides in environmental samples especially in seawater. Caesium normally exists and migrates in monovalent state in an aquatic environment. The determination of ^{137}Cs in the ocean can be applied to monitor a release of any radiological incident in the pacific region [1]. The classical ammonium molybdophosphate (AMP) pre-concentration method has been used routinely to determine ^{137}Cs in seawater. This analysis passed the proficiency test exercise under the project "IAEA proficiency test exercise on Marine benchmark study on the possible impact of the Fukushima radioactive releases in the Asia-Pacific Region for Caesium Determination in Sea Water (RCA RAS/7/021)" [2]. Routinely seawater samples are collected from fields and brought to our laboratory. The samples are acidified to pH 1.6 and followed by adsorption of ^{137}Cs on AMP as the scavenger. The precipitate samples are filtrated and dissolved with 10 M NaOH before gamma-ray measurement using HPGe detector. However the radiochemical method is laborious and time consuming. Moreover the method limits number of samples that can be brought to laboratory due to 20 L seawater per sample to be collected. A simple method to determine ^{137}Cs in seawater has been developed based on hexacyanoferrate ion exchanger. Hexacyanoferretes have advantage of being selective of Cs (excluding Na and K). Moreover it is easily to prepare in three forms i.e. intrinsic, supported and composite [3]. Terada [4] and Su [1] used silica gel as supporting bed for Cu-hexacyanoferrate (in the following text abbreviation CuHCF) which exhibited stability in seawater when compared with other transit element such as Co, Ni and Zn [5]. Our work used also CuHCF but on cotton cartridge filters supporting bed which can be prepared easily and performed with large volume samples for high detection of contained Cs. The traditional AMP technique and developed method based on CuHCF are compared and described in this paper.

2. Experiment

2.1. Seawater sample description
Local sea water at Sriracha Fishery Research Station, Thailand (13°11' 03.8"N, 100°55' 43.7"E) which contained natural ^{137}Cs was used for both of AMP pre-concentration and CuHCF filter methods. Seawater salinity was 35 ppt and its pH was 7.8.

2.2. AMP pre-concentration method

2.2.1. Reagents and reference solution for AMP pre-concentration method. AMP, CsCl, HCl and NaOH used were analytical grade. The ^{137}Cs certified reference solution used to prepare calibration source was purchased from Eckert and Ziegler Isotope Product.

2.2.2. Sample preparation and analysis for AMP pre-concentration method. The AMP-concentration method was developed from Hirose and Phaopeng's work [6-7]. Two hundred litre seawater samples in the field were pumped through 1 μm filter into plastic containers. The samples were acidified with HCl to pH 1.6. Then 2.6 g of CsCl (caesium carrier) and 40 g of AMP were added. The samples were stirred for one hour and left overnight to allow the precipitate to settle. The precipitate samples were centrifuged and washed with 1 M HCl twice. The precipitate samples were dissolved with 10 M NaOH. The solutions were heated to remove ammonia. The precipitation step was repeated. The second precipitate samples were dissolved with 10 M NaOH. The solutions were transferred to calibrated containers (the same shape and type as prepared calibration source i.e. 100 ml transparent polypropylene bottle) and made up to the same volume as the calibration source i.e. 20 ml. The pictures of AMP pre-concentration method are shown in figure 1.

2.2.3. Calibration source preparation for AMP pre-concentration method. The ^{137}Cs certified reference solution which contained 0.385±0.002 Bq of ^{137}Cs activity, was spiked into 20 ml of 1 M HCl in the calibrated container (100 ml transparent polypropylene bottle).

Figure 1. The AMP pre-concentration method (a) Filtrated seawater samples, (b) Precipitating AMP in the seawater samples, (c) Samples after precipitation and (d) Calibration source and samples for gamma-ray measurement.

2.3. *CuHCF filter method*

2.3.1. Reagents and standard for CuHCF filter method. $K_4[FeCN_6]$ and $Cu(NO_3)_2$ used were analytical grade. The ^{137}Cs certified reference solution used to prepare the calibration source was purchased from Eckert and Ziegler Isotope Product.

2.3.2. CuHCF filter preparation. The 1 μm cotton wound cartridge filters which had stainless steel core were used. The filters were soaked in a $K_4[FeCN_6]$ solution (containing 10 g of $K_4[FeCN_6]$/300 ml of deionised water) for two days. The filters were removed from the solution and were dried in an oven at 80°C. The dried filters were submerged in $Cu(NO_3)_2$ solution (containing 10 g of $Cu(NO_3)_2$/300 ml of deionised water) for at least one day. The filters were removed from the solution and rinsed with deionised water. The washed filters were again dried at 80°C. The dried filters were packed in plastic bags and ready to be used as Cs ion exchanger.

2.3.3. CuHCF filter method. Sets of cartridge filter composed of three filters i.e. pre filter cotton wound cartridge filter and the two CuHCF filters in series in filter housing were fabricated. The various amount of seawater samples were pumped through the filter sets with flow rate of 240 L hr^{-1}. The CuHCF filter samples contained Cs were packed in plastic bag and labelled "A" and "B" as the first and second CuHCF filters respectively and brought back to the laboratory. The both filters were dried and ashed at 400°C for 15 hours. The ash samples were transferred into calibrated containers (the same shape and type as the prepared calibration source i.e. 100 ml transparent polypropylene bottle). The pictures of CuHCF filter method are shown in figure 2.

2.3.4. Calibration source preparation for CuHCF filter method. The ^{137}Cs certified reference solution which contained 0.943±0.005 Bq of ^{137}Cs, was spiked into ashed CuHCF filter in the calibrated container (100 ml transparent polypropylene bottle).

2.3.5. Calculation of ^{137}Cs collection efficiency and ^{137}Cs activity concentration. The ^{137}Cs collection efficiency onto the ion exchanger of each sample was calculated from the activity of filter A (front filter) and B (back filter) described below [1].

$$E = \frac{[F] - [B]}{[F]} \qquad (1)$$

When E is the absolute collection efficiency. *[F]* and *[B]* are the activity of ^{137}Cs in the front and back filters respectively.

And the final ^{137}Cs activity concentration in seawater samples is referred to equation (2).

$$A = \frac{[F]}{E \times V} \qquad (2)$$

When A is the final ^{137}Cs activity concentration in seawater samples. V is the volume of seawater samples.

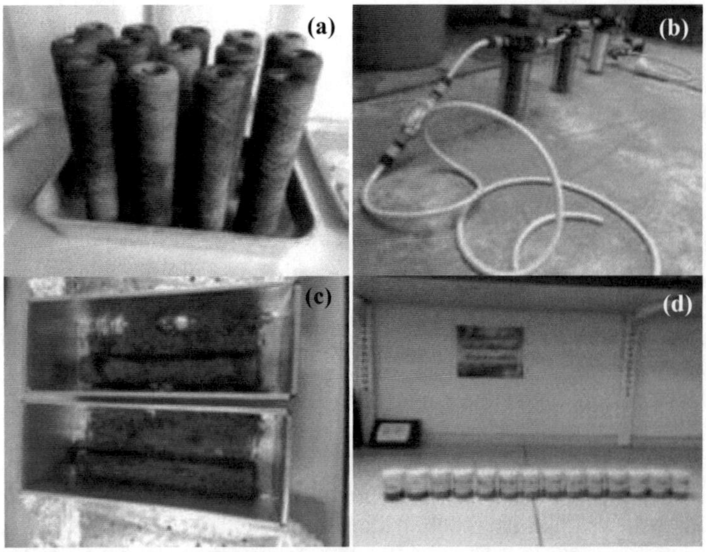

Figure 2. The CuHCF filter method (a) Prepared CuHCF filters, (b) CuHCF filter set for ^{137}Cs absorption in seawater, (c) CuHCF samples after ashing and (d) Calibration source and samples for gamma-ray measurement.

2.4. Counting equipment and measurement

Gamma-ray spectrometry system, coaxial HPGe detector (CANBERRA) with MAESTRO was used. The detector was shielded by low background lead. The prepared calibration sources for both AMP solution and CuHCF ash were used to calibrate counting efficiencies at 661.657 keV. Then the both AMP solution samples and CuHCF ash samples in calibrated containers were measured with the same gamma-ray spectrometry system.

3. Results and discussion

The results of AMP pre-concentration and CuHCF filter methods are compared and discussed in this section. The method to determine activity concentration of ^{137}Cs in seawater samples using AMP pre-concentration method passed the proficiency test exercise, "IAEA proficiency test exercise on Marine benchmark study on the possible impact of the Fukushima radioactive releases in the Asia-Pacific Region for Caesium Determination in Sea Water (RCA RAS/7/021)", for accuracy, precision and trueness criteria with -1.00 % bias [2]. The same method was applied to determine ^{137}Cs in seawater in this analysis and result is shown in table 1. It obtained 0.026 mBq L^{-1} of the detection limit. The result of three samples was precise with 2.55 % relative standard deviation (RSD).

Table 1. AMP pre-concentration method result.

Sample volume (L)	Sample	Activity (mBq L^{-1})
200	sample 1	0.782±0.059
	sample 2	0.748±0.058
	sample 3	0.749±0.059
	average	0.760±0.059

For the CuHCF filter method, the parallel results with various amounts of seawater volume i.e. 200 L, 300 L and 600 L are presented in table 2. The detection limits of ^{137}Cs activity were 0.016 mBq L^{-1}, 0.010 mBq L^{-1} and 0.005 mBq L^{-1} for sample volumes 200 L, 300 L and 600 L respectively.

IOP Conf. Series: Journal of Physics: Conf. Series **860** (2017) 012011 doi:10.1088/1742-6596/860/1/012011

Table 2. CuHCF filter method results.

Sample volume (L)	Sample	Efficiency of Cs collection	Activity (mBq L^{-1})
200	sample 1	79.05	0.837±0.080
	sample 2	52.08	0.698±0.101
	average	65.57	0.768±0.091
300	sample 1	57.58	0.901±0.077
	sample 2	89.86	0.831±0.060
	average	73.72	0.866±0.069
600	sample 1	68.33	0.889±0.057
	sample 2	74.31	0.762±0.050
	average	71.32	0.826±0.079

The Cs collection efficiencies of each repeated sample were not stable. This can be effected by combined sources i.e. unequal quality of prepared CuHCF filters (different amount of ion exchanger onto cotton wound in each cartridge), some loss of CuHCF and unstable flowrate of seawater during filtration [1]. However the repeated samples with the different Cs collection efficiencies gave quite similar ^{137}Cs activity concentration. Their repeated activity concentrations had 12.80 %, 5.72 % and 10.87 % RSD for 200 L, 300 L and 600 L respectively. The RSD of the three results were quite high as expected due to the low-level activity concentration and only two repeated samples used.

The average activity concentrations were close together i.e. 0.768 mBq L^{-1}, 0.866 mBq L^{-1} and 0.826 mBq L^{-1} which were 0 %, + 12.76 % and + 7.55 % bias for 200 L, 300 L and 600 L respectively as referred to the AMP pre-concentration method result from the PT exercise [2]. Even the minimum and maximum activity concentrations i.e. 0.698 mBq L^{-1} for sample 2 200L and 0.901 mBq L^{-1} for sample 1 300L, obtained bias in the range of – 9.16 % to + 18.55 %. It can be therefore concluded this method gave the accepted result as the PT exercise judged ± 20% bias for ^{137}Cs determination was accepted for accuracy. This also means that the minimum 200 L seawater would be adequate to perform the CuHCF filter method and therefore 300 L and 600 L seawater is not necessary. However it should be noted that the CuHCF filter method gave higher bias than those of AMP pre-concentration method due to the unstable filter quality and flowrate during filtration.

4. Conclusion

The simple method to determine ^{137}Cs in seawater based on CuHCF scavenger was compared with the classical AMP pre-concentration method. The developed CuHCF filter method gave accepted results as the ^{137}Cs activity concentration in seawater samples obtained % bias within the range of ± 20 i.e. - 9.16 to + 18.55 when compared with the AMP method as referred to the IAEA proficiency test exercise RCA RAS/7/021. In addition, it was found the minimum 200 L seawater can be adequate to perform the CuHCF filter method as well as higher volumes i.e. 300 L and 600 L. This means the method can be easily and shortly performed at sampling fields within an hour when using 200 L sample with flowrate of 240 L hr^{-1}. However some concern in the CuHCF method was found surrounding the quality of unstable prepared CuHCF filters since the repeated samples had significantly different Cs collection efficiencies. Also the different Cs collection efficiency could be improved by constantly regulating and controlling flowrate during filtration. Perhaps this method can be optimised to obtain higher accuracy and Cs collection efficiency by improving quality of prepared CuHCF filters and lowering the flowrate during filtration.

References

[1] Su C C, Huh C A and Chen J C 2000 A rapid method for the determination of ^{137}Cs in seawater *Terr Atmos Ocean Sci.* **11** pp 753–764

[2] Harms A V, Osvath I and Osborn D 2016 *IAEA-RML-2015-02 Proficiency Test for Determination of Radionuclides in Seawater* (Vienna: IAEA)

[3] Teng H C, Sun C H and Chen C L 1999 The selective ion exchanger hexacyanoferrates in radioactive waste water management *Iner Quarterly.* **33** pp 80–9

[4] Terada K, Hayakawa H, Sawada K and Kiba T 1970 Silica gel as a support for inorganic ion-exchangers for the determination of caesium-137 in natural waters *Talanta.* **17** pp 955-63

[5] Gaur S 1996 Determination of Cs-137 in environmental water by ion-exchange chromatography *J Chromatogr A.* **733** pp 57–71

[6] Hirose K, Aoyama M, Igarashi Y and Komura K 2008 Improvement of ^{137}Cs analysis in small volume seawater samples using the Ogoya underground facility *J. Radioanal. Nucl. Ch.* **276** pp 795–798

[7] Phaopeng N 2013 *The Measurement of Cs-137 Radioactivity in Seawater and the Oyster (Crassostrea belcheri) from the lower Gulf of Thailand for Radiation Dose Assessment in the General Public* (Bangkok: Kasetsart University)

IOP Conf. Series: Journal of Physics: Conf. Series **860** (2017) 012012 doi:10.1088/1742-6596/860/1/012012

Investigation of radon level in air and tap water of workplaces at Thailand Institute of Nuclear Technology, Thailand

P Sola[1], U Youngchuay, S Kongsri and A Kongtana

Thailand Institute of Nuclear Technology, Ongkharak, Nakhon Nayok, 26120 Thailand

[1] Corresponding author e-mail: ph_solar@hotmail.com, phachirarats@tint.or.th

Abstract. Thailand Institute of Nuclear Technology (TINT) has continuously monitored radiation exposure and radionuclide in workplaces specifically radon gas to estimate effective dose for workers. Radon exposure is the second leading cause of lung cancer in the world. In this study, radon in air and tap water at building no. 3, 7, 8, 9 and 18 on Ongkharak site of TINT have been measured for 5 years from 2012 to 2016. Radon level in air and tap water were investigated on 83 stations (workplaces) and 54 samples, respectively. Radon concentrations in air and tap water were measured by using the pulsed ionization chamber (ATMOS 12 DPX). Indoor radon concentrations in air were in the range of 12-138 Bq.m^{-3} with an average value of 30.13±17.05 Bq.m^{-3}. Radon concentrations in tap water were in the range of 0.10 to 2.89 Bq.l^{-1} with an average value of 0.51±0.55 Bq.l^{-1}. The results of radon concentrations at TINT were below the US Environmental Protection Agency (US EPA) safety limit of 148 Bq.m^{-3} and 150 Bq.l^{-1}, for, air and tap water, respectively. The average effective dose for TINT's workers due to indoor radon exposure was approximately 0.20±0.11 mSv.y^{-1}. The value is 100 times less than the annual dose limit for limit occupational radiation worker defined by the International Commission on Radiological Protection (ICRP). As a result, the TINT's workplaces are radiologically safe from radon content in air and tap water.

1. Introduction

Radon (Rn-222) is a colorless, odorless gas, a naturally occurring radioactive gas formed in the ground by the radioactive decay of uranium that presents in all rocks, soils and groundwater. Radon is the most important radio isotope from the environment in the viewpoint of human radiation exposure. People who inhale high level of radon and its progeny increase risk of developing lung cancer. Radon gas can enter a building and can be trapped indoors, thus the hazardous level for people can be elevated.

TINT has continuously monitored radiation exposure and radionuclide in workplaces specifically radon gas to estimate effective dose for workers. Radon concentration in workplaces (77 stations) at TINT was previously surveyed in 2012-2015 for determining level of the average radon concentrations and the annual effective dose in workplaces [1]. The average radon concentration and the annual effective dose for workers in workplaces at TINT were estimated to be 30.50 Bq.m^{-3} and 0.19 mSv.y^{-1} respectively [1]. In this study, the radon measurement in 2016 was analyzed and combined with the previous radon data in 2012 to 2015 for obtaining higher accuracy and reliability. Besides, additional radon concentrations in 54 workplaces tap water were determined. The aim of this study is to investigate the radon exposure level for worker at TINT's workplaces via radiation measurement of air and tap water.

2. Experiments

The radon concentration in air and tap water were investigated at building no. 3, 7, 8, 9 and 18 on Ongkharak site of TINT. The measurement has performed continually for 5 years between 2012 to 2016. The stations for measurements were selected by considering the workplaces where people always work such as laboratory, office, meeting room and basements, etc. The water samples were collected directly from hydrant at particular workplaces. Radon level in air was investigated on 83 stations (workplaces) and 54 tap water samples. Radon concentrations in both air and tap water was measured by the pulsed ionization chamber (ATMOS 12 DPX), showed in Figure 1. The measurement time for each air and tap water samples was 2 hours.

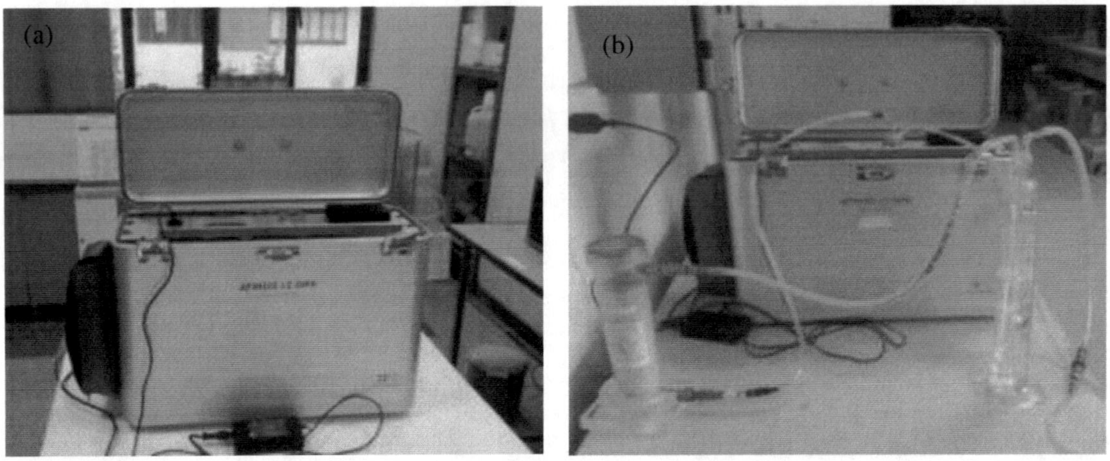

Figure 1. The equipment set up for the measurement of radon concentrations in (a) air and (b) tap water [2].

The values of indoor radon from ATMOS 12 DPX were compared with the safety limit (148 $Bq.m^{-3}$) recommended by the US Environmental Protection Agency (US EPA, 1992) [3]. The indoor radon concentration ($Bq.m^{-3}$) from the detector were used to calculate of an annual effective dose (*H*) by using equation (1) which was introduced by United Nation Scientific Committee on the effects of atomic radiation (UNSCEAR, 2000) [4] as follows:

$$H = C \times E \times F \times T \times D \tag{1}$$

Where H is the annual effective dose ($mSv.y^{-1}$), C is radon concentration ($Bq.m^{-3}$), E is equilibrium factor (=0.4) [5], F is the occupancy factor (5.85 h/day or 1500 h/year excluding primary industry, so, 1500/8760=0.17 [6], T is hours in a year (8760 h y^{-1}), and D is the dose conversion factor (9.0×10^{-6} mSv ($Bq.m^{-3}.h^{-1}$)).

The annual effective dose ($mSv.y^{-1}$) was used for the total effective dose (H_{total}) calculation of TINT worker by the following equation [6]:

$$H_{total} = H_{dwelling} + H_{outdoor} + H_{workplace} \tag{2}$$

Where $H_{workplace}$ is the effective dose calculated from the radon concentration at indoor workplace ($mSv.y^{-1}$), $H_{dwelling}$ is the effective dose calculated from the indoor radon concentrations in dwellings in Thailand which is equal to 20 Bq m^{-3} or 0.40 mSv y^{-1} [7-8]. $H_{outdoor}$ is an effective dose calculated from the radon concentration at outdoor environment which is between 5-10 Bq m^{-3} or about 0.058 mSv y^{-1}[6]. The annual effective dose for workers were compared with the safety limit of 20 mSv.y^{-1} for the occupational worker, regulated by International Commission on Radiological Protection (ICRP, 1991) [9].

The quantity of each water sample collected from particular workplaces was about 350 ml. The water sample was then introduced into degassing cylinder for 2 hours measurement per sample. The radon concentration (C_a, Bq.m^{-3}) from the detector was used for determining the radon concentration in tap water (C_w, Bq.l^{-1}) by using the following equation:

$$C_a = \frac{\alpha_1}{\alpha_1 + \alpha_2} \frac{V_w}{V_a} C_w \left[e^{-\lambda t} - e^{-(\alpha_1 + \alpha_2 + \lambda)t} \right] \quad (3)$$

Where λ is the radon decay constant (min^{-1}), t is measurement time (min), V_w is the water volumes (l) and V_a is the air volumes (m^{-3}). The transfer factors α_1 and α_2 are parameters of the transferred radon gas from the water to air, and air to water, respectively. The evaluations of the transfer factors, α_1 and α_2 were 0.0845 min^{-1} and 0.00813 min^{-1}[10], respectively. Radon concentrations in tap water were finally compared with the safety limit of 150 Bq.l^{-1}, recommended by the US Environmental Protection Agency (US EPA, 1991) [11].

3. Results and discussions

The levels of indoor radon concentration from each workplace are showed in Figure 2.

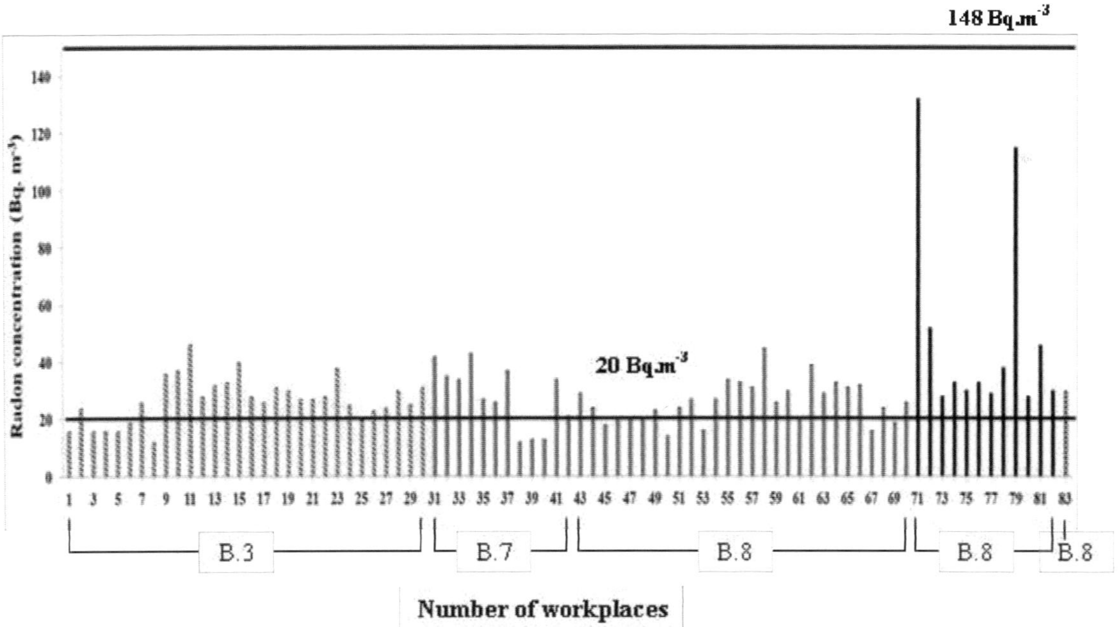

Figure 2. Level of indoor radon concentration at TINT compared with the safety limit (upper line, 148 Bq.m^{-3}) and indoor radon of dwelling in Thailand (lower line, 20 Bq.m^{-3}), ( ), ( ), ( ), ( ), and ( ) refered to radon measurement at building no. 3, 7, 8, 9, and 18, respectively.

The levels of indoor radon concentration from each workplace found in the range of 12 to 138 Bq.m^{-3}, with an average of 30.13±0.13 Bq.m^{-3}, which were within the safety level of 148 Bq.m^{-3}, recommended by the US EPA (1992), however, many workplaces found radon concentration in higher amount than indoor radon of dwelling in Thailand (20 Bq.m^{-3}) [7-8]. Sample no. 71 and 79 collected at the basement laboratories of building no. 9 showed extremely high radon concentration.

The annual average and total effective dose of TINT's workers at particular building were summarized in Table1. Besides, the radiation safety limits were added the table for data comparison.

Table1. The calculation results of average radon concentration, average annual effective dose and total effective dose from exposing radon at workplaces of TINT, including the radiation safety limits.

No. of building	Average Rn concentration (Bq.m^{-3})	Average annual effective dose (mSv. y^{-1})	Total effective dose of workers (mSv.y^{-1})
3	27.00±7.90	0.17±0.05	0.63±0.05
7	28.08±11.21	0.18±0.07	0.64±0.07
8	26.07±7.30	0.16±0.46	0.62±0.46
9	34.70±8.23*	0.31±0.09*	0.71±0.10*
18	30.00±4.00**	0.19±0.03**	0.65±0.48**
Average	31.17±17.05	0.20±0.06	0.66±0.04
Safety limit	148 (US EPA, 1992)	1 for public, 20 for the occupation worker	

* data obtained by excluding sample no. 71 and 79
** refers to random error obtained by one station measurement

An annual effective dose in workplaces were in the range of 0.08 to 0.83 mSv.y^{-1}, which the maximum value found in building no. 9, because the building has two basements with limit of ventilation. The average annual effective dose from exposing radon and total effective dose of workers at TINT's workplaces were 0.20±0.06 and 0.66±0.04 mSv.y^{-1}, respectively. All workplaces showed lower radon concentration than the safety limits of 20 mSv.y^{-1} for occupational worker. However, it is a recommendation for every workplace that the regular ventilation is needed for avoiding of unnecessary radon up taking.

The levels of radon concentrations in tap water at TINT workplaces and the safety limit of 150 Bq.l^{-1}, recommended by the US EPA (1991) are showed in Figure 3.

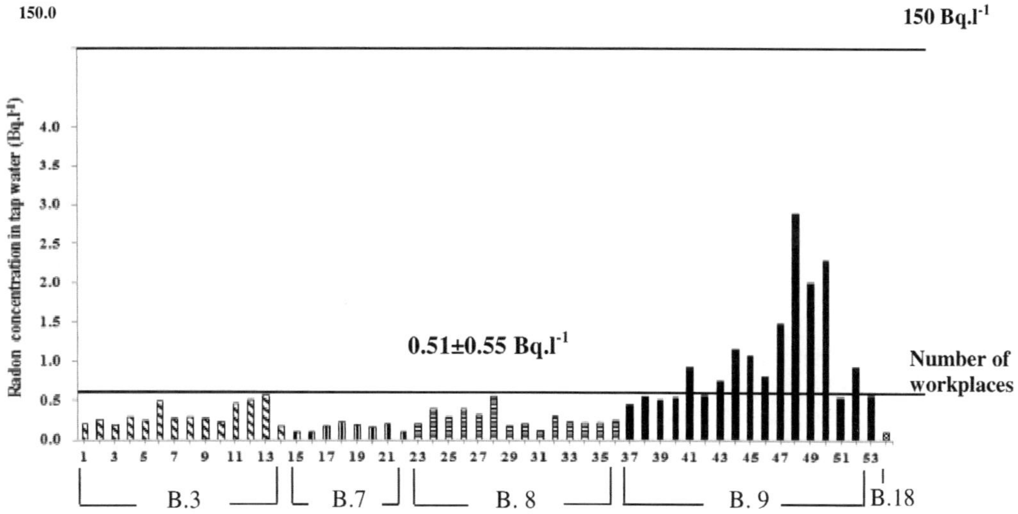

Figure 3. Radon concentration in tap water at TINT compared with the safety limit (upper line, 150 Bq. l^{-1}) and an average of radon concentration in tap water (lower line, 0.51±0.55 Bq. l^{-1}), ( ), ( ), ( ), ( ), and ( ) refered to radon measurement at building no. 3, 7, 8, 9, and 18, respectively.

The levels of radon concentrations in tap water were in the range of 0.10 to 2.89 Bq.l^{-1}, with an average of 0.51±0.55 Bq.l^{-1}.The radon concentration in tap water collected from building no. 9 showed significantly high concentration than samples from other buildings (no. 3, 7, 8 and 18). However, radon concentrations in all tap water samples are below the safety limit of 150 Bq.l^{-1}. General recommendation to routinely used groundwater workplaces is that the clarifier system may be needed for reducing the radon concentration in groundwater.

4. Conclusions

Thailand Institute of Nuclear Technology (TINT) has been continuously monitored radiation exposure and radionuclides in workplaces include radon gas to estimate effective dose for workers. In this study, the results of radon concentration investigation in air and tap water at building no. 3, 7, 8, 9 and 18 on Ongkharak site during 2012 to 2016 were reported. Radon levels in air and tap water were investigated on 83 stations (workplaces) and 54 samples, respectively. Radon concentrations in air and tap water were measured by the pulsed ionization chamber (ATMOS 12 DPX). The indoor radon concentrations at TINT workplaces were in the range of 12-138 Bq.m^{-3}, with an average of 30.13±17.05 Bq.m^{-3} (exclude building no.9). All indoor radon concentrations remain within the safety level (148 Bq.m^{-3}), recommended by the US EPA. However, it was noteworthy that some workplaces showed significantly high radon concentration than value of the indoor radon concentration of dwelling in Thailand (20 Bq.m^{-3}). Indoor radon at workplaces no. 71 and 79 (building no. 9) were extremely higher than other workplaces because they are basement laboratories with limit of ventilation. The average annual effective dose and total effective dose of TINT's workers were about 0.20±0.06 and 0.66±0.04 mSv.y^{-1}, respectively. As a result, the TINT workplaces have indoor radon concentration about 100 times lower than the safety limits (20 mSv.y^{-1}) for the occupational worker. Radon concentrations in tap water from all buildings were found in the range of 0.08 to 0.83 Bq. l^{-1} and an average was 0.51±0.55 Bq.l^{-1}. The building no. 9 showed relatively high radon concentration in tap water. However, all tap water samples contain very low radon concentration when compared with the safety limit (150 Bq.l^{-1}). In addition, if any workplaces regularly use groundwater, the clarifier may be necessary to reduce radon concentration.

References

[1] Sola P, Kongsri S, Youngchuay U and Kongtata A 2016 Indoor radon concentration survey in workplaces at Thailand Institue of Nuclear Technology *Pure and Applied Chemistry Int., Conf., (PACCON 2016), 9-11 February 2016, BITEC, Bangkok, Thailand*

[2] Sola P, Srisuksawad K, Loaharojanphand S, O-Manee A, Permnamtip V, Issarapan P & Thummagarun L 2012 Radon concentration in air, hot spring water, and bottled mineral water in one hot spring area in Thailand *J Radioanal Nucl Chem.* Doi: 10.1007/s10967-012-2359-9

[3] US EPA 1992 National emission standards for hazardous air pollutants; Emission standards for radon emissions from phosphogypsum stacks; Final rule (40CFR61) *Federal Register* **57** 23305

[4] United Nations Scientific Committee on the Effects of Atomic Radiation 2000 Sources and effects of ionizing radiation *2000 Report to the General Assembly with Annex B: Exposure from Natural Source of Radiation* (New York: UNSCEAR)

[5] Tokonami S, Furukawa M, Shicchi Y, Sanada T, Yamada Y 2003 Characteristics of radon and its progeny concentration in air-conditioned office building in Tokyo *Radiation Protection Dosimetry.* **106** 71

[6] Oikawa S, Kanno N, Sanada T, Abukawa J, Higuchi H 2006 A Survey of indoor workplace radon concentration in Japan *J. Environ. Radioactivity* **87** 23

[7] Boonyaprapa S, Wanabongse P, Cheepsattayakorn A, Saeung S, Sola B and Bovornkitti S 2008 Residential Radon Exposure and Lung Cancer: A Survey in Chiang Mai Province *J. Health Systems Res.* **2** 460

[8] Moontep C, Suparom K, Sola B, Chantarachote W, Wanabongse P and Bovornkitti S 2008 Residential Radon Exposure and Lung Cancer: A Survey in Khon Kaen Province *Thammasat Medical J* **8** 455

[9] ICRP 1991 *Annals of the ICRP; 1990 Recommendation of the Int. Commission on Radiation Protection* (*ICRP Publication 60)* (Oxford: Pergamon Press)

[10] Salih I, Pettersson H, Lund E (2000) Determination of ^{222}Rn and ^{226}Ra in water using a large volume ionisation chamber *J Environ Radioact* **48** 235

[11] US EPA 1991 National primary drinking water regulations: Radionuclides (Proposed Rule) *Federal Register* **56** 33050

IOP Conf. Series: Journal of Physics: Conf. Series **860** (2017) 012013 doi:10.1088/1742-6596/860/1/012013

Determination of stable cesium and strontium in rice samples by inductively coupled plasma mass spectrometry

W Srinuttrakul[1,3] and S Yoshida[2]

[1] Thailand Institute of Nuclear Technology (Public Organization), 9/9 Mu 7, Sai Mun, Ongkharak, Nakhon Nayok 26120, Thailand
[2] National Institutes for Quantum and Radiological Sciences and Technology, Anagawa 4-9-1, Inage-ku, Chiba-shi, Chiba 263-8555, Japan

[3] Corresponding author e-mail: wannees@tint.or.th

Abstract. For long-term radiation dose assessment models, food ingestion is one of the major exposure pathways to human. In general, the stable isotopes can serve as analogues of radioisotopes. In this study, rice samples were collected from 30 paddy fields in Si Sa Ket, Yasothon and Roi Et in the northeast of Thailand in November 2014. The concentrations of stable cesium (Cs-133) and strontium (Sr-88) in polished rice were determined by inductively coupled plasma mass spectrometry (ICP-MS). The standard reference material of rice flour (NIST 1568a) with spiked Cs and Sr was used to validate the analytical method. The concentration of Cs in polished rice from Si Sa Ket, Yasothon and Roi Et was 0.158 ± 0.167 mg kg^{-1}, 0.090 ± 0.117 mg kg^{-1} and 0.054 ± 0.031 mg kg^{-1}, respectively. The concentration of Sr in polished rice from Si Sa Ket, Yasothon and Roi Et was 0.351 ± 0.108 mg kg^{-1}, 0.364 ± 0.215 mg kg^{-1} and 0.287 ± 0.102 mg kg^{-1}, respectively. Comparison of the results with Japanese data before the Fukushima Di-ichi nuclear power plant accident showed that the concentrations of both Cs and Sr for Thai rice were higher than those for Japanese rice.

1. Introduction

Radioactive cesium (Cs-134; $T_{1/2}$ = 2.06 years, Cs-135; $T_{1/2}$ = 2.3 x10^6 years and Cs-137; $T_{1/2}$ = 30.17 years) and radioactive strontium (Sr-90; $T_{1/2}$ = 28.1 years) [1] have long half-lives and they play an important role to human because of their potential to harm the human body. Signs and symptoms of acute toxicity from external and internal exposure to high levels of radioactive cesium and radioactive strontium are similar to those observed in cases of high exposure to ionizing radiation. Depending on the radiation dose, symptoms may include those typical of acute radiation syndrome (vomiting, nausea, and diarrhea), skin and ocular lesions, neurological signs, chromosomal abnormalities, compromised immune function, and death. Limited human and animal data indicate that exposure to radioactive cesium can result in adverse effects such as reduced fertility, abnormal neurological development, genotoxicity, and possibly cancer [2]. Since radioactive strontium will accumulate in bone, bone itself and the soft tissues nearby may be damaged by radiation emitted over time. Radioactive strontium may cause cancer as a result of damage to the DNA in cells. An increase in leukemia was reported in one foreign population who swallowed relatively large amounts of Sr-90 (and other radioactive materials) in river water contaminated by a nuclear weapons plant. In the case of a breathing exposure, leukemia and cancers of the bone, nose, and lung were reported in animal studies. In addition, skin and bone cancer were reported in animals that received radiation at high doses to the skin [3]. In general, isotopes of the same element behave similarly in terms of their physical properties and chemical reactivity. Therefore soil to plant transfer factor of stable isotopes can serve as analogues of radioisotopes, e.g., stable Cs for Cs-137 and stable Sr for Sr-90 [4]. Because of very low concentration of radioactive materials in the environment of most countries in the world, the concentration of stable element related to long-lived radionuclides was studied. For long-term radiation dose assessment models, food ingestion is one of the major exposure pathways to human [5]. Rice (*Oryza sativa*) is a diet staple for almost half of the world's population including Thailand. Rice is the most important economic crop of Thailand. Thailand is one of the world's biggest rice producers and exporter. Consequently, safety assessment for Thai is required to consider rice as the critical food. However the data of stable cesium and strontium in Thai rice are limited. Therefore, the levels of these elements in rice should be investigated.

Several methods have been widely used for the determination of elemental concentrations in rice samples including graphite furnace atomic absorption spectrometry (GFAAS) [6], x-ray fluorescence (XRF) [7], instrumental neutron activation analysis (INAA) [8-9], inductively coupled plasma optical emission spectrometry (ICP-OES) [8, 10] and inductively coupled plasma mass spectrometry (ICP-MS) [5, 11]. The Cs and Sr contents in rice samples analyzed by various techniques have been reported. Gonzálvez *et al* [12] studied the concentration of Sr in rice samples from Spain including Valencia, Tarragona, Murcia and Extremadura, Japan, Brazil and India by ICP-OES. Nadimi-Goki *et al* [13] investigated the Sr content in rice samples from Italy by ICP-OES. The concentrations of Cs and Sr in rice samples from Jamaican market were analyzed by INAA reported by Antoine *et al* [14]. The Cs and Sr compositions in rice samples collected from Japan were determined by ICP-MS studied by Uchida *et al* [5] and Tsukada *et al* [11]. Cheajesadagul *et al* [15] reported the Cs and Sr levels in Thai rice and foreign (France, India, Italy, Japan and Pakistan) rice samples analyzed by high resolution ICP-MS. Among the techniques widely used, ICP-MS is the most powerful technique for elemental analysis with the advantage of simultaneous determination of multi-element, wide linear range, low detection limit and rapid analysis.

In this study, rice samples were collected from 30 paddy fields in Si Sa Ket, Yasothon and Roi Et in the northeast of Thailand in November 2014. The concentrations of Cs and Sr in the polished rice (or white rice) samples were determined by inductively coupled plasma mass spectrometry (ICP-MS). The obtained results of Cs and Sr concentrations of Thai rice were compared with available literature data.

2. Materials and methods

2.1 Reagents and standards

All the chemical reagents used were supra-pure analytical grade. Water (>18 MΩ cm) treated with a Milli-Q water system (Millipore Corporation; Billerica, MA, USA) was used throughout the research. The standard solutions were purchased from Agilent Technologies, (New Haven, CT, USA). Standard reference material used to validate the method was rice flour (NIST 1568a) from National Institute of Standards and Technology (NIST; MD, USA).

2.2 Sample collection and preparation

Rice samples were collected from 30 paddy fields in Si Sa Ket, Yasothon and Roi Et in the northeast of Thailand in November 2014. Rice grains (2-3 kg) were harvested from the paddy field. After dehulling and polishing, the polished rice or white rice samples were obtained. All polished rice samples were dried in an electric oven at 50-60°C and ground to a fine powder by the high speed blender (1093 Cyclotec Sample Mill, Hoganas, Sweden). The rice powders were dried again in an oven at 50-60°C till constant weight, kept in polyethylene containers and stored in a desiccator until chemical analysis.

All dry rice samples (0.5 g) were digested in PTFE vessels on a hot plate with 65% HNO_3 and 48% HF. After digestion, the samples were evaporated to dryness and the residues were dissolved in 65% HNO_3 and 30% H_2O_2. After dryness again, the residues were dissolved with 2% HNO_3 as described by Ogiyama *et al* [10] with slight modifications. Sample digestions were done in triplicate.

2.3 Determination of Cs and Sr by ICP-MS

After diluting the digested samples to a suitable concentration, stable Cs and Sr in rice samples were determined by ICP-MS (Agilent 7900, Agilent Technologies, USA). The instrumental operating conditions for the determination of the elements are summarized in Table 1. Indium (In) was used for internal standard to compensate for change in analytical signal during the operation. Standard solutions were prepared from Agilent multi-element solution and used to get calibration curves. The quantification was based on at least 5 point external calibrations. Rice flour (NIST 1568a) with spiked Cs and Sr from Agilent multi-element solution was used to validate the analytical procedure.

Table 1. ICP-MS operating conditions.

Parameter	Condition
Nebulizer	Mira Mist
Spray chamber	Scott type
Sampling cone	Nickel
Skimmer cone	Nickel
RF power	1550W
Carrier gas flow	0.75 L min^{-1}
Dilution gas flow	0.25 L min^{-1}
Spray chamber temperature	2 °C
Nebulizer pump	0.1 rps

3. Results and discussion

3.1 Method validation

Measurement of calibration standard solutions provided linear response with correlation coefficients of 1.000 as shown in Table 2. The limit of detection (LOD) of Cs and Sr determined by ICP-MS was evaluated from three times the standard deviation of ten replicate blank measurements. The limit of detection (in mg kg^{-1}) was obtained from LOD (in (μg L^{-1}) multiplied by the overall dilution factor given in Table 2 which is adequate for the determination of Cs and Sr concentrations in rice sample. The limit of quantification (LOQ) of Cs and Sr calculated from ten times the standard deviation of ten replicate blank determinations was also presented in Table 2.

Table 2. Equation of calibration curve, LOD and LOQ in ICP-MS measurement.

Element	Cs	Sr
Analytical mass	133	88
Regression equation	y = 0.4826x + 0.0081	y = 0.4199x + 0.0143
Regression, R^2	1.000	1.000
LOD (μg L^{-1})	0.015	0.212
LOQ (μg L^{-1})	0.051	0.709
LOD (mg kg^{-1})	0.0015	0.021

Since there is no data of Cs and Sr in the standard reference material of rice flour (NIST 1568a), the standard reference material of rice flour spiked with Cs and Sr was included in the digestion and analytical procedure for quality control. The recoveries of Cs and Sr were summarized in Table 3. The result showed good agreement between techniques. The recoveries for reference materials spiked with Cs and Sr were within acceptable limits of ±15%. Precision, expressed as the relative standard deviation (RSD) of the three independent measurements was within 5%.

Table 3. Analytical results of rice flour (NIST 1568a) spiked with Cs and Sr obtained by ICP-MS (mean ± standard deviation for $n = 3$).

Spiked concentration (mg kg^{-1})	Recovery (%)
Cs	
0.025	88.2 ± 1.7
0.05	96.6 ± 0.1
0.25	96.6 ± 2.8
0.5	97.4 ± 1.1
0.75	97.1 ± 2.2
1.0	97.3 ± 1.2
Sr	
0.05	90.1 ± 0.9
0.1	93.4 ± 0.7
0.5	93.7 ± 0.4
1.0	94.2 ± 1.2
1.5	95.8 ± 0.9
2.0	95.9 ± 2.1

3.2 Stable Cs and Sr in rice samples

Summary Statistics of Cs and Sr concentrations in rice samples determined by ICP-MS were presented in Table 4. All rice samples had the Sr content higher than the Cs content. Both Cs and Sr showed the arithmetic mean concentration higher than the geometric mean concentration for all rice location. The arithmetic mean concentration of Cs was highest in rice sample from Si Sa Ket and lowest in rice sample from Roi Et. The concentration of Cs in polished rice from Si Sa Ket, Yasothon and Roi Et was 0.158 ± 0.167 mg/kg, 0.090 ± 0.117 mg/kg and 0.054 ± 0.031 mg/kg, respectively. The arithmetic mean concentration of Sr was highest in rice sample from Yasothon and lowest in rice sample from Roi Et. The concentration of Sr in polished rice from Si Sa Ket, Yasothon and Roi Et was 0.351 ± 0.108 mg/kg, 0.364 ± 0.215 mg/kg and 0.287 ± 0.102 mg/kg, respectively. The maximum/minimum concentration ratio of Sr was less than 10 showing the narrow distribution range for all location. Whereas, the wide distribution range of Cs was found with the maximum/minimum concentration ratio of 57, 25 and 20 for rice samples from Si Sa Ket, Yasothon and Roi Et, respectively. The concentrations of both Cs and Sr observed from this work were higher than the previous work reported by Cheajesadagul *et al* [15] possibly due to the different year of rice cultivation [16] and the polishing rate or milling ratio for rice preparation [17]. Tsukada *et al* [18] reported that 7% and 10% of the total stable Cs were found in polished rice and bran, respectively. 0.6% and more than 99% of the total Sr were observed in polished rice and in the non-edible parts [19].

The data of Cs and Sr from this study were also compared with the previous data as shown in Table 5. Both Cs and Sr data of Thai rice were higher than those of Japanese rice. Although the Cs concentration of Thai rice was higher than that of Jamaican rice, the Sr concentration of Jamaican rice was higher than that of Thai rice about 10 times. The Sr content of Thai rice was also lower than Spanish rice and Chinese rice. The levels of Cs and Sr in rice samples depend on the topography, soil properties and soil conditions [4, 20].

The estimated daily intake (EDI) of elements resulting from rice consumption was calculated from the elemental content multiplied by the average per capita daily consumption of rice. The daily average rice consumption in Thailand was about 233 g which is equivalent to yearly amount of about 85 kg per person [22]. The EDI of Cs and Sr for Thai rice consumption was 0.024 and 0.078 mg day^{-1}, respectively. The EDI for male and female (in Table 6) was evaluated from the body weight of 69 and 57 kg, respectively. The EDI of Sr for male and female from this study was lower than the minimal risk level (MRL) of stable Sr for intermediate-duration oral exposure (2.0 mg kg^{-1}day^{-1}) [3]. The MRL

Table 4. Summary Statistics of Cs and Sr concentrations in rice samples determined by ICP-MS.

Location	Cs (mg kg^{-1})	Sr (mg kg^{-1})
Roi Et		
Number of sample	10	10
Minimum	0.005	0.182
Maximum	0.107	0.521
Max./Min.	20	3
Median	0.053	0.260
Arithmetic mean	0.054	0.287
Geometric mean	0.042	0.274
Si Sa ket		
Number of sample	10	10
Minimum	0.010	0.231
Maximum	0.557	0.501
Max./Min.	57	2
Median	0.133	0.312
Arithmetic mean	0.158	0.351
Geometric mean	0.086	0.337
Yasothon		
Number of sample	10	10
Minimum	0.015	0.107
Maximum	0.381	0.825
Max./Min.	25	8
Median	0.038	0.313
Arithmetic mean	0.090	0.364
Geometric mean	0.051	0.312

Table 5. Cs and Sr concentrations in rice samples from different countries.

Sample	Cs (mg kg^{-1})		Sr (mg kg^{-1})	
	Mean ± SD[a]	Range	Mean ± SD[a]	Range
Thai rice (this study) (*n*=30)	0.101 ± 0.123	0.005 - 0.557	0.334 ± 0.149	0.107 - 0.825
Japanese rice (*n*=37) [5]	0.003[b]	0.0004 - 0.017	0.072[b]	0.014 - 0.14
Japanese rice (*n*=20) [11]	0.0017 ± 0.0017	0.00045 - 0.0065	0.070 ± 0.017	0.044 - 0.12
Jamaican rice (*n*=1) [14]	<0.004	No data	3.16	No data
Spanish rice-Valencia (*n*=52) [12]	No data	No data	0.9[b]	<0.2 - 3.7
Spanish rice-Murcia (*n*=11) [12]	No data	No data	1.5[b]	1.1 - 3.0
Chinese rice (*n*=18) [21]	No data	No data	1.582 ± 0.434	0.932 - 2.253

[a]Mean ± standard deviation
[b]Mean

Table 6. Estimated daily intake (EDI) of Cs and Sr for Thai rice consumption.

Element	EDI (mg day^{-1})	EDI (mg kg^{-1} day^{-1})	
		Male	Female
Cs	0.024	0.0003	0.0004
Sr	0.078	0.0011	0.0014

represents an estimate of daily human exposure that is likely to be without an appreciable risk of adverse health effects. Thus the consumption of Thai rice would be safe for Sr. There is no data of oral or inhalation MRLs for stable Cs due to lack of available data, therefore it was not possible to compare the results.

4. Conclusion

Stable Cs and Sr in rice samples collected from the paddy field in the northeast of Thailand were determined by ICP-MS. The ICP-MS technique proved to be accurate and reliable to analyze the Cs and Sr in rice samples. The obtained results would be valuable to be the database of Thai rice and further fundamental knowledge on the intake of Cs and Sr through food ingestion. For long-term radiation dose assessment models, these data would be useful to evaluate the transfers of Cs and Sr from soil to rice which they would be served as analogues of radioisotope information. Therefore the determination of stable Cs and Sr in the paddy soil samples with the same region of rice cultivation will be studied.

Acknowledgments

We gratefully acknowledge the staff of Agricultural Extension Office for their assistance in cooperation with the farmers. We wish to express our thanks to the farmers for providing their rice samples. Mr. S. Photong and Ms. T. Suksawasdi are acknowledged for their help in sample collection and preparation.

References

[1] International Atomic Energy Agency 2009 *Quantification of Radionuclide Transfer in Terrestrial and Freshwater Environments for Radiological Assessments: IAEA-TECDOC-1616* (Vienna: IAEA) p 130-3
[2] ATSDR. 2001c. Toxicological profile for cesium. Atlanta, GA: Agency for Toxic Substances and Disease Registry. U.S. Department of Health and Human Services. Public Health Service.
[3] ATSDR. 2001e. Toxicological profile for strontium. Atlanta, GA: Agency for Toxic Substances and Disease Registry. U.S. Department of Health and Human Services. Public Health Service.
[4] Uchida S, Tagami K, Shang ZR and Choi YH 2009 *J. Environ. Radioactiv.* **100** 739–45
[5] Uchida S, Tagami K and Hirai I 2007 *J. Nucl. Sci. Technol.* **44** 628-40
[6] Parengam M, Judprasong K, Srianujata S, Jittinandana S, Laoharojanaphand S and Busamongkol A 2010 *J. Food Compos. Anal.* **23** 340–5
[7] Fleming DEB, Foran KA, Kim JS and Guernsey JR 2015 *Appl. Radiat. Isotopes* **104** 217–23
[8] Srinuttrakul W and Busamongkol A 2014 *Energy Procedia* **56** 85 – 91
[9] Laoharojanaphand S, Busamongkol A, Permnamtip V, Judprasong K and Chatt A 2012 *J. Radioanal. Nucl. Ch.* **294**, 323-327
[10] Ogiyama S, Tagami K and Uchida S 2008 *Nucl. Instrum. Meth.* B **266** 3625–32
[11] Tsukada H, Hasegawa H, Takeda A and Hisamatsu S 2007 *J. Radioanal. Nucl. Ch.* **273** 199–203
[12] Gonzálvez A, Armenta S and de la Guardia M 2011 *Food Chem.* **126** 1254–60
[13] Nadimi-Goki M, Wahsha M, Bini C, Kato Y, Vianello G and Antisari LV 2014 *J. Geochem. Explor.* **147** 200-214
[14] Antoine JMR, Hoo Fung LA, Grant CN, Dennis HT and Lalor GC 2012 *J. Food Compos. Anal.* **26** 111–21
[15] Cheajesadagul P, Arnaudguilhem C, Shiowatana J, Siripinyanond A and Szpunar J 2013 *Food Chem.* **141** 3504–9
[16] Huang Y, Tong C, Xu F, Chen Y, Zhang C and Bao J 2016 *Food Chem.* **192** 873–78
[17] Okuda M, Joyo M, Tokuoka M, Hashiguchi T, Goto-Yamamoto N, Yamaoka H and Shimoi H 2012 *J. Biosci. Bioeng.* **114** 600-5
[18] Tsukada H, Hasegawa H, Hisamatsu S and, Yamasaki S 2002 *J. Environ. Radioactiv.* **59** 351-63

IOP Conf. Series: Journal of Physics: Conf. Series **860** (2017) 012013 doi:10.1088/1742-6596/860/1/012013

[19] Tsukada H, Takeda A, Takahashi T, Hasegawa H, Hisamatsu S and Inaba J 2005 J. Environ. Radioactiv. **81** 221–231
[20] Kelly S, Baxter M, Chapman S, Rhodes C, Dennis J and Brereton P 2002 *Eur. Food Res. Technol.* **214** 72-78
[21] Wei YH, Zhang JY, Zhang DW, Luo LG and Tu TH 2014 *Food Chem.* **159** 507–11
[22] National Statistical Office and Office of Agricultural Economics of the Kingdom of Thailand 2012 *Food insecurity assessment at national and subnational levels in Thailand, 2011*. Bangkok, Thailand.

IOP Conf. Series: Journal of Physics: Conf. Series **860** (2017) 012014 doi:10.1088/1742-6596/860/1/012014

Determining the bio-based content of bio-plastics used in Thailand by radiocarbon analysis

T Ploykrathok and S Chanyotha[1]

Department of Nuclear Engineering, Faculty of Engineering, Chulalongkorn University, Bangkok, Thailand

[1] Corresponding author e-mail: supitchacu18@gmail.com

Abstract. Presently, there is an increased interest in the development of bio-plastic products from agricultural materials which are biodegradable in order to reduce the problem of waste disposal. Since the amount of modern carbon in bio-plastics can indicate how much the amount of agricultural materials are contained in the bio-plastic products, this research aims to determine the modern carbon in bio-plastic using the carbon dioxide absorption method. The radioactivity of carbon-14 contained in the sample is measured by liquid scintillation counter (Tri-carb 3110 TR, PerkinElmer). The percentages of bio-based content in the samples were determined by comparing the observed modern carbon content with the values contained in agricultural raw materials. The experimental results show that only poly(lactic acid) samples have the modern carbon content of 97.4%, which is close to the agricultural materials while other bio-plastics types are found to have less than 50% of the modern carbon content. In other words, most of these bio-plastic samples were mixed with other materials which are not agriculturally originated.

1. Introduction

Nowadays, the usage of plastic is increased due to the increase in population. Materials or equipments made from certain kinds of plastic can be recycled, while some cannot. Normally, plastic that has major uses in everyday life is disposable or is used only once then tossed away. This leads to the increasing problems of plastic garbage management because plastic garbage is quite difficult to eliminate from the environment [1]. Leaving or burying plastic garbage, the current approaches for spontaneous degradation, could take a very long duration up to hundreds of years. Moreover, certain kinds of plastic are hardly self-degradable and must be processed to degrade them. The processes, e.g., burning, can result in environment pollution [2]. The critical solution for this plastic material management is to make the plastic self-degradable in a short duration so that it does not pollute the environment. The production of bio-plastics is one of the approaches to solve the problems. Bio-plastics are made from biodegradable agriculture products, for instance, sugarcane and tapioca, which can be degraded in a short period of time and are environmentally friendly [3]. Most importantly, these bio-plastic materials are made from plants that can be grown as many as desired, i.e. renewable, unlike the petrochemical plastics made from hydrocarbon-based fossil materials that take very long time to produce and cause pollution in the production processes. This is the main reason why bioplastics are increasingly gaining attention in the market. However, the amount of agricultural materials contained in the bio-plastics product can be specified by the amount of modern carbon in bio-plastics. Therefore, this research aims to determine the modern carbon in bio-plastic commercially available in Thailand market using the carbon dioxide absorption method.

Bio-based content analysis is one of the methods of bio-plastic testing. The bio-based content is defined as the ratio of the weight of the bio-based carbon in material to that of the total organic carbon in the product [4]. The value obtained from the analysis is the ratio of the combination between bio-plastic and petroleum plastic contents. If the plastic comes from 100% bio-plastic, the C-14 amount would be equal to the natural C-14 ratio. If the plastic content comes entirely from petroleum, there would be no C-14. In this present study, we aimed to examine the bio-plastic materials in Thailand in order to determine the bio-based content using the carbon absorption method.

In the present study, the plastic samples from the disposable plastic container produced in Thailand were analysed for C-14 ratio. This analysis method can be applied to all carbon-based products and was approved by American Society for Testing and Material (ASTM) No. ASTM 6866. Furthermore, there are many studies that proved that this method is simple, accurate and achievable [5,6,7]. The analysis method employed can be either accelerator mass spectrometry (AMS), benzene synthesis or carbon absorption. In the previous study by Noakes et al. [8], all three methods were evaluated and it was found that the most accurate method was AMS and that the benzene synthesis and carbon absorption methods had lower accuracy, respectively. The accuracy of all analysing methods, however, was acceptable. Culp et al. compared the accuracy and precision of C-14 between natural products and bio-based materials analysed by AMS and LSC. Results showed that degree of naturalness accurate and precise ^{14}C measurement is achievable by both methods [9]. The analysing method in this present study was performed using carbon absorption because of the simplicity of the process and the variation was acceptable. In the study by Molnar et al. [10], which fixed the parameters in order to minimize the variation of the carbon absorption analysing method, the minimized variation is found to be around 2 percent. Each of the three methods has its own advantages and disadvantages. However, when comparing the duration and cost of analysis, the carbon absorption method is the most preferred method in the industrial application.

2. Experimental

In this work, carbon absorption method was used for sample preparation. The process of this method [1] is as follows:

2.1. Samples and standard

Plastic samples in this present study are divided into two groups, which are plastic granules and commercial bio-plastic products. Plastic granules are divided into six types. The first four are bio-plastics, i.e. poly(lactic acid) (PLA), polybutylene succinate (PBS) type 1 and 2, thermoplastic elastomers and bio-TPE, and the other two are petroleum-based plastics i.e., polybutylene adipate-co-terephthalate (PBAT) and polypropylene (PP). Among nine samples for commercial bio-plastic products, five are Thai products (BP1-5), two are imported products (BPA from America, BPB from Australia), and the other two are research laboratory plastic products (BPX1, BPX2) with well-known composition which were added to this experiment as the reference for the efficacy of the test. The known standard materials were sugar with adjusted strength of the radiation as the Oxalic II (49960c) which is the reference for the substance that has natural C-14 amount before 1950 and marble as the background radiation reference.

2.2. Carbon dioxide preparation from samples

Plastic samples were in a solid form, so they had to be transformed into carbon dioxide gas before analysis using oxygen combustion bomb. The samples were combusted in high pressure of around 30 atm in the oxygen-provided combustion tank and electricity was used to ignite the combustion. The samples were prepared by transferring 8-10 grams of plastics into the container in the high pressure tank, then the electric circuit was completed by connecting the electrode on the tank lid to the plastic samples in the tank. The air in the tank was expelled using oxygen gas. The pressure was adjusted by the oxygen pump before the combustion started. After combustion, carbon dioxide gas from the samples was contained in the tank.

2.3. Carbon dioxide purification

Carbon dioxide from the combustion step had to be purified to eliminate the ions and the contamination by transporting the gas through the following 3 solutions:

- Potassium iodide and iodine (KI/I_2) for phosphorus, nitrogen, and sulfur oxides filtration.
- Silver nitrate solution ($AgNO_3$) for halide and acid precipitation.
- Potassium dichromate in sulfuric acid ($K_2Cr_2O_7/H_2SO_4$) for carbon monoxide (CO) and sulfur dioxide (SO_3) extraction.

The whole purifying system must be vacuumed before letting the carbon dioxide gas in. Moreover, after decontamination, the gas then passed through a container containing the solution between liquid nitrogen and alcohol in order to dehumidify by condensation. Then, the gas would be transformed into asolid form after passing through liquid nitrogen at the back of the system. All carbon dioxide was transported into the storage container.

2.4. Carbon dioxide absorption

Carbon dioxide was absorbed in the scintillation cocktail before undergoing modern carbon analysis by LSC. The scintillation cocktail was the compound between the Carbosorb and Permafluor scintillator in the equal ratio. In this experiment, 10 ml of Cabosorb was mixed with 10 ml of Permafluor. The cocktail was carefully prepared in the nitrogen atmosphere to avoid exposure of carbon dioxide to the air. The prepared cocktail was then injected into the glass tube connected to the gas absorption system in the suitable amount compared with the carbon dioxide sample. The system must be vacuumed before carbon dioxide injection, because the cocktail can absorb the carbon dioxide remaining in the gas absorption system. The absorption proceed was started by releasing the carbon dioxide gas from the storage container into the vacuum system in a preferred amount. Secondly, the cocktail-filled glass tube was connected to the system by open the valve on the top of the glass tube and the carbon dioxide gas was pumped to circulate in the system for 20 minutes. The carbon dioxide-adsorbed cocktail was then weighted for the adsorbed gas around 2.5-2.9 grams after subtraction the cocktail's weight. The cocktail was transferred to the glass container for further analysis [11].

2.5. Carbon-14 analysis using liquid scintillation counter

The suitable counter is the low strength radioactive counter which is normally used in the measurement of naturally occurring radioactive material (NORM). The liquid scintillation counter measures the energy released from the radioactive substance which activates the fluorochrome to the excited state. The fluorochrome then emits the energy in the visible spectrum and this light can be measured by photomultiplier tubes. This instrument is able to measure the alpha ray and low-energy beta ray due to the high sensitivity. The efficiency of the measurement can be increased when using the cocktail combined with the sample [12].

2.6. Bio-based content determination

The calculation of bio-based content was performed by comparing the C-14 amount to that of the standard samples. All measurements were adjusted for the carbon isotope (δ13) which is C-13/C-12, the stable isotopes in all samples before bio-based content calculation. The standard value of VPBD was used [13]. The calculated values were then calculated for the modern carbon ratio from the equation 1 [14]:

$$pMC = \frac{A_{sn}}{A_{on}} * 100 \quad (1)$$

where

pMC	=	Percent modern carbon
A_{sn}	=	Normalized sample activity
A_{on}	=	Normalized standard activity (Oxalic acid)

The obtained pMC values were then used to calculate the bio-based content as shown in the equation 2 [14]:

$$\text{Bio-based content} = \left(\frac{pMC_{sample}}{pMC_{natural}}\right)*100 \qquad (2)$$

where pMC_{sample} is the percent modern carbon of samples while $pMC_{natural}$ is percent modern carbon of the raw agriculture material, representing the natural C-14.

3. Results

Percent modern carbon of the raw agriculture material represented the natural C-14 amount and the substance was used in the bio-based content calculation. In the present study, the sugar from sugarcane and cassava which are biomass materials, were selected to be the C-14 references. Table 1 shows that percent modern carbon values of both biomass materials were around 112 percent, and this value was set to be the reference for the 100 percent bio-based product substance [15]. Then, this modern carbon value of the agriculture product was compared to the values of other plastics to calculate the bio-based content. The efficiency of the liquid scintillation counter measurement was around 67 percent.

Table 1. Percent modern carbon and bio-based content of biomass materials

Biomass material	pMC (*N=3*)	*p*	Bio-based content (*N=3*)	*p*
Sugarcane	112.15	2.18	100.00	2.75
Cassava	112.05	2.49	99.91	4.03

The bio-based content of each sample is shown in Table 2. Results show that the bio-based content of PLA is 97.36 percent, which is closest to the content from the agriculture product, while that of .Type 1 and 2 PBS and Bio-TPE is in the range of 40-50 percent. From the experiment, PLA is the only bio-plastic that its bio-based content is as high as that of the natural products. On the other hand, PBAT and PE resulted in bio-based content around 1-3 percent. All five bio-plastic products in Thailand have very low bio-based content around 2-7 percent. Bio-plastics from Thai bio-plastic laboratory have higher bio-based content at 52 and 56 percent. However, two imported bio-plastics display over 90 percent of bio-based content.

Table 2. Bio-based content of each bio-plastics

Bio-plastics	Bio-based content (%) (*N=3*)	*p*
PLA	97.36	3.90
PBS1	39.60	1.79
PBS2	45.22	1.95
Bio-TPE	46.53	1.90
PBAT	2.44	0.88
PP	1.91	0.82
BP1	4.78	1.04
BP2	6.30	1.13
BP3	6.11	1.02

BP4	2.93	1.60
BP5	4.55	0.97
BPX1	51.96	2.21
BPX2	55.68	2.39
BPA	90.42	3.74
BPB	91.93	3.62

4. Conclusions and discussion

In this study, the modern carbon in bio-plastic was determined using the carbon dioxide absorption method to indicate how much agricultural materials are contained in the bio-plastic products commercially available in Thailand market. The results show that PLA was the only bio-plastic that was made entirely from agriculture material without synthetic plastic. This bio-plastic can be recognized as the whole bio-based content plastic. PBS and Bio-TPE were found to have less than 50 percent of bio-based content, thus the major composition of these materials are synthetic plastics. PBAT and PE products, which are petroleum-based synthetic plastics, showed a very low bio-based content and the measured values were comparable to the background radiation reference. While five Thai bio-plastic products were found to have very low (less than 10 percent) bio-based content, which is in contrast to the imported bio-plastic products that had more than 90 percent of bio-based content. This shows that bio-plastic products produced in Thailand have a very low composition of bio-based materials.

To validate the method of our analysing protocol, it is necessary to have a known standard reference of bio-plastic. For this study, imported BPA and BPB samples were used as references, with each having its bio-based content (%) shown on its label. This reference served as the known composition standard. The results of measurement value from BPA and BPB samples differed from the value stated on the label about 8.8 percent. This variation was comparable to the previous study by Culp Ret.al [9] and was less than the variation of the standard testing protocol which is 15 percent. In addition to protocol validity testing by comparing with the composition label, this study was also approved by the formulator of the bio-plastic produced for the research group. It was approved that the sample of the plastic container (BPX1) had the bio-based content of 51.96 ± 2.21 percent. Therefore, our protocol yielded the acceptable measurement value. The advantages of our protocol are the minimal analytical time which allows multi-type product analysis and the lower cost of analysis.

References

[1] Leja K and Lewandowicz G 2010 Polymer biodegradation and biodegradable polymers - a review*Pol. J. Environ. Stud.* **19** p 255

[2] Ammala A, Bateman S, Dean K, Petinakis E, Sangwan P, Wong S, Yuan Q, Yu L, Patrick C and LeongK H 2011 An overview of degradable and biodegradable polyolefins *Prog. Polym. Sci.* **36** p 1015

[3] Zheng Y, Yanful E K and Bassi A S 2005A review of plastic waste biodegradation *Crit Rev Biotechnol.* **25** p 243

[4] Standard Test Method for Determining the Biobased Content of Solid, Liquid and Gaseous Samples Using Radiocarbon Analysis Designation: D 6866-08 (Pennsylvania: ASTM international)

[5] JouR M , MacarioK D, CarvalhoC, DiasR S, BrumM C, CunhaF R, FerreiraC G and ChancaI S 2015 Biogenic fraction in the synthesis of polyethylene terephthalate*Int. J. Mass spectrom.***388** p 65

[6] Nagakawa Y,Yunoki S and Saito M 2014 Liquid scintillation counting of solid-state plastic pellets to distinguish bio-based polyethylene*Polym. Test.* **33** p 13

[7] Norton G A, Hood D G and Devlin S L 2007 Accuracy of radioanalytical procedures used to determine the biobased content of manufactured products *Bioresource Technol.* **98** 1052

[8] Norton G A and Devlin S L 2005 Determining the modern carbon content of biobased

IOP Conf. Series: Journal of Physics: Conf. Series **860** (2017) 012014 doi:10.1088/1742-6596/860/1/012014

products using radiocarbon analysis *Bioresource Technol.* **97** p 2084
[9] Culp R, Noakes J, Cherkinsky A, G.V. Ravi Prasad and Dvoracek D. 2013 A decade of AMS at the University of Georgia *Nucl. Instrum. Methods Phys. Res., Sect.* B **294** p 46
[10] Molnar M, Nagy S, Svingor E and Svetlik I 2006 *LSC 2005, Advances in Liquid Scintillation Spectrometry* (Arizona: University of Arizona)
[11] Canducci C, Bartolomei P, Magnani G, Rizzo A, Piccoli A, Tositti L, Esposito M 2013 Upgrade of The CO_2 Direct Absorption Method For Low-Level 14C Liquid Scintillation Counting *Radiocarbon* **55** p 260
[12] Schonhofer F 1995 Liquid scintillation spectrometry in environmental measurements *Sci Total Environ* **173/174** p 29
[13] Verkouteren R M and Klinedinst D B 2004 *Value Assignment and Uncertainty Estimation of Selected Light Stable Isotope Reference Materials* (Washington: Standard Reference Materials)
[14] Stuiver M and Polach H A 1977 *Discussion: Reporting of 14C Data* (Arizona: Radiocarbon) p 355
[15] Society of the Plastics Industry 2012 *Understanding Biobased Carbon Content* (Washington: Bioplastics Council)

Acknowledgments
The authors would like to acknowledge the funding support provided through the 90th Anniversary of Chulalongkorn University Ratchadaphiseksomphot Endowment Fund.

IOP Conf. Series: Journal of Physics: Conf. Series **860** (2017) 012015 doi:10.1088/1742-6596/860/1/012015

A study of residence time distribution using radiotracer technique in the large scale plant facility

S Wetchagarun[1], C Tippayakul, A Petchrak, K Sukrod and P Khoonkamjorn

Reactor Center, Thailand Institute of Nuclear Technology (Public Organization), Bangkok 10900, THAILAND

[1] Author to whom any correspondence should be addressed: saensuk@tint.or.th

Abstract. As the demand for troubleshooting of large industrial plants increases, radiotracer techniques, which have capability to provide fast, online and effective detections to plant problems, have been continually developed. One of the good potential applications of the radiotracer for troubleshooting in a process plant is the analysis of Residence Time Distribution (RTD). In this paper, the study of RTD in a large scale plant facility using radiotracer technique was presented. The objective of this work is to gain experience on the RTD analysis using radiotracer technique in a "larger than laboratory" scale plant setup which can be comparable to the real industrial application. The experiment was carried out at the sedimentation tank in the water treatment facility of Thailand Institute of Nuclear Technology (Public Organization). Br-82 was selected to use in this work due to its chemical property, its suitable half-life and its on-site availability. NH_4Br in the form of aqueous solution was injected into the system as the radiotracer. Six NaI detectors were placed along the pipelines and at the tank in order to determine the RTD of the system. The RTD and the Mean Residence Time (MRT) of the tank was analysed and calculated from the measured data. The experience and knowledge attained from this study is important for extending this technique to be applied to industrial facilities in the future.

1. Introduction

Chemical, and petrochemical industries are technically complex and generally operates continuously [3]. The radiotracer technique is competitive with other inspection methods in troubleshooting inspection and process analysis of such systems due to its sensitivity and ability to be performed as the process is in operation. Residence Time Distribution (RTD) method using radiotracer has been applied extensively in industry to optimize processes, solve problems, improve product quality, save energy and reduce pollution. The RTD and the mean residence time (MRT) are parameters relevant to the performance of chemical reactors, influencing both the throughput and the quality of the product [1]. As the application of RTD method using radiotracer are mostly applied on large scale industrial plants, knowledge and experience on performing the test and analysis on "larger" than laboratory scale is needed. Therefore, research and hands on work on large industrial-like system shall be conducted to develop the techniques and engineering the operation equipment to be suitable for large and harsh environments in the chemical/petrochemical processes. This paper discusses the experiment at Thailand Institute of Nuclear Technology (TINT) in attempt to develop and analyze RTD and MRT of a large scale system using radiotracer technique. The experiment was conducted at the water treatment facility of TINT at Nakhon Nayok. This paper provides the setup details of the experiment. The measured data obtained from the experiment were processed using the recent developed software called preRTD to determine RTD and MRT of the system.

2. Experimental Description

The experiment was carried out for the RTD of the sedimentation tank in the water treatment facility of TINT, Nakhon Nayok. This system was a “larger than a laboratory scale” which could be considered to be comparable to systems in an industrial plant. The measured radiation signals obtained from the experiment were further processed for the RTD and MRT of the sedimentation tank using the preRTD program, originally developed in this work. The details of the experimental setup and the calculation methods are as follows.

2.1. Experimental Setup

The experimental setup consists of three main parts: 1) the water treatment facility, 2) the radiotracer injection and detection system, and 3) the liquid radiotracer. The water treatment facility consists of water supply tanks, a sedimentation tank, an underground storage tank, and pumps and piping systems. The diagram of the experimental setup and the configuration of the sedimentation tank are as shown in Figure 1 and Figure 2, respectively.

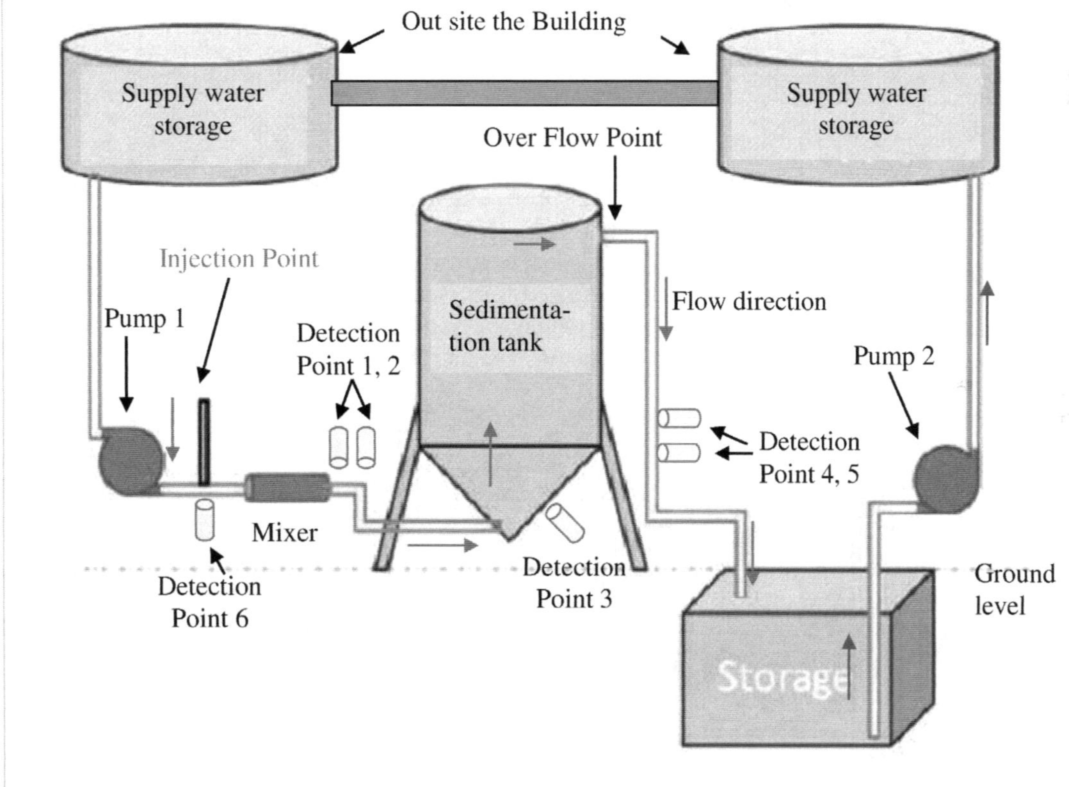

Figure 1. Layout of the experimental setup

IOP Conf. Series: Journal of Physics: Conf. Series **860** (2017) 012015 doi:10.1088/1742-6596/860/1/012015

Figure 2. Sedimentation tank configuration

The radiotracer, Br-82 (approximately 2 R/h at contact), in the form of NH_4Br aqueous solution was injected into the test system at the injection point as shown in Figure 3 . Six NaI detectors were placed along the pipeline and at the sedimentation tank to measure the radiation signals (see Figure 3). The first two detectors were located at the inlet of the sedimentation tank as redundancy to measure signal of the radiotracer at the inlet. The third detector was setup at the tank wall in order to detect the radiotracer in the tank. The fourth and fifth detectors were set, as redundancy system, at the outlet of the tank for the RTD measurement of the radiotracer inside of the tank. The last detector, detector #6, was located close to the injection point to ensure the injection of the radiotracer into the test system.

Figure 3. Injection and detection points

2.2. Calculation method

The radiation signals obtained from the experiment were further processed to determine RTD and MRT of the system of interest, the sedimentation tank. The measured signals were input into the preRTD program for computing RTD and MRT of the tank. The signal treatments of the experimental response curve in the preRTD program include background correction, radiation decay correction, signal smoothing, and normalization of the area of experimental tracer curve. After the treatments, the system RTD and MRT were computed from [1], [2]

$$RTD = E(t) = \frac{n_c(t)}{\int_0^\infty n_c(t)dt}$$

$$MRT = \int_0^\infty tE(t)dt$$

Where $n_c(t)$ is the corrected count rate after the signal treatments.

3. Results and Discussion

The radiation signals obtained from all detectors are shown in Figure 4. At the beginning, the signals from all detectors were approximately constant due to the background radiation, influenced from the radiotracer source, at each measured locations. The levels of the background radiation depend on the distance between the measured point and the radiation source at the injection point. The level of radiation from detectors #1-#5 behaved similarly as they increased when the radiotracer reached the detection points and decrease as the radiotracer passed by. The signal from the detector #6 behaved differently as it only decreased during the injection time. This result may come from the configuration of the system at the injection point (valve and bends of the system) that caused significant accumulation of the radiotracer around that location. The accumulative radiotracer near the detection point cause high radiation level almost at all time for the detector #6.

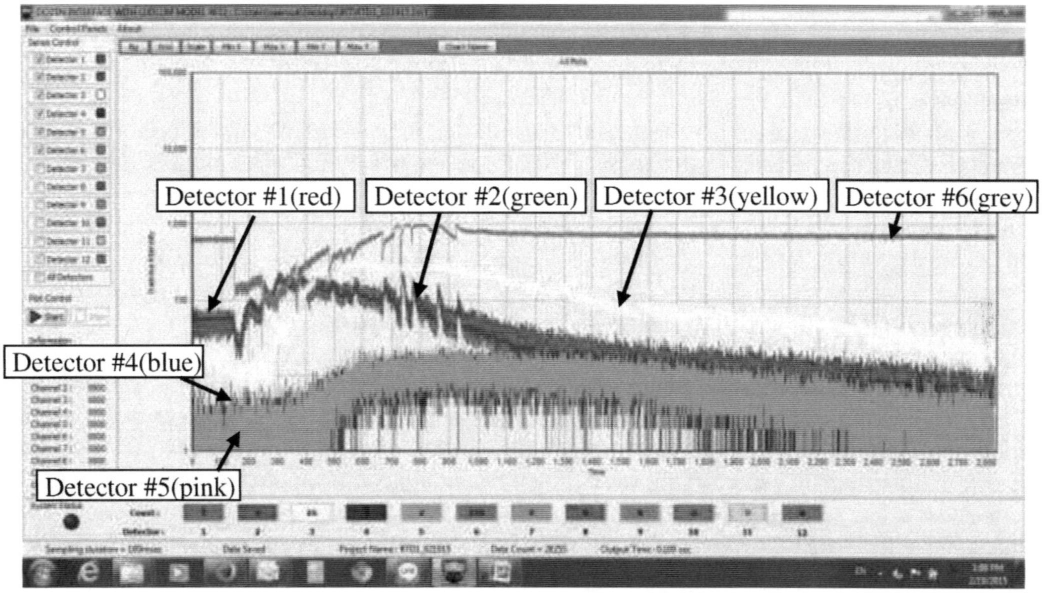

Figure 4. Radiation signals from six NaI detectors

The signals from detector #4 and #5, which were measured at the outlet of the sedimentation tank, can be used to calculate the RTD and MRT of the system. Signals from detector#2 (inlet signal) and from detector#4 (outlet signal) were selected for RTD and MRT analysis. The signals were treated with background and radiation decay correction. Unwanted noise was filtered out by median filter with n=500. The signals after treatments by the preRTD program were shown in Figure 5. The calculated MRT of the sedimentation tank is approximately 16 minute.

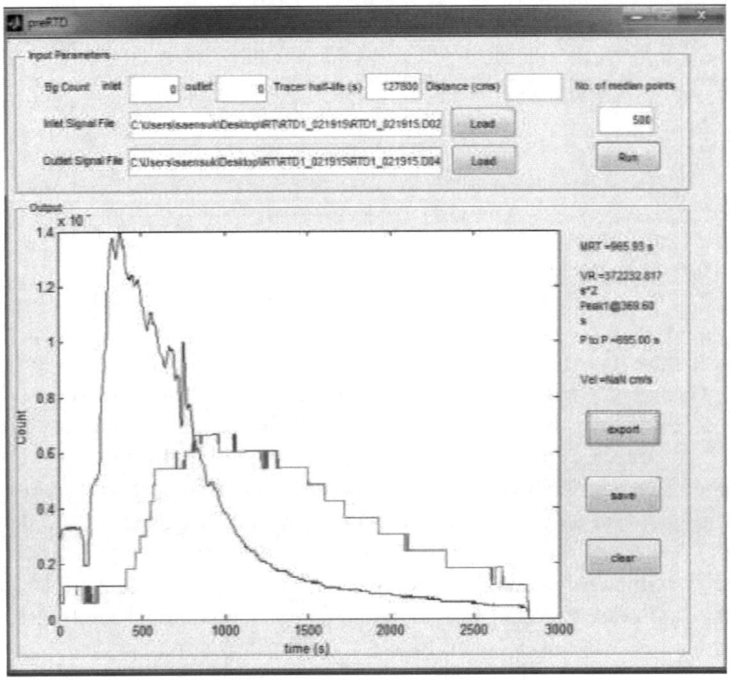

Figure 5. Corrected signals from preRTD program

4. Conclusions

The study of RTD in a large scale plant facility using radiotracer technique was performed at TINT water treatment facility. Br-82, in the form of NH_4Br aqueous solution, was used as a radiotracer in this study. The MRT of the system of interest, the sedimentation tank, computed by preRTD program was approximately 16 minute. As the plant is quite large, the high dose rate radiotracer has to be employed in this work. In order to perform the experiment safely and successfully, various aspects have to be considered at all steps including before, during, and after the experimental time. These include preparation for the experimental procedure, production and transportation of the high dose rate radiotracer, remote radiotracer injection system and detection setup, data collecting process, data analyzing, radiation protection procedure, emergency preparedness, site preparation, and site clearing after completing the experiment. The success and safe operation of this experiment raises the confidence of applying this technique to the large scale facility. The experience and knowledge attained from this study is important for extending the technique to be applied to industrial facilities in the future.

References

[1] International Atomic Energy Agency 2008 *Radiotracer Residence Time Distribution Method for Industrial and Environmental Application: material for education and on-the-job training for practitioners of radiotracer technology, Training course series No.31* (Vienna: International Atomic Energy Agency)

[2] International Atomic Energy Agency 2004 *Radiotracer Applications in Industry - A Guidebook, Technical Reports Series 423* (Vienna: International Atomic Energy Agency)

[3] International Atomic Energy Agency 2001 *Radiotracer technology as applied to industry – Final report of a co-ordinated research project 1997-2000, IAEA-TECDOC-1262* (Vienna: International Atomic Energy Agency)

IOP Conf. Series: Journal of Physics: Conf. Series **860** (2017) 012016 doi:10.1088/1742-6596/860/1/012016

Validation of ^{182}Ta, ^{54}Mn and ^{46}Sc Radioactivity Measurements in Blue Topaz Using Gamma Spectrometric Analysis

A Maneewong[1,3], J Channuie[1], K Pangza[1], N Jangsawang[1], T Charoennam[1] and T Chokesiriwiriyakul[2]

[1] Thailand Institute of Nuclear Technology (Public Organization), 9/9 Moo 7, Saimoon, Onkharak, Nakhon-Nayok, 26120, Thailand
[2] Department of Physics, Faculty of Science, Kasetsart University, 50 Ngam Wong Wan Rd., Chatuchak, Bangkok, 10900, Thailand

[3] Corresponding author e-mail: mwchate@tint.or.th

Abstract. Radioactivity is one of the most discussed topics surrounding blue topaz, which is commonly treated by neutron and/or electron irradiation from near-colorless initiating material. The detection of residual radioactivity after irradiation was usually done by the analytical method of gamma-ray spectroscopy. Therefore, using of validated methods is important for an analytical procedure to show performance and qualification of the method. This work describes the methods for validation of gamma spectrometric analytical procedure in the determination of some gamma emitters in blue topaz samples. Repeatability, reproducibility and minimum detectable activity were the main validation parameters.

1. Introduction

Blue topaz is one of the most popular colored gemstones in the jewelry market today, especially fashion-forward pieces and production lines. In nature, topaz is most common in transparent colorless and pale yellow colors. Blue topaz is achieved by a combination of irradiation and heat treatment. Sky Blue topaz (a lighter shade of blue) is the result of treatment of colorless topaz with either gamma rays from Cobalt 60 sources or electrons produced by electron accelerators, followed by a heat treatment. London Blue (a medium to dark grayish blue) and Swiss Blue (a deep blue) were produced by using the conjunction of neutron from a nuclear reactor, electron from a linear accelerator and heat treatment. In most cases electron accelerators at 10 MeV is used for treatment of topaz and these facilities cannot produce radioactivity in topaz. Also, if the electrons power is increased to 20 MeV, blue topaz can be activated for days or a few weeks. On the other hand, topaz treated with neutrons in a nuclear reactor turns radioactive from the nuclides of trace-element impurities [1, 2], such as Fe, Mn, Co, Zn, Sb, Ta, Cs, Sc, and Tb [3-5]. The decay rates of nuclides created by irradiation depend on the individual nuclides. The length of decay to be anywhere from hours to days to a few years. For this reason the health authorities have issued very strict rules for measuring of the activity in the topaz before they can be released from these irradiation facilities. The detection of residual radioactivity after irradiation was usually done by the analytical method of gamma-ray spectroscopy.

Quality control in gamma-ray spectrometry is an important process of operational techniques and activities that are used to fulfill the requirements of a quality system based on ISO/IEC 17025. It requires monitoring of the performance of the spectrometer, the analysis procedure and the consistency of the results. Therefore, using validated methods is important for an analytical procedure to show performance and qualification of the method. This work describes the methods for validation of gamma spectrometric analytical procedure in the determination of some gamma emitters in blue topaz samples. Repeatability, reproducibility and minimum detectable activity were the main validation parameters.

2. Material and method

Gamma spectrometry based on hyper-pure germanium detectors (HPGe) (Fig. 1) was used to measure the residual radioactivity of the irradiated blue topaz samples. The HPGe detector has a relative efficiency of 25% and full width at half maximum (FWHM) of 1.80 keV for ^{60}Co gamma energy line at 1332 keV and operated with Canberra Genie 2000 software for gamma acquisition and analysis. The gamma ray spectrum represent the radioactivity of specific nuclides in irradiated topaz. The measurements were semi-quantitative, obtained for 3,600 seconds of counting time.

Figure 1. HPGe detector system was installed since 2015 at the Gems Irradiation Center, Thailand Institute of Nuclear Technology (TINT).

Four samples of cutting topaz (weighing ~10.0 g.) of pegmatite origin from Nigeria were studied. The samples were irradiated with 5.94 x 10^{11} n cm^{-2} s^{-1} neutron flux in Thai Research Reactor 1/Modification 1 (TRR-1/M1) on March 28, 2014, and removed from the reactor on December 4, 2014. The decay time about 13 month till the experiment day.

Description of the analytical procedure and spectrometer calibration for the determination of the radioactivity is shown in Fig. 2.

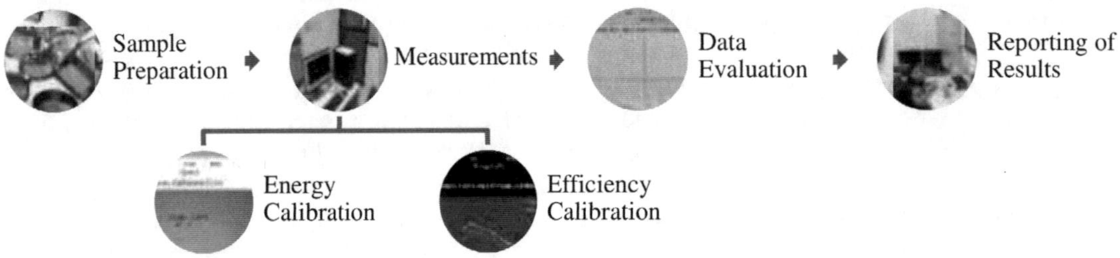

Figure 2. Analytical procedure for the determination of radioactivity of blue topaz samples.

3. Results and discussion

3.1. Energy Calibration

The energy calibration of the HPGe detector system using the radioisotope calibration sources of ^{133}Ba, ^{137}Cs, ^{60}Co, with characteristics described in their calibration certificates reference. The energy range of interest, 50 - 1332 keV, was used for this calibration. All the samples (including standards) were counted directly in the same geometry and counting time. The centroid of photopeak of calibration sources were obtained the peak positions of gamma ray energy lines exhibited in Table 1.

Table 1. Relation between peak position and energy lines of calibration source.

Radioisotope	Gamma ray Energy [keV]	Channel number
^{133}Ba	302.85	905
	356.02	1064
^{137}Cs	661.66	1977
^{60}Co	1173.24	3505
	1332.50	3982

Taking into account above results, the calibration curve is described by the following second degree polynomial function:

$$\text{Energy [keV]} = a_0 + (a_1 \times \text{Channel}) + (a_2 \times \text{Channel}^2) \quad (1)$$

where a_0 = -0.3987, a_1 = 0.3351, $a_2 = -8\times10^{-8}$. These coefficients values are from the fitting curve represented in Fig. 3. The integral nonlinearity is < 0.023% in the interval of 50 - 1332 keV, the differential nonlinearity is < ±0.1%.

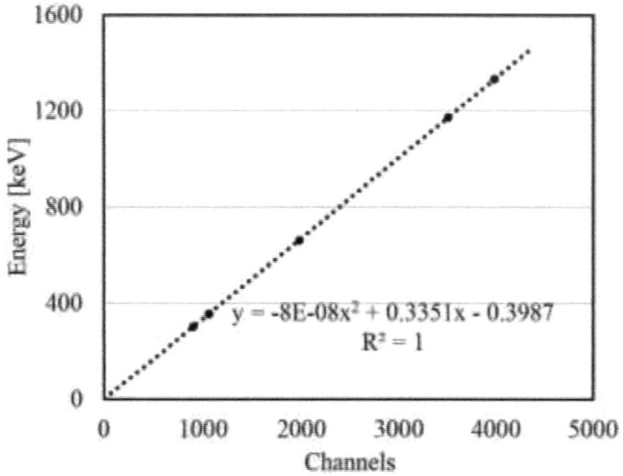

Figure 3. The energy calibration curve of the HPGe detector system.

3.2. *Efficiency Calibration*

The source used for efficiency calibration was the europium mixed sources. The main gamma ray energy, the emission rate, the half-life and the certified activity, are listed in Table 2. The calibration sources were positioned at 2.54 cm distance from the end cap of the detector. All of the measured gamma spectra were analyzed using Genie 2000 program. After measurement of calibration sources, the experimental efficiencies in each dominant energy between range 100 and 1596 keV were calculated. The experimental efficiency at energy E_γ for a given set of measuring conditions are:

$$\varepsilon_{E_\gamma} = \frac{N_{E_\gamma}}{A_n I_{E_\gamma} t} \quad (2)$$

Where N_{E_γ} is the net area in the region of interest of the full-energy peak corresponding to E_γ energy photons emitted by a radionuclide with a known activity, A_n, I_{E_γ} is the probability of emitting a gamma photon and t is the counting time. Efficiency curves were constructed from these calculated full-energy-peak efficiencies as shown in Fig. 4.

Table 2. The characteristics of the europium mixed sources.

Peak No.	Energy [keV]	Eu-152 (Photons/s)	Eu-154 (Photons/s)	Eu-155 (Photons/s)
1	40.1 (SmKa)	5370		
2	60.0			105
3	86.5			2950
4	105.3			2060
5	121.8 –doublet	2640		
	123.1		3920	
6	244.7	695		
7	344.3	2460		
8	591.7			
9	723.3		447	
10	873.2		1170	
11	1004.8		1710	
12	1274.5		3350	
13	1408.0	1930		
14	1596.5		180	
	Activity (Becquerels, Bq)	9260	9710	9520
	Activity (Microcuries)	0.250	0.262	0.257
	Half-Life (Days)	4.96E+03	3.14E+03	1.74E+03
	Decay Constant (Days^{-1})	1.40E-04	2.21E-04	3.98E-04

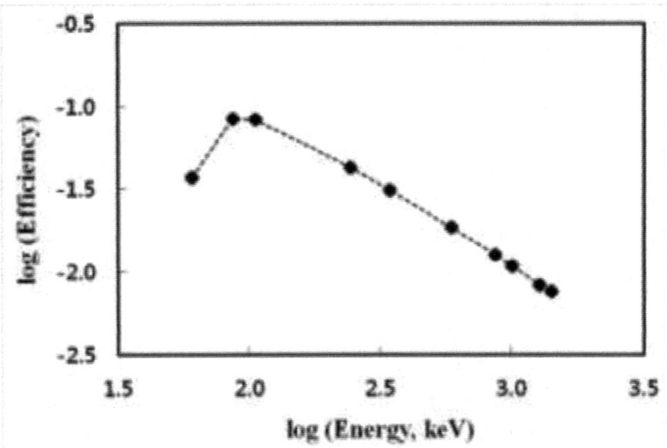

Figure 4. Efficiency calibration curve obtained for the reference geometry.

In Fig. 4, the efficiency curve was exhibited in log of efficiency vs. log of energy scales. In the 60-1408 keV range shows two regions of different behavior because distinct attenuation and absorption processes dominate. At below 100 keV energies the efficiency rises rapidly, but above a few hundred keV the efficiency decreases monotonically. For calculating activity of radioisotope ^{182}Ta, ^{46}Sc and ^{54}Mn in irradiated topaz can use the efficiency curve above 100 keV, which showed the linear curve. Therefore, the efficiency fitting curve is described by a linear function as follow:

$$y = b_1x^2+b_2x+b_3 \tag{3}$$

where y is logarithm of efficiency, x is the logarithm of energy and b_1, b_2 and b_3 are the fitting parameters, b_1 = -0.1698, b_2 = - 0.0509, and b_3 = - 0.2813, from the Fig. 4. The efficiency used for calculating activity of ^{182}Ta, ^{46}Sc and ^{54}Mn are mentioned in Table 3.

Table 3. The efficiency used for calculating activity.

Isotope	Energy [keV]	Branching ratio	Efficiency
^{54}Mn	834.89	0.9999	0.0132
^{46}Sc	889.28	0.9999	0.0124
^{182}Ta	1189.05	0.1623	0.0090
^{137}Cs	661.66	0.9436	0.0168
^{60}Co	1173.24	0.9985	0.0092
	1332.50	0.9998	0.0080

3.3. Performance characteristics of test method

3.3.1. Activity Accuracy. The activity accuracy obtained after the efficiency calibration by setting the measuring conditions as used in measuring efficiency. The results are presented in the Table 3.

Table 4. Activity accuracy.

Radionuclide	Point source code	Calculated actual activity [Bq]	Measured activity [Bq]	Accuracy [%]
^{137}Cs	D-63-10	26229.60	27287.30	95.97
^{60}Co	D-63-9	4821.80	4640.66	96.24

3.3.2. Minimum Detectable Activity (MDA). The calculation of Minimum Detectable Activity for ^{182}Ta, ^{54}Mn and ^{46}Sc, at the 95% confidence level, are based on Currie's derivation [6], with simplified formulation as follows:

$$\text{MDA (Bq/unit } wt) = \frac{2.71+4.66(\sigma)}{T \cdot EFF \cdot I \cdot wt} \tag{4}$$

where σ is the standard deviation of the background collected during the counting time, *(T*, in sec*)* over the energy range of interest (ROI), *EFF* is the efficiency at the energy of interest, *I* is the Branching Ratio and *wt* is sample weight. The MDA values of measuring nuclides are shown in Table 5.

Table 5. MDA values.

Radionuclide	Energy [keV]	Background count at ROI [s^{-1}]	MDA [Bq/gram]
^{54}Mn	834.89	0	0.028
^{46}Sc	889.28	0	0.030
^{182}Ta	1189.05	0	0.256

3.3.3. The internal comparison. This part is most important for our measuring method. The internal comparison is a valuable part of validation, which having three participant and their various spectrometer systems for measuring one sample of blue topaz for this study. The results obtained by the three laboratories are presented in Table 6, indicating with high accuracy in comparable activity values. However, the activity values of ^{46}Sc and ^{182}Ta at R&D NAA Lab show lower than those at GIC labs. The difference in efficiency calibration method may take effect in high energy rang of gamma ray (more than 800 keV).

Table 6. The internal comparison results. (GIC stands for Gems Irradiation Center)

Radionuclide	Measured activity [Bq]		
	GIC Ongkharak* (Point source calibration)	GIC Bangkhen (Point source calibration)	R&D Group NAA Lab (ISOCS theoretical calibration)
^{54}Mn	23.19	21.67	21.44
^{46}Sc	449.78	460.10	371.38
^{182}Ta	69.80	66.21	50.57

*GIC Ongkharak is the main lab, and the other two labs are used for comparison.

4. Conclusion

The validation of the results and of the calibration method was made in a series of measurements to check the values of the parameters declared in the HPGe detector technical documentation, using certified point sources and respecting the initial measuring conditions and also participating at three intercomparisons with the internal laboratory (GIC Lab. in Bangkhen branch and R&D Group NAA Lab). The results obtained by GIC Lab. in Ongkharak branch validate the measuring method, emphasizing the fact that our gamma spectrometric system with HPGe detector is properly calibrated in energy and efficiency, respecting the measurement geometry. The sample size of irradiated topaz used in this test is ~10 gram which may comparable to ~3.0 cm^3. Therefore, the difference in measurement geometry still have to find the solution. We will concentrate more in this issue.

References

[1] Crowningshield R 1981 *G&G.* **17(4)** 215.
[2] Ashbaugh C E 1988 *G&G.* **24(4)** 196.
[3] Foord E E, Jackson L L, Taggart J E, Crock J G and King T VV 1988 *Mineral Rec.* **26** 69.
[4] Northrup P A and Reeder R J 1994 *Am. Mineral.* **79** 1167
[5] Helal A I, Zahran N F, Gomaa M A M and Salama S 2006 *Physics & Protection Conference*, Beni Suef - Fayoum, Egypt, November 13–15, 447
[6] Currie L A 1968 *Anal. Chem.* **40** 586

INAA Application for Trace Element Determination in Biological Reference Material

D P D Atmodjo, S Kurniawati, D D Lestiani and N Adventini[1]

Center for Applied Nuclear Sains and Technology, National Nuclear Energy Agency, Tamansari No.71, Bandung, 40132, Indonesia

[1] Corresponding author e-mail: djoko_prakoso@batan.go.id

Abstract. Trace element determination in biological samples is often used in the study of health and toxicology. Determination change to its essentiality and toxicity of trace element require an accurate determination method, which implies that a good Quality Control (QC) procedure should be performed. In this study, QC for trace element determination in biological samples was applied by analyzing the Standard Reference Material (SRM) Bovine muscle 8414 NIST using Instrumental Neutron Activation Analysis (INAA). Three selected trace element such as Fe, Zn, and Se were determined. Accuracy of the elements showed as %recovery and precision as %coefficient of variance (%CV). The result showed that %recovery of Fe, Zn, and Se were in the range between 99.4-107%, 92.7-103%, and 91.9-112%, respectively, whereas %CV were 2.92, 3.70, and 5.37%, respectively. These results showed that INAA method is precise and accurate for trace element determination in biological matrices.

1. Introduction

Increasing knowledge and advancement in technology demand that researchers improve research method in getting accurate and precise research results. One of the field studies that should have good quality control (QC) is the study of biological samples, because many of them are used in the health sciences and toxicology. Elemental analysis on research related to toxicology and health must describe the results as close as possible to the actual content in the sample studied [1]. This is of particular importance when small differences in concentration values may help better understanding of the role played by the trace elements under investigation [2]. Determination change to its essentiality and toxicity of trace element requires suitable method that can yield accurate results, which implies that a good Quality Control (QC) procedure should be performed. Neutron Activation Analysis (NAA) is a nuclear analytical technique that has high accuracy in determining the elements both qualitatively and quantitatively, and has the ability to measure many elements simultaneously, has high sensitivity and is capable of analyzing the elements of major, minor and traces [3,4]. NAA is based on the measurement of induced radioactivity of the sample, formed in the reaction of atomic nuclei by neutron particles [5]. Some of the elements that have thin boundaries between essentiality and toxicity in the body are Fe, Zn and Se, so they require a set of data analysis that are controlled and selective and should be verified by quality assurance procedures. Therefore, laboratory performing elemental analysis must always perform the quality control of the data analysis results. To ensure the quality of the test results, which is also as one of the requirements in ISO/IEC 17025:2005, method validation activity using reference material is performed. Reference materials are materials or substances that have homogeneous and stable properties and serve as standards of measurement or in testing a sample [6,7]. Reference materials play a very important role in validating the accuracy of the data; especially for testing laboratory, in this case is the Laboratory of Center for Applied Nuclear Science and Technology (PSTNT) that uses reference materials as validation of results from testing a sample. Reference materials aim to facilitate the accurate testing of the whole system testing during the development or application of an analytical method. There are several types of reference materials, among which are the Reference Material (RM), Certified Reference Material (CRM), and Standard Reference Material (SRM). One of SRM that been used for validating the NAA method in PSTNT laboratory, particularly for biological matrices samples was SRM NIST 8414 Bovine Muscle, that is produced by NIST (National Institute of Standards and Technology, USA) [8]. In this study, method validation of NAA in determining the elements in biological matrices sample was done by measuring the concentration of target elements in SRM NIST 8414, then comparing with the certificate value for its accuracy and precision. QC chart was made for evaluating the performance of analytical method during routine use. The results of these activities were expected to show that the result data in the field of biological research using nuclear analytical techniques, especially generated by our laboratory are accurate.

2. Methodology

2.1. NAA relative method

NAA relative method was applied for trace element determination in SRM NIST 8414 Bovine Muscle. In this method, the standards and the samples were irradiated together in order to obtain the same conditions. Therefore, that the levels of the elements in the sample can be calculated by comparing the activity of the sample (A spl) with the known activity of prepared standard (A std). A level of elements in the sample is calculated by the following equation [9]:

$$Wx = \frac{A\,spl}{A\,std} \times Wstd \qquad (1)$$

Wx and *Wstd* are weight of interested elements of sample and standard, respectively.

2.2. Preparation of NAA standards

Preparation of NAA standards was carried out by pipetting 100 μl of ICP multi-elemental standard solution (E-Merck) then dripping into the polyethylene vial. The filled vials were then dried under the infrared lamp and sealed by heating. Masses of elements in the prepared standard are shown in table 1.

Table 1. Elemental mass of prepared standards.

Elements	Mass (μg)
Fe	59.9
Se	10.2
Zn	10

2.3. Preparation of biological reference materials

SRM NIST 8414 Bovine muscles biological reference material samples were weighed as much as 25 mg, placed in the 0.273 mL polyethylene vial, and then sealed by heating the end of the lid. After checking for leakage possibility, the samples were ready for irradiation. Leakage of the vial can affect the reference material and cause contamination of the research reactor irradiation facility.

2.4. Irradiation and counting of biological reference material.
Biological reference material NIST 8414 Bovine muscle was irradiated in a multipurpose research reactor, PRSG G. A. Siwabessy located in Serpong, Banten province with neutron flux of ~10^{13} $n.cm^{-2}.s^{-1}$ for 120 minutes with cooling time until 4 weeks to get the long half-life elements such as Fe, Zn and Se. Counting was performed using Gamma spectrometer equipped with HPGe detector (Canberra) in PSTNT laboratory for 5000 seconds, to form the characteristic energy peak that shows the elements Fe, Zn and Se.

2.5. Quality control and data analysis.
The SRM was used as quality control assessment of data validity. The results of SRM analysis were compared with its certificate value and evaluated its accuracy and precision by %Recovery and %CV calculation. Net areas from Genie 2000 were used to calculate the concentrations of elements in the samples using a comparative method. Further evaluation of validation results was carried out by plotting it on a diagram of control. Data on control charts were evaluated statistically using the average value and standard deviation with the following conditions [10]:

- Upper warning limit = mean + 2SD
- Lower warning limit = mean - 2SD
- Upper control limit = mean + 3SD
- Lower control limit = mean - 3SD

After statistical evaluation, accuracy and precision were then calculated. Accuracy testing was done by comparing the results obtained with the value of the certificate and commonly describes as %Recovery, which is expressed in the following equation:

$$\%Recovery = \frac{Value_{measurement}}{Value_{certificate}} \times 100\% \qquad (2)$$

Precision is the ability to produce consistent measured value. Precision is measured as the standard deviation or coefficient of variance (CV). Parameters precision in the validation of methods, expressed in %CV as shown in the following equation:

$$\%CV = \frac{SD}{Mean} \times 100\% \qquad (3)$$

Where CV is coefficient of variance, SD is standard deviation and Mean is mean concentration.

The results of accuracy (%Rec) and precision (%CV) were evaluated by observing the acceptability criteria of accuracy results required by the AOAC guidelines for single laboratory validation, and of precision required by Horwitz equation, as shown in the following table 2 and 3.

Table 2. Acceptance criteria for accuracy [11].

Concentration	*Recovery* (%)
100%	98-101
10%	95-102
1%	92-105
0.1%	90-108
0.01%	85-110
10 ppm	80-115
1 ppm	75-120
10 ppb	70-125

Table 3. Acceptance limits of precision by Horwitz equation [11].

Concentration	Horwitz (%RSD)
100%	2
10%	2.8
1%	4
0.1%	5.7
100 ppm	8
10 ppm	11.3
1 ppm	16
100 ppb	22.6
10 ppb	32
1 ppb	45.3

According to the table 2 and 3, it can be stated that the analytical method observed is valid if the validation data, meet the acceptance values listed in the tables 2 and 3 above. If it does not meet the acceptance criteria then it will be necessary to evaluate the whole measurement activity beginning with the sample preparation to data processing, otherwise it should be considered that the methods used do not correspond to the reference materials used.

3. Results and Discussion

NAA has been applied in PSTNT laboratory for several samples including biological matrices. Data quality assurance was performed in this laboratory. The easiest and really effective way to carry out data QA controls is to analyze using the same procedure, standard reference materials with well-known and established data on the trace elements under investigation. If correct results are obtained, the overall analytical performance meets the precision and accuracy requirements for trace analysis [12]. Accuracy and precision of NAA method for Fe, Se and Zn was tested by analyzing SRM NIST 8414 Bovine muscle. Results are shown in table 4. Table 4 shows that Fe has an average concentration of 72.9±2.13 ppm. The results obtained from measurements of the 5 samples of SRM gives standard deviation of 2.13 ppm. The values of the measurements were still within the range of the certificate value; 71.20±9.20 ppm, with %Recovery were in the range of 99.4 until 107%, and the %CV was 2.92%. The results of subsequent measurements are described in a diagram control as shown in figure 1. Figure 1 shows that the analysis results of SRM NIST 8414 using NAA method from repeated measurements lie within the mean±2SD. With all the results of testing that has been done, it can be ascertained that the measurement of Fe element content contained in NIST SRM 8414 using NAA method is valid. Zn was also analyzed for the parameters of validation, as shown in table 4. Mean concentration of Zn was 140 ppm, with the range values were 132-146 ppm. The measurements values were still within the range of certificate value; 142.0±14.0 ppm, with the %Recovery were in the range of 92.7-102.75%, and the value of %CV was 3.70%. Performance evaluation of NAA method of test results Zn showed by its control diagram in figure 2.

Table 4. The results of the validation using SRM 8414 Bovine muscle.

No	Sample	Fe		Zn		Se	
		Measured (mg/kg)	Accuracy (%Rec)	Measured (mg/kg)	Accuracy (%Rec)	Measured (mg/kg)	Accuracy (%Rec)
1	SRM BM A	73.7±2.92	104	132±2.15	92.7	0.075±0.008	99.1
2	SRM BM 1	70.8±1.48	99.4	141±0.60	99.6	0.075±0.006	98.7
3	SRM BM 2	72.4±7.14	102	146±1.43	103	0.085±0.009	112
4	SRM BM 3	71.5±2.49	100	140±0.44	98.5	0.079±0.007	104

5	SRM BM B	76.2±3.73	107	141±0.65	99.2	0.076±0.010	99.95
	Mean	72.9±9.04	102	140±2.76	98.5	0.078±0.02	103
	Certificate	71.2±9.20		142±14.0		0.076±0.010	
	%CV	12.4		1.97		23	

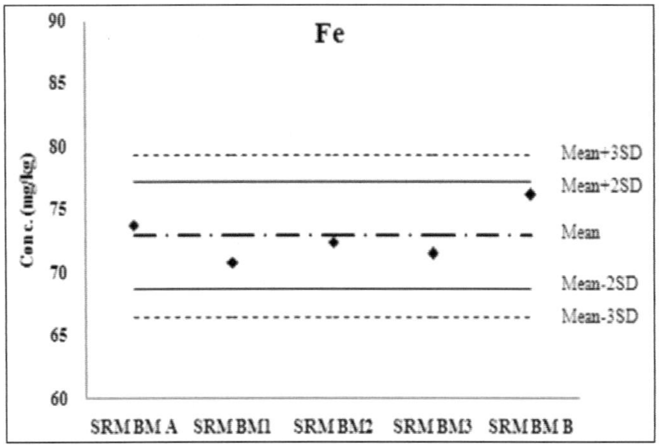

Figure 1. QC chart of Fe in SRM NIST 8414 Bovine muscle.

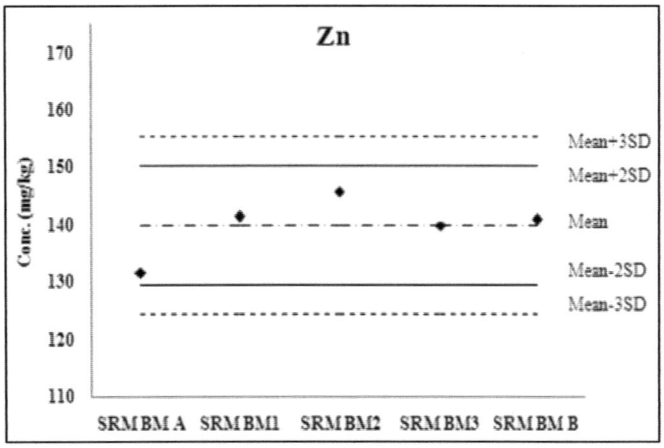

Figure 2. QC chart of Zn in SRM NIST 8414 Bovine muscle.

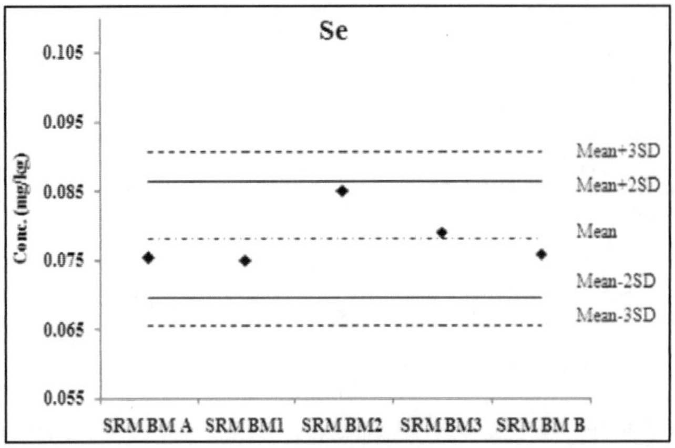

Figure 3. QC chart of Se in SRM NIST 8414 Bovine muscle.

Diagram control for Zn as shown in figure 2, present the range of measurement results for SRM NIST 8414 using NAA between 130 until 150 ppm and still within the range of mean±2SD values. The control diagram and the results of accuracy and precision of Zn measurements contained in SRM NIST 8414 using NAA showed valid results. Validation method of NAA for Selenium was also performed by analyzing SRM NIST 8414. Selenium is an element that usually quite difficult to be analyzed using the NAA. Table 4 shows the element Se has an average concentration of 0.078 ppm, with the results values were in the range of 0.075 until 0.085 ppm, and standard deviation from 5 measurements was 0.004 ppm. This value was still within the range certificate value that is 0.076±0.010 ppm, with the value of %Recovery range between 98.68 until 111.84%, and the %CV was 5.37%. QC chart for Se is shown in figure 3, and the concentration of 5 measurements was still in the mean±2SD, it means that the performance of this method is good. Based on of both accuracy and precision test, Se determination in SRM NIST 8414 using NAA showed good results. The accuracy and precision obtained from this study were affected by several factors such as weighing procedure of the Fe, Se and Zn in the prepared standards for NAA relative method and stability of the neutron flux during irradiation. From these results, it can be concluded that the NAA method in determining of Fe, Zn and Se in SRM NIST 8414 are valid, can thus be applied with accuracy and precision on similar biological matrices samples.

4. Conclusion

The results obtained from testing the accuracy (%Recovery) and precision (%CV) of the Fe, Zn and Se contents in SRM NIST 8414 Bovine muscle are within the range of their acceptance criteria, and are strongly supported by the repeated measurements as described in the control diagram. It can be stated that the NAA method has high accuracy and precision for the determination of elements observed in the SRM Bovine muscle, thus NAA method is valid and can be applied to determine trace elements in similar biological matrices samples.

References

[1] Gallorini M 1995 Standard reference materials and data quality assurance: lesson from the analysis of trace elements *Toxicology Letters* **77** 209-212

[2] Rennert O M and Chan W Yee 1984 *Metabolism of Trace Metals in Man* Vol I and II (CRC Press Inc., Boca Raton, Fla)

[3] Greenberg R R, Bode P and Fernandes E A D N 2001 Neutron activation analysis: A primary method of measurement: A review *Spectrochimica Acta Part B: Atomic Spectroscopy* **66** 193–241

[4] Bhatia C, Byun S H, Chettle D R, Inskip M J and Prestwich W V 2015 A neutron activation technique for manganese measurements in humans *Journal of Trace Elements in Medicine and Biology* **31** 204–208

[5] Mulyaningsih T R, Rukihati and Istanto 2008 Analysis of essential elements in Indonesian commercial tea by neutron activation analysis (in Indonesian) *Proceeding of National Seminar NAA* Bandung 22 October

[6] Venelinov T, Sahuquillo A 2006 Optimizing the uses and the costs of reference materials in analytical laboratories *Trends in Analytical Chemistry* **25** 5

[7] Kantasubrata J 2005 *Method Validation (in Indonesian)* (Lembaga Ilmu Pengetahuan Indonesia)

[8] NIST Certificate of Analysis Standard Reference Material 8414 Bovine Muscle

[9] Lestiani D D, Santoso M and Adventini N 2009 Application of Neutron Activation Analysis in Characterization of Environmental SRM Samples (in Indonesian) *Indo. J. Chem.* **9(2)** 231–235

[10] Santoso M 2007 *Repeatability and Reproducibility: The training modules of K0 neutron activation analysis method validation (in Indonesian)* (Center for Education and Training National Nuclear Energy Agency Jakarta Indonesia)

[11] Gonza´lez A G and Herrador M A 2007 A practical guide to analytical method validation, including measurement uncertainty and accuracy profiles *Trends in Analytical Chemistry* **26(3)** 227-238

[12] Mummatsu Y and Parr R M 1985 *Survey of Current Available Reference Materials for Use in Connection with the Determination of Trace Elements in Biological Materials IAEA/RL/28* (International Atomic Energy Agency Vienna)

A calculation of relative efficiencies of a GEM-based neutron detector using different solid neutron converters

A Rittirong, S Sripreeprem and K Saenboonruang[1]

Department of Applied Radiation and Isotopes, Faculty of Science, Kasetsart University, Chatuchak, Bangkok, 10900, Thailand

[1] Corresponding author e-mail: fscikssa@ku.ac.th

Abstract. A Gas Electron Multiplier (GEM) detector is one of the most promising particle detectors nowadays due to its excellences in high rates of detection capability, good spatial resolution, and flexibility in designs. In addition to typical applications of ionizing particle detection, the GEM detector could also be modified to detect thermal neutrons by coating suitable solid neutron converters to its GEM drift cathode. Although, wide selections of materials are available to be used as neutron converters, most suitable ones must have high neutron-absorption-cross-section properties as more chances of nuclear interactions between thermal neutrons and their nuclei are desirable. Another factor that plays important roles to detection efficiencies in addition to types of materials is thicknesses of coating layers. Although thicker layers mean higher numbers of nuclei available for interactions with neutrons, ionizing particles produced after the interactions have less chances of successfully penetrate thick layers to ionize gas molecules for detection. Consequently, an optimum thickness for each material is crucial for GEM-based thermal neutron detector and must be carefully determined. This article aims to illustrate in-depth calculations to find these optimum thicknesses and their corresponding relative efficiencies for different solid neutron converters including ^{6}Li, ^{nat}Li, ^{10}B, ^{nat}B, ^{113}Cd, ^{nat}Cd, ^{149}Sm, and ^{nat}Sm, using a simulation and actual data of GEM's properties. Basic information of GEM detector and neutron detection, simulations, efficiency calculations, results, and discussion will be thoroughly reported in this article.

1. Introduction

1.1. Gas Electron Multiplier (GEM) detector

A Gas Electron Multiplier (GEM) detector is a gaseous charge amplification detector invented by F. Sauli in 1997 [1,2]. Since the invention, the GEM detector has been developed and used in great numbers of nuclear and high energy particle researches including the large size GEM for Super Bigbite Spectrometer (SBS) polarimeter for Hall A 12 GeV program at Jlab [3], large size GEM for future PREX-like experiments [4], a triple-GEM for CMS muon detector upgrade, and a triple-GEM tracker for KLOE experiment [5]. In addition to scientific researches, the GEM detector is also utilized in various applications including low energy x-ray monitor for burning plasmas in tokamaks [6] and Single-Photon Emitted Computed Tomography (SPECT) combined with x-rays Computed Tomography (CT) in medical applications [7].

The standard triple-GEM detector is a cascade of

IOP Conf. Series: Journal of Physics: Conf. Series **860** (2017) 012018 doi:10.1088/1742-6596/860/1/012018

- GEM drift cathode: a thin aluminized polyimide film placed topmost of the cascade with the most negative voltage to guide ionized electrons towards the GEM readout.
- GEM foils: a thin polyimide film coated by metal clads (usually copper) on both sides of the film. A GEM foil is perforated with a regular matrix of 70-μm holes. These holes are closely spaced with a pitch of ~140 μm from one center of the hole to centers of the closest neighboring holes. The first GEM foil is placed 3 mm below the GEM drift cathode, while all GEM foils are spaced 2 mm from each other. The third GEM foil is placed 2 mm above the GEM readout.
- GEM readout: in most applications, the GEM readout consists of two sets of thin wires running perpendicular to each other (XY configuration). However, the GEM readout could also be designed as arrays of micro-sized square pads depending on intended applications. The GEM readout is usually connected to a common ground (highest voltage) of electronic circuits. Amplified numbers of electrons are transferred through connectors that connect to data acquisition system (DAQ) for data processing. Major components of a standard GEM detector are shown in figure 1.

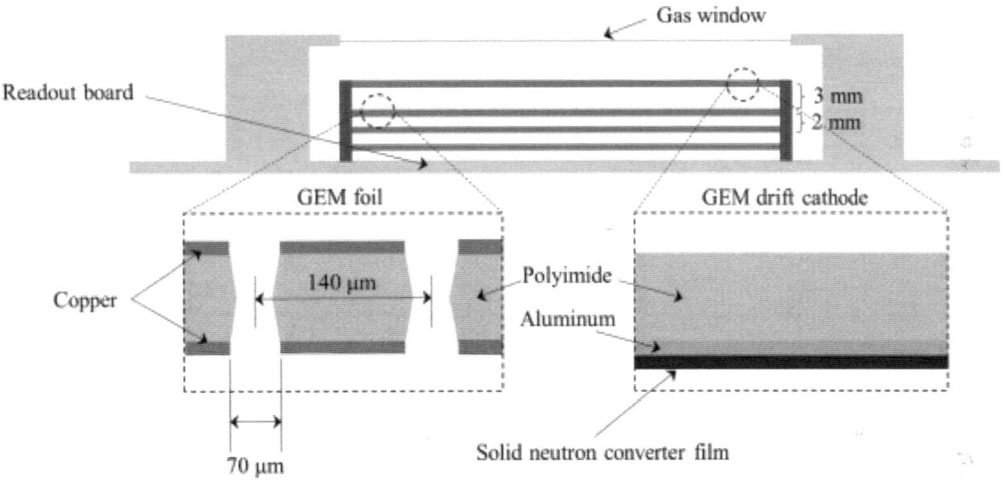

Figure 1. Side views of a complete assembly of a GEM detector, a GEM foil with magnified GEM holes, and a GEM drift cathode, which is coated by a solid neutron converter film for thermal neutron detection.

Since GEM's properties heavily rely on properties of gas used, selection of appropriate gas flow is important to operate the detector. In principle, a pure noble gas such as helium or argon could be used, however, in order to increase stability of the detector, a quenching gas such as CO_2 or C_4H_{10} is usually added to the gas mixture. For general applications, a gas mixture of Ar/CO_2 (70:30) is normally used. Another important factor needed to be considered is the gas flow rate. Although there was a report that high gas flow rates (up to 10.5 L/hr) did not have significant effects on a standard 10 cm x 10 cm GEM's properties [8], the gas flow rate should be high enough to flush 10 times of a volume of a GEM detector such that new gas could sufficiently replace depleted gas for better detection. For example, a standard 10 cm x 10 cm GEM prototype needs at least 3.0 L/hr of gas flow rates to efficiently operate the detector.

1.2. GEM-based thermal neutron detector

Although GEM detector is known for their excellences in ionizing particle and low-energy photon detection, it could be modified by coating its GEM drift cathode with a high-neutron-absorption-cross-section converter [9] as shown in figure 1 for thermal neutron detection. Criteria for selecting a

IOP Conf. Series: Journal of Physics: Conf. Series **860** (2017) 012018 doi:10.1088/1742-6596/860/1/012018

suitable solid neutron converter depends on intended applications, however, major points that are crucial for consideration are:

- Neutron-absorption-cross-section of a material. Since neutrons are electrically neutral, they do not directly ionize gas molecules, leading to extremely low detection efficiency. In order to increase the efficiency, a process to convert neutrons to secondary ionizing particles is required. This process could be done via nuclear reactions between thermal neutrons and nuclei of neutron converters as shown in figure 2.

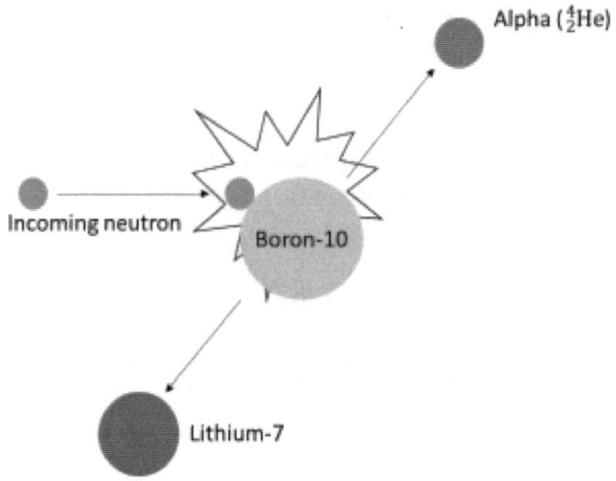

Figure 2. Nuclear interaction between an incoming thermal neutron and a ^{10}B nucleus, which produces an alpha particle and ^{7}Li traveling in opposite directions.

Examples of commonly used high-neutron-absorption-cross-section converters are ^{6}Li, ^{nat}Li, ^{10}B, ^{nat}B, ^{113}Cd, ^{nat}Cd, ^{149}Sm, and ^{nat}Sm. Although ^{6}Li, ^{10}B, ^{113}Cd, and ^{149}Sm isotopes have much higher neutron-absorption-cross-section values than their natural composites, they require sophisticated procedures to enrich, which could significantly increase complications and costs of manufacturing. Equations showing interactions between thermal neutrons and selected nuclei are shown in equations (1.1)-(1.4).

$$^{6}Li + n \longrightarrow {}^{3}H\ (2.75\ MeV) + \alpha\ (2.05\ MeV) \quad (1.1)$$

$$^{10}B + n \longrightarrow {}^{7}Li\ (0.84\ MeV) + \alpha\ (1.47\ MeV) \quad (1.2)$$

$$^{113}Cd + n \longrightarrow {}^{114}Cd + \gamma\ (0.558\ MeV) \quad (1.3)$$

$$^{149}Sm + n \longrightarrow {}^{146}Nd\ (1.66\ MeV) + \alpha\ (8.96\ MeV) \quad (1.4)$$

- Thickness of a coating layer. Thicker layers of neutron converters mean more numbers of available nuclei to interact with neutrons per unit area, leading to better detection ability. However, too thick layers could inversely reduce efficiencies as secondary ionizing particles produced from the interactions could not penetrate through the converter layer to ionize gas molecules for detection. Consequently, an optimum thickness of each coating layer must be thoroughly calculated in order to maximize detection ability.
- Types/energies of secondary ionizing particles produced. As shown in equations (1.1)-(1.4), secondary ionizing particles produced depend on types of neutron converters. For ^{6}Li, ^{10}B, and ^{149}Sm, alpha particles (α) are produced, while gamma (γ) is produced in the case of ^{113}Cd. Generally, an alpha particle has extremely short ranges in matter (usually in micrometer

ranges) due to its relatively high electric charge. On the other hand, gamma is an electromagnetic wave that could travel much further in matter due to their less chances of interactions and low Linear Energy Transfer (LET). These properties of secondary ionizing particles must be taken into considerations when selecting neutron converters.

As several factors could play roles in defining GEM-based thermal neutron detection, a careful and thorough determination must be performed in order to maximize detection efficiencies. This article aims to report details of calculations and analysis to find optimum thicknesses for different solid neutron converters: ^{6}Li, ^{nat}Li, ^{10}B, ^{nat}B, ^{113}Cd, ^{nat}Cd, ^{149}Sm, and ^{nat}Sm. Results will be useful for all GEM-based neutron detector in both basic and applied researches.

2. Methodology

To calculate relative efficiencies (ε_{rel}) for each solid neutron converter, two main parts that are needed to be considered are reaction rates per unit area (*RR*) of neutrons and nuclei of neutron converters, and corrections (κ) from trajectories of emitted ionizing particles (κ_{θ}) and GEM's efficiency to detect that particular ionizing particles (κ_{eff}). The equation that relates ε_{rel}, *RR*, and κ is shown in equation (2.1).

$$\varepsilon_{rel} \propto RR \times \kappa \tag{2.1}$$

2.1. Reaction rates per unit area (RR)

For a uniform neutron flux on an active area of GEM detector, reaction rates per unit area (*RR*) of the interactions between incoming thermal neutrons and nuclei of solid neutron converters are a function of incoming neutron flux (Φ), number of solid neutron converter's nuclei per unit area (*N*), and neutron-cross-section of solid neutron converter (σ) as shown in equation (2.2).

$$RR = \Phi \times N \times \sigma \tag{2.2}$$

However, N could be expressed as

$$N = n \times t \tag{2.3}$$

$$N = \left(\frac{\rho}{M}\right) \times t \tag{2.4}$$

where *n* is the nuclear density, *t* is the thickness of the solid neutron converter's layer, ρ is the mass density, and *M* is the molecular mass of the solid neutron converter's nuclei. As a result, when considering only factors from designs and properties of neutron converter materials, *RR* is related to ρ, *M*, *t*, and σ as

$$RR \propto \left(\frac{\rho}{M}\right) \times t \times \sigma \tag{2.5}$$

Values of ρ, M, and σ for each type of solid neutron converters are shown in table 1.

Table 1. Values of mass density per unit area (ρ), molecular mass (*M*), and neutron-cross-section (σ) for each solid neutron converter.

Element	ρ (g/cm^3)	*M* (g)	σ (10^{-24} cm^2)
^{6}Li	0.535	6.00	940.4
^{nat}Li	0.535	6.94	70.5
^{10}B	2.460	10.00	3835.9
^{nat}B	2.460	10.81	767.8
^{113}Cd	8.650	113.00	20600.4
^{nat}Cd	8.650	112.41	2520.5
^{149}Sm	7.353	149.00	42080.4
^{nat}Sm	7.353	150.36	5922.5

2.2. Corrections (κ)

2.2.1. Trajectories of emitted ionizing particles (κθ). As nuclear interactions occur, emitted ionizing particles could travel in any direction with equal probability. However, only particles that travel downwards with distances shorter than its ranges are able to penetrate through solid neutron converter layers and ionize gas molecules for detection.

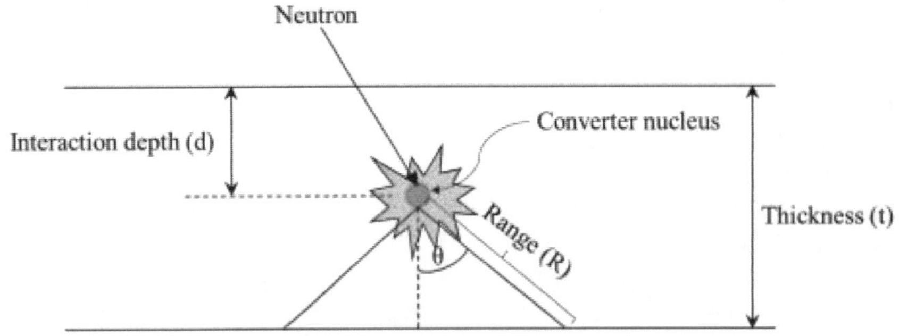

Figure 3. Trajectory of emitted ionizing particles. Only particles that travel within 2θ are able to penetrate through solid neutron converter layer and ionize gas molecules for detection.

As shown in figure 3, only particles that are emitted within 2θ are considered to be active particles for detection. Thus, a correction factor for trajectory of emitted ionizing particles (κ_θ) could be calculated as

$$\kappa_\theta = {}^{2\theta}/_{2\pi} = {}^{\theta}/_{\pi} \tag{2.6}$$

To calculate θ, ranges of emitted ionizing particles (R) and average depth of interactions (d) must be determined. Since different types/energies of ionizing particles have different values of R depending on types of materials encountered, simulations must be done to find R. To serve this purpose, a software called SRIM-2013 [10] was used and results of simulations are shown in figure 4 and table 2.

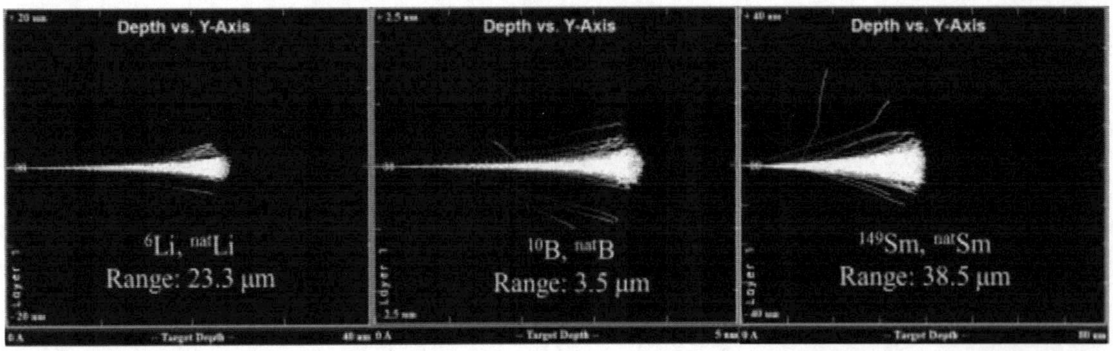

Figure 4. Simulations using SRIM-2013 to find ranges of alpha particles emitted from interactions between neutrons and ^{6}Li/^{nat}Li (left), ^{10}B/^{nat}B (middle), and ^{149}Sm/^{nat}Sm (right) nuclei respectively.

In the case of ^{113}Cd and ^{nat}Cd, γ with the energy of 0.558 MeV is emitted. Since gamma has low Linear Energy Transfer (LET), its ranges are much longer than the thickness of materials ($R>>t$) used in this efficiency calculations.

Table 2. Ranges of emitted ionizing particles for all solid neutron converters.

IOP Conf. Series: Journal of Physics: Conf. Series **860** (2017) 012018 doi:10.1088/1742-6596/860/1/012018

Element	Emitted particles	Energy (MeV)	Ranges (μm)
$^{6}Li/^{nat}Li$	α	2.05	23.3
$^{10}B/^{nat}B$	α	1.47	3.5
$^{113}Cd/^{nat}Cd$	γ	0.558	>>t
$^{149}Sm/^{nat}Sm$	α	8.96	38.5

As shown in equation 2.5, *RR* is directly proportional to *t*, i.e. numbers of interaction increase linearly with *t*. By assuming that numbers of incoming neutrons are large and *t* is much smaller than ranges of neutrons in the materials, the average interaction depth (*d*) could be simplified to be at the midpoint of the materials, i.e. $d = t/2$. Although the actual interaction depth could vary throughout the materials, this simplification is sufficient for relative efficiency calculation. Hence, θ could be expressed as

$$\theta = \arccos\left(\frac{t}{2R}\right) \tag{2.7}$$

and consequently,

$$\kappa_\theta = \frac{arccos(^{t}/_{2R})}{\pi} \tag{2.8}$$

2.2.2. GEM's efficiency (κeff). GEM has different detection efficiencies for different types of ionizing particles. In the case of alpha particles, the efficiency is almost 100% [11], while only 4% in the case of gamma [12]. These reported efficiencies were measured using standard gas mixtures of Ar/CO_2 (70:30), while the gas flow rates did not significantly affect detection efficiency of GEM as shown in [8]. Thus, the overall corrections (κ) could be expressed as

$$\kappa = \kappa_\theta \times \kappa_{eff} \tag{2.9}$$

where $\kappa_{eff} = 1.00$ for ^{6}Li, ^{nat}Li, ^{10}B, ^{nat}B, ^{149}Sm and ^{nat}Sm, while $\kappa_{eff} = 0.04$ for ^{113}Cd and ^{nat}Cd.

3. Results and discussion

Using (2.5) and data in table 1, *RR* as a function of *t* could be calculated. Results of relative *RR* using the value from ^{149}Sm at $t = 40$ μm as a reference for normalization is shown in figure 5.

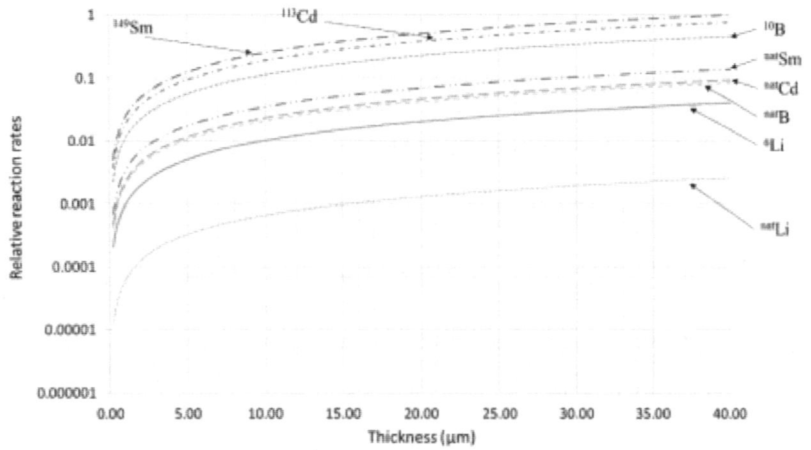

Figure 5. Relative reaction rates (*RR*) of nuclear interactions between thermal neutrons and ^{6}Li, ^{nat}Li, ^{10}B, ^{nat}B, ^{113}Cd, ^{nat}Cd, ^{149}Sm, and ^{nat}Sm using the value from ^{149}Sm at the thickness of 40 μm for normalization.

As shown in figure 5, relative *RR* for ^{149}Sm and ^{113}Cd are much higher than other nuclei. These behaviors are mainly due to their much higher σ. However, when factoring in corrections κ_θ and κ_{eff}, ε_{rel} behaves differently from results of relative *RR*. Results of ε_{rel} with *t* varying from 0 to 40 μm and from 0 to 8 μm are shown in figure 6 and figure 7 respectively.

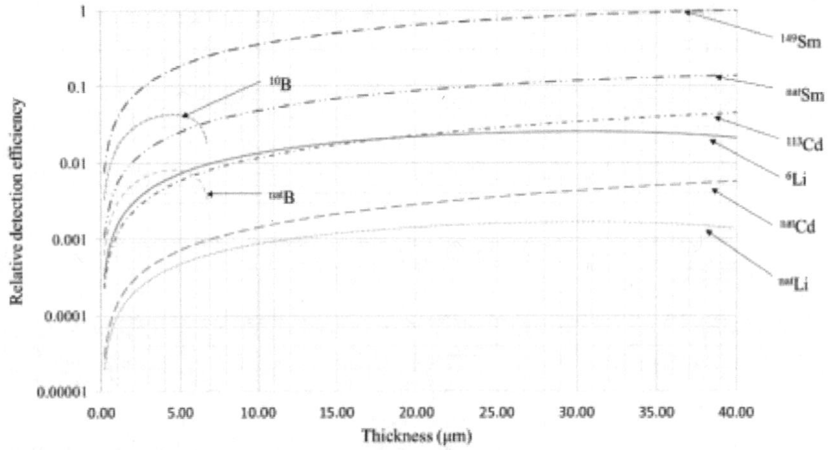

Figure 6. Relative efficiency (ε_{rel}) of GEM-based neutron detection using ^{6}Li, ^{nat}Li, ^{10}B, ^{nat}B, ^{113}Cd, ^{nat}Cd, ^{149}Sm, and ^{nat}Sm as solid neutron converters (thicknesses of coating layers from 0 to 40 μm). Results are normalized using the value from ^{149}Sm at the thickness of 40 μm as a reference.

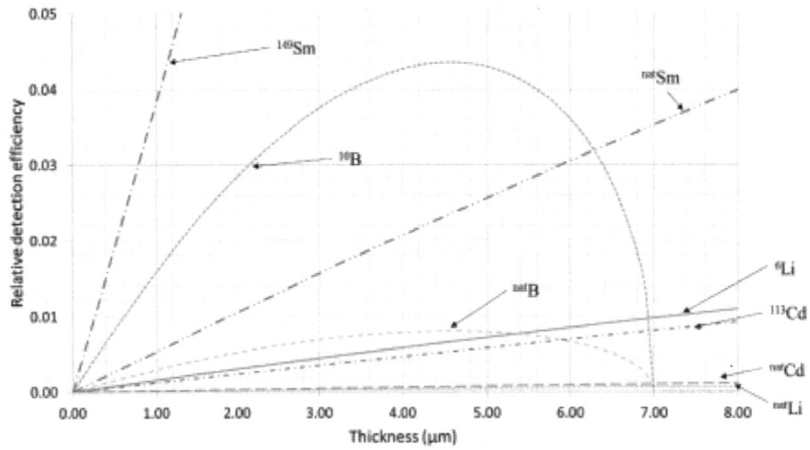

Figure 7. Relative efficiency (ε_{rel}) of GEM-based neutron detection using ^{6}Li, ^{nat}Li, ^{10}B, ^{nat}B, ^{113}Cd, ^{nat}Cd, ^{149}Sm, and ^{nat}Sm as solid neutron converters (thicknesses of coating layers from 0 to 8 μm). Results are normalized using the value from ^{149}Sm at the thickness of 40 μm as a reference.

As shown in figure 6, a solid neutron converter made of ^{149}Sm has the highest ε_{rel} due to its high σ, relatively longer *R* of its 8.96-MeV emitted alpha particles (~38.5 μm), and almost perfect alpha detection efficiency of GEM detector. On the other hand, ε_{rel} of ^{113}Cd is significantly reduced due to much lower (only 4%) gamma detection efficiency of the GEM detector. Furthermore, the importance of thickness is greatly emphasized in the case of ^{10}B and ^{nat}B, where both types of nuclei have their highest ε_{rel} at the thickness of ~4.5 μm and decrease substantially to zero at the thickness of ~7 μm.

These behaviors are due to their 1.47-MeV emitted alpha particles having low values of *R* (~3.5 μm) where thicker coating layer absorbs most of alpha particles and reduces ε_{rel}.

4. Conclusions

The GEM-based thermal neutron detector could be used to detect neutrons with different efficiencies depending on types/thicknesses of solid neutron converters. For this study, data of high-neutron-absorption-cross-section nuclei including ^{6}Li, ^{nat}Li, ^{10}B, ^{nat}B, ^{113}Cd, ^{nat}Cd, ^{149}Sm, and ^{nat}Sm along with actual GEM properties are used to calculate relative efficiencies. Results show that, ^{149}Sm has the highest detection efficiency, mostly due to its highest neutron-cross-section, longer ranges of its 8.96-MeV emitted alpha particles, and almost perfect alpha detection efficiency of GEM detector. Information from this study would be useful for all GEM-based neutron detection researches as researchers could utilize an optimum thickness for each selected solid neutron converter to maximize neutron detection efficiency.

Acknowledgement

This research is supported in part by the Graduate Program Scholarship from the Graduate School, Kasetsart University. We would also like to acknowledge supports from Thailand Research Fund (TRF), the grant number TRG5780292, the Department of Applied Radiation and Isotopes, Faculty of Science, Kasetsart University, and Kasetsart University Research and Development Institute (KURDI).

References

[1] Sauli F 1996 A new concept for electron amplification in gas detectors *Nucl. Instr. and Meth. A.* **386** 531—34
[2] Sauli F 2004 Progress with the gas electron multiplier *Nucl. Instr. and Meth. A.* **522** 93-98
[3] Gnanvo K, *et al.* 2015 Large Size GEM for Super Bigbite Spectrometer (SBS) polarimeter for Hall A 12 GeV program at Jlab *Nucl. Instr. and Meth. A.* **782** 77—86
[4] Saenboonruang K and Liyanage N 2015 Q^2 Measurement and Challenges in PREX *Kasetsart J. (Nat. Sci.)* **49** 277—87
[5] Sauli F 2016 The gas electron multiplier (GEM): Operating principles and applications *Nucl. Instr. and Meth. A.* **805** 2—24
[6] Murtas F 2013 Applications of triple GEM detectors beyond particle and nuclear physics *Proc. Int. Conf. on Micro Pattern Gaseous Detectors (Zaragoza, Spain)* (IOP Science) C01058
[7] Tsyganov E, *et al.* 2008 Gas Electron Multiplying detectors for medical applications *Nucl. Instr. and Meth. A.* **597** 257—65
[8] Saenboonruang K, *et al.* Effects of High Gas Flow Rates on the Standard 10 cm x 10 cm GEM prototype *Chiang Mai J. Sci.* **43**(4) 876—83
[9] Park S H, Kim Y K and Kim J K 2005 Neutron detection with a GEM *IEEE Trans. Nucl. Sci.* **52** 1689—92
[10] Ziegler J F Interactions of ions with matter; Available from: http://www.srim.org/. Accessed: 2016-07-14.
[11] Charpak G, Benaben P, Breuil P and Peskov V 2008 Detectors for alpha particles and X-rays operating in ambient air in pulse counting mode or/and with gas amplification *J. Instrum.* **3** P02006
[12] Lee S, Jung J H and Lee R 2014 Research of Efficiency for Gas Electron Multiplier Detector to Monitor Low Energy Gamma-Ray and Beta-Ray *Progress. Med. Phys.* **25**(2) 95—9

IOP Conf. Series: Journal of Physics: Conf. Series **860** (2017) 012019 doi:10.1088/1742-6596/860/1/012019

The current status of the Gas Electron Multiplier (GEM) research at Kasetsart University, Thailand

P Kumpiranon, K Kulasri, A Rittirong and K Saenboonruang[1]

Department of Applied Radiation and Isotopes, Faculty of Science, Kasetsart University, Chatuchak, Bangkok, 10900, Thailand

[1] Corresponding author e-mail: fscikssa@ku.ac.th

Abstract. During the past decade, Gas Electron Multiplier (GEM) detectors have been greatly developed and utilized in numbers of applications including advanced nuclear and particle researches, medical imaging, astrophysics, and neutron detection for national security. Our GEM research group at the Department of Applied Radiation and Isotopes, Faculty of Science, Kasetsart University, Thailand, realized in its excellent properties/potentials and started extensive researches on GEM detectors. To build a strong foundation on our research group, two 10 cm × 10 cm triple GEM detectors were characterized on their important properties including absolute gains and detection uniformity. Moreover, to widen applications of the GEM detector, our group had modified the GEM detector by introducing either solid or gaseous neutron converters to the detector so that the detector could effectively detect neutrons. These modifications included coating a thin film of ^{10}B and ^{nat}B to the GEM drift cathode for thermal neutron detection and flowing a gas mixture of He/CO_2 (80:20 and 70:30) and $C_4H_{10}/He/CO_2$ (7:70:23) for fast neutron detection. Results showed that the modified GEM-based neutron detector could detect both types of neutrons with different relative efficiencies and gains depending on thicknesses and types of neutron converters. This article discusses basic knowledge of the GEM detector, construction and testing procedures, results, and discussion.

1. Introduction

Radiation detection has become one of the most important tools in scientific researches and industrial applications nowadays. Great numbers of advanced nuclear and particle physics researches have utilized and relied their precision and accuracy on radiation detection instruments. Examples are vertical drift chambers for the Hall A high-resolution spectrometers at Jefferson Lab [1], combinations of imaging calorimeters, plastic scintillator, and position sensitive sensors in LHCf at the CERN Large Hadron Collider [2], and a fast multi-parameter data acquisition system for microbeam analyses [3]. Not only scientific researches but also industrial and medical applications have benefited from the fast-developed radiation detection techniques including the detection of titanium ion release from dental and orthopedic implants [4], high performance silicon-based extreme ultraviolet (EUV) radiation detector for next-generation 13.5 nm wavelength lithography [5] and neutron detectors for soil moisture and its relevance to the landmine [6].

One particular interesting radiation detector named Gas Electron Multiplier (GEM) detector, in which Sauli F [7] invented in 1997, has gained popularity amongst researchers and has been proposed to replace former position-sensitive detectors such as the large size GEM for Super Bigbite Spectrometer (SBS) polarimeter for Hall A 12 GeV program at Jlab [8] and large size GEM for future

PREX-like experiments [9]. The GEM detector is a gaseous particle detector consisting of a cascade of a GEM drift cathode, GEM foils, and a readout board. The GEM drift cathode is a thin aluminized polyimide foil placed topmost of the cascade with the most negative voltage, while the GEM foils are made of 50 μm polyimide foils coated by thin copper plates on both sides of the foils. The GEM foils are perforated with a regular matrix of holes with 70 μm diameter and a pitch of 140 μm between two adjacent holes. In a standard triple-GEM detector, three GEM foils are approximately 2 mm spaced from each other and 3 mm below the GEM drift cathode. For a readout board, several designs are possible to use depending on intended applications. However, in most applications, a standard 10 cm × 10 cm GEM readout board consists of two sets of 512 thin wires running perpendicular (X and Y) to each other. These wires are connected to data acquisition system (DAQ) for data processing. Schematic drawing of the GEM detector is shown in figure 1.

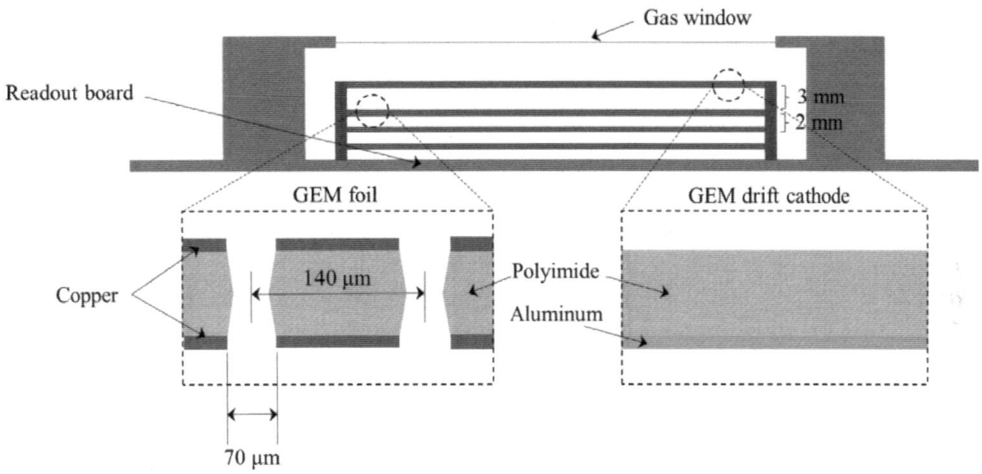

Figure 1. Schematic drawing of a GEM detector showing side views of a complete detector, a GEM foil and a GEM drift cathode.

Since GEM detector relies on ionizations of gas molecules from incoming ionizing radiation, types of gas flow play major roles in defining GEM properties. In principle, a pure noble gas such as argon (Ar) could be used to operate the detector, however, in order to increase stability, a gas mixture with a quencher is usually used. For example, a mixture of Ar/CO_2 in a ratio of 70:30 is usually used in most applications where Ar acts as the main ionized gas while CO_2 acts as a quencher.

Although the GEM detector is well known for its excellent properties when detecting ionizing particles and low-energy photons [10], the detector could be modified by adding suitable neutron converters such as high neutron absorption cross-section materials (^{3}He or ^{10}B) or high energy transferred materials (^{1}H or ^{4}He) such that the detector is able to detect neutrons. The two main methods involving neutron detection are:

- Nuclear reaction between thermal neutrons and high neutron absorption cross-section nuclei.
- Elastic scattering of fast neutrons of light nuclei.

Each method of nuclear reaction and interaction are shown in figure 2.

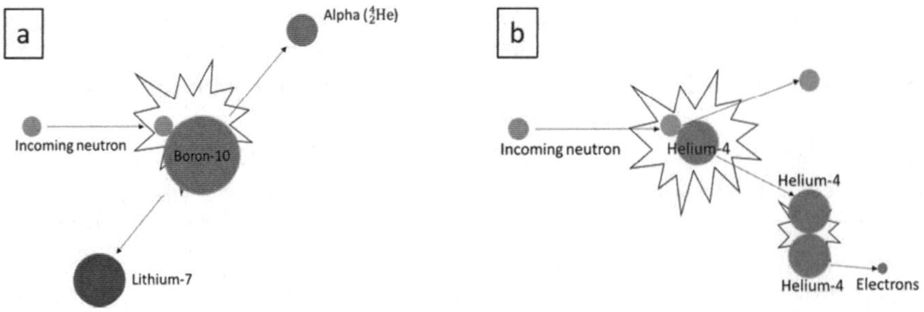

Figure 2. [a] The mechanism of a nuclear reaction between a thermal neutron and a ^{10}B nucleus, which produces a ^{7}Li nucleus and a 1.47 MeV alpha particle traveling in opposite directions and [b] the mechanism of an elastic scattering of a fast neutron off a ^{4}H nucleus where neutron energy is transferred to the recoiled ^{4}H nucleus.

In fact, neutron detection through nuclear reaction has better detection efficiencies than detection through elastic scattering due to its much larger neutron cross-section [11], however, the procedures to modify detector are considered more complicated as moderators and neutron converters must be introduced into the detection system [12] [13]. On the other hand, detection through elastic scattering is relatively simpler as the only modification required is to flow ^{1}H- or ^{4}He-contained gas mixtures through the detector, while other components remain the same.

Our GEM research group has emphasized on building a strong foundation on the first GEM research group in Thailand by investigating on important properties of the GEM detector including absolute gains and detection uniformity. Moreover, our group has also widened possible uses of the GEM detector by attempting to improve neutron detection ability by using either solid or gaseous neutron converters to detect fast neutrons and thermal neutrons.

2. Materials and methods

2.1. GEM construction and initial tests

Two complete sets of 10 cm × 10 cm triple GEM detectors were ordered from a Gas Detection Development (GDD) group at CERN. Once receiving all components, GEM foils were tested in a N_2-filled sealed box for current leakage. An acceptable GEM foil should have current leakage amplitudes less than 200 nA when supplied by a voltage of 500 V. All GEM foils used in the research must pass this requirement of current leakage tests. Then, GEM detectors were put together in a class-1000 cleanroom at the Thailand Institute of Nuclear Technology (TINT) and tested for possible gas leaks, discharged frequencies, and preliminary 5.9 keV x-ray detection. During the initial tests, a gas mixture of Ar/CO_2 (70:30) was flowed through the detector with a constant and optimized flow rate of 3.0 L/hr measured by a direct-reading flow meter (Cole-Parmer Model 3204776). A high voltage of -4100 V was supplied to the GEM detector via a voltage divider circuit shown in figure 3. Signals collected at the readout board were sent to a preamplifier (Canberra Model 2006), an amplifier (Ortec Model 590A), an oscilloscope (Hantek DSO 5202B), and a counter and timer (Canberra Model 2071A) for signal processing and analysis.

2.2. Absolute gain measurement

To measure absolute gains for different voltages supplied, signal currents from 5.9 keV x-ray detection were measured using a pico-ammeter that had a 20 fA current resolutions with the supplied voltages varied from -3900V to -4300V in 50 V increments.

From the equation

$$I = R \times N \times G \times e \quad (1)$$

where I is the detection current, R is the count rate of the 5.9 keV x-ray from ^{55}Fe photon source, N is the number of primary electrons, G is the gain of the detector, and e is the charge of an electron (e = 1.6×10^{-19} C), interested quantities could be calculated, given that other variables are accurately measured.

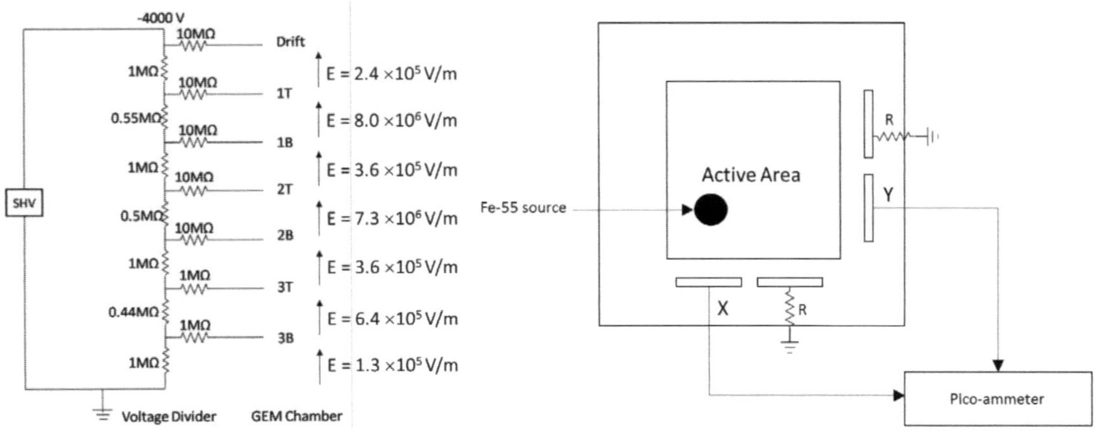

Figure 3. (Left) voltage divider circuit used during the tests and (right) diagram of setup for gain measurement.

In order to obtain G, values of I, R, and N must be carefully measured and evaluated. To measure R, the count rate of 5.9 keV x-ray at the counting plateau region was measured, while N could be estimated using the average energy dissipation per ion-pair (*w-value*) of the gas mixture Ar/CO_2 (70:30), which approximately equals 27.8 eV [14]. Assuming that only photoelectric effect occurs during the interaction between x-ray and gas molecules, N would be approximately 200 electrons.

2.3. Uniformity measurement
To test uniformity of the 10 cm × 10 cm triple GEM detector, the active area was divided into 36 equally spaced positions (6 rows and 6 columns). ^{241}Am source, which emitted 59 keV secondary gamma, was placed 0.5 cm above each position and 3 minute counting time were measured with a constant supplied voltage of -4100V and a constant gas flow rate of 3.0 L/hr. After completing all 36 positions, counts for each position were plotted using the OriginPro software to produce a contour of uniformity.

2.4. Thermal neutron detection
Since thermal neutron detection is most efficient with the addition of high neutron absorption cross-section nuclei, ^{10}B and ^{nat}B, which have neutron absorption cross-section of 3835.9 barns and 767.8 barns respectively, were selected and coated onto the GEM drift cathode. In the case of ^{10}B, the coating was performed by Richter Precision, Inc. (USA) using Tribo-Kote S-X1 technique [15]. However, the manufacturer could only coat with a thickness of 1 μm of ^{10}B. On the other hand, ^{nat}B was coated onto the drift cathode using an electron beam evaporation technique with the thicknesses of 2.5, 3.5, and 4.5 μm. Attempts to coat ^{nat}B thicker than 4.5 μm resulted in cracks along the film surface and could not be used. Setup for the thermal neutron test is shown in figure 4(a).

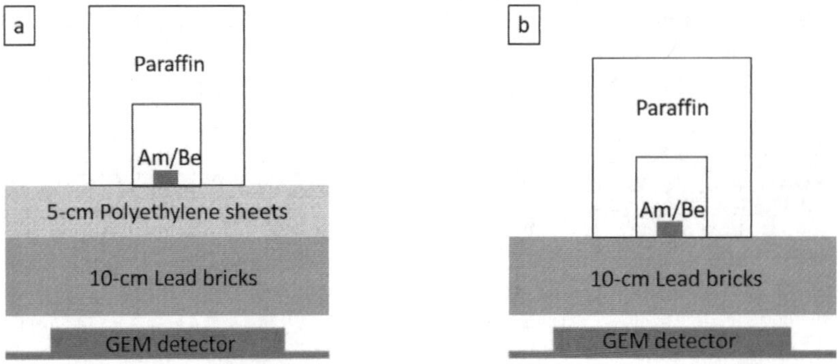

Figure 4. Setups of the GEM detector for [a] thermal neutron detection and [b] fast neutron detection.

From figure 4, the polyethylene (PE) sheets were used to thermalize fast neutrons from ^{241}Am/Be, while 10 cm lead bricks were used to shield gamma. The thickness of PE sheets was varied and optimized at 5 cm for the highest yield of thermal neutrons measured using ^{3}He neutron detector. Also, to ensure the least counts of gamma and other types of ionizing radiation contributed to neutron counts, the GEM detector itself was used to measure backgrounds and subtracted from all neutron detection counts. A gas mixture of Ar/CO_2 (70:30) was flowed through the GEM detector to measure count rates and signal amplitudes, where biased voltages were varied from -3,900V to -4,400 V in 50 V increments. To determine the average thermal neutron count rate and signal amplitude for each voltage, 10 sets of neutron measurement at each 5 minutes counting time were performed. Counts and amplitudes of signals larger than threshold level were used to find average values. After completing all required measurement, a 1 μm ^{10}B-coated drift cathode was replaced by a 2.5, 3.5, and 4.5 μm ^{nat}B-coated drift cathodes respectively and all above procedures were repeated.

2.5. Fast neutron detection

Since fast neutron detection is most efficient with the addition of light nuclei such as ^{4}He due to their large energy transfer during the scattering, gas mixtures of ^{4}He/CO_2 (80:20 and 70:30) and ^{4}He/CO_2/C_4H_{10} (70:23:7) were flowed through the GEM detector. The gas flow rates for all gas mixtures were kept constant at 3.0 L/hr throughout the measurement. To measure count rates and signal amplitudes for all three gas mixtures, the biased voltages were varied from -3900 to -4400 V in 50 V increments. For each high voltage bias setting, 10 sets of neutron measurement at each 50 minutes counting time were performed in order to find average count rate and signal amplitude. Proper background counts from cosmic radiation measured by the GEM detector without a neutron source presented were measured and subtracted from measured signal counts to ensure only neutron data was included in the calculations.

3. Results and discussion

3.1. Absolute gain measurement

In order to calculate gains (G), the count rate of 5.9 keV x-ray from ^{55}Fe source at plateau region (R) must be measured. As shown in figure 5, R was approximately 670 cps. From equation (1) along with current (I) values shown in figure 6(a), absolute gains of the GEM detector for each supplied voltage were calculated. As shown in figure 6(b), the absolute gains exponentially increased as the supplied voltages increased due to stronger electric fields inside GEM holes, resulting in higher electron multiplication.

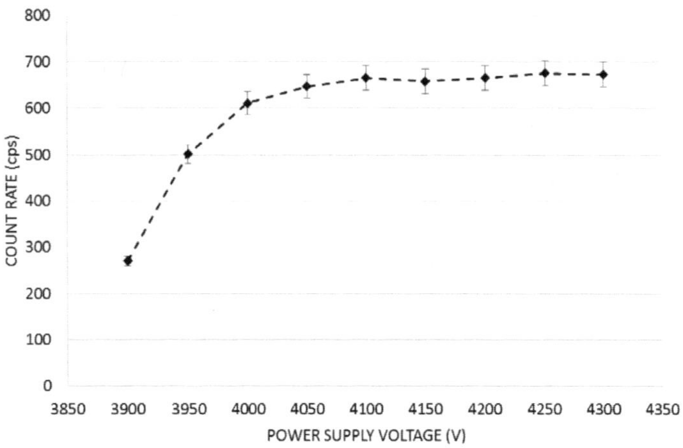

Figure 5. Count rates of 5.9 keV x-ray from ^{55}Fe showing the rate at plateau region was approximately 670 cps.

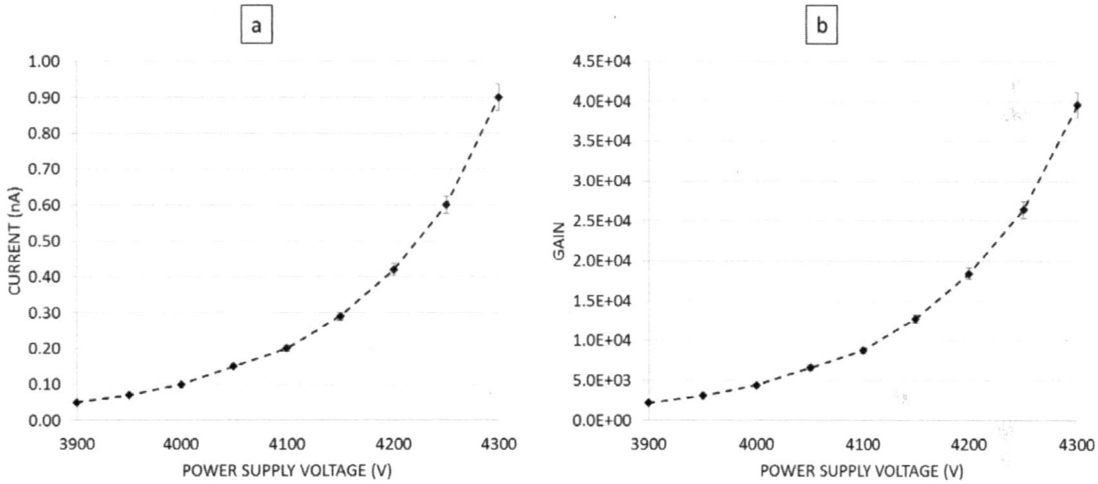

Figure 6. [a] Current measurement for each supplied voltage and [b] absolute gains for each supplied voltage.

4. Uniformity measurement

Figure 7 shows the uniformity of the GEM prototype using ^{241}Am as a gamma emitter. Areas near the center of the active area had higher relative efficiencies compared with areas near edges. This behavior was expected since incoming particles or ionized electrons occurred near edges had possibilities to drift out of the active area, lowering its overall sensitivities and signal amplitudes. However, if considering areas with at least 1 cm away from edges, the relative sensitivities were different less than 20% from each position.

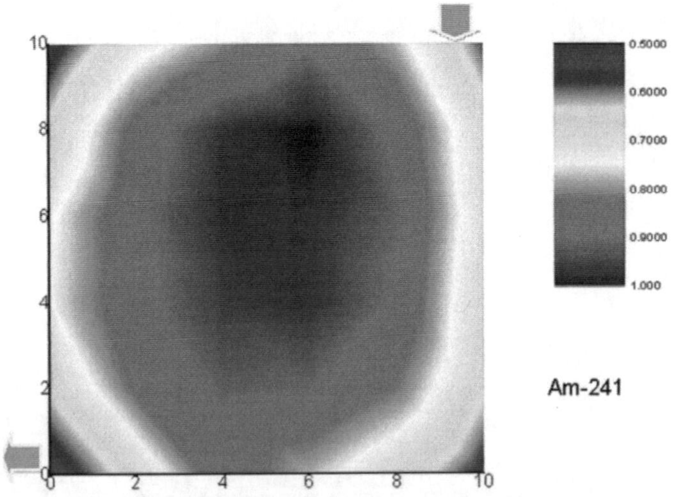

Figure 7. The detection uniformity of the GEM prototype. Higher relative efficiencies are clearly shown at the center of the detector

5. Thermal neutron detection

Figure 8(a) shows that thermal neutron count rates increased as supplied voltages increased and reached plateau when the voltage was approximately -4,200 V. Also, the setup, which the cathode was coated by 1 μm ^{10}B, had the highest count rates compared to any other thickness of ^{nat}B. This is because, despite only 1 μm thick, ^{10}B has the value of neutron absorption cross-section approximately 5 times higher than ^{nat}B, leading to more nuclear reactions between thermal neutrons and boron nuclei. However, as thicknesses of ^{nat}B increased, count rates of detection also increased as more ^{nat}B nuclei were available for nuclear reactions. Figure 8(b) shows that signal amplitudes increased as supplied voltages increased. The increase in the amplitudes was caused by stronger electric fields inside GEM holes, resulting in more ability to multiply electrons.

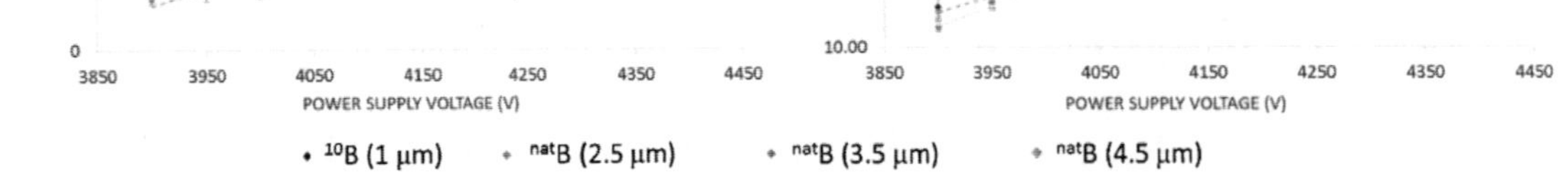

Figure 8. [a] Count rates and [b] signal amplitudes of the GEM-based thermal neutron detector using different solid neutron converters.

6. Fast neutron detection

As shown in figure 9(a), count rates increased as biased voltages increased and reached plateau region when the high voltage was approximately -4300 V, while flowing a gas mixture of $^{4}He/CO_2$ (80:20) yielded slightly higher count rates amongst all three gas types. This behavior was due to the highest concentration of light nuclei, specifically ^{4}He, in $^{4}He/CO_2$ (80:20), which the energy transfer from incoming fast neutrons to recoil nuclei was larger. However, when considering stability of signals, it was found that the gas mixture of $^4He/CO_2/C_4H_{10}$ (70:23:7) gave more stable signals as less discharged frequencies and electrical noises were observed.

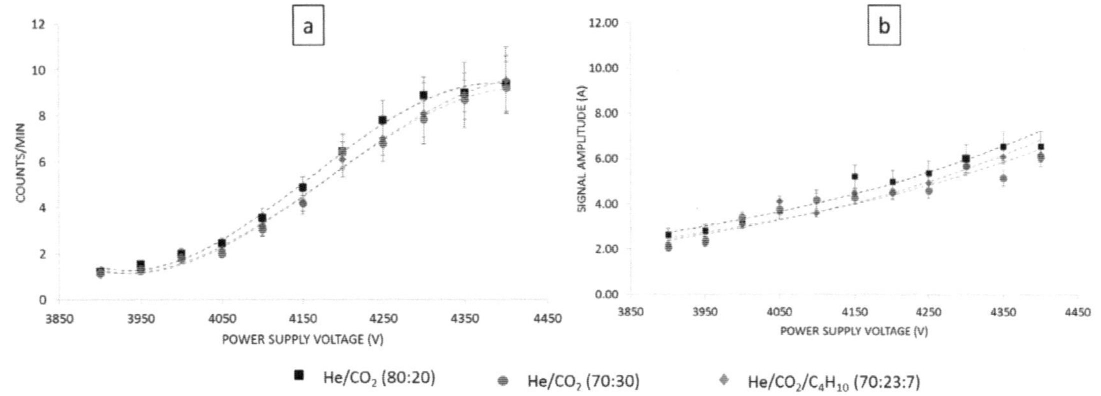

Figure 9. [a] Count rates and [b] signal amplitudes of the GEM-based fast neutron detector using different gaseous neutron converters.

7. Conclusions

Our GEM research has conducted both basic and applied researches on the GEM detector. In order to strengthen our understanding on GEM's characteristics, absolute gains and detection uniformity of the detector were investigated. Results showed that gains of the GEM detector increased exponentially with biased voltages due to stronger electric fields inside GEM holes, while relative detection sensitivity at the center of the active area had the highest detection ability compared to areas near edges. Furthermore, our research group has investigated on the possible uses of the GEM detector for neutron detection. By introducing a $^{10}B/^{nat}B$-coated GEM drift cathode, the GEM detector was able to detect thermal neutrons. On the other hand, fast neutrons could be detected by flowing gas mixtures of $^4He/CO_2$ (80:20 and 70:30) and $^4He/CO_2/C_4H_{10}$ (70:23:7). However, fast neutron detection had much lower relative efficiencies than thermal neutron detection due to smaller neutron cross-sections of ^4He compared to $^{10}B/^{nat}B$ nuclei. These results are significant and useful for our research on GEM development and also for references in future GEM-related applications.

Acknowledgements

We would like to acknowledge financial supports from Thailand Research Fund (TRF), the grant number TRG5780292, and Kasetsart University. We would also like to thank Thailand Institute of Nuclear Technology (TINT) and the National Electronics and Computer Technology Center (NECTEC) for equipment and facility supports.

References

[1] Fissum K G, *et al.* 2001 Vertical drift chambers for the hall a high-resolution spectrometers at Jefferson lab *Nucl. Instr. and Meth. A.* **474**(2) 108–31
[2] Adriani O, *et al.* 2008 The LHCf detector at the CERN Large Hadron Collider *J. Instrum.* **3** S08006
[3] Sakai T, *et al.* 1998 Development of a fast multi-parameter data acquisition system for microbeam analyses *Nucl. Instr. and Meth. B.* **136-138** 390–94
[4] Ektessabi A M, *et al.* 1994 Application of micro beam PIXE to detection of titanium ion release from dental and orthopedic implants *Int. J. PIXE.* **4** 81
[5] Shi L, *et al.* 2009 High performance silicon-based extreme ultraviolet (EUV) radiation detector for industrial application *Proc. 35th Annual Conference of IEEE* (Porto, IEEE) p 1877–82

IOP Conf. Series: Journal of Physics: Conf. Series **860** (2017) 012019 doi:10.1088/1742-6596/860/1/012019

[6] Obhodas J, *et al.* 2004 The soil moisture and its relevance to the landmine detection by neutron backscattering technique *Nucl. Instr. and Meth. B.* **213** 445–51

[7] Sauli F 1996 A new concept for electron amplification in gas detectors *Nucl. Instr. and Meth. A.* **386** 531–34

[8] Gnanvo K, *et al.* 2015 Large size GEM for Super Bigbite Spectrometer (SBS) polarimeter for Hall A 12 GeV program at Jlab *Nucl. Instr. and Meth. A.* **782** 77–86

[9] Saenboonruang K and Liyanage N 2015 Q^2 measurement and challenges in PREX *Kasetsart J. (Nat. Sci.)* **49** 277–87

[10] Saenboonruang K 2015 Recent developments in GEM-based neutron detectors *J. Phys. Conf. Ser.* **611** 012016

[11] Crane T W and Baker M P 1991 Neutron detectors: passive nondestructive assay of nuclear materials (Nuclear Regulatory Commission) p 379

[12] Prasad K R, Dighe P M and Alex M 2000 Boron-lined neutron detectors with gamma discrimination for reactor applications *Indian. J. Eng. Mater. S.* **7** 21–4

[13] Lacy J L, *et al.* 2011 Boron-coated straws as a replacement for ^{3}He-based neutron detectors *Nucl.* Instr. *and Meth. A.* **652** 359–63

[14] Zecchin M 2014 Characterization of a triple-GEM detector prototype for the CMS muon spectrometer upgrade with GEM detectors *PhD thesis* (Universitas Bruxellensis)

[15] Richter Precision Inc. [cited 2016 July 7]. Available from: http://www.richterprecision.com/glossary.html

Gamma response characterizations of optically stimulated luminescence (OSL) affects personal dosimetry

S Monthonwattana[1,3], J Esor[1], T Rungseesumran[1] and A Intang[2]

[1] Thailand Institute of Nuclear Technology (Public Organization), Nakorn Nayok, Thailand
[2] Office of Atoms for Peace, Bangkok, Thailand

[3] Corresponding author e-mail: sahakanm@tint.or.th

Abstract. Optically Stimulated Luminescence (OSL) is the current technique of personal dosimetry changed by Nuclear Technology Service Center instead of Thermoluminescence dosimetry (TLD) because OSL has more advantages, such as repeat reading and elimination of heating process. In this study, OSL was used to test the gamma response characterizations. Detailed OSL investigation on personal dosimetry was carried out in the dose range of 0.2 – 3.0 mSv. The batch homogeneity was 7.66%. R^2 value of the linear regression was 0.9997. The difference ratio of angular dependence at ± 60° was 8.7%. Fading of the reading was about 3%.

1. Introduction
Radiation protection theory is initially limited for occupational and public exposure. Therefore, the measuring exact radiation dose is very important. At present, optically stimulated luminescence (OSL) is the technique for radiation dose assessment. It has been variously used in radiation dosimetry fields, such as space dosimetry, medical dosimetry, environmental dosimetry and personal dosimetry. This technique is a related phenomenon in which the luminescence is stimulated by the absorption of optical energy, rather than thermal energy [1].

The objective of this study is to evaluate the batch homogeneity, dose linearity, angular dependence and fading of OSL for gamma response characterizations.

2. Materials and Methods

2.1. Gamma radiation
OSL dosimeters were irradiated with ^{137}Cs gamma rays by Secondary Standard Dosimetry Laboratory (SSDL) of Office of Atom for Peace (OAP), delivered doses of 0.2, 0.4, 0.6, 0.8, 1.0, 1.5, 2.0 and 3.0 mSv.

2.2. Calibration
The InLight Automatic reader (Landauer, Glenwood, IL,USA) was calibrated with calibrate set of Landauer calibration facility. It is directly traceable to the National Institute of Standards and Technology (NIST).

2.3. Radiation dosimeter
A commercial InLight Automatic reader, which uses OSL technology, measures radiation exposure with aluminum oxide doped with carbon (Al_2O_3:C). The read out process uses the green light of 532 nm wavelength produced by light emitting diode (LED) arrays to stimulate the detectors, and the blue

light of 420-440 nm wavelength emitted by the OSL material is detected and measured by the photomultiplier tube (PMT) of the OSL reader, as shown in figure 1. The amount of light released during optical stimulation is directly proportional to the absorbed radiation dose and the intensity of stimulation light. A dose calculation algorithm is applied for determination of personal equivalent dose.

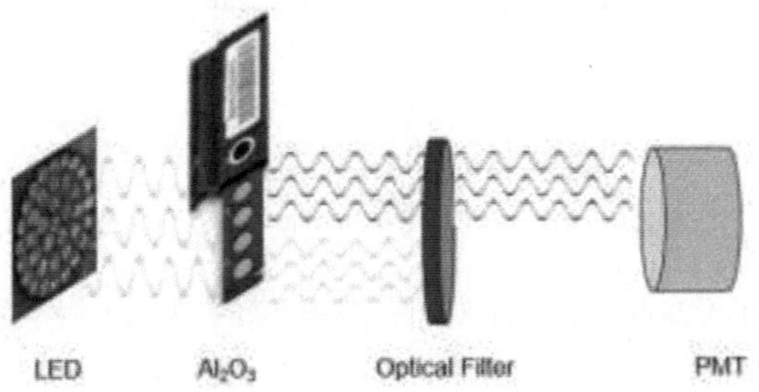

Figure 1. The read out process (Landauer, Inc.).

2.4. Batch homogeneity

Twenty-four dosimeters were used for testing the batch homogeneity. All dosimeters were irradiated with 2.0 mSv. The batch homogeneity of evaluated personal equivalent dose was calculated from

$$\text{Batch homogeneity (\%)} = ((E_{max}-E_{min})/(E_{max}))\times 100 \quad (1)$$

where E_{max} and E_{min} are the maximum and minimum evaluated values of the batch, respectively. The evaluated value for any one dosimeter in a batch shall not differ from the evaluated value of any other dosimeter in the batch by more than 30% [2].

2.5. Dose linearity

Ninety-six dosimeters were used for testing the dose linearity and divided into eight groups of twelve. The personal equivalent dose was calculated from the average dose with the evaluation of Hp (10), at a depth equivalent to 1,000 mg/cm^2 or about 1.0 cm deep below the ICRU phantom surface.

2.6. Angular dependence

Twenty–one dosimeters were used for testing the angular dependence and divided into seven groups of three. The dosimeters were mounted on a 30 × 30 × 15 cm PMMA ISO water phantom and were exposed in vertical centerline of the phantom face to the proper angle for irradiation from seven different angles (-60°, -45°, -30°, 0°, 30°, 45° and 60°). The personal equivalent dose was calculated from the average dose. The relative response value was obtained by normalized the response for each angle to the response at an incidence angle of 0°.

2.7. Fading of the reading

Thirty-six dosimeters were used for testing the fading of the reading and divided into nine groups of four. The effect of storage time over the period of 1-365 days after irradiation were investigated. The dosimeters were irradiated with 3.0 mSv and were kept in the dark. The personal equivalent dose was calculated from the average dose.

3. Results and Discussion

3.1. Batch homogeneity

The result of batch homogeneity test is shown in figure 2. The batch homogeneity was 7.66%, complying with the acceptance limit and the error was less than ± 5%.

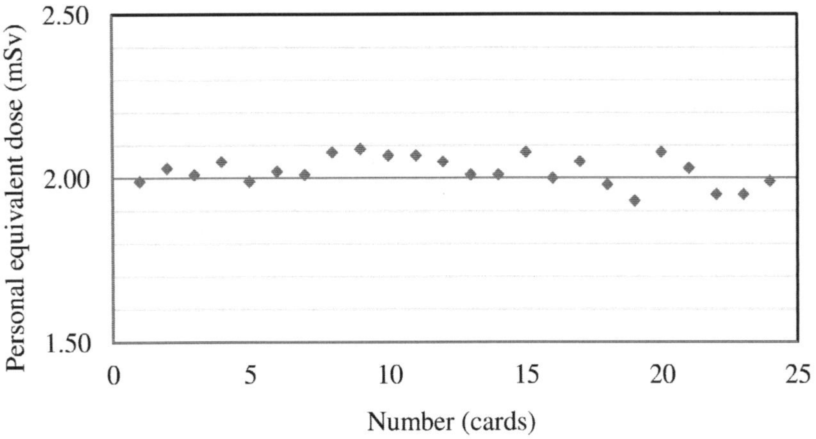

Figure 2. The result of batch homogeneity test.

3.2. Dose linearity

The dose linearity of OSL as a function of radiation dose was given in figure 3. It is shown that the reported dose increases linearly with the delivered dose ranging from 0.2 – 3 mSv. The dosimeter shows a very high linearity indicated by the R^2 value of linear regression is 0.9997 ($P < 0.0001$). The function of linear fitting is $Y = 0.9994X + 0.0135$ where X and Y stand for personal equivalent dose (mSv) from SSDL and the reader, respectively.

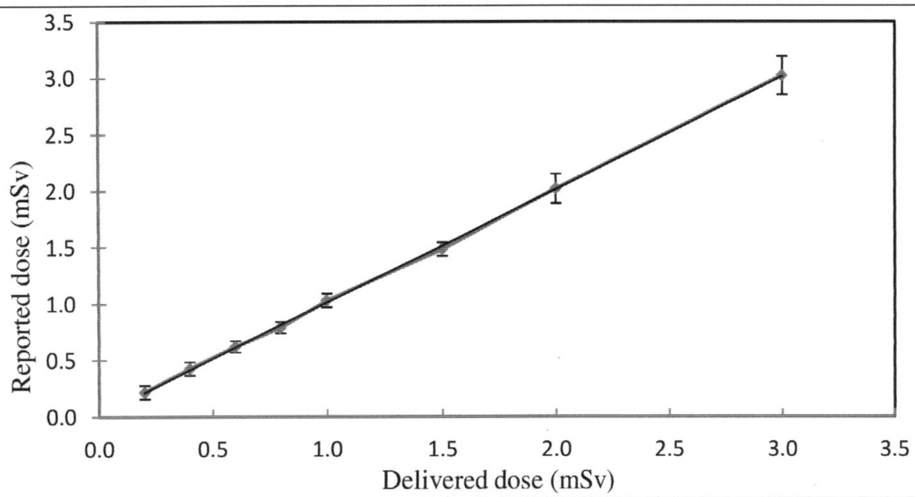

Figure 3. The reported dose as a function of delivered dose. The R^2 value of linear regression is 0.9997 ($P < 0.0001$).

IOP Conf. Series: Journal of Physics: Conf. Series **860** (2017) 012020 doi:10.1088/1742-6596/860/1/012020

3.3. Angular dependence

The angular dependence of dosimeters was measured by irradiated with 2.0 mSv from seven different angles. The results of radiation incident angles of -60°, -45°, -30°, 0°, 30°, 45° and 60° were 0.92, 0.97, 0.99, 1.00, 0.99, 0.96 and 0.91, respectively. Figure 4 shows the data of relative response of OSL. The difference ratio of angular dependence at ± 60° was 8.7%.

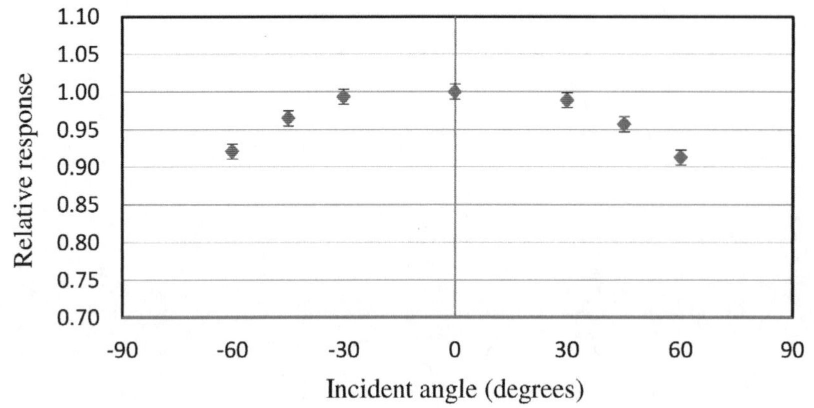

Figure 4. The relative response as a function of incident angle.

3.4. Fading of the reading

For these measurements the dosimeters were irradiated and readings were begun within nine different periods (1, 7, 30, 60, 90, 120, 150, 180 and 365 days) after irradiation. The result of fading test is shown in figure 5. Fading of the reading was about 3%.

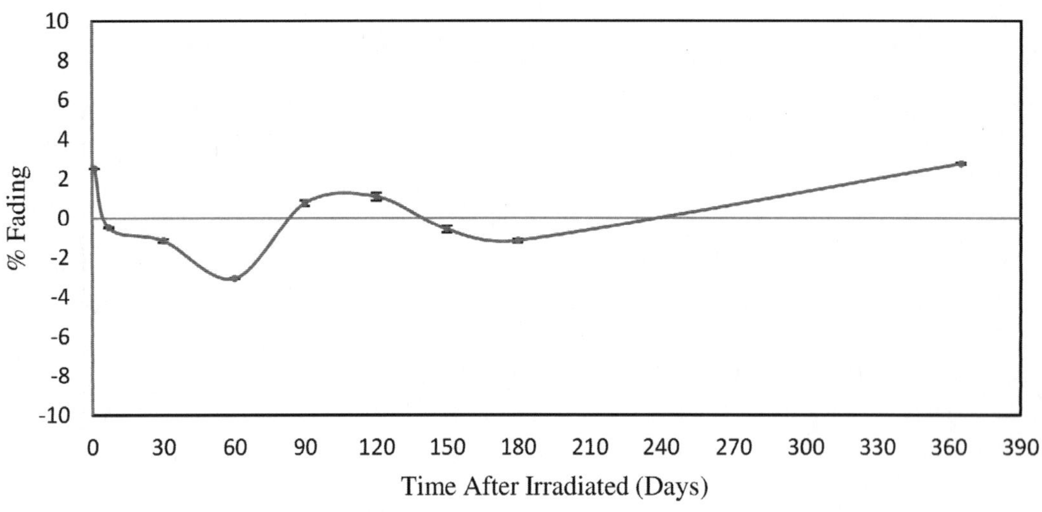

Figure 5. Fading of the reading.

IOP Conf. Series: Journal of Physics: Conf. Series **860** (2017) 012020 doi:10.1088/1742-6596/860/1/012020

4. Conclusions

In this study show that OSL dosimeters can response to the radiation doses with high precision and accuracy. It may be useful for occupational and public exposure. Nevertheless, personal dosimeters should be aligned in the vertical direction because they measure the exact dose value at this alignment [3]. Fading of the reading was about 3% over different time periods of up to 365 days after irradiation. These results are in accordance with the previous results that fading is generally quoted 3-5% per year [1].

Acknowledgement

The authors wish to thank Office of Atoms for Peace for the ^{137}Cs standard source.

References

[1] Yukihara E G and McKeever S W S 2011 *Optically Stimulated Luminescence Fundamentals and Applications* (Physics Department, Oklahoma State University)

[2] Perks C A and Passmore C N 2006 *The InLight dosimeter – comparison with IEC 61066* (Landauer Europe and Landauer Inc.)

[3] Lim C S, Lee S B and Jin G H 2011 Performance of optically stimulated luminescence Al_2O_3 dosimeter for low doses of diagnostic energy X-rays *Appl. Radiat. Isotop.* **69** 1486–9

IOP Conf. Series: Journal of Physics: Conf. Series **860** (2017) 012021 doi:10.1088/1742-6596/860/1/012021

Investigation of the response characteristics of OSL albedo neutron dosimeters in a ^{241}AmBe reference neutron field

T Liamsuwan[1,3], S Wonglee[1], J Channuie[1], J Esoa[2] and S Monthonwattana[2]

[1] Nuclear Research and Development Division, Thailand Institute of Nuclear Technology, Ongkharak, Nakorn Nayok, Thailand
[2] Nuclear Technology Service Center, Thailand Institute of Nuclear Technology, Ongkharak, Nakorn Nayok, Thailand

[3] Corresponding author e-mail: thiansinl@tint.or.th

Abstract. The objective of this work was to systematically investigate the response characteristics of optically stimulated luminescence Albedo neutron (OSLN) dosimeters to ensure reliable personal dosimetry service provided by Thailand Institute of Nuclear Technology (TINT). Several batches of InLight® OSLN dosimeters were irradiated in a reference neutron field generated by the in-house ^{241}AmBe neutron irradiator. The OSL signals were typically measured 24 hours after irradiation using the InLight® Auto 200 Reader. Based on known values of delivered neutron dose equivalent, the reading correction factor to be used by the reader was evaluated. Subsequently, batch homogeneity, dose linearity, lower limit of detection and fading of the OSLN dosimeters were examined. Batch homogeneity was evaluated to be 0.12 ± 0.05. The neutron dose response exhibited a linear relationship (R^2=0.9974) within the detectable neutron dose equivalent range under test (0.4-3 mSv). For this neutron field, the lower limit of detection was between 0.2 and 0.4 mSv. Over different post-irradiation storage times of up to 180 days, the readings fluctuated within ±5%. Personal dosimetry based on the investigated OSLN dosimeter is considered to be reliable under similar neutron exposure conditions, i.e. similar neutron energy spectra and dose equivalent values.

1. Introduction

Reliable evaluation of personal dose equivalent is important for occupational radiation protection and radiation monitoring. Thailand Institute of Nuclear Technology (TINT) provides a personal dosimetry service based on optically stimulated luminescence (OSL) dosimeters for gamma rays, beta rays, x-rays and neutrons. The OSL dosimeters are regularly tested in reference photon fields to ensure reliable photon dose equivalent evaluation, while type testing of the dosimeters in reference neutron fields could not be conveniently done in the past due to a lack of neutron reference fields in Thailand. Until recently, TINT has set up a radioactive neutron irradiator for generating neutron reference fields for detector testing and research purposes [1], making it possible to test the performance and reliability of the OSL dosimeters for neutron dose equivalent evaluation.

The objective of this work was to investigate the response characteristics of InLight® optically stimulated luminescence albedo neutron (OSLN) dosimeters (Nagase Landauer, Ltd) in a reference neutron field and assess their reliability for neutron personal dosimetry service. For this investigation, the accuracy of evaluated neutron dose equivalent, batch homogeneity, dose linearity, lower limit of detection and fading of the OSLN dosimeters were examined.

2. Materials and Methods

2.1. OSLN dosimeters

OSL is a process in which a material pre-exposed to ionizing radiation emits light after appropriate optical stimulation. The light signal is proportional to the absorbed dose. Therefore, OSL dosimetry can be used for assessment of occupational dose equivalent. The personal dosimetry service at TINT is based on InLight® OSL/OSLN dosimeters (Nagase Landauer, Ltd). Each dosimeter has four sensors made of aluminum oxide doped with carbon (Al_2O_3:C). The sensors are placed in a plastic holder, covered with an open window, plastic and metal filters for measurement of Hp(0.07), Hp(3) and Hp(10) (shallow, lens-of-eye and deep dose, respectively), as shown in Figure 1. E2 is the only neutron-sensitive element of the OSLN dosimeter.

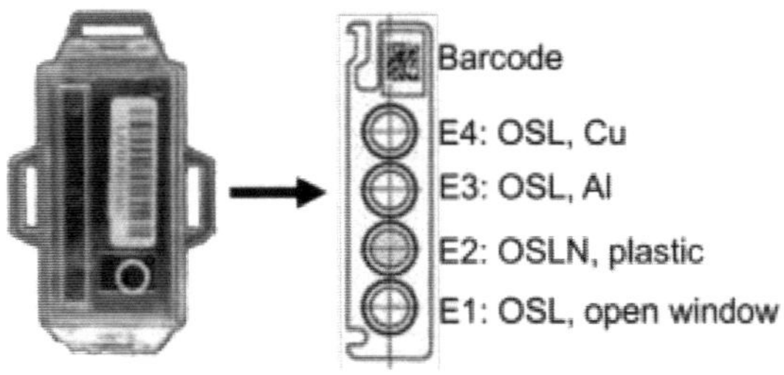

Figure 1. InLight® OSLN dosimeter (Nagase Landauer, Ltd).

When the Al_2O_3:C crystal is exposed to ionizing radiation, free electrons and holes are produced. The free charge carriers will be eventually trapped in impurity states. The trapped electrons can be released by optical stimulation with green light of 532 nm wave length produced by a light emitting diode (LED), as illustrated in left panel of Figure 2. The emitted blue light signal (420-440 nm wave length) is measured by the photomultiplier tube (PMT) of the InLight® Auto 200 Reader (Landauer, Inc.). Subsequently, a dose calculation algorithm is applied for determination of dose equivalent received by the wearer. Since neutrons are indirectly ionizing radiations and the OSL material is not directly sensitive to neutrons, one of the Al_2O_3:C sensors of the OSLN dosimeter (E2) is coated with $^6LiCo_3$. Neutron absorption by ^6Li produces a tritium (^3H) and an alpha particle (^4He), which can cause ionization and generate free charge carriers in the Al2O3:C crystal, as shown in right panel of Figure 2.

Figure 2. The OSL process (left) and the principle of albedo-type OSLN dosimeters.

In the presence of detected neutron signal, the response on the open-window element is used by the reader for estimating the photon contribution on the neutron-sensitive element. The neutron dose is calculated by applying a correction factor (chosen by the operator) to the net neutron signal. Shallow and deep photon doses of the mixed radiation field are calculated from the response of the Cu element, while the signal ratio of the Al element and the Cu element is used as an independent variable for selecting the function for calculating the lens-of-eye photon dose. Since neutron is strongly penetrating radiation, neutron shallow, lens-of-eye and deep doses are set equal.

2.2. Neutron irradiation

Several batches of the OSLN dosimeters were randomly selected for neutron irradiation. For each batch, 4 OSLN dosimeters were placed on a water phantom (30 cm width × 30 cm height × 15 cm thick) facing the center of the neutron beam, as shown in Figure 3. The reference neutron field was produced by TINT's 50 Ci ^{241}AmBe neutron irradiator [1]. Neutron irradiation was done at 100 cm distance from the source with the rectangular beam of 26.7 cm x 26.7 cm and the beam homogeneity of 10.6% (standard deviation of the average neutron dose equivalent rate of the field). $H_p(10)$ for neutrons at the irradiation position was determined to be 2,102 μSv/h, based on Monte Carlo simulation and measurement with a moderated neutron spectrometer [1], with the agreement of 13% between both methods.

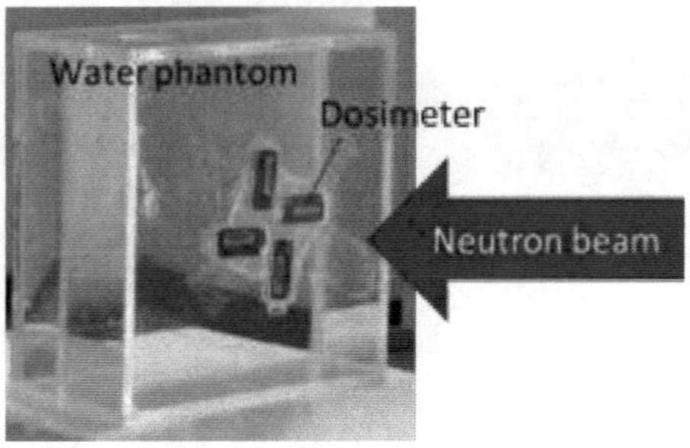

Figure 3. Irradiation set-up

2.3. The response characteristics of the OSLN dosimeters

OSL signals were read using the InLight® Auto 200 Reader before and after irradiation, typically 24 hours after irradiation. The corrected reading corresponded to the difference between the readings post- and pre-irradiation. The dose calculation algorithm implemented in the Neutron Mode of the reader was applied to the corrected reading for evaluation of received neutron dose equivalent. The average dose equivalent for each batch represented its evaluated value.

To test the accuracy of evaluated values, 3 batches of the OSLN dosimeters were irradiated with ~1-3 mSv neutrons. The correction factor for the Neutron Mode of the reader was calculated from the ratio of delivered to evaluated neutron dose equivalent and the average correction factor was applied throughout this study.

The reliability of neutron personal dosimetry based on the OSLN dosimeter was investigated in terms of batch homogeneity, linearity of the dose-response relationship and fading characteristic. A work instruction for type testing of OSL dosimeters [2], which is based on IEC 61066:2006 [3], was used as the reference of the acceptance tests.

Batch homogeneity was defined as:

$$\text{Batch homogeneity} = (E_{max}-E_{min})/E_{max} \tag{1}$$

where E_{max} and E_{min} are the maximum and minimum evaluated values of each batch. The batch homogeneity should not exceed 0.3 [2].

To test the linearity of the OSLN dosimeter response, 7 batches of the OSLN dosimeters were randomly selected for irradiation with ~0.2-3 mSv neutron dose equivalent. Evaluated values were plotted against delivered neutron dose equivalent. Subsequently, linear regression was applied to find the relationship between evaluated and delivered values. The R^2 of linear regression should be at least 0.995 for the dosimeters to be qualified for use [2].

IOP Conf. Series: Journal of Physics: Conf. Series **860** (2017) 012021 doi:10.1088/1742-6596/860/1/012021

Finally, to investigate the fading property of the OSLN dosimeters, 6 batches were irradiated with 1 mSv neutrons. The exposed dosimeters were kept for 1, 7, 30, 60, 90 and 180 days post irradiation, respectively. Fading at any time point after irradiation should not exceed ±15% [2].

3. Results and Discussion

Table 1 shows evaluated values compared with delivered neutron dose equivalent before implementing the new correction factor in the InLight® Auto 200 Reader. The correction factor for the Neutron Mode of the reader was calculated to be 1.82 ± 0.06, which was used for evaluating neutron dose equivalent in the other tests throughout this study. The batch homogeneity was 0.12 ± 0.05, complying with the acceptance limit of ≤0.3.

Table 1. Delivered and evaluated neutron dose equivalent before implementation of the new correction factor (CF).

Delivered neutron dose equivalent (mSv)	Evaluated values (mSv)	Correction factors
1.03	0.56 ± 0.07	1.84
2.05	1.17 ± 0.10	1.75
3.08	1.65 ± 0.09	1.86
	Average	1.82 ± 0.06

By irradiating the OSLN dosimeters with ~0.2-3 mSv neutrons, the relationship between evaluated and delivered neutron dose equivalent was obtained, as shown in Figure 4. The evaluated value for 0.2 mSv was found to be 60% lower than the delivered dose equivalent. Therefore, the result of this dose level was excluded from the linearity test and it was concluded that the lower limit of detection of the OSLN dosimeters should be in the range of 0.2-0.4 mSv. Linear regression was applied to the data of 0.4-3 mSv delivered neutron dose equivalent, yielding an excellent R^2 value of 0.9974, which complied with the acceptance limit of ≥0.995. The slope of the graph was 0.9813, i.e. close to 1, indicating good consistence between evaluated and delivered dose equivalent.

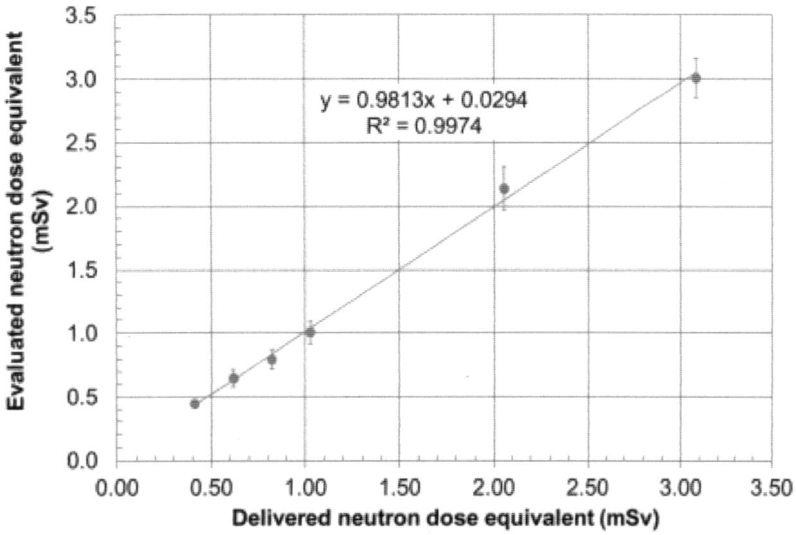

Figure 4. Dose-response of the OSLN dosimeters.

Six batches of the OSLN dosimeters were used in the fading test over the period of 1-180 days after irradiation. The result of the fading test is shown in Figure 5, indicating that the evaluated values fluctuated over time in the range of -4.75 to 4.25%, which were within the acceptance limit of ±15%.

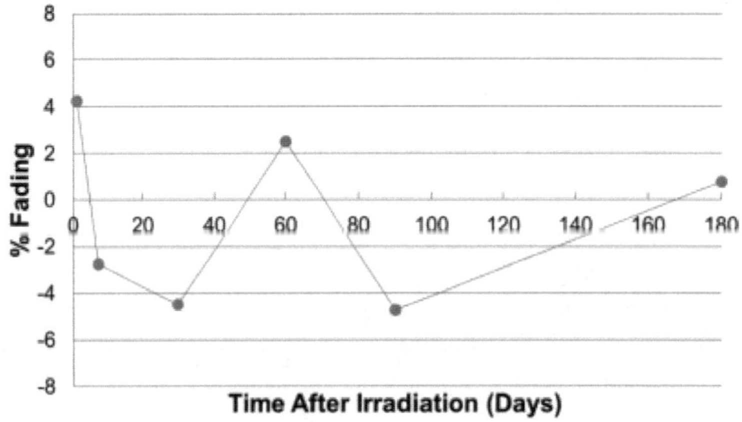

Figure 5. Fading of the OSLN dosimeters. Percentage fading corresponds to percentage discrepancy of the evaluated value from the delivered neutron dose equivalent

4. Conclusions

The InLight® OSLN dosimeters as used for the personal dosimetry service of TINT were tested with a reference neutron field generated by the in-house 50 Ci ^{241}AmBe neutron irradiator. Using the Neutron Mode of the InLight® Auto 200 Reader, a correction factor of 1.82 ± 0.06 was required for accurate evaluation of personal neutron dose equivalent. The evaluated values exhibited excellent linearity and good agreement with delivered neutron dose equivalent in the detectable neutron dose equivalent under test from ~0.4 to 3 mSv. The lower limit of detection of the OSLN dosimeters was found to be in the range of 0.2-0.4 mSv. Fading of OSL signals did not exceed ±5% over the different time periods of up to 180 days post irradiation. From this study, personal dosimetry based on the investigated OSLN dosimeters is acceptably reliable under similar neutron exposure conditions, i.e. similar neutron energy spectra and dose equivalent values. Since the response of OSLN dosimeters depends highly on neutron energy spectra, the correction factor used by the reader should be determined for specific workplace neutron fields. Nevertheless, linearity of the dose-response relationship and fading of the OSLN dosimeters are expected to be independent of neutron fields and should be similar to the results presented in this paper.

References

[1] Liamsuwan T, Channuie J and Ratanatongchai W 2015 Characterization of neutron calibration fields at the TINT's 50 Ci americium-241/beryllium neutron irradiator *J. Phys. Conf. Ser.* **611** 012004

[2] Monthonwattana S and Esoa J 2014 *Work Instruction: Type testing for OSL dosimeters (WI-NSC-29), Thailand Institute of Nuclear Technology*

[3] International Electrotechnical Commission 2006 *Thermoluminescence dosimetry systems for personal and environmental monitoring (IEC 61066: 2006)*

The development of practice manual for LSC based on job analysis in radiation measurement and analysis

W H Shin[1], T J Park

Korean Association for Radiation Application, Seoul, 04790, Korea

[1] Corresponding author e-mail: swh0410@ri.or.kr

Abstract. Radiation technology is closely related to the industrial growth and the creation of employment in Korea. The techniques as radiation or/and radioactivity measurement, and the practical skills achieving a higher level analysis are required. In this study, practice manual for liquid scintillation counter were developed by job analysis. Raw data applied in job analysis are collected by on/off line survey by 420 workers employed in KOREA. Importance-priority analysis was performed to make duties and competency unit that consists of knowledge, skills as each task. Refined data was reviewed by expert who experienced actual duties on site. Classification was conducted by focus group interview to deduct duties and competency unit. From the radiation devices in measurement and analysis, liquid scintillation counter was preferentially selected because of the high demands for training. Investigation of build-up status to liquid scintillation counter in KOREA was conducted. Then technical specification and operating procedure of 2 main devices were analyzed and integrated by practice manual. Duties and competency unit were applied to integrated materials respectively. To validate effectiveness, test curriculum was designed by the advanced course to workers who engaged in radiation measurement and analysis. The developed manual is structured to take advantage of test training. This manual will be a practical handbook that can improve the knowledge, skills of radiation workers in Korea.

1. Introduction

Job analysis is a crucial process to identify and determine in definite the particular job duties and requirements and relative importance of these practical duties for a given job. An important concept of Job analysis is that the analysis process is conducted to job, not the person. The purpose of job analysis is to establish guideline and procedure such as training, selection, compensation, and performance appraisal. Through the process of job analysis, more practical curriculum and training program can develop for workers to reflect industrial demands. Additionally, job analysis can be used in development of training contents, assessment tests to measure effectiveness of training, equipment to be used in delivering the training, and methods of training [1, 2].

Hence, variety methodologies are suggested and developed to reflect industrial demands such as national competency standards in Korea. National competency standards are currently operating in more than 130 countries worldwide. A successful country is typically British. It has a history of 60-70. Australia's vocational education and basic skills are based on the NCS. These educational methods are based on the job analysis because there are good resources for curriculum development and operations corresponding to the practical industrial demands. In particular, the techniques explained as radiation or/and radioactivity measurement, and the highly practical skills/knowledge achieving a higher level analysis are required. The tasks of radiation measurement and analysis have specialty related industrial characteristic on site. It is necessary to compensate for the gap between the operated curriculum and job on site consistently [3, 4].

But now, systematic job analysis in radiation measurement and analysis are not conducted so the difference between the theoretical training and field training was more going on. This gap has had a negative impact in industrial development and reducing job competency of related workers on site. The study, reflects the recent social interest in job analysis including standard practice manual of liquid scintillation counter, presents ad detailed practice manual that is a reference for work sites concerning troublesome that may occur in radiation fields.

2. Materials and method

2.1 Demand survey on workers

The training needs survey was conducted targeting 420 professionals engaged in the field to perform a job analysis of the radiation industry. The purpose of the survey was to determine training status and labor supply and demand realities and needs of radiation-related industries. Convenience Sampling was extracted to radiation technology-related institutions from total population. The research was conducted online survey by 420 samples. The survey was used to a structured questionnaire. Survey results were utilized for statistical analysis SPSS 21.0 program. To validate and design radiation measurement training curriculum, in-depth interviews were conducted by 15 experts. Table 1 indicated the procedure and method of demand survey step by step.

Table 1. Demand survey procedures and method

Procedure		Method
Step 1	Setting objective	Identifying radiation-related industrial labor supply and demand situation Identify your needs and radiation-related industries Education Status
Step 2	Survey design	Population: convenience sampling (radiation technology-related agencies and academia, and industry workers) Sample size: online (370 sample), offline (50 sample) Survey type: on / off-line survey Questionnaire type: structured questionnaire (closed, open, etc.) Items array: Similar question the purpose community form
Step 3	Survey conducted	Survey period: on-line('15.2~'15.3) Off-line('15.3)
Step 4	Result analysis	Analysis Method(Frequency Analysis, Ratio Analysis, reliability analysis, analysis IPA) Statistical analysis program : SPSS 21.0

2.2 Job analysis of radiation measurement and analysis

Through the Human Resource Development Council, 8 technical fields are suggested in radiation industry. Hence, validation of organization was conducted by demand survey. Importance-priority analysis was performed to make duties and competency unit that consists of knowledge, skills and attitude about total radiation measurement and analysis. Refined data was reviewed by 15 experts who experienced duties about radiation measurement and analysis. Classification procedure was conducted by focus group interview to integrate various opinions. Duties and competency unit were deducted by regular expert verification [5, 6, 7]. Radiation measurement and analysis field was selected for the rank three. Since the equipment education is suitable for radiation measurement and analysis training, we analyzed detailed job analysis of radiation measurement and analysis. Various measuring devices have been derived to develop practice manual. Then, additional demand survey is conducted. It established the equipment deployment plan.

2.3 Analysis of installed equipment on site

From the radiation devices for use in measurement and analysis, a practical manual for liquid scintillation counter was preferentially selected because of the high demands for training. Construction status of liquid scintillation counter in KOREA was investigated to develop standard manual for applicable in radiation field. After examining the equipment that is most often distributed in the domestic priority, we performed a comparison and analysis of the technical specifications of the equipment survey. Figure 1, 2 indicated the composition of liquid scintillation counter about each manufacturer.

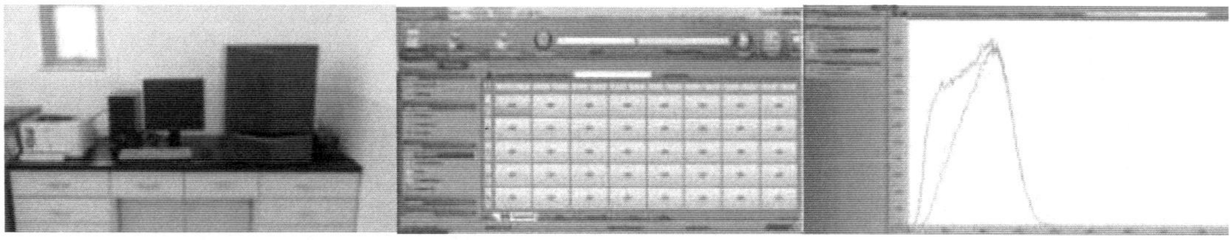

(a) Main system (b) MicroWin 300SL S/W (c) Analysis results in display

Figure 1. HIDEX 300SL LSC System

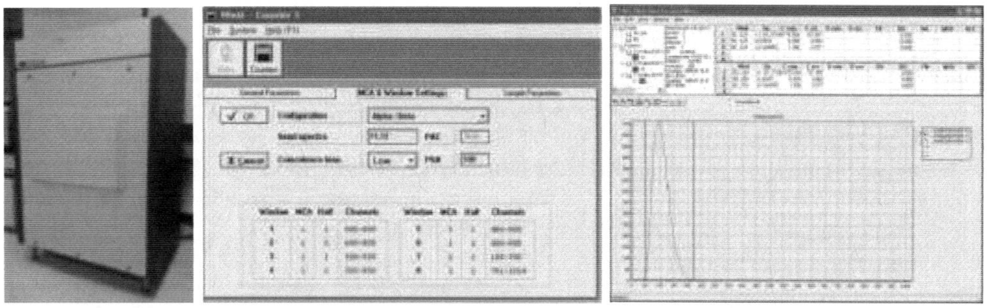

(a) Main system (b) WinQ detection S/W (c) Easy view spectrum analysis S/W

Figure 2. Quantulus 1220 LSC System

2.4 Development of practice manual

Liquid scintillation counter is utilized after dissolving the sample to a detector using a liquid scintillator dissolved fluorescent substance in a solution, and equipment for measuring the radiation emitted by the sample, it can be measured by dissolving a sample in a scintillator better the efficiency tritium (H-3), C-14, Ra-226, such as that often used in low-energy beta nuclide analysis [8]. Therefore, beta nuclide analysis training using liquid scintillation counter should provide essential incumbent to radiation, this liquid scintillation counter equipment practice should also be established. Equipment preliminary research was carried out and expert validation, evaluation and selection of suitable devices for training. Through Findings demonstrated to be the PerkinElmer and HIDEX showed that delivery of liquid scintillation counter from two manufacturers, two of PerkinElmer's Quantulus HIDEX's 300SL models and liquid scintillation counting two types of equipment most often distributed in Korea

[9, 10, 11, 12, 13]. Examine the technical specifications of the two devices, and do the comparison and analysis on the educational aspects. To design test education curriculum, we established 1 liquid scintillation counter through technical comparison. However, the development has been reflected in the equipment manuals and the operating procedures of the two companies. Detailed manual contents are consists of results that based on job analysis of radiation measurement and analysis.

3. Results and discussion

3.1 Reliability analysis of the demand survey

Utilizing SPSS 21.0 in order to ensure the reliability of the survey data about the demand investigation was carried out reliability analysis. Confidence coefficient for radiation-related manpower supply and demands are 0.786 and the confidence coefficient value about the radiation educational status and needs for appeared to .750. We conducted validation of reliability to the survey questionnaire and reliable factor for this result is shown below in Table 2.

3.2 Survey analysis and implications

It was performed to investigate the training needs of workers and their job demand survey showed 28% of response rate. Table 3 shows response results of the demand survey to worker in radiation field. As you can see from the survey results, most of the workers are hoping for specialized courses. Workers are engaged in routine task so prefer the short-term course. Education time in 1 day is preferred 6 to 8 hour for education efficiency. Also, curriculum is the best considerations to participate actively. Other areas of highest needs in radiation technology is "safe use and handling" sector ranked first, "Operation" field position 2, "Measurement Analysis" appears as the field is a priority 3. It was investigated to primarily educational technology sector that appears to be the "Measurement Analysis" field and "equipment maintenance" areas. In this results are reflected test education curriculum

3.3 Determination of definition

A job definition for the radiation measurement and analysis was defined as follows :

"Professionals to the measurement and analysis of radioactive materials through the practical verification of radiation measuring and understanding the principles and operating procedures and systems, sample preparation process and the results analysis by the radiation source is required"

The job definition and the knowledge, skills, were obtained through the expert focus group interview. Job descriptions are applied practice manual development in liquid scintillation counter. Table 4 shows that essential competency results as definition of job in Radiation measurement and analysis

Table 2. The results of demand survey (Providing training).

Survey results	Confidence coefficient (α)	Target people	Number of respondents	Response rate
Radiation workforce supply and demand status	78.6%	1,500	420	28%
Radiation Education Status and Requirements	75.0%	1,500	420	28%

Table 3. Response results of the demand survey.

Question	Response material
Do you think you need a specialized educational institution with facilities and equipment acquired for radiation technologies?	YES (90.5%) NO (9.5%)
What you want in a place where education and training?	Company (21.4%) Outside experts (78.6%) ETC (0%)
How long take Job training for the workers?	3Days (52.4%) 5Days (8.3%) 2Weels (8.3%) 1Month~2Month (131%) ETC (17.9%)
What is appropriate time for training in one day?	Under 2 Hours (15.5%) Under 2~4 Hours (13.1%) Under 4~6 Hours (29.8%) Under 6~8 Hours (40.5%) Above 8 Hours (1.2%)
What are the considerations to select education institution?	Curriculum (50.0%) Quality of instructor (7.1%) Period (6.0%) Famous (1.2%) COST (6.0%) Location (29.8%)

Table 4. Essential competency results as definition of job (Radiation measurement and analysis).

Knowledge	Radiation measurement system Understanding of devices for radiation measurement and analysis
skills	Pretreatment of the sample Radiation devices and operating systems Radionuclide factor and spectral analysis capabilities Environmental radiation analysis capabilities Analytical measuring sword / calibration capabilities

3.4 Development of liquid scintillation counter practice manual

Derived job description and definition can improve the job skills. On the basis of the outputs were designed training manuals and content. To enhanced practical education contents, standardized operating manual of the latest equipment being used in the country. Table 5 shows the contents of the standard configuration of the operating manual for practical training liquid scintillation counter. Figure 3 shows the outputs of the handbook type for developed manual. Based on the demand survey and job analysis results, liquid scintillation counter test curriculum was designed a short-term education program. Table 6 shows that organized main contents of liquid scintillation counter practice manual

IOP Conf. Series: Journal of Physics: Conf. Series **860** (2017) 012022 doi:10.1088/1742-6596/860/1/012022

Table 5. Organized main contents of liquid Scintillation counter practice manual

Chapter	Subject	Num. of page
1	Outline of LSC	1
1-1	Principle of measurement	1
1-2	Objective	2
2	Organization of LSC system	2
2-1	Hidex 300SL LSC	2
2-2	Quantulus 1220 LSC	3
3	Procedure of pretreatment as nuclide	4
3-1	H-3(Water, air sample)	4
3-2	C-14(Water, air sample)	12
4	Procedure of operation	21
4-1	Hidex 300SL LSC (Operation, Program performing, turn off)	21
4-2	Quantulus 1220 LSC (Operation, Program performing, turn off)	35
Reference	-	42

Figure 3. The Handbook of developed liquid scintillation counter practice manual

IOP Conf. Series: Journal of Physics: Conf. Series **860** (2017) 012022 doi:10.1088/1742-6596/860/1/012022

Table 6. Organized main contents of liquid Scintillation counter practice manual

Schedule	Main contents	Time(h)
1st Day	LSC principles and basic theories Commentary Understanding the radioactivity measured using LSC LSC measurement definitions and descriptions Beta nuclide analysis of new technologies	6
2nd Day	Sample preparation & LSC Tips Radioactivity measured using LSC Working Questions and Answers and Discussion	6
3rd Day	Using the LSC analysis of urine samples	3

4. Conclusion

Radiation measurement and analysis are need to essential education through a systematic analysis of the actual duties performed in the field due to the peculiarity that has its own duties. Many developed countries have changed the job training courses focused field training in the curriculum of the existing theories centered. It is necessary to develop job analysis-based training manual on radiation measurement and analysis because there are conducted the role of the primary jobs in the radiation industry. Among them, practical manual for liquid scintillation counter that can measure low-energy beta radiation effectively was selected and developed preferentially. Integrated manual for liquid scintillation counter are based on job analysis in radiation measurement and analysis. There are continuous necessity to the practical manual to encourage the activation of professional education and employment. It needs to additional onsite check continuously because developed competencies are reified to meet on-site demands. This manual will be utilized as practical handbook that can improve the knowledge, skills of radiation workers in Korea.

References

[1] L.Usha, B Arthi 2014 *Job Analysis for Changing a Workplace* (INDIAN JOURNAL OF APPLIED RESEARCH) Vol(4) Issue(9) pp 293-295

[2] Mark A W, Winston B, Jr.,Shanan G G and George M A 2013 *The Handbook of Work Analysis : The Methods, Systems, Applications and Science of Work Measurement in Organizations* (Routledge) p 1, pp 41-79?

[3] Jo J Y and Oh H J 2013 *The introduction direction and implementation strategies of the national qualifications framework (NQF)* (The HRD Review) **Vol(16)** Issue(3)

[4] Ham N H, Kim J J, Yu O G 2016 *A Study on the development and Operation of a Curriculum Based on National Competency Standards (NCS)* (The Journal of the Korea Contents Association) pp 694-711.

[5] Ahn S H 2014 *Development of Standardized Clinical Practice Manual for Trainees Dental Hygiene* (Kyuing-Hee University)

[6] Hoang G S, Ahn I J, Lee J K, Noh Y H 2008 *A Study on the Tasks of Public Librarians based on Job Analysis* (The Journal of the Korea Library Information Association) pp.407-427

[7] Kim S G, Hong J S 2002 *Articles : New Approaches to Job Analysis* (International Management Research) pp223-244

[8] Charles J Passo 1994 *Handbook of Environmental Liquid Scintillation Spectrometry* (Packard)

[9] C T Peng 1977 *Sample preparation in liquid scintillation counting* (Amersham Corporation)

[10] Instrument Manual 2002 *wallac 1220 Quantulus Ultra Low Level Liquid Scintillation Spectrometer* (PerkinElmer)

[11] Owner's handbook 2014 *Hidex 300 SL automatic liquid scintillation counter* (HIDEX)
[12] User Manual 2005 *wallac Easy View Spectrum analysis program for 1220 Quantulus^{TM} and System 1400 DSA Liquid Scintillation Counters* (PerkinElmer)
[13] User Manual 2004 *wallac WinQ Windows software for controlling Wallac 1220 Quantulus^{TM} for software* (PerkinElmer)

Acknowledgments
The research was supported by National Research Foundation of Korea Grant funded by the Ministry of Science, ICT and Future Planning of Korea (NRF-2016-003822)

Macro elemental analysis of food samples by nuclear analytical technique

W Y N Syahfitri[1], S Kurniawati, N Adventini, E Damastuti and D D Lestiani

Center For Applied Nuclear Science and Technology, National Nuclear Energy Agency (BATAN), Tamansari 71 Bandung, Indonesia 40132

[1] Corresponding author e-mail: worosyahfitri99@gmail.com

Abstract. Energy-dispersive X-ray fluorescence (EDXRF) spectrometry is a non-destructive, rapid, multi elemental, accurate, and environment friendly analysis compared with other detection methods. Thus, EDXRF spectrometry is applicable for food inspection. The macro elements calcium and potassium constitute important nutrients required by the human body for optimal physiological functions. Therefore, the determination of Ca and K content in various foods needs to be done. The aim of this work is to demonstrate the applicability of EDXRF for food analysis. The analytical performance of non-destructive EDXRF was compared with other analytical techniques; neutron activation analysis and atomic absorption spectrometry. Comparison of methods performed as cross checking results of the analysis and to overcome the limitations of the three methods. Analysis results showed that Ca found in food using EDXRF and AAS were not significantly different with p-value 0.9687, whereas p-value of K between EDXRF and NAA is 0.6575. The correlation between those results was also examined. The Pearson correlations for Ca and K were 0.9871 and 0.9558, respectively. Method validation using SRM NIST 1548a Typical Diet was also applied. The results showed good agreement between methods; therefore EDXRF method can be used as an alternative method for the determination of Ca and K in food samples.

1. Introduction

Many studies have demonstrated that the consumption of foods, such as fruits and vegetables is associated with human health improvement. Thus, individuals who eat five or more daily servings of fruits and vegetables have decreased risk of a wide variety development of cancer types, particularly those of the gastrointestinal tract. Other studies have shown that fruits and vegetables alleviate the effects of Alzheimer's disease, diabetes, obesity and metabolic, due to the bioactive components in these foods [1]. Fruits and vegetables are valuable sources of these minerals. Micronutrients in human diet are essential nutrition required by the human body in a small amount. They are needed in maintaining normal physiological function and involved in many metabolism processes [2,3]. An adequate intake of these minerals is essential to prevent deficiency-related diseases. The risk of nutritional deficiency, and of associated pathologic conditions, depends on a wide range of factors, including the magnitude of dietary intake, processing practices, the presence of substances that could reduce or increase mineral bioavailability and the physiologic and health condition of the individual [4,5].

Micronutrients consist of vitamin and mineral. Minerals play an important role in chemical, biological, biochemical, metabolic, catabolic and enzymatic reactions and involved in many metabolism processes in the living cells of plants, animals and human beings and they have great significance due to their tendency to accumulate in the vital human organs over prolonged periods of time [6,2,3]. A mineral is an element which is needed in an amount of more than 100 mg per day like calcium (Ca) and potassium (K) and they are referred as macro minerals. Calcium is very important in human nutrition due to its direct relationship with bone mass and teeth, it plays a major regulatory role in numerous biochemical and physiological processes, while adequate of both K and Na intake are related to the development of hypertension and potassium is the principle cation in the extra and intracellular fluid [4,7]. They regulate nerves and muscle function. If these essential elements removed from diet, consistent and reproducible impairment of physiological function will occur. The deficiency of these elements results from a combination of poor availability and low intake. All the nutrient elements primarily supplied through diet. For the entire elements essential for metabolism, each of them has a range of intake over which their supply is adequate for the body. However, beyond this range, deficiency and toxic effects are observed [8,9].

Multi elemental methods for food analysis are increasingly attractive and often necessary due to increasing needs for information that is more detailed and for simpler, lower cost analytical methods that are reliable. For this purpose, a wide range of diverse instrumental techniques are available. Atomic Absorption Spectrometry (AAS) is a technique for measuring quantities of chemical elements present by measuring the absorbed radiation by the chemical element of interest. This is done by reading the spectra produced when the sample is excited by radiation [10]. Atomic absorption spectrophotometric technique has specificity, sensitivity, precision, simplicity and relatively low cost per analysis [11]. However, this technique implies a prior total destruction of the matrix by mineral acids, which may lead to problems of contamination by the reactants employed or disturbances of the measured concentration by element losses due to incomplete solubilisation and evaporation; and sample matrix effects must be carefully avoided for certain elements. Moreover, the methods of matrix destruction strongly depend on the chemical composition of the sample and on the element to be determined [12].

Instrument neutron activation analysis (INAA) is a non-destructive, versatile, sensitive, multi-element analytical technique with a very low detection limits that can be used for the investigation of samples [13]. Therefore, INAA is a very useful technique for performing both qualitative and quantitative multi-element analysis of major, minor and trace elements in samples from various fields of scientific or technical interest [14]. It is also having easier samples preparation, no need to dissolve the samples, and less matrices effects. However, NAA is high cost and requires access to a nuclear reactor [15,16] and relatively spent more time to analysis some elements with medium-long lived half time. These made the NAA were not quite suitable to be used for routine analysis. Due to the easier samples preparation, multi-elements and relatively fast analysis time, EDXRF is a potential nuclear analytical technique in the determination of elemental contents in food.

The application of X-Ray Fluorescence spectrometry in the determination of the mineral profile of foods has been reviewed with regard to the tremendous possibilities of these methodologies to provide fast, environmentally friendly alternative methods for evaluating the presence of essential and toxic elements or, at least, to provide screening of the mineral composition of foods and their safety for human nutrition [17]. The use of Energy Dispersive X-Ray Fluorescence spectrometry (EDXRF) for direct and multi-elemental analysis has increased over the last few years. For example; the application of XRF for food analysis such as Mg, P, S, Cl, K, Ca, Mn, Fe, Cu, and Zn in fruits, vegetables, grain products and trace elemental characterization of some food crustacean tissue samples have been carried out [15,18].

This paper described the use of EDXRF method for food analysis and the aim of this work was to demonstrate the applicability of EDXRF for food analysis especially its macro elemental contents. The analytical performance of non-destructive EDXRF was compared with other analytical techniques; neutron activation analysis and atomic absorption spectrometry. This paper reported the concentrations of Ca and K in fruits and vegetables. In addition, the quality control procedures were applied to ensure the precision and accuracy of EDXRF, NAA and AAS methods in food analyses by analyse standard reference materials from the U.S. National Institute of Standards and Technology (NIST).

2. Material and Method

2.1. Chemical and reagent

Stock solutions (Ca and K) of certified atomic absorption references (4000 mg/L) and HNO_3 for trace element analysis, Inductively Coupled Plasma (ICP) standard, standard reference materials, SRM NIST 1548a Typical Diet, deionized water made by filtration through a Millipore-Q system.

2.2. Sample preparation

Each type of food was weighed individually (only edible portion), and then mashed using Titanium blade-blender. Collected food samples were spinach, beans, papaya, start fruit, guava, mango, mangosteen, banana, salak (Indonesian fruit) and watermelon. Food samples that have been smoothed and homogenized were weighted and placed into a small tube, then stored in the freezer until frozen. The tubes filled with frozen samples were then attached to the freeze dryer. Freeze drying process was carried out at -85°C under vacuum for 2x24 hours until the samples dried and had constant weight. Dried samples then refined into fine powder using a Teflon mortar and pestle, and then placed into polyethylene container.

2.3. Analytical procedure for AAS

A 0.5 g powder food sample was dissolved by adding 2.5 mL deionized water, 7.5 mL concentrated HNO_3 and digested using microwave digestion for 20 minutes at 160°C and power 1000 W. The same steps were also applied to SRM NIST 1548a Typical Diet. Measurement of samples was done by flame atomic absorption spectrometer GBC Avanta P. The wavelength observed was 422.70 nm for Ca. The dissolved sample then diluted with deionized water into 25 mL flask and transferred to polyethylene container. The calibration optimum working range for Ca is 0.01-4 mg/L.

2.4. Analytical procedure for NAA

As much as 40-50 mg dried samples powder were placed into 0.3 mL polyethylene vial and then sealed by heating. Samples were then irradiated along with Standard Reference Material (SRM) NIST 1568a Typical Diet and standard 21 µg of K for 15 minute at thermal neutron flux of 10^{13} $n.cm^{-2}.s^{-1}$ in rabbit system facilities of multipurpose reactor the G.A. Siwabessy, Serpong. Samples were let to decay for 1 day then measured for 1000 s using a gamma spectrometer with high resolution of HPGe detector. The spectrum was observed using Genie 2000 software [19].

2.5. Analytical procedure for EDXRF

One gram of food samples and reference materials SRM NIST 1648a Typical diet was placed in a disposable plastic cup with 26 mm diameter which was assembled with Mylar; then pressed until the surface was homogeneous. No binder material was applied. The samples were placed in sample holders and loaded into the MiniPal 4 EDXRF spectrometer with two replicates (n=2). Sample were irradiated by X-ray generated from Rh tube (maximum power 9 W, window 75µm Be, maximum high voltage 30 kV, maximum current 300 µA, cooling medium air). Filters was not used in this application because Ca and K were considered as light elements. Matrix effects were corrected by using standards that have similar/approach to the sample matrices. Corrected were also done by applying correction factor from the results of the SRM. Optimum measurement parameters of elements were obtained in the previous study, presented in Table 1 [20].

Table 1. Optimum measurement parameters for the measurement of macro elements in food by Minipal 4 spectrometer [20].

Parameter	Information
Voltage (kV)	14
Current (μA)	220
Medium or atmosphere	Air
Filter Material	None
Measurement time (s)	900

2.5.1. Lower limit detection (LLD) EDXRF. Lower Limit Detection of X-ray fluorescence method is particularly applicable to the quantitative and qualitative analysis of low concentrations of elements in a wide range of samples as well as allowing the analysis of elements at higher concentrations in limited quantities of materials. LLD of EDXRF was also measured in this study. The LLD was calculated using the following formula

$$LLD = \frac{3}{m}\left(\frac{R_b}{t_b}\right)^{1/2}$$

Where t_b is the time spent counting on the background, Rb is count of background and m is sensitivity [21].

2.6. Quality control

The standard reference material (SRM) was used as quality control assessment of data accuracy and precision. The results of SRM analysis were compared with its certificate value and evaluated its accuracy and precision by % recovery and % Coefficient of Variation (CV) calculation.

3. Result and Discussion

The quality control of analytical results of food samples using AAS, NAA and EDXRF was evaluated by analyzing SRM NIST 1548a Typical Diet. The comparison of measured values with the certified values as well as the accuracy and precision of each method that was describe as % recovery and %CV, were provided in Table 2.

Table 2. Quality control assessment using SRM NIST 1548a Typical Diet.

Reference	Methods	Elements	Result mg/kg	Certified Value (mg/kg)	% Recovery	% CV
Typical Diet (SRM 1548a)	AAS	Ca	1929±127	1967 ± 113	98	6.6
		K	6983±156	6970 ± 125	100	2.2
	NAA	Ca	1967±145	1967 ± 113	100	7.4
		K	6754 ± 155	6970 ± 125	97	2.3
	EDXRF	Ca	1820 ± 90	1967 ± 113	93	4.9
		K	6375 ± 192	6970 ± 125	91	3.0

Table 2 showed that the results obtained were in good agreement with the certified values. The accuracies of the analysis were in the range of 91-100%; while the analytical precisions, described as %CV were in the range of 2.2 to 7.4%. Both analytical accuracy and precision were can be accepted according to AOAC international guidelines, 90-108%. The measured value for K is out of certified value. However, this result was still in a good agreement with laboratory quality policy that referred to the AOAC guidelines, which requires the results of recoveries on verification methods for the concentration range of 0.1% was 90-108%. While analytical precision, described as % CV were in the range 2.2-7.4% which are fit in its acceptance level of RSD ≤ 15 % [22,23,24]. Therefore, it can be concluded that the three methods used for food samples analysis were reliable and accurate.

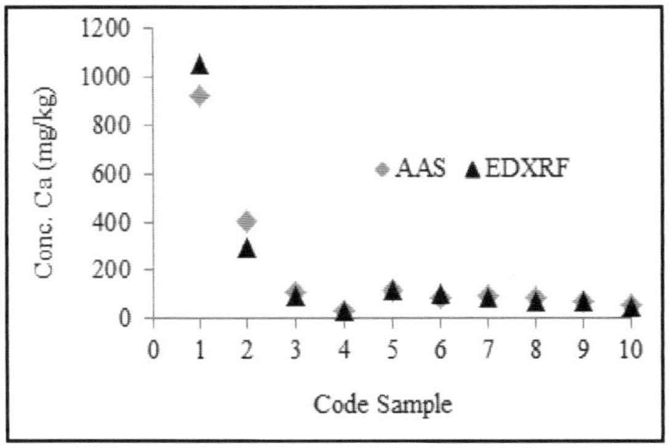

Figure 1. The results of Ca content (wet wt basis) in vegetables and fruits by AAS and EDXRF.

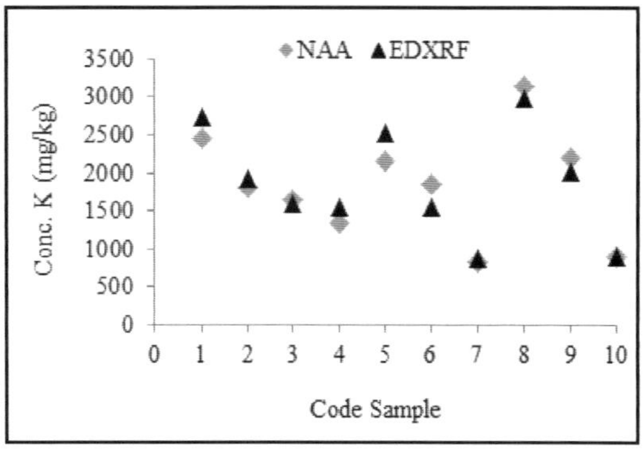

Figure 2. The results of K content (wet wt basis) in vegetables and fruits by NAA and EDXRF.

Table 3. Ca and K content in this study, Spain fruits, marketed internationally and USA.

Elements	Fruits/	Mineral Content mg/kg (wet wt basis)				
	Vegetable	This Study	LLD	Spain [1]	Markets International [1]	USA [9]
Ca	Papaya	97.8-108.0	0.11	190±90	243±50	-
	Starfruit	34.9-36.0	0.08	20±2	26±5	-
	Mango	89.7-103.0	0.19	90±25	64±1	-
	Spinach	927-1059.0	0.17	-	-	1800±420
K	Papaya	1600-1644	0.44	2840±38	1840±130	-
	Starfruit	1343-1541	0.42	1410±150	1450±240	-
	Mango	1551-1848	0.78	1380±210	1670±130	-
	Spinach	2446-2728	0.34	-	-	3600±570

Figure 1 and Figure 2 showed that the content of Ca and K respectively from vegetables (spinach and beans) and fruits samples (papaya, starfruit, guava, mango, mangosteen, banana, salak (Indonesian fruit) and watermelon). The results of EDXRF were close to the results obtained by NAA as well as AAS. The Ca concentrations ranged between 36.0-927mg/kg by AAS, while EDXRF got range between 34.9-1059 mg/kg. The K content of the NAA and EDXRF methods were in the range of 834-3152 mg/kg and 869-2993 mg/kg, respectively.

Concentrations of Ca and K in fruits in this study and results from other countries are summarized in Table 3. Generally, the Ca and K content in fruits and vegetables from this study were in the same range of other countries results. The Ca and K concentrations in the vegetable foodstuffs in our study were less than the values recorded in earlier study from USA. This can be due to the mineral and trace element contents of plants were known to be affected by the cultivar of plant, soil conditions and weather conditions during the growing, the use of fertilizers and the ripeness of the plants at harvest time [25].

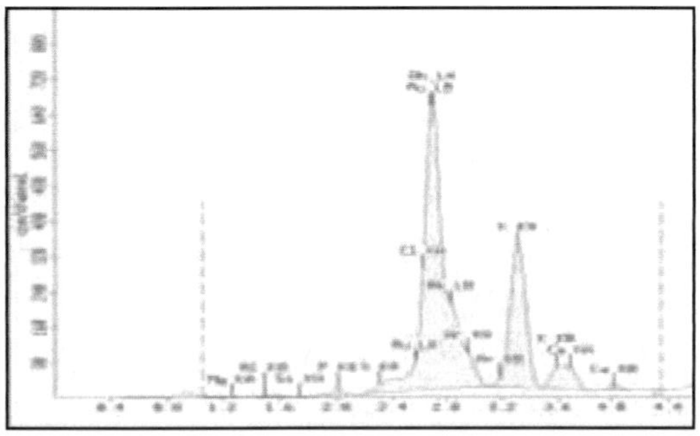

Figure 3. The spectrum of mango sample by EDXRF.

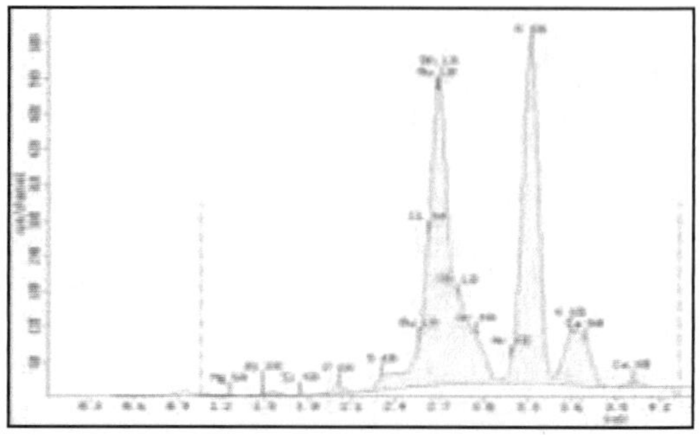

Figure 4. The spectrum of papaya sample by EDXRF.

Figure 3-4 showed EDXRF spectrum of the two samples; dried fruits mango and papaya. Mango and papaya have water content 80.51 and 88.88%, respectively. The analytical results of non-destructive EDXRF were compared with the results obtained by other analytical techniques; neutron activation analysis and atomic absorption spectrometry for analysis of Ca and K content in food samples. Comparison of methods were performed to ensure the analytical results of EDXRF and to overcome the limitations of the three methods. Statistical results of t-test and Pearson correlation of Ca and K in food using EDXRF-AAS and EDXRF-NAA were shown in Table 4 .

Table 4. Results of t-test, content elements Ca and K using EDXRF, NAA and AAS.

Elements	p-value (accept 0.05)	Pearson correlation
Ca	0.9887	0.9871
K	0.6575	0.9558

Table 4 shown that the p-value for both Ca and K are more than 0.05, which means that the results of Ca measurements for fruits and vegetables using EDXRF and AAS were not significantly different**.** Pearson correlation results for Ca and K were 0.9871 and 0.9558, which showed that there were good correlations between EDXRF and AAS results of Ca content as well as EDXRF and NAA results of K content in food samples. In general, the comparison showed a good agreement between methods; therefore, the EDXRF method can be used as an alternative method for the determination of Ca and K in food samples and can be as complementary method with NAA and AAS methods. However, EDXRF method has advantages over the AAS method for elemental analysis in food samples that EDXRF method does not involve the complex preparation stage such as in NAA and AAS methods. It has good accuracy up to the levels of trace elements. It is also simultaneous or multiple elements analysis, fast and cost-effective. Furthermore, it provides a fairly uniform detection limit across a large portion of the periodic table and is applicable to a wide range of concentrations from a 100% to few parts per million (ppm). Detection limits for Ca and K used EDXRF Minipal 4 were 0.08-0.19 and 0.34-0.78 mg/kg, respectively. Its main disadvantage is that analyses are generally restricted to elements heavier than fluorine [14].

The XRF method used in these analyses exhibited consistent accuracy, with well-defined precisions that were proportional to element concentrations. The consistency was promoted by automated analysis of samples and data, which avoided operator judgements, adjustments, and other interactions beyond sample preparation. Potential contamination and recovery problems related to chemical dissolution were avoided by direct analysis of the solid, lyophilized sample pellets. By reanalysing samples non-destructively, actual analytical precisions were measured separately from homogeneity-related variations among replicate sample aliquots. The resulting analyses suggested that most of the fruit and vegetable materials were sufficiently homogeneous to be well represented by the measured mean mineral concentrations.

4. Conclusion

The EDXRF spectrometry was used for the determination of Ca and K contents in fruits and vegetables and evaluated its accuracy, precision and its conformity with other analytical techniques; NAA and AAS. The EDXRF method was quite accurate and precise to determine Ca and K content in fruits and vegetables. Prior notification should be taken to the limit of quantification of each XRF spectrometer to be used. Good comparability with NAA and AAS showed by statistical results (p-value) confirming no significant difference between EDXRF and both NAA/AAS methods. Therefore, and with its advantages of easier sample preparation and relatively fast analysis made the EDXRF potential to be used as an alternative and complementary analytical method to determine Ca and K in food matrix samples.

References

[1] Alvarez M B, Andrade C D, Haro A, Olalla M, Seiquer I and Henare J A 2016 *Journal of* Food Composition and Analysis 48 34-40
[2] Pascal L, Andre B, Fernando E and Angela ML 2006 *Food Chemictry* 95 644-652
[3] Endah D, Widya A, Muhayatun S, Syukia K and Woro YNS 2011 *J.Atom Indonesia* **37** 62-70
[4] Montserrat B A, Cristina D A, Ana H, Manuel O, Isabel S and Jose A R H 2016 *J. Of Food Composition and Analysis* **48** 34-40
[5] Vrhovnik P, Dolenec M, Serafimovski T, Tasev G and Arrebola J P 2016 *Science of the Total Environment* **559** 204-211
[6] Vishal M and Vishwa 2012 *APCBEE Procedia* **2** 135-140
[7] Mofida WAE and Khairia MAAA 2005 *JKAU: Sci.* **17** 117-129
[8] Edet OA 2006 *Journal of Nutrition* **5** 198-202
[9] Alplasan A, Osman S, Birol E and Mehmet E 2008 *Spectroscopy* **22** 57-62
[10] Rocío G and Armando AB 2012 *Atomic Absorption Spectrometry* (Intech) p 258
[11] Sadia A, Feroza H W, Mahmood A, Muhammad H S W, Syed A T and Abdul W 2015 *Alexandria Journal of Medicine* **51** 19-23
[12] Marijan N, Peter K, Janez S, Radojko J and et al 2008 *Spectrochimica Acta Part B.* **63** 1240-47
[13] Supalak K, Wanne S, Phachirarat S and Arporn B 2016 *Energy Procedia* **89** 361-365
[14] Hayder S N, Pegah M K and Eid M E 2013 *J. Chem.Chem. Eng* **7** 803-807
[15] Kirk K N, Arthur W M, Lisa S W and Vern C R 1991 *Journal of Food Composition and Analysis.* **4** 39-51
[16] Ravisankar R, Naseeruthen A, Chandrasekaran A, Satya N B, Kanagasabapthy K V, Prasad Mvr and Satpathy K K 2014 *Journal of Radiation Research and Applied Sciences.* **7** 44-54
[17] Miguel D L G and Salvador G 2015 *Handbook of Mineral Elements in Food*, First Edition (Spain: John Wiley & Sons. Ltd) p 285
[18] Anil M, T R Rautray, V Vijayan, Rajeeb K M and S K Dey 2007 *Aquaculture* **270** 552-558
[19] Endah D, Widya D A, Muhayatun S, Syukria K and Woro YNS 2011 Atom Indonesia **37** 62-70
[20] Woro Y N S, Endah D, Syukria K and Diah D L 2015 *Proc.of National Conference on Science and Nuclear Technology* (Bandung: Batan) pp 323-329
[21] Selma MHAJ 2009 *Journal of Kirkuk University Scientific Studies* **4** 12-19
[22] Anonymous 2002 *Guidelines For Single Laboratory Validation* (AOAC International)
[23] Anonymous 2011 *Guideline on Bioanalytical Method Validation* (United Kingdom: European Medicines Agency) p 8
[24] Gustavo G and Angeles HM 2007 *Trends in Analytical Chemistry.* **26** 227-238

[25] Ekholm P, Reinivuo H, Mattila P, Pakkala H and Koponen J 2007 *Journal of Food Composition and Analysis* **20** 487-495

Acknowledgements
The authors would like to extends our thanks to Dr. Muhayatun S. for the time and guidance, Djoko Prakoso DA., Indah Kusmartini and Dyah Kumalasari for helping in analysis processes and also to other staffs for or their valuable contribution.

IOP Conf. Series: Journal of Physics: Conf. Series **860** (2017) 012024 doi:10.1088/1742-6596/860/1/012024

Effect of the Scattering Radiation in Air and Two Type of Slap Phantom between PMMA and the ISO Water Phantom for Personal Dosimeters Calibration

N Kamwang[1], T Rungseesumran[1,3], D Saengchantr[1], S Monthonwattana[1] and V Pungkun[2]

[1] Thailand Institute of Nuclear Technology (public Organization), Nakhon Nayok, 26120 Thailand
[2] Office of Atoms for Peace, Bangkok, 10900 Thailand

[3] Author to whom any correspondence should be addressed: thitir@tint.or.th

Abstract. The calibration of personal dosimeter to determine the quantities of the personal dose equivalent, $H_p(d)$, is required to be placed on a suitable phantom in order to provide a reasonable approximation to the radiation backscattering properties as equivalent as part of body. The dosimeter which is worn on the trunk usually calibrated with slap phantom which recommended in ICRU 47 with dimension of 30 cm (w) x 30 cm (h) x 15 cm (t) PMMA slab phantom to achieve uniformity in calibration procedures, on the other hand the International Organization for Standardization (ISO), ISO 4037-3, proposed the ISO water slap phantom, with PMMA walls, same dimension but different wall thickness (front wall 2.5 mm and other side wall 10 mm thick) and fill with water. However, some laboratories are still calibrating a personal dosimeter in air in term of ambient dose equivalent, $H^*(d)$. This research study the effect of the scattering radiation in two type of those slap phantoms and in air, to calibrate two type of OSL (XA and LA) and electronic personal dosimeters. The X-ray and Cs-137 radiation field with the energy range from 33 to 662 keV were used. The results of this study will be discussed.

1. Introduction

The personal dose equivalent, $H_p(d)$, be known in standardization of radiation protection to define as the operational quantity for individual monitoring of external exposure in ICRU standard tissue at a depth d, in the body of an exposed person. ICRU report 47 recommended depth, d, being 10 mm for strongly penetrating radiation and 0.07 mm for weakly penetrating radiation. ICRU 47 also recommended the 30 cm x 30 cm x 15 cm PMMA slab phantom to achieve a uniformity in calibration procedures[1], on the other hand the International Organization for Standardization (ISO), in ISO 4037-3, has proposed the ISO water slap phantom, which consists of a 30 cm x 30cm x 15 cm water phantom with PMMA walls (front wall 2.5 mm thick and other side wall 10 mm thick)[2]. But at the same time some laboratories were still calibrate it in air, in terms of ambient dose equivalent, $H^*(d)$. The accuracy and precision of the calibration of the personal dosimeters are important for the radiation protection, so the calibration laboratory should calibrate those dosimeters in the appropriate condition in order to obtain the proper results for user.

This study focused on the effect of the scattering radiation in air and two type of slap phantom between PMMA and the ISO water phantom by comparison of the calibration results from two type of

IOP Conf. Series: Journal of Physics: Conf. Series **860** (2017) 012024 doi:10.1088/1742-6596/860/1/012024

OSL dosimeters (XA and LA) and also the calibration results from electronic personal dosimeters in the energy range of the x-ray (33 to 118 keV) and Cs-137 (662 keV).

2. Materials and Methods

The study was performed by the cooperation of Dosimetry Calibration Laboratory of Thailand Institute of Nuclear Technology (TINT) for ^{137}Cs gamma radiation field and the Secondary Standard Dosimetry Laboratory of Office of Atoms for Peace (OAP) for the X-ray narrow spectrum series (N40 - N150). Two types of the OSL dosimeters (manufacturer by Landauer), XA (used in X-Ray, Gamma and Beta measurement) and LA (used in X-Ray, Gamma, Beta and Neutron measurement) type, and the Electronic personal dosimeters (EPDs) (model RAD-60s, manufacturer by RADOS) were used in the measurement. OSL holder is covered by 1 mm thick PVC, whereas the EPD device had a plastic case and aluminium sheet inside.

Before starting the calibration procedure, the OSL dosimeters were carefully selected in order to unify the accuracy and precision of the dosimeters. 85 XAs and 75 LAs type were irradiated at 1.0 mSv for $H_p(10)$ in the ^{137}Cs gamma radiation field, then select the OSLs that had the error less than ±5.0% to be used in this study. Finally, 67 XAs and 51 LAs of the OSL dosimeters were selected.

2.1. Calibration procedure 1: ISO water slab phantom

This procedure based on the use of ISO water slab phantom, which is recommended by ISO 4037-3 as a phantom for personal dosimeter. The phantom was placed at a distance of 1 meter measured from the radiation source. The Dosimeters were carefully installed on the surface of phantom to ensure that all dosimeters were irradiated with the same quality of radiation beam. In each irradiation batch, 6 XAs and 6 LAs of the OSL dosimeter were irradiated at the same time on the phantom while 6 EPDs could be irradiated at once.

The dosimeters were irradiated at 1.0 mSv for Personal Dose Equivalent ($H_p(10)$) (calculated from the reference Air Kerma Rate) using the ^{137}Cs gamma radioactive source (662 keV) and the X-ray narrow spectrum series (33, 48, 65, 83, 100 and 118 keV). The radiation qualities that used in this study are showed in Table 1.

The reading dose of the OSLs were read out by OSL automatic reader (InLight Reader, manufacturer by Landauer) at the Laboratory of Radiation Dose Measurement and Assessment of TINT, while the EPDs were read directly from their indicator after the irradiated.

Table 1. The radiation qualities used for irradiated XAs and LAs type of the OSL dosimeter and the Electronic personal dosimeters (EPDs).

Radiation Quality	Series	Effective Energy (keV)	Air Kerma Rates (Gy/h)	Conversion Coefficients	
				$H_p(10)/K_a$ (Sv/Gy)	$H^*(10)/K_a$ (Sv/Gy)
X-ray	N-40	33	7.180E-03	1.22	1.17
	N-60	48	6.245E-03	1.71	1.58
	N-80	65	6.294E-03	1.89	1.73
	N-100	83	6.027E-03	2.21	1.71
	N-120	100	6.269E-03	1.81	1.64
	N-150	118	1.167E-02	2.24	1.58
S-Cs-137	-	662	7.577E-03	1.21	1.20

2.2. Calibration procedure 2: PMMA slab phantom

The setting up procedure of this experiment was the same as stated in procedure 1, but the ISO water slab phantom was replaced by PMMA slab phantom, (recommended by ICRU 47). The radiation qualities, in terms of $H_p(10)$, were the same as indicated in Table 1.

2.3. Calibration procedure 3: "in air" condition

The setting up procedure of this experiment was the same as stated in procedure 1 and 2. The 52 cm (w) x 35 cm (h) x 11 cm (t) foam sheet was selected to attach the dosimeters and called as a "in air" condition. The dosimeters were irradiated at 1.0 mSv for Ambient Dose Equivalent ($H^*(10)$) with the radiation qualities as indicated in Table 1.

The statistical difference of the reading results of each calibration procedures were examined by a "two-sample t-stat model method". It indicates whether or not the difference between two groups' averages of the calibration procedures. The statistical analysis, "t-value" and the "t-critical" of a t-test, was used in this study. The following criteria were used to justify the different; (a) if "t-value" less than the "t-critical" means the results are not statistically significant different, and (b) if the "t-value" is greater than the "t-critical" means the reading results are statistically significant different.

3. Results

The Figure 1, 2 and 3 show the results of the irradiation of OSLs (XA and LA type) and EPDs on the ISO water slab phantom (calibration procedure 1), PMMA slab phantom (calibration procedure 2), and "in air" condition (calibration procedure 3), respectively.

The results showed that the reading dose of OSLs type XA and LA on the PMMA slap phantom produced the scattering radiation at nearly the same of ISO water slap phantom. The highest scattering from PMMA was observed at energy of 83 keV from X-ray. The lowest scattering radiation was observed in the "in air" irradiation. The main reason of the difference was the material density of each phantoms. The density of PMMA slap phantom, ISO water slap phantom and foam sheet are 1.17–1.20 g/cm^3, 1 g/cm^3 and 0.01–0.02 g/cm^3, respectively. On the other hand, the scattering radiation of EPDs on all phantoms had few differences. The effective energy from 33 to 65 keV presented that the reading dose were not difference on each phantom, whereas at the effective energy from 83 to 662 keV showed a few difference reading dose, "in air" condition presented the lowest reading dose than the ISO water slap phantom and PMMA slap phantom.

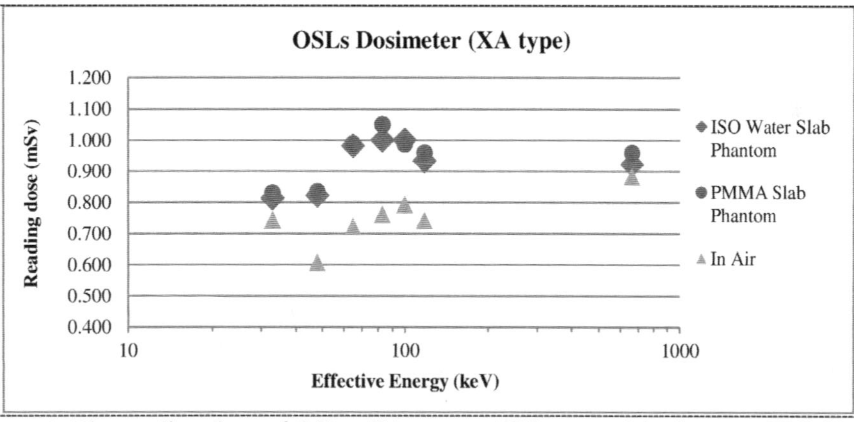

Figure 1. The reading dose of OSLs, XA type on ISO water slab phantom, PMMA slab phantom and "in air" condition.

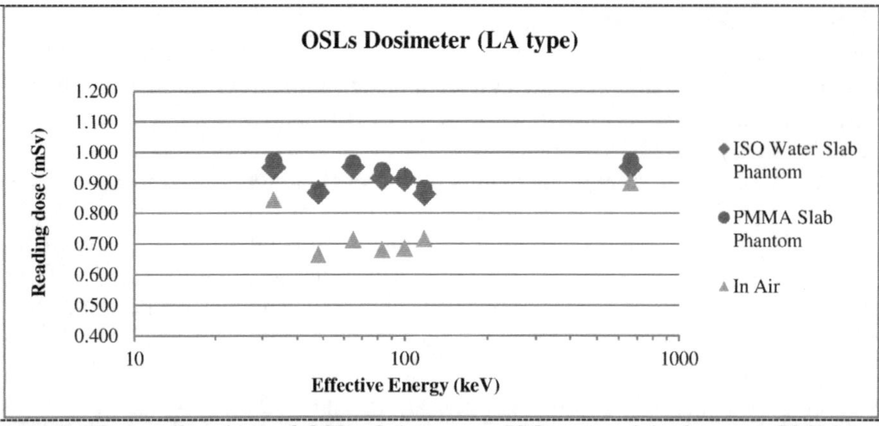

Figure 2. The reading dose of OSLs, LA type on ISO water slab phantom, PMMA slab phantom and "in air" condition.

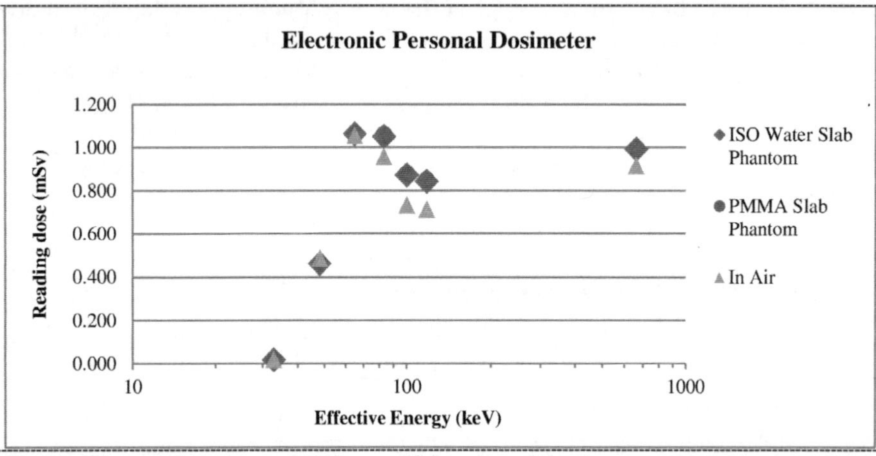

Figure 3. The reading dose of EPDs on ISO water slab phantom, PMMA slab phantom and "in air" condition.

A justification value of two-sample t-stat model with the significant level of 0.05 were used for analysis the reading doses in this experiment and are shown in Table 2, 3 and 4.

Table 2. The statistical analysis results between PMMA slab phantom & ISO water slab phantom of OSLs, XAs and LAs type dosimeter and the Electronic personal dosimeters (EPDs).

Eff. Energy	XA		LA		EPD	
keV	t-value	t-critical	t-value	T-critical	t-value	t-critical
33	2.113	2.228	1.320	2.228	0.488	2.228
48	0.529	2.447	0.223	2.228	0.173	2.228
65	0.233	2.228	1.620	2.228	0.168	2.228
83	1.533	2.228	2.179	2.228	0.417	2.228
100	0.373	2.228	0.187	2.228	0.337	2.228
118	0.793	2.228	0.627	2.228	0.360	2.228
662	1.838	2.228	0.865	2.365	0.606	2.228

Table 3. The statistical analysis results between PMMA slab phantom & "in air" condition of OSLs, XAs and LAs type of the OSL dosimeter and the Electronic personal dosimeters (EPDs).

Eff. Energy	XA		LA		EPD	
keV	t-value	t-critical	t-value	T-critical	t-value	t-critical
33	6.731	2.228	6.031	2.228	0.752	2.228
48	9.349	2.447	8.768	2.228	1.373	2.228
65	18.024	2.228	10.062	2.228	0.440	2.228
83	5.317	2.447	9.456	2.447	5.020	2.228
100	5.225	2.228	6.112	2.228	9.243	2.228
118	7.187	2.228	4.762	2.228	9.193	2.228
662	3.532	2.228	3.414	2.228	10.376	2.228

Table 4. The statistical analysis results between ISO water slab phantom & "in air" condition of OSLs, XAs and LAs type dosimeter and the Electronic personal dosimeters (EPDs).

Eff. Energy	XA		LA		EPD	
keV	t-value	t-critical	t-value	T-critical	t-value	t-critical
33	4.663	2.228	4.554	2.228	1.275	2.228
48	20.764	2.228	6.203	2.447	1.588	2.228
65	9.013	2.228	9.766	2.228	0.278	2.228
83	6.930	2.228	6.663	2.365	4.984	2.228
100	7.198	2.228	4.961	2.228	8.796	2.228
118	8.624	2.228	4.712	2.228	8.346	2.228
662	1.847	2.228	1.668	2.228	10.836	2.228

4. Conclusion and Discussion

By comparing the reading dose from OSLs, XA and LA type from the experiment between the PMMA phantom and ISO water phantom, the results were "not significantly different in statistic point of view", in contrast the reading dose "in air" condition from OSLs were "significantly different in statistic point of view" from the others (PMMA and ISO water phantoms). The reason of the difference was the calibration of OSL reader, the OSL reader which used in this study was calibrated in term of the personal dose equivalent, $H_p(10)$[3]. However, the reading results of "in air" condition was the ambient dose and it was regarded in term of ambient dose equivalent, $H^*(10)$). Thus the reading results were lower than the PMMA phantom and ISO water phantom results.

The reading dose of EPDs on the solid phantom, water phantom and "in air" condition were "not significantly different in statistic point of view" at the effective energy of the x-ray from 33 to 65 keV. However, this study showed that the reading dose at the effective energy from 83 to 662 keV were "significantly different in statistic point of view". The reason to explain this behaviour is because of the EPD case contained aluminium sheet inside, therefore the low scattering radiation energy could not reach to the detector of EPD. Moreover, the results of EPDs reading dose at the effective energy 33 and 48 keV of the x-ray showed the under estimation result. The manufacturer company was aware of this behaviour, hence it is recommended in the equipment specification that the effective energy of photon should be more than 60 keV[4].

It is a matter of facts that selection of phantom is very important for the calibration of the personal dosimeter in order to determine the radiation qualities for human. It is not recommended for the calibration the personal dosimeter in term of $H^*(10)$. In practical, the personal dosimeter calibration using PMMA phantom is more comfortable than water phantom. However, in order to harmonize and unify the proficiency testing, the accredited laboratories have adopt the water phantom for personal dosimeter calibration, as recommended by the ISO. Nevertheless, the results of this study showed that the effect of the scattering radiation between the PMMA phantom and ISO water phantom are not significantly different. The comparison of calibration procedure recommended by ICRU and ISO are not implying any important differences in former dose estimation. Thus, the dosimetry calibration laboratory can use both phantoms for calibration in term of $H_p(10)$.

References

[1] ICRU Report 47, 1992, *Measurement of Dose Equivalents from External Photon and Electron Radiations*
[2] ISO 4037-3, 1999, *Calibration of area and personal dosimeters and the measurement of their response as a function of energy and angle of incidence*
[3] User manual of Landauer, Inc., *InLight® Basic dosimeter & InLight® Auto 200 Reader*
[4] User manual of LAURUS Systems, Inc., *RAD-60 Personal alarm dosimeter*

Acknowledgments

The authors would like to express our sincerely thanks and appreciation to the Secondary Standard Dosimetry Laboratory (SSDL) of Office of Atoms for Peace (OAP) for their support on the X-ray narrow spectrum series irradiator for this study. We also would like to extend our thankful to the Laboratory of Radiation Dose Measurement and Assessment of TINT for supporting all OSLs relevant apparatus that used in this study.

IOP Conf. Series: Journal of Physics: Conf. Series **860** (2017) 012025 doi:10.1088/1742-6596/860/1/012025

Modelling of a proton spot scanning system using MCNP6

O Ardenfors[1,5], A Dasu[2,3], M Kopeć[4] and I Gudowska[1]

[1] Medical Radiation Physics, Department of Physics, Stockholm University, Stockholm, Sweden
[2] The Skandion Clinic, Uppsala, Sweden
[3] Department of Medical and Health Sciences, Linköping University, Linköping, Sweden
[4] Department of Nuclear Energy, Faculty of Energy and Fuels, University of Science and Technology, Krakow, Poland

[5] Corresponding author e-mail: oscar.ardenfors@fysik.su.se

Abstract. The aim of this work was to model the characteristics of a clinical proton spot scanning beam using Monte Carlo simulations with the code MCNP6. The proton beam was defined using parameters obtained from beam commissioning at the Skandion Clinic, Uppsala, Sweden. Simulations were evaluated against measurements for proton energies between 60 and 226 MeV with regard to range in water, lateral spot sizes in air and absorbed dose depth profiles in water. The model was also used to evaluate the experimental impact of lateral signal losses in an ionization chamber through simulations using different detector radii. Simulated and measured distal ranges agreed within 0.1 mm for R_{90} and R_{80}, and within 0.2 mm for R_{50}. The average absolute difference of all spot sizes was 0.1 mm. The average agreement of absorbed dose integrals and Bragg-peak heights was 0.9%. Lateral signal losses increased with incident proton energy with a maximum signal loss of 7% for 226 MeV protons. The good agreement between simulations and measurements supports the assumptions and parameters employed in the presented Monte Carlo model. The characteristics of the proton spot scanning beam were accurately reproduced and the model will prove useful in future studies on secondary neutrons.

1. Introduction

The interaction properties of protons enabling substantial energy depositions in the Bragg peak at well-defined depths have established proton radiation therapy as a suitable option for treating tumors with considerable sparing of the surrounding normal tissues [1-3]. In recent years there has been an increase in the use of spot scanning technique for the delivery of proton therapy using magnets to laterally scan the target by one narrow proton beam at a time and using different nominal energies to adjust the range of protons to deliver the required dose to the whole target volume.

The properties of proton beams can vary substantially between different proton facilities and should be characterized to allow for an independent validation of the performance of the treatment planning system (TPS). In order to characterize a proton beam a verification of different physical quantities such as depth and lateral dose distribution is essential. This verification can be performed by modelling the proton interactions, the beam transport through the beamline and the patient using Monte Carlo (MC) simulations as this is generally the most accurate tool for such studies [3, 4]. To accurately model a proton beam, the characteristics of the beam must be established through benchmarking between MC simulations and corresponding measurements performed at the proton facility. As the primary protons undergo scattering and nuclear interactions within the patient, scattered and secondary particles can give rise to a low dose envelope surrounding the primary proton beam causing a broadening of the beam spatial distribution (commonly denoted 'halo' contribution). The effect of this low-dose envelope can lead to an increase in total dose deposition when irradiating large tumor volumes [5, 6] and its influence on the results should hence be accounted for when comparing simulations or measurements with dose distributions calculated by the TPS.

The low production of secondary particles in the treatment nozzle of spot scanning beams generally allows for excluding full treatment head simulation when modelling the beam characteristics for in-field assessment [7, 8]. Secondary particles are instead predominantly produced within the patient [9-11] and much concern exists with respect to the neutron contribution. The transport of these particles is most accurately modeled using a benchmarked MC code in which evaluated nuclear data tables and nuclear models are employed. The latest version of the Monte Carlo N-Particle transport code (MCNP6) developed at the Los Alamos National Laboratory, USA, implements evaluated nuclear data files from thermal energies up to 150 MeV and additional libraries and physics models for higher energies to simulate neutron transport [12]. This makes it a highly suitable MC code for simulation of neutrons and other secondary particles in the energy range of proton therapy [2, 13-15]. The implementation of MCNP for proton dose calculations has been previously validated against other MC codes such as GEANT4 and FLUKA [14, 16-18].

Simulations of secondary neutrons require an accurate beam transport model with proper beam characteristics benchmarked with experimental measurements. The aim of this work was to create a MC model using the code MCNP6 (ver. MCNP6.1.1 released in 2014) based on input parameters obtained from measurements performed at beam commissioning. The model was used to benchmark simulations of absorbed doses against measurements to allow for future studies of the production of secondary neutrons doses using the advantages of MCNP6. The MC model was also used to simulate the influence of lateral spread and beam halo with regard to signal losses when performing measurements using a commercially available plane parallel ionization chamber (IC).

2. Material and Methods

2.1. Measured data

All measurements were performed on a dedicated spot scanning proton facility with a maximum cyclotron energy of 230 MeV from IBA (Ion Beam Applications, Louvain-La-Neuve, Belgium) recently installed at the Skandion Clinic, Uppsala, Sweden. Integral depth dose (IDD) curves in water for nominal proton energies between 60 and 226 MeV in steps of 5 MeV were obtained from measurements performed at commissioning using a PTW 34070 Bragg peak plane-parallel IC with an active radius of 4.08 cm (PTW-Freiburg, Freiburg, Germany). The IC was irradiated in water with a single spot beam with a fixed energy at a time. The output air ionization signal from the IC was converted to that corresponding to dose to water using stopping power ratio values calculated using equation B.13 in TRS-398 [19]. This conversion corresponds to an uncertainty of 1% in a clinical proton beam [19].

Beam profile measurements in air were carried out in energy intervals of 10 MeV from 60 to 226 MeV using a Lynx scintillation detector with a pixel resolution of 0.5 mm x 0.5 mm (IBA dosimetry, Schwarzenbruck, Germany). The detector was irradiated at five positions along the beam axis for each energy: -19 cm, -10 cm, 0 cm, +10 cm and +20 cm (positions relative to isocenter). The nozzle exit was located at approximately -53 cm relative to isocenter. Standard deviations of the beam Gaussian spatial distribution (denoted spot sizes henceforth) in the x- and y-direction ($\sigma_{x,y}$) were obtained from the measured beam profiles in air for all positions and energies.

Absorbed dose assessment was carried out for nominal energies from 60 to 226 MeV in 5 MeV intervals through measurements using two ionization chambers calibrated in terms of absorbed dose to water at the National Metrology Laboratory for ionising radiation. A cylindrical FC65-G chamber (IBA dosimetry, Schwarzenbruck, Germany) was used for nominal energies above 150 MeV and a plane-parallel Roos electron chamber (PTW-Freiburg, Freiburg, Germany) was used for lower energies. The chamber was irradiated with a homogeneous 10 cm x 10 cm field using a spot separation of 2.5 mm. The absorbed dose at reference depth, D_{cal}, was determined and equation (1) was used to calculate a standardization factor at the reference depth, D_{ref}, in units of Gy·mm^2/MU relating the number of monitor units (MU) to the dose determined at the reference depth. The reference depth in water was 3 cm for beams with nominal energies above 150 MeV. For lower energies the reference depth was at 2 cm to prevent the chamber being placed at the proximal rise of the depth dose curve.

$$D_{ref} = D_{cal} \times \frac{SAD_{factor}}{MU_{factor}} \qquad (1)$$

where MU_{factor} is the number of MU delivered per proton spot multiplied with the number of spots per mm^2. The source-axis distance factor (SAD_{factor}) corrects for the distance between reference depth, d_{ref}, and position of the MU reference chamber in the treatment nozzle, d_{MU}. The SAD_{factor} is calculated with equation (2), where SAD_x=1830 mm and SAD_y=2203 mm are the distances from isocenter to the virtual source position on the two axes.

$$SAD_{factor} = \frac{(SAD_x - d_{ref}) \times (SAD_y - d_{ref})}{(SAD_x - d_{MU}) \times (SAD_y - d_{MU})} \qquad (2)$$

The standardization factor at reference depth, D_{ref}, was then used to convert the IDD curves measured for single spot irradiations under non-reference conditions, IDD_{spot}, into units of Gy·mm^2/MU using equation (3).

$$IDD_{abs} = IDD_{spot} \times \frac{D_{ref}}{IDD_{spot}^{ref-depth}} \qquad (3)$$

where $IDD_{spot}^{ref-depth}$ is the value on the IDD curve for the spot irradiation at the reference depth (2 or 3 cm depending on energy) and IDD_{abs} is the measured IDD curve converted into units of Gy·mm^2/MU.

2.2. Modelling of integral depth dose and lateral beam size

Proton beams of energies from 60 MeV to 226 MeV in steps of 5 MeV were individually simulated and modelled. The ionization energy losses of the protons in water were evaluated in MCNP6 using the Bethe-Bloch formula with a mean excitation potential of 75.3 eV. The number of multiple scattering points was set to maximum by decreasing the energy spacing to minimum with the MCNP parameter efac set to 0.99 [20]. The nuclear interactions were simulated using the Bertini intranuclear cascade model [21]. The proton source was defined using the -41 entry on the source definition card (SDEF) corresponding to a built-in Gaussian spatial distribution. The standard deviations of the Gaussian distributions were defined using the energy specific $\sigma_{x,y}$-values obtained from the spot size measurements at -19 cm using the Lynx detector.

The beam divergence was characterized by x- and y-components which can vary depending on the different properties of the beamline such as the placement of the steering dipole magnets. For the beamline employed in this work the x-component of the spot sizes was constantly increasing with distance (divergent) for all energies whereas the y-component slightly converges before isocenter and then diverges for energies above 170 MeV. The beam divergence was simulated using an energy dependent distribution with three components: x divergence, y divergence and no divergence. For nominal energies above 170 MeV the x- and y-divergence was simulated as a cone in which the cosine angle was calculated from the difference in spot size between the -19 cm and +20 cm Lynx measurements. For lower nominal proton energies the y-divergence was set to zero. The largest divergence angle was 2.3 degrees for the x-component at 60 MeV.

The energy distribution of the primary proton beam was assumed to be normally distributed around a nominal energy with an energy-specific standard deviation, σ_E. The final value of σ_E for each initial proton energy was obtained through an iterative process in which the nominal proton energy was held constant and σ_E was varied. The IDD curves of each specific proton energy were then evaluated with regard to the spread of the Bragg peak and the height of the plateau. When further changes in σ_E did not improve the results the process was completed. The standard deviation of the energy distribution decreased with increasing energy from 0.8% (60 MeV) to 0.3% (226 MeV). As illustrated in Figure 1, a low energy tail was added to the Gaussian shaped energy distribution to account for the change in proton energy distribution due to primary protons scattered in the treatment nozzle. The magnitude of the low energy tail in the probability distribution function (PDF) was determined individually for each primary proton energy through repeated simulations and evaluations of the IDD curves at shallow depths. The value varied between 1.0×10^{-4} and 1.4×10^{-4} depending on the primary proton energy. The choice to use a constant value for the low energy tail was based on the assumption that the fluence of secondary protons originating from the treatment head exit is rather constant with regard to energy as shown by Grassberger et al. [8].

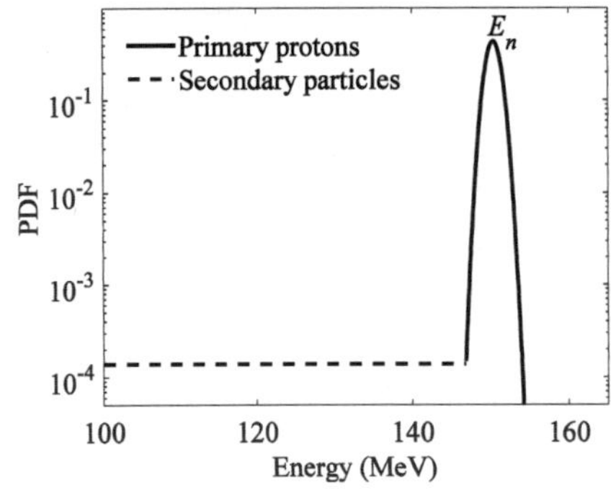

Figure 1. Probability distribution function (PDF) of proton energy distribution with nominal energy (E_n) of 150.4 MeV and energy standard deviation (σ_E) of 0.6%.

The IDD simulations were performed by scoring energy depositions in 0.1 mm thick water cylinders with radius of 4.08 cm for primary proton energies from 60 to 226 MeV in steps of 5 MeV. The simulated IDD curves were evaluated against the measured IDD curves with regard to the distal ranges R_{90}, R_{80} and R_{50}. The statistical uncertainties of all cylinder tallies were less than 0.005 for all simulated energies.

The beam profile simulations in air were simulated using an energy deposition mesh tally scoring the deposited proton energy. The mesh grid was placed at -19 cm, -10 cm, 0 cm, +10 cm and +20 cm (distances relative to isocenter) and the proton source was placed at -19 cm. Mesh sizes were defined as equal to the pixel resolution of the Lynx detector (0.5 mm x 0.5 mm). All spot sizes were calculated and compared with the corresponding spot sizes determined from the beam profiles measured with the Lynx detector.

2.3. Absorbed dose benchmarking simulations

Absorbed dose simulations were carried out using the same setup as for the reference dosimetry measurements. Thus, a water phantom was irradiated with a 10 cm x 10 cm field of 1681 proton spots with a spot separation of 2.5 mm and each proton source was defined according to the IDD simulations. The F6 tally output from MCNP6 (in MeV per unit mass per initial proton) at reference depth in the water phantom, $IDD_{MCNP}^{ref-depth}$, was calculated and used to convert the IDD curves from the single spot simulations into units of Gy·mm^2/MU with equation (4).

$$IDD_{abs}^{MCNP} = IDD_{spot}^{MCNP} \times \frac{D_{ref}}{IDD_{MCNP}^{ref-depth}} \tag{4}$$

where IDD_{abs}^{MCNP} is the simulated IDD curve in units of Gy·mm^2/MU, D_{ref} is the standardization factor calculated from measured data and IDD_{spot}^{MCNP} is the IDD curve simulated for single spots under non-reference conditions. The statistical uncertainty of $IDD_{MCNP}^{ref-depth}$ was less than 0.005 for all simulated energies.

The calculated IDD curves normalized to MU were evaluated against IDD curves derived from measurements with regard to the absolute height of the Bragg Peaks and the total dose deposition defined as the integral of the absorbed dose curves.

2.4. Lateral signal loss simulations

Signal losses due to lateral spread when performing measurements with the 4.08 cm radius PTW ionization chamber were investigated through simulations of IDD curves using an increased detector radius of 25 cm. Simulations using larger radii ensured that a radius of 25 cm was sufficient to account for all signal contribution from laterally scattered protons. The ratios between the IDD curves from the small radius (4.08 cm) simulations and the full signal simulations (25 cm radius) were calculated to evaluate signal losses at different depths for proton energies between 60 and 226 MeV in steps of 5 MeV.

3. Results

3.1. Integral depth dose in water and lateral beam size in air

Ranges derived from simulated and measured IDD agreed within 0.1 mm for R_{90} and R_{80}, and within 0.2 mm for R_{50}. Calculated differences between simulated and measured R_{90}, R_{80}, and R_{50} for all energies are presented in Figure 2a. Measured spot sizes of the x-component in isocenter ranged between 7.2 mm (60 MeV) and 2.9 mm (226 MeV), corresponding values of the y-component were 7.2 mm and 2.8 mm. The largest spot size was measured for 60 MeV at +19 cm corresponding to 8.1 mm and 7.6 mm for the x- and y-component respectively. The difference in spot size between simulations and measurements for both the x- and y-component was relatively constant for the five positions along the beam axis with the largest average difference found at +20 cm. The average difference in spot size for the five positions was 0.1 mm for both the x- and the y-component. The largest difference of the x-component was -0.6 mm for 60 MeV at +20 cm and the corresponding value for the y-component was -0.7 mm for 65 MeV at -19 cm. The simulated spot sizes were based on the measurements at -19 cm and the reason for the difference between measurements and simulations at this position was due to the beam divergence. The percentage difference in spot sizes (simulated-measured) of the x-component are presented in Figure 2b where it is seen that all spot sizes agree within 10% (similar results for the y-component).

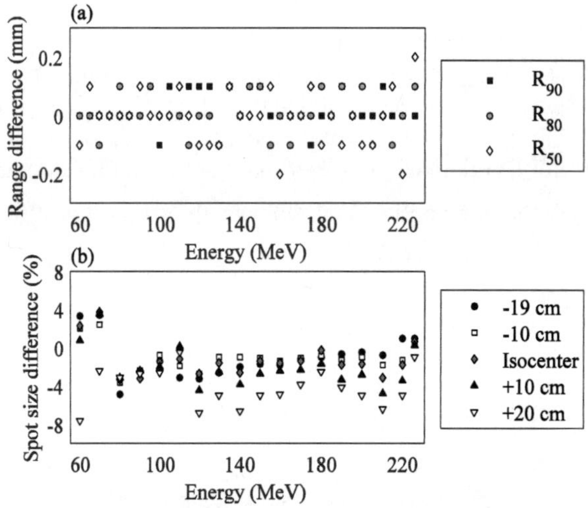

Figure 2. (a) Calculated differences in range between normalized IDD curves from simulations and measurements for proton energies in the range 60-226 MeV. (b) Percentage differences between the x-component of the spot sizes from simulations and measurements.

3.2. Absorbed dose benchmarking simulations

Absolute IDD curves in units of Gy·mm^2/MU derived from measurements and simulations using the standardization factors, D_{ref}, for energies between 60 and 226 MeV are presented in Figure 3. The average absolute difference of the IDD curve integrals derived from simulations and measurements for all energies was 0.8% and the maximum difference was 2%. The largest difference in the height of the Bragg-peak was found at 145 MeV corresponding to 2.1% (1.8 Gy·mm^2/MU) and the average absolute difference of all energies was 0.9%.

3.3. Lateral signal loss simulations

Ratios between MCNP6 simulations using a tally radius of 4.08 cm and 25 cm are presented in Figure 4 for a selection of representative energies. The signal loss of the 4.08 cm radius PTW IC increased with incident proton energy. The maximum signal loss was approximately 0.2%, 1.5%, 3.4% and 6.7% for the incident proton energies 120, 160, 190 and 226 MeV, respectively.

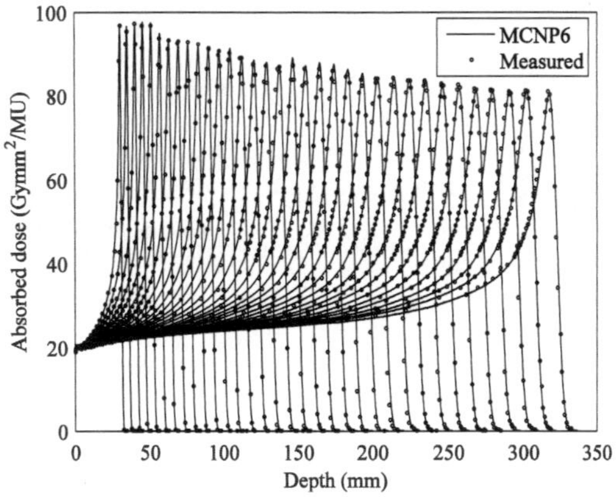

Figure 3. Absolute IDD curves for nominal beam energies in the range 60–226 MeV derived from measurements and MCNP6 simulations in water using the absorbed dose values obtained from reference measurements performed with calibrated ionization chambers.

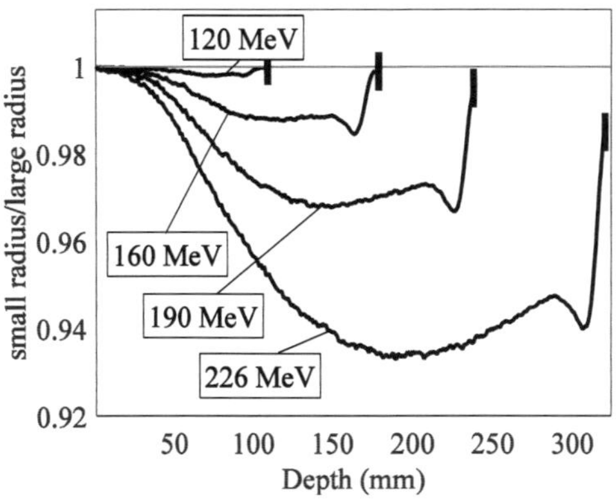

Figure 4. Calculated ratios between simulations using 4.08 cm radius and 25 cm radius for a selection of energies. The vertical markers correspond to the Bragg peak of the corresponding energy.

4. Discussion and Conclusions

In this work, the latest version of MCNP (MCNP6.1.1) has been used to model the characteristics of a dedicated proton spot scanning facility. All modelling parameters were based solely on measurements performed during commissioning and recognized assumptions regarding the energy distribution of the proton beam. The MCNP6 model has been benchmarked against measurements with regard to absorbed dose IDD normalized to the number of MU employed.

The energy distribution of the initial proton beam was defined using energy specific PDF defined with energy factors modifying the nominal primary proton energy and a Gaussian distribution defined with energy specific standard deviations. The low energy tail in the PDF of the energy distribution modelling the contribution from secondary particles scattered in the treatment head nozzle (mainly protons) resulted in an increase in MCNP6 output at shallow depths of roughly 0.5% for lower proton energies and about 2.5% for higher energies. Thus, simulations performed without including the low energy tail in the PDF caused an underestimation of the normalized IDD curves at shallow depths in comparison to measured IDD profiles (especially for higher energies). This result does not necessarily mean that the PDF employed in this work is a full representation of the actual energy distribution of the protons exiting the treatment head but that it is a rather valid simplification supported by previous findings [8]. It should be noted that the exclusion of geometrical modelling of the treatment nozzle itself could lead to underestimations of doses to regions outside the primary proton field [8]. However, considering that most scattered radiation in treatment-like situations originates from within the patient when irradiated with a spot scanning beam [9-11] the low energy tail employed in this work is a reasonable simplification for modelling of the secondary particle contribution originating from the treatment nozzle. The agreement between the integrals of the measured and simulated IDD curves in Figure 3 supports this assumption.

The spatial distribution of the beam was simulated using standard deviations and beam divergence derived from measurements. The average differences in spot size all correspond to a full width at half maximum of less than that of the resolution of the Lynx detector (0.5 mm). The largest average difference in spot sizes between simulations and measurements was found at +20 cm relative to isocenter. Considering that the patient, or phantom, is usually positioned so that the surface is located before the isocenter this discrepancy will most likely have very little impact on spot sizes inside a medium.

The results from the simulations of IDD curves using different IC radii presented in Figure 4 confirmed previously reported results that a limited detector width can lead to an underestimation in detected energy deposition of several percent for higher energies [22-24]. The largest effect of the limited detector width was seen in the distal half of the particle track and is a result of protons undergoing elastic scattering on hydrogen and nonelastic scattering on oxygen resulting in large scattering angles [25]. The trajectories of the scattered protons naturally cause an increase in signal loss with depth as they travel away from the central beam axis. However, the relative contribution of the forward directed unscattered protons increase at the proximal rise of the IDD curves as the ratios in Figure 4 converges towards unity. High energy protons are more likely to scatter in low angles when undergoing elastic scattering with the oxygen nucleus in the water phantom [1, 8, 26] and consequently a large fraction of the scattered protons will have a range rather close to that of the primary protons. The small discontinuity of the ratios observed before the Bragg peaks in Figure 4 is due to the Coulomb-nuclear interference effects described by Gerstein et al. [27]. This effect causes a local maximum of the proton differential cross section resulting in a higher yield of secondary protons for a specific angle and thus an increase in signal loss for the IC with limited diameter. The rather large signal losses showed that the low dose envelope originating from nuclear interactions of the primary protons expands over distances larger than the radius of the measurement IC and that this effect increases with incident proton energy. This emphasizes the need to quantify these signal losses and consequently correct the measured data when used as input for treatment planning systems.

The agreement between MCNP6 simulations and measurements supports the assumptions and input parameters employed in this work. The Monte Carlo model provided accurate representation of the beam characteristics with regard to absolute dose in water, range in water and lateral spot sizes in air. The ability to simulate accurate absolute dose IDD normalized to number of MU will prove useful for future studies regarding secondary dose assessment with MCNP6.

Acknowledgements
The authors would like to thank the physicists involved in the beam commissioning at the Skandion Clinic for providing depth dose curves, beam profile details and reference dosimetry data.

This work was partially supported by the Swedish Radiation Safety Authority under the contract SSM2016-2425. The authors declare no conflict of interest.

References

[1] Wilson R R 1946 *Radiology* **47** 487-91
[2] Paganetti H and Palmans H 2012 *Proton Therapy Physics*, ed H Paganetti (Boca Raton, Florida: CRC Press/Taylor & Francis) chapter 1, 7, 9
[3] Newhauser W D and Zhang R 2015 *Phys. Med. Biol.* **60** 155–209
[4] Paganetti H 2014 *Br. J. Radiol.* **87** 20140293
[5] Pedroni E, Scheib S, Böhringer T, Coray A, Grossmann M, Lin S and Lomax A *2005 Phys. Med. Biol.* **50** 541–61
[6] Li Y, Zhu R X, Sahoo N, Anand A and Zhang X 2012 *Phys. Med. Biol.* **57** 983–97
[7] Grevillot L, Bertrand D, Dessy F, Freud N and Sarrut D 2011 *Phys. Med. Biol.* **56** 5203–19
[8] Grassberger C, Lomax A and Paganetti H 2015 *Phys. Med. Biol.* **60** 633–45
[9] Schneider U, Agosteo S, Pedroni E and Besserer J 2002 *Int. J. Radiat. Oncol. Biol. Phys.* **53** 244–51
[10] Zacharatou Jarlskog C, Lee C, Bolch W E, Xu X G and Paganetti H 2008 *Phys. Med. Biol.* **53** 693–717
[11] Hultqvist M and Gudowska I 2010 *Phys. Med. Biol.* **55** 6633–53
[12] Goorley T et al. 2016 *Ann. Nucl. Energy.* **87** 772-83
[13] Taddei P J, Mirkovic D, Fontenot J D, Giebeler A, Zheng Y, Kornguth D, Mohan R and Newhauser W D 2009 *Phys. Med. Biol.* **54** 2259–75
[14] Titt U, Bednarz B and Paganetti H 2012 *Phys. Med. Biol.* **57** 6381–93
[15] Bielajew A F 2013 *Monte Carlo techniques in radiation therapy*, ed J Seco and F Verhagen (Boca Raton, FL: CRC Press/Taylor & Francis) chapter 1
[16] Randeniya S D, Taddei P J, Newhauser W D and Yepes P 2009 *Nucl. Technol.* **168** 810–14
[17] Newhauser W D, Rechner L, Mirkovic D, Yepes P, Koch N C, Titt U, Fontenot J D, and Zhang R 2013 *Radiat. Meas.* **58** 37–44
[18] Lee C C, Lee Y J, Tung C J, Cheng H W and Chao T C 2014 *Radiat. Phys. Chem.* **95** 302–4
[19] International Atomic Energy Agency 2006 *TRS 398-Absorbed Dose Determination in External Beam Radiotherapy: An International Code for Practice for Dosimetry based on Standards of Absorbed Dose to Water ver. 12* (Vienna: IAEA)
[20] *MCNP6 User's Manual Version 1.0* 2013 ed Pelowitz D B Los Alamos National Laboratory report LA-CP-13-00634
[21] Bertini H W 1963 *Phys. Rev.* **131** 1801–21
[22] Sawakuchi G O, Mirkovic D, Perles L A, Sahoo N, Zhu X R, Ciangaru G, Suzuki K, Gillin M T, Mohan R and Titt U 2010 *Med. Phys.* **37** 4960–70
[23] Anand A, Sahoo N, Zhu X R, Sawakuchi G O, Poenisch F, Amos R A, Ciangaru G, Titt U, Suzuki K, Mohan R and Gillin M T 2012 *Med. Phys.* **39** 891-900
[24] Clasie B, Depauw N, Fransen M, Gomà C, Panahandeh H R, Seco J, Flanz J B and Kooy H M 2012 *Phys. Med. Biol.* **57** 1147–58
[25] Gottschalk B, Cascio E W, Daartz J, Wagner M S 2015 *Phys. Med. Biol.* **60** 5627–54
[26] Braunn B, Boudard A, David J C, Koning A J, Leprince A, Leray S and Mancusi D 2015 *Eur. Phys. J. Plus* **130** 153
[27] Gerstein G, Niederer J and Strauch K 1957 *Phys. Rev.* **108** 427–32

IOP Conf. Series: Journal of Physics: Conf. Series **860** (2017) 012026 doi:10.1088/1742-6596/860/1/012026

A correlation of longterm effects and radiation quality in the progeny of bystander cells after microbeam radiations: The experimental study of radiotherapy for cancer risk mitigation

N Autsavapromporn[1,2,7], T Konishi[2], C Liu[2], I Plante[3], T Funayama[4], N Usami[5], EI Azzam[6] and M Suzuki[3]

[1] Divison of Radiation Oncology, Department of Radiology, Faculty of Medicine, Chiang Mai University, Chiang Mai, Thailand
[2] National Institute of Radiological Sciences (NIRS), National Institutes for Quantum and Radiological Science and Technology (QST), Inage-ku, Chiba, Japan
[3] Wyle Science Technology & Engineering, NASA Johnson Space Center, Houston, Texas, USA
[4] Takasaki Advanced Radiation Research Institute (TARRI), Quantum Beam Sciences Directorate, National Institutes for Quantum and Radiological Sciences and Technology (QST), Takasaki, Gunma, Japan
[5] Photon Faculty, Institute of Material Structure Sciences (IMSS), High Energy Accelerator Research Organization (KEK), Tsukuba, Ibaraki, Japan
[6] New Jersey Medical School Cancer Center-Rutgers, The State University of New Jersey, New Jersey, USA

[7] Corresponding author e-mail: narongchai.a@cmu.ac.th

Abstract. The goal of this study is to investigate the role of radiation quality and gap junction intercellular communication (GJIC) in the propagation of delayed stressful effects in the progeny of bystander human skin fibroblasts cultures (NB1RGB). Briefly, confluent NB1RGB cells in the presence and absence of gap junction inhibitor (AGA) were exposed to ionizing radiation (IR) with a different linear energy transfer (LET) either 5.35 keV X rays (LET ~6 keV/μm) or 18.3 MeV/u carbon (LET ~103 keV/μm) microbeam radiations. Following 20 populations post-irradiation, the progeny of bystander NB1RGB cells were harvested and assayed for several of biological endpoints. Our results showed that expression of stressful effects in the progeny of bystander cells is dependent on LET. The progeny of bystander cells exposed to low-LET X rays showed the persistence of oxidative stress and it was correlated with the increased mutant fraction. Such effect were not observed after high-LET carbon ions. Interestingly, inhibition of GJIC mitigated the toxic effects in the progeny of bystander cells. Together, the results contribute to the understanding of the fundamental radiation biology relating to the high-LET carbon ions to mitigate cancer risk after radiotherapy. Furthermore, GJIC be considered as a critical mediator in the bystander mutagenic effect.

1. Introduction

It is well established that the biological effects of low-and high-linear energy transfer (LET) of ionizing radiation (IR) such as DNA damage, chromosome aberration, cell death and mutation typically occurs in directly or hit-irradiated cells. However, there is a significant evidence suggesting that similar effects can also be seen in non-irradiated bystander cells, a phenomenon known as the non-targeted effects which include bystander effects, genomic instability and others [1-9]. Bystander effect refers to the phenomenon that the induction of biological effects in non-irradiated bystander cells whose nucleus has not been directly traversed by IR [10]. Genomic instability is found when a cell population surviving after IR, shows an increased rate of chromosome aberrations, gene mutations, the delayed cell death in the progeny of cells surviving irradiation and a hallmark of cancer cells [11,12].

In recent years, heavy ions radiotherapy such as high-LET carbon ion is one of the most effective treatment for cancer patients compared to conventional radiotherapy (such as X rays and γ rays). By delivering higher doses to tumour volume and sparing healthy normal tissues through manipulation of the Bragg peak area, resulting in the increased therapeutic ratio, and a higher relative biological effectiveness (RBE) in cell killing and induction of DNA damage [13-17]. However, the underlying bystander responses and genomic instability mechanisms after high-LET carbon ions are still poorly understood and investigated. In this context, previous studies have suggested that the bystander effects dependence of LET and radiation dose [10]. Although, a little known about the late effects in the progeny of bystander cells for many generations after exposed to high-LET carbon ions. To address some of these issues, we investigated the role of radiation quality and gap junction intercellular communication (GJIC) in the propagation of delayed stressful effects in the progeny of bystander normal cells after low-LET X ray and high-LET carbon ions.

2. Material and method

2.1. Cell culture

Low passage (6-9) normal human skin fibroblasts (NB1RGB) were grown in Eagle's minimum essential medium (MEM) contacting kanamycin (60 mg/L), supplemented with 10% fetal bovine serum (FBS). They were maintained in 37 °C humidified incubators in an atmosphere of 5% CO_2 in air. At the time of irradiation, the cells were confluent, thus allowing direct communication through gap junctions; approximately 94% of the cells were in the G_1/G_0 phase as analyzed by a flow cytometer (data not shown). For all irradiation experiments carried out at the two different facilities, the same passage of cell, the same cell culture medium and the same experimental protocol were used [10, 12].

2.2. X ray microbeam irradiation

Monoenergetic 5.35 keV X ray microbeam irradiation (LET ~6 keV/μm) was performed at the BL-27B station in the Photon Factory Synchrotron, High Energy Accelerator Research Organization (KEK), Tsukuba, Japan. The beam characteristics, biological irradiation procedures, and dosimetry have been described previously [10, 12, 18].

Cells (5×10^6 cell/dish) were cultured on microbeam dishes with 8 μm-thick kapton polyimide film on the bottom, and cultured for 1 day before irradiation to form confluent monolayers. Irradiation was carried out by using the 256 (16 × 16)-cross-strip method in the center of the microbeam dishes (containing MEM growth medium). At each spot, a dose of 0.4 Gy was delivered within a beam area of 20 μm × 20 μm, so that around 0.036-0.144% of the total cells attached on the bottom of the microbeam dishes were hit by IR [10, 12]. The irradiated and control samples were carried out at room temperature under ambient conditions, and the exposure time was less than 10 min. Under these conditions, the X ray energy was mostly absorbed by the targeted area within the cell nucleus and/or cytoplasm. The absorbed dose D (in Gy) for NB1RGB cells was calculated from the exposure, X (in C/kg) according to the following equation (1);

$$D = X \times W/e \times (\mu/\rho)_{NB1RGB} / (\mu/\rho)_{air} \qquad (1)$$

Where μ/ρ is the mass energy deposition coefficient (m^2/kg); W, the energy required to generate an ion pair (~34 eV) and e, the elementary electric charge (~1.6022×10^{-19} C). The values of μ/ρ in soft tissue for NB1RGB cells and air were 35.2 and 33.5 cm^2/g, respectively [18]. The energy of the X ray was 5.35 keV and the exposure rate was ~7.7×10^{-3} C kg^{-1} s^{-1} (~0.0103 Ckg^{-1}) as measured with an AXUV-100 absolute XUV silicon photodiode.

2.3. Carbon ion microbeam irradiation

Irradiation experiments with 18.3 MeV/u carbon ion microbeam (LET ~103 keV/μm) were performed at the Takasaki Ion Accelerator for Advanced Radiation Application (TIARA), Takasaki Advanced Radiation Research Institute, Quantum Beam Sciences Directorate, National Institutes for Quantum and Radiological Sciences and Technology (QST) (Takasaki, Japan). The beam characteristics and dosimetry of carbon ion-microbeam irradiations have been described elsewhere [10, 12, 19]. The preparation of cell samples have been described in detail previously. During irradiation, the medium was removed and the dishes were covered with Kapton polyimide film in order to maintain hydration during irradiation and to prevent microbiological contamination which was less than 8 min. The direction of beam was from top to bottom, and a total of 256 (16 × 16) points within a 15 × 15 mm^2 square area in the center of the microbeam dishes were irradiated using a microbeam spot of 20 μm in diameter. Within each spot, 8 traversal of carbon ions (Fluence = 2.55×10^6 cm^{-2} equivalent to 0.42 Gy) were delivered in the cell monolayer [10, 12]. Therefore, under these conditions, the vast majority of the cells (~99.8%) could be considered as cells not targeted by a primary particle (i.e. bystander cells). As a consequence, the contribution of the effects induced in hit cells to the overall response is negligible. The irradiation time was less than 8 min. Immediately after irradiation, 2 mL of fresh growth medium was added to the microbeam dishes, and the cultures were incubated at 37°C in 95% air/ 5% CO_2 atmosphere until harvesting. The absorbed dose (Gy) for NB1RGB cells with the consideration of a traversal of a number of carbon ions through a single target cell (Table 1) was calculated based on the fluence (number of ion particles/cm^2) and the following equation (2):

$$\text{Dose (Gy)} = \text{Fluence (ions/cm}^2) \times \text{LET (keV/}\mu\text{m)} \times 1.602 \times 10^{-9} \qquad (2)$$

2.4. Inhibition of gap junction communication

18-α-Glycyrrhetic acid (AGA), a reversible inhibitor of gap junction communication, was dissolved in dimethyl sulfoxide (DMSO) and added to cells at a concentration of 50 μM at 30 mins prior to irradiation. The cells were incubated in the presence of AGA until they were trypsinized. Under this protocol, AGA did not alter the micronucleus formation or *HPRT* colonies of NB1RGB cells [10, 12].

2.5. Micronucleus formation assay

The fraction of micronucleated cells in the exposed cultures was examined using the cytokinesis block technique [20]. Briefly, cells were subcultured by seeding ~3×10^4 cells in chamber flask and grown in the presence of 2 μg/mL cytochalsin B. The cultures were maintained at 37 °C for 72 h in order to obtain an optimum frequency of binucleated cells. Then the cells were rinsed in phosphate buffered saline (PBS), fixed in ethanol, stained with Hoechst-33342 solution (1 μg/mL in PBS) and viewed under a fluorescence microscope. At least 1000 cells were examined for each data point in each individual experiment, and only micronuclei in binucleated cells were considered for analysis [10, 12].

Table 1. Physical properties of microbeam at KEK and TIARA [10, 12, 19].

Institute	Types of microbeam	Energy	LET (keV/μM)	absorbed dose (Gy) to a target cell	Maximum range of δ-rays (μm)
KEK	X rays	5.35 keV	6	0.37	0.01
TIARA	Carbon ions	18.3 MeV/u	103	0.42	35

2.6. Mutation assay

The mutant frequencies were assessed at the HYPOXANTHINE-GUANINE PHOSPHORIBOSYL-TRANSFERASE (*HPRT*) locus by selection of cells resistant to 6-thioguanine (6TG), [21]. Cells were suspended by trypsinization and cultured in 75 cm^2 plastic flaks containing nonselective medium

(MEM with 10% FBS). The cells were cultured for 8-10 weeks until the population doubling number reached 20 in order to allow for phenotypic expression of the mutation. The cells in each flask were seeded in 10 new P-100 plastic dishes at a density of 1×10^5 cells/dish and cultured with selective medium supplemented with 40 μM of 6TG. The colonies were fixed and stained with 20% methanol and 0.2% crystal violet after 14 days of incubation. The frequency of mutation was calculated based on the number of 6-TG resistant colonies scored per 10^6 surviving cells and the cloning efficiencies at the time of seeding.

2.7. Oxidative stress assay (protein oxidation)

Protein carbonyl levels (an index of protein oxidation) [22] were determined by immunoblotting using the OxyBlot Oxidized Protein Detection Kit. Briefly, when proteins are oxidized by reactive oxygen species (ROS), some amino acids are modified, generating carbonyl groups. The latter can react with 2,4-dinitrophenyl hydrazine (DNPH), which in turn is recognized by anti-2,4-dinitrophenylhydrazone (DNP) antibodies on immunoblots. Carbonyl groups were derivatzed with DNPH using 20 μg protein samples denatured with 6% sodium dodecyl sulfate (SDS). Negative controls were reacted with a derivatization-control solution. After incubation at room temperature for 20 min, a neutralization solution was added to each tube and samples were analyzed by SDS polyacrylamide gel electrophoresis (SDS-PAGE), blotted onto nitrocellulose membranes, reacted with anti-dinitrophenylhydrazone antibody, and visualized by standard immune techniques.

2.8. Statistical analysis

All graph presented in the results section are the representatives of three independent experiments (n = 3), and the data were presented as mean standard errors (SE). Comparisons between treatment groups and respective controls were performed using Pearson's χ^2-test (micronucleus formation) and student's *t*-test (*HPRT* mutation assay). A p value of < 0.05 between groups was considered to be statistically significant.

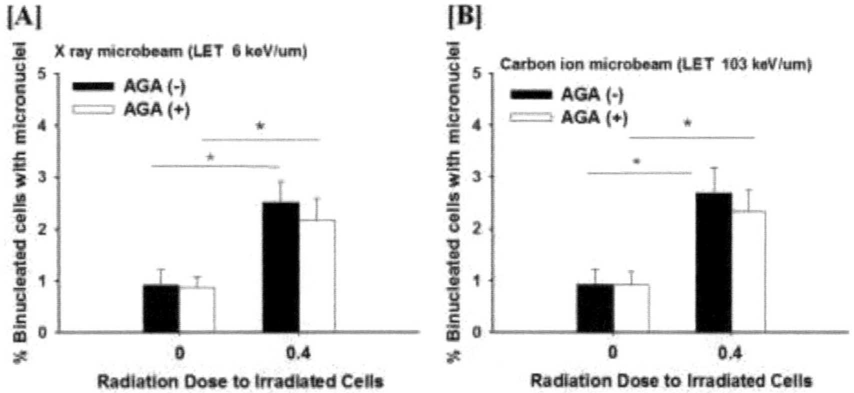

Figure 1. The frequency of micronucleus formation in the bystander NB1RGB cells were exposed to 0.4 Gy from X ray (A) and carbon ion microbeam (B) in the presence and absence of gap junction inhibitor (AGA) [10]. All values are presented by means ± SE; $*p < 0.05$.

3. Result and discussion

Second cancers are one of the leading causes of mortality for cancer survivors after receiving radiotherapy treatment. In this context, during cancer radiotherapy, normal tissues are also irradiated along with the tumor. The transmission of stressful effects from irradiated tumor/normal cells to bystander normal cells, and the persistence of such effects in their progeny would have profound implication for long term health risks of the exposure to radiotherapy such as the emergence of secondary cancer induction in normal cells [6-7, 11-12]. Therefore, it is important to elucidate the risk associated with various ionizing radiation type and radiation quality of the current radiotherapies use. However, the correlation between different types of radiation and cancer risk is not yet clear. The objective of this study is to assess the relationship between the non-targeted effects (early bystander responses and the late effects of bystander response) and secondary cancer risk through gap junction communication and oxidative metabolism in normal human bystander NB1RGB cells after low-LET X rays (LET ~6 keV/μm) and high-LET carbon ion (LET ~103 keV/μm) microbeam irradiation.

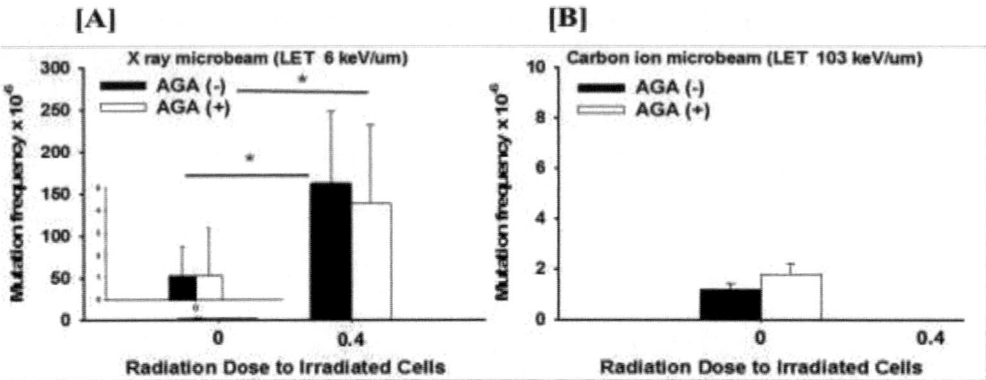

Figure 2. *HPRT* mutant fraction in the progeny of bystander cells after microbeam irradiation with X rays (A) and carbon ions (B) at mean absorbed dose of 0.4 Gy in the presence and absence of gap junction inhibitor (AGA). Note that the radiation dose shown in the table refers to the absorbed dose by the irradiated cells [12]. All values are presented by means ± SE; $*p < 0.05$. Note that the radiation dose described in the Figure refers to the absorbed dose by the irradiated cancer cells.

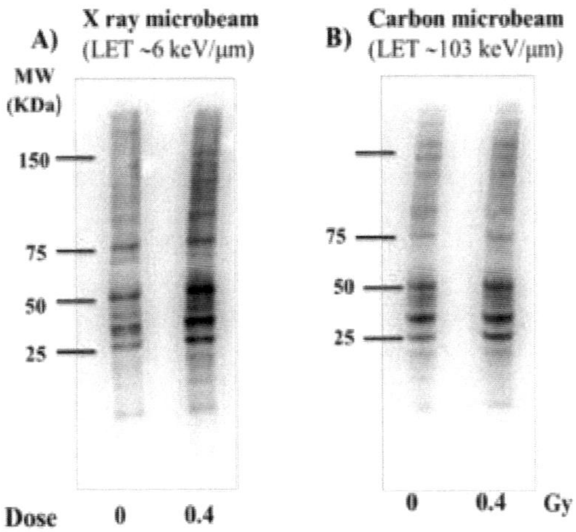

Figure 3. Persistence of oxidative stress (protein oxidation) in the progeny of bystander NB1RGB cells after microbeam irradiation with X rays (A) and carbon ions (B) is dependent on the quality of radiation. Note that the radiation dose shown in the table refers to the absorbed dose by the irradiated cells [12].

As shown in Figure 1, an increase in MN formation induced by low-LET X rays (Fig. 1A) was similar to that induced by high-LET carbon ions (Fig. 1B), implying similar magnitude of bystander response. However, it should be note that the cell culture medium was removed from the microbeam dishes during irradiation with carbon ions but not in X ray microbeam. Therefore, the potential contribution of culture medium-mediated bystander response during the irradiation was not accounted for carbon ion microbeams. This could explain a similar bystander cell killing in low LET X ray and high LET carbon ion microbeams at dose 0.4 Gy [10]. Based on previous studies, we can conclude that the bystander responses dependence with LET (~6-1260 keV/μm) and radiation dose (0-1.2 Gy), and these results fit well with the work presented in refs 23 and 24. Interestingly, incubation of cells with AGA show decrease (n = 3, $p > 0.05$) in micronucleus formation among bystander cells. These data indicate that the damaging effects in bystander cells is amplified by gap junctional communication shortly after microbeam irradiations.

Next, we investigated the role of radiation quality or LET and GJIC in the propagation of stressful effects in the progeny of bystander NB1RGB cells (following 20 population doublings that occurred in ~8-10 weeks). Figure 2 shows the *HPRT* mutation frequencies in the progeny of bystander cells induced by low-LET X rays, but not high-LET carbon ion microbeam, suggesting an absence of risk of induction of gene mutation that can lead to carcinogenesis [25]. Taken together, these results suggest that the delayed effects (genomic instability) occurred in the progeny of bystander cells after exposed to low-LET radiation. Moreover, treatment with AGA resulted in the decrease but not significantly different (n = 3, $p > 0.05$) of induced *HPRT* mutant fraction in the progeny of bystander cells. Therefore, these data indicate that gap junction communication plays a major role of a bystander mutagenic effect.

Furthermore, to examine whether oxidative stress influence the mutagenic effect in the progeny of bystander normal cells, we investigated the levels of protein carbonylation induced by reactive oxygen species (ROS). Whereas protein carbonylation was increased in the progeny of bystander cells exposed to low-LET X rays (Figure 3) and this data is consistent with the induction of *HPRT* mutation (Figure 2). In contrast to low-LET X rays, it is interesting result to note that the levels of protein carbonylation changes only slightly decrease after high-LET carbon ion is used. The result show that it is consistent with the lack of *HPRT* mutant fraction in the progeny of bystander cells (Figure 2). Overall, the bystander mutagenic effect in the progeny of bystander cells might affect from the production of ROS after receiving irradiation.

Based on the results obtained in this study, we proposed a model for bystander effects and the late effects in the progeny of bystander cells (genomic instability)-induced by low-and high-LET microbeam irradiation. Figure 4 is a schematic diagram showing the physical direct cell-cell contact via gap junctional-mediated bystander response by directly irradiated cells may transmit to bystander cells, and the propagation of stressful effects in the bystander cells is LET-dependent. Futuremore, the progeny of bystander cells that undergo genomic instability and cancer development greatly depend on the quality of radiation, induction of mutation, oxidative stress and other, especially in the high-LET radiation where small effects are expected [26-27].

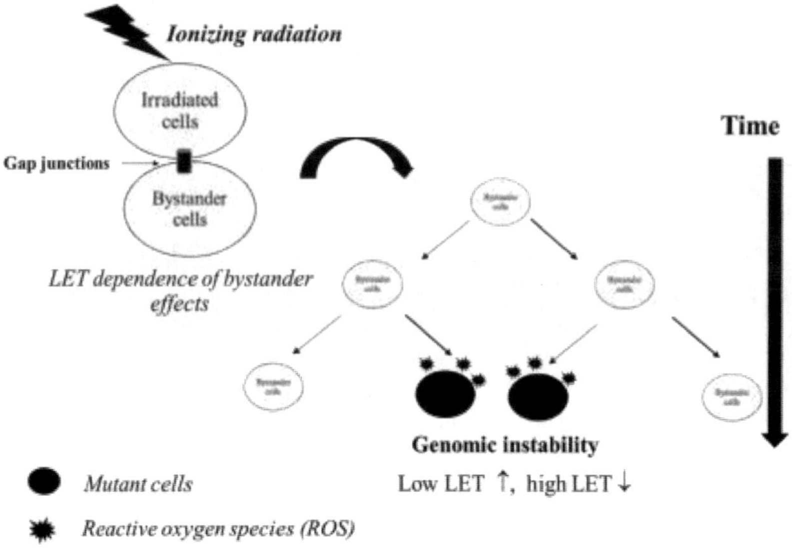

Figure 4. Model for non-targeted effects of low-and high-LET microbeam irradiation.

4. Conclusion

Result of the current study suggest that non-targeted effects of low-LET radiation in the bystander normal cells can increase genetic changes and may ultimately lead to carcinogenesis, but not high-LET carbon ion microbeam. These findings clearly indicate that the impact of second cancer risks is depending on the LET value in the progeny of bystander normal cells. In addition, the results highlights that the important role of gap junctional communication and oxidative metabolism in the propagation of the stressful effect in the bystander cells and their progeny. Future studies are needed to elucidate the role of long term effects after high-LET carbon ion irradiation in *vivo* and to develop biological measures to minimize the potential risk of radiotherapy-induced second cancer in cancer survivors.

References

[1] Nagasawa H and Little JB 1992 *Cancer Res.* **52** 6394
[2] Azzam EI, de Toledo SM, Gooding T and Little JB 1998 *Radiat. Res.* **150** 497
[3] Mothersill C and Seymour CB 1997 *Int. J. Radiat. Biol.* **71** 421
[4] Zhou H, Suzuki M, Randers-Pehrson G, Vanais D, Chen G, Trosko JE, Waldren CA and Hei TK
2001 *Proc. Natl. Acad. Sci. USA* **98** 14410
[5] Azzam EI, de Toledo SM and Little JB 2001 *Proc. Natl. Acad. Sci. USA* **98** 473
[6] Morgan WF 2003 *Radiat. Res.* **159** 567
[7] Morgan WF 2003 *Radiat. Res.* **159** 581
[8] Prise KM and O'Sullivan JM 2009 *Nat. Rev. Cancer* **9** 351.
[9] Yang H, Asaad N and Held KD 2005 *Oncogene* **24** 2096.
[10] Autsavapromporn N, Suzuki M, Funayama T, Usami N, Plante I, Yokota Y, Mutou Y, Ikeda H, Kobayashi K, Kobayashi Y, Uchihori Y, Hei TK, Azzam EI and Murakami T 2013 *Radiat.*

Res. **180** 367
[11] Buonanno M, de Toledo SM, Pain D and Azzam EI 2011 *Radiat. Res.* **175** 405
[12] Autsavapromporn N, Plante I, Liu C, Konishi T, Usami N, Funayama T, Azzam EI, Murakami T and Suzuki M 2015 *Int. J. Radiat. Biol.* **91** 62
[13] Oike T, Sato H, Noda S-e and Nakano T G 2016 *Front. Oncol.* **6** 139
[14] Imaoka T, Ishii N, Kawaguchi I, Himma-Takeda S, Doi K, Daino K, Nakanishi I, Tagami K, Kokubo T, Morioka T, Hosoki A, Takabatake M and Yoshinaga S 2016 *Int. J. Radiat. Biol.* **92** 289
[15] Ando K, Koike S, Ohmachi Y, Ando Y and Kobashi G 2014 *Int. J. Radiat. Biol.* **90** 1119
[16] Manem VSK, Kohandel M, Hodgson DC, Sharpe MB and Sivaloganathan S 2015 *Int. J. Radiat. Biol.* **91** 209
[17] Fernandez-Palomo C, Mothersill C, Bräuer-Krisch E, Laissue J, Seymour C, Schültke E 2015 *PLos ONE* **10** e0119924
[18] Maeda M, Usami N, Kobayashi K 2008 *J. Radiat. Res.* **49** 171
[19] Funayama T, Wasa S, Yokota Y, Fukamoto K, Sakashita T, Taguchi M, Kakizaki T, Hamada N, Suzuki M, Furusawa Y, Watanabe H, Kiguchi K and Kobayashi Y 2008 *J. Radiat. Res.* **49** 71
[20] Fenech M, Morley AA 1985 *Mutat. Res.* **147** 29
[21] Tsuruoka C, Suzuki M, Furusawa Y, Anzai K, Okayasu R 2009 *Biolog. Sci. Space* **23** 55
[22] Dalle-Donne I, Giustarini D, Colombo R, Rossi R, Milzani A 2003 *Trends. Molec. Med.* **9** 169
[23] Wakatsuki M, Magpayo N, Kawamura H and Held KD 2012 *Int. J. Radiat. Oncol. Bio. Phys.* **84** 103
[24] Anzenberg V, Chanairamani S, Coderre JA 2008 *Radiat. Res.* **170** 467
[25] Fournier C, Winter M, Zahnreich S, Nasonova E, Melnikova L and Ritter S 2007 *Radiother. Oncol.* **83** 277
[26] Fernandez-Palomo C, Schültke E, Bräuer-Krisch E, Laissue JA, Blattmanna H, Seymour C and Mothersill C 2016 *Health Phys.* **111** 149
[27] Lam RKK, Fung YK, Han W and Yu KN 2015 *Int. J. Mol. Sci.* **16** 2591.

Acknowledgments
The authors would like to acknowledge National Institutes for Quantum and Radiological Science and Technology (QST) and High Energy Accelerator Research Organization (KEK) for providing the experimental instruments. We would like to thank Drs. Takeshi Murakami, Yasuhiko Kobayashi, Katsumi Kobayashi, Yasuko Mutou, Yuichiro Yokota and their colleagues for their support during the experiments. This research is partially supported by grants from Japan Society of Promotion of Sciences (JSPS), National Institutes for Quantum and Radiological Science and Technology (QST); The Higher Education Research Promotion and National Research University (NRU) Project of Thailand; Faculty of Medicine Research Fund, Chiang Mai University; and Chiang Mai University Research Fund for lung cancer study, Chiang Mai, Thailand.

IOP Conf. Series: Journal of Physics: Conf. Series **860** (2017) 012027 doi:10.1088/1742-6596/860/1/012027

Overview summary of clinical heavier-ion progress in Japan

N Matsufuji

National Institute of Radiological Sciences, National Institutes for Quantum and Radiological Science and Technology, 9-1, Anagawa-4, Inage-ku, Chiba-shi, Chiba 263-8555 Japan

Email: matsufuji.naruhiro@qst.go.jp

Abstract. Swift ion beams such as carbon has unique characteristics suitable for treating deep-seated tumours. In Japan, carbon-ion radiotherapy was started in 1994 at Heavy Ion Medical Accelerator in Chiba (HIMAC) at National Institute of Radiological Sciences and more than 10,000 patients have been treated by Aug. 2016. Clinical outcomes show superior efficacy of carbon ions even against radioresistant tumour while keeping the quality of life at high level, and also the usefulness of hypofractionated irradiation down to the completion of the course of lung-cancer treatment in 1 day. During the decades, the improvement of hardware and software technology such as 3D scanning technique, superconducting rotating gantry or biology model have been carried out aiming at further optimized ion-beam radiotherapy as well as reducing the cost of the facility. The developed technology has been transferred to the following facilities. As of 2016, 5 carbon ion radiotherapy facilities are in operation in Japan.

1. Introduction

Elevating energy loss of ion beams traveling in a matter toward its range end, known as Bragg curve, is attractive characteristics when thinking about delivering dose locally to deep-seated tumour in a patient's body while sparing surrounding normal tissues. The characteristics gets more significant on heavier ions. In case of carbon ions for example, the straggling of the beam is reduced to be about 1/3 of proton beam at certain depth. This realizes more localized dose delivery specific to tumour. In addition to that, for ions in proximity to carbon, their RBE (relative biological effectiveness) gradually increases as the energy loss increases, *i.e.,* from entrance to the Bragg peak. This increase in RBE tends to enhance the peak-to-plateau ratio of the Bragg curve from biological viewpoint. Here, it should be noted that this peak-to-plateau ratio gets rather diminished in reversal for further heavier ions because of increasing frequency of projectile fragmentation and overkill effect. Therefore radiotherapy with certain ion species such as carbon (hereafter ion-beam RT or C-ion RT) is considered advantageous for therapeutic application as a method to achieve both high clinical effect on tumour and low toxicity on normal tissues at the same time.

However, in order to deliver carbon ions to 30 cm depth in water for example, it is necessary to accelerate the carbon ions up to about 400 MeV/n which is more than 70 % of the light speed. That is why though the original idea of the ion-beam RT was proposed by R. Wilson in 1940s, it was necessary to wait for the development of powerful accelerator to make it possible. The first ion-beam RT was accomplished by Lawrence Berkeley Laboratory (LBL) [1] with Bevatron synchrotron accelerator dedicated for physics study. When the pioneering clinical trials at the LBL came to the end in 1993, Japan and Germany ran with the ball respectively. In Japan, world's first accelerator dedicated to ion-beam RT, HIMAC (Heavy Ion Medical Accelerator in Chiba) [2], was established at the National Institute of Radiological Sciences (NIRS) in 1994. Soon after, German first experimental ion-beam RT facility was built at the Gesellschaft für Schwerionenforschung mbH (GSI) in Darmstadt and started clinical trials in 1997 [3]. As of 2016, 10 ion-beam facilities have been in operation in the world mainly in Asia and Europe. This paper introduces the progress of C-ion RT in Japan.

2. Outline of the HIMAC project

The project of HIMAC was started in 1984 under "Comprehensive 10-years Strategy for Cancer Control" lead by Government of Japan. The HIMAC project was aimed at establishing the optimum ion-beam RT modality. Figure 1 shows the bird's eye view of the HIMAC facility. The main synchrotron ring is doubled and can be operated independently. HIMAC has 3 treatment rooms (room A, B and C) in the building. Rooms A and C are equipped with one vertical or horizontal port, respectively while room B has both ports. In addition, experimental rooms for physics and biology are prepared and open for fundamental research with ion beams. HIMAC supplies only carbon ion beams to treatment rooms for therapeutic purpose in daytime, and various ion beams ranging from H to Xe up to 800 MeV/n for the experimental research in night time.

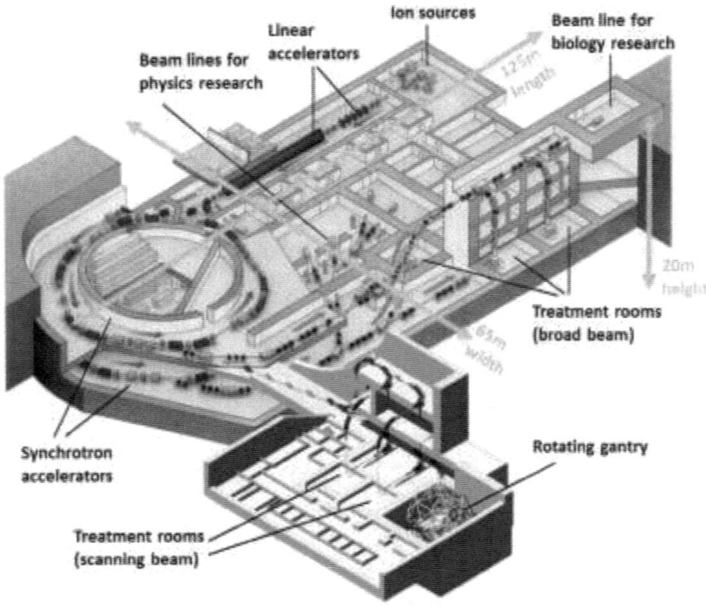

Figure 1. Bird's eye view of HIMAC facility.

Clinical trials of the C-ion RT was initiated in June of 1994. Figure 2 shows the annual accrual and patient distribution statistics treated at HIMAC as of March 2016. The first majority of patients treated so far is those suffered from prostate cancer, followed by those with solid tumours occurred at various sites. The annual number of treated patients gradually increases. In these days about 1,000 patients are treated annually and the total number exceeds 10,000 by August 2016. Most of clinical trials have been promoted to clinical practice by the approval by Government of Japan as the "advanced medical technology" in 2003. Under this category, 3.14 M JPY is charged per treatment. Since 2016, C-ion RT for inoperable bone and soft tissue sarcoma has been covered by Japanese national health insurance.

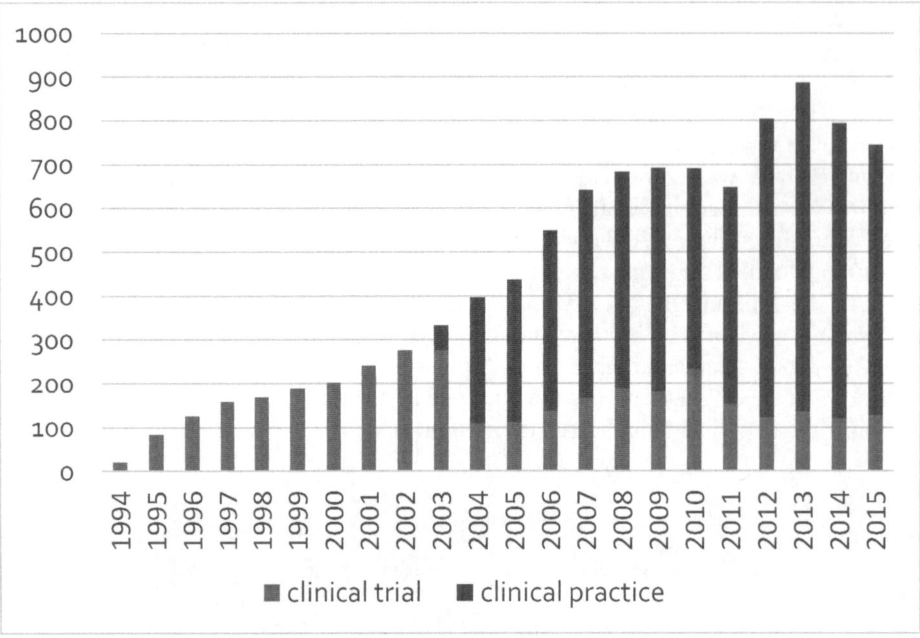

Figure 2. Annual accrual of patients treated at HIMAC in 1994 to 2015 (in fiscal year).

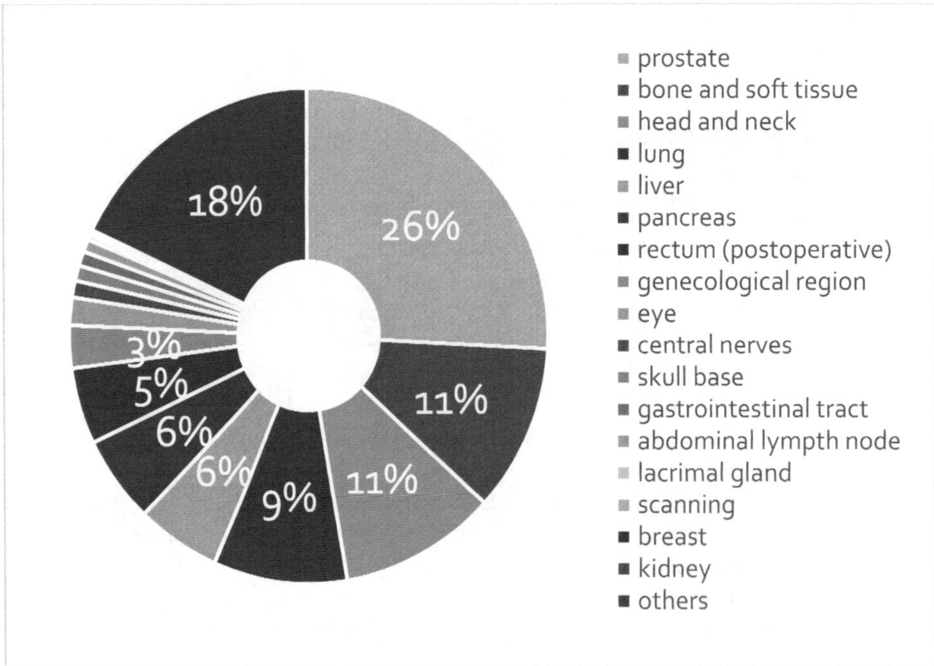

Figure 3. Distribution of patients treated at HIMAC by Feb. 2016.

One of the theme on clinical trials at HIMAC is to exploit the possibility of hypofractionation. In general, fractionated irradiation is preferred in radiotherapy as a method to spare normal tissue surrounding tumour by enhancing the difference in repair capacity between tumours and normal tissues. The reduction of the number of fractionation, hypofractionation, could be therefore regarded potentially hazardous. However, if hypofractionation can be performed in safe, it is beneficious for patient's reducing the burden. In case of C-ion RT, due to previously mentioned superior peak-to-plateau ratio, dose deposited to normal tissue can be kept under tolerable limit even in hypofractionation scheme.

In each fraction schedule, dose escalation study was carried out in order to find the optimum dose level to prescribe by carefully observing the normal tissue response. In case of the most frequently-treated prostate cancer [4] for example, the number of fractionation was gradually reduced from 20 in 5 weeks at the beginning to 12 in 3 weeks. The ultimate is the non-small cell lung cancer treatment. The treatment with carbon ions was carried out in 18 fractions in 6 weeks at the beginning but now completed by successive 4 portal irradiations given within one hour [5]. Throughout these clinical experiences [6], it was confirmed that normal tissue response is tolerable in most cases and QOL (quality of life) after the treatment has been kept at high level even under hypofractionated irradiation.

3. Improvement of C-ion RT

While carrying out the C-ion RT routinely, HIMAC has been kept on updating as a research facility. A part of the hardware and software progress is introduced below.

3.1. Scanning beam delivery

The therapeutic ports of HIMAC was originally configured as a broad-beam irradiation method [7] to deliver the carbon-ion beam to entire tumour volume. In this method, a pencil-like beam extracted from the accelerator is at first circulated by a pair of dipole wobbler magnets. Then this ring-like beam passes through a thin metal foil as a scatterer inserted downstream of the wobbler magnets. Due to the scattering there, at the isocenter located about 10 m downstream from the scatterer, beam size is broadened up to 200 mm in diameter with uniformity better than 95 %. Regarding the axial direction, this dish-like beam is spread out up to 150 mm in water so that the spread out Bragg peak (SOBP) covers the maximum target thickness by passing through a ridge filter made of tapered aluminium blades lined in parallel. The resultant laterally and axially broadened beam is finally shaped by a patient collimator made of brass or a multileaf collimator made of iron, and a compensator made of polyethylene manufactured to match the beam to the cross section and distal plane of the individual target, respectively. The irradiation field shaped by the broad-beam method is characterized by the steadiness in time and space which is favourable especially treating tumours moving with respiration [8]. In case of lung or liver cancer treatment, the respiration phase is detected by a LED light attached to the patient, and the beam is delivered only at the expiration phase.

On the other hand, the rigid characteristic of the broad-beam method is disadvantageous from the viewpoint of flexibility. A few days of the leading time necessary for manufacturing the collimator or compensator makes it difficult to adapt to the change in tumour shape. In addition, due to the constant SOBP width irrespective to the difference in tumour thickness point by point, some portion of normal tissues inevitably receives full dose.

In order to overcome the problems, the development of the 3D scanning method [9] was started in 2007. In the method, the pencil-like beam extracted from the accelerator is directly delivered to appropriate position in the target. A pair of scanning magnets is used to control the lateral position while the axial stopping point is adjusted by the accelerator by changing the beam energy dynamically within one extraction phase from 430 MeV/n down to 56 MeV/n in 250 steps.

In general, it is regarded difficult to apply the scanning method for the irradiation to the moving target because the interplay between the scanning with narrow beam and target motion can spoil the dose uniformity easily. The 3D scanning technique developed at HIMAC tries to overcome this problem by a combination of gating and rescan feature [10]. The gating technique for scanning irradiation [11] has been updated from those used in the broad-beam irradiation. During the irradiation, the position of the target is frequently monitored on X-ray transmission image. The scanning irradiation takes place selectively when the target comes within a planned area in the expiration phase where the respiration motion is minimized. The small movement within the expiration phase is statistically smeared out by swiftly repeating the scanning irradiation for several times. In order to make the rescan possible, the scan speed reaches to 100 mm / 1 ms to the irradiation field size of 220 mm square.

The scanning irradiation ports have been installed to a vertical and horizontal port of newly built room E and F in an annex building in 2011. After pilot study successfully completed for 11 patients with stationary tumour, the scanning irradiation has been carried out even for moving tumour together with the broad-beam ports.

3.2. Rotating gantry

Currently all the irradiation ports in existing treatment rooms at HIMAC is fixed to either vertical or horizontal direction. Due to the limitation, sometimes a patient is tilted up to 20 degrees by a couch to deliver the beam from necessary direction. This can be a cause of uncomfortableness for the patient. Moreover, because positioning of the patient takes place for each portal irradiation, it is inevitable to position the patient again for multi-portal irradiation. The geometrical difference due to the deformation of the patient's organ by the repositioning makes it difficult to perform the intensity-modulated ion-beam RT, which is analogous to the intensity-modulated radiotherapy (IMRT) technique in X-ray RT and demands high precision and accuracy on the geometry of the target under current fixed port configuration. From the viewpoint, a rotating gantry for C-ion RT is under development at HIMAC [12]. The rotating gantry configuration is essential for the IMRT machine and getting popular even for proton therapy, however, challenging for C-ion RT due to the large $B\rho$ value necessary for magnets to bend the therapeutic carbon-ion beam in a relativistic energy range. HIT (Heidelberg Ion Therapy Centre) [13] in Germany has realized the world's first rotating gantry applicable for C-ion RT. The huge construction has about 600 t of rotating weight with 20 m in diameter. The size of the gantry at HIMAC is reduced to be comparable to proton gantry, *i.e.*, 200 t in weight and 11 m in diameter by making use of compact superconducting magnets. Some of the magnets are designed as a hybrid of dipole and quadrupole in one magnet in order to further save the geometrical space. The 3D scanning technology is integrated to the rotating gantry. The gantry is under commissioning, and will be enrolled in C-ion RT by the end of 2016.

3.3. Biological model

It is requisite to estimate the changing RBE value of carbon ions quantitatively and precisely to conduct the C-ion RT effectively and safely. LET (linear energy transfer) has been widely used as an index of the radiation quality of the beam in estimating the RBE of radiation. In initiating the C-ion RT at HIMAC, the RBE value of human salivary gland (HSG) cells were chosen as the reference biological endpoint. The RBE value at 10 % clonogenic survival level of the HSG cells was experimentally studied by changing the LET of the carbon ions step by step. The derived dose-survival relationship at each LET was fitted with commonly-used LQ model $S=\exp(-\alpha D-\beta D^2)$ where S and D is the survival probability and the absorbed dose, respectively and α and β are the fitting parameters as a representative of the radiosensitivity for the beam. The derived α and β were tabulated as a function of LET and looked up in designing the ridge filter to form the SOBP beam [7].

This original approach was practically useful for estimating the RBE of therapeutic carbon-ion beams: in case of non-small cell lung cancer treatment for example, the clinical RBE values estimated by the model and verified with the clinical observation were both about 2.4 [14]. However, the model was not suitable for causal understanding between the incident radiation quality and observed radiosensitivity. From the viewpoint, the original model was updated to Microdosimetric Kinetic Model (MKM) [15] when starting the scanning irradiation at HIMAC. MKM was originally developed by Hawkins [16] as an update of microdosimetric dual radiation action theory [17]. The model starts with an energy imparted to a micrometre scale "domain". The number of lesions produced in the domain is associated with the energy deposition there, and the cell survival probability is given from Poisson statistics when none of the domains in a cell nucleus develop any single lesion. The final cell survival probability also shows the LQ dependency on the macroscopic absorbed dose while the parameter α can be interpreted with the microdosimetric energy information. MKM is tuned to harmonize with the original model [18] and now routinely utilized in a treatment planning system for both broad-beam and scanning irradiation [19].

4. Spreading of C-ion RT in Japan

Soon after HIMAC, Hyogo Ion Beam Medical Centre (HIBMC) [20] project was started. HIBMC offers both proton and C-ion RT in one machine first in the world since 2003 and 2004, respectively. In addition to the fixed vertical and horizontal ports, HIBMC has one 45 degree oblique port and two rotating gantries for proton beam in 4 treatment rooms in total.

Upon the superior clinical outcome of C-ion RT through the clinical trials and successive clinical practices at HIMAC, development of hospital-based compact C-ion RT machine by optimizing related technology developed at HIMAC was funded in the third decade of the cancer control strategy started in 2004 to spread this modality. In this framework, the GHMC (Gunma University Heavy Ion Medical Centre) project [21] was started in 2006. The accelerator ring was optimized solely for carbon ion beam up to 400 MeV/n, and matured broad-beam technique was adopted there. Together with additional optimizations and improvements, this compact model became about 1/3 of HIMAC in size and cost (33 G JPY to 12 G JPY). GHMC has been carried out C-ion RT since 2010. SAGA HIMAT (Saga Heavy Ion Medical Accelerator in Tosu) [22] also adopted the similar design and has started the clinical operation in 2013.

The newly developed techniques have been transferred to the successive facilities. i-ROCK [23], C-ion RT facility at Kanagawa Cancer Centre has started the C-ion RT in the end of 2015 with the respiratory gating 3D scanning technique. Ion-beam RT facility in Osaka is under construction for the open in 2018. The facility will be equipped with the scanning technique applicable for respiratory motion. The latest C-ion RT facility project at Yamagata University will open in 2019 with the superconducting rotating gantry with 3D scanning technique.

In order to accelerate the spreading further, study for further reduction of the facility cost has been in progress [24]. At the same time, cost-effectiveness analysis on C-ion RT has been in progress. The number of studies is still limited, however, the result is promising that C-ion RT can be superior to conventional modalities in the cost effectiveness, too [25].

5.Conclusion

C-ion RT has been carried out since 1994 in Japan. In more than two decades of experience, C-ion RT has been proven to be advantageous especially in treating radioresistant tumours such as bone and soft-tissue sarcoma, as well as in safely completing the course of the therapy even in short period as hypofractionation regime [26]. The high QOL and high clinical effectiveness of the C-ion RT will be advantageous in terms of cost effectiveness and contribute to sustainable development of the society. At the same time, research and development has been continuously carried out in order to further improve the ion-beam RT and further reduce the facility cost.

Acknowledgments
The Author wishes to acknowledge all the members at NIRS HIMAC.

References

[1] Castro J R 1995 *Radiat. Environ. Biophys.* **34** 45-8
[2] Hirao Y *et al* 1992 *Nucl. Phys.* A **538** 541-50
[3] Kraft G 2000 *Nucl. Instr. and Method in Phys. Res. Sect.* A **454** 1-10
[4] Tsuji H, Yanagi T, Ishikawa H, Kamada T, Mizoe J, Kanai T, Morita S and Tsujii H 2005 *Int. J. of Radiat. Oncol. Biol. Phys.* **63** 1153-60
[5] Karube M, Yamamoto N, Nakajima M, Yamashita H, Nakagawa K, Miyamoto T, Tsuji H, Fujisawa T and Kamada T 2016 *Int. J. of Radiat. Oncol. Biol. Phys.* **95** 542-8
[6] Tsujii H and Kmada T 2012 *Jpn. J. of Clin. Oncol.* **42** 670-85
[7] Kanai T *et al* 1999 *Int. J. of Radiat. Oncol. Biol. Phys.* **44** 201-10
[8] Minohara S, Kanai T, Endo M, Noda K and Kanazawa M 2000 *Int. J. of Radiat. Oncol. Biol. Phys.* **47** 1097-1103
[9] Furukawa T *et al* 2010 *Med. Phys.* **37** 5672-82
[10] Furukawa T *et al* 2010 *Med. Phys.* **37** 4874-9
[11] Mori S *et al* 2016 *Int. J. of Radiat. Oncol. Biol. Phys.* **95** 258-66

[12] Iwata Y *et al* 2013 *Nucl. Instr. and Method in Phys. Res. Sect.* B **317** 793-7
[13] Combs S E, Jäkel O, Haberer T and Debus J 2010 *Radioth. Oncol.* **95** 41-44
[14] Matsufuji N *et al* 2007 *J. Radiat. Res.* **48** A81-6
[15] Kase Y, Kanai T, Matsumoto Y, Furusawa Y, Okamoto H, Asaba T, Sakama M and Shinoda H 2006 *Radiat. Res.* **166** 629-38
[16] Hawkins R B 1994 *Radiat. Res.* **140** 485-96
[17] Kellerer M and Rossi H H 1978 *Radiat. Res.* **75** 471-88
[18] Inaniwa T, Kanematsu N, Matsufuji N, Kanai T, Shirai T, Noda K, Tsuji H, Kamada T and Tsujii H 2015 *Phys. Med. Biol.* **60** 3271-86
[19] Inaniwa T, Furukawa T, Kase Y, Matsufuji N, Toshito T, Matsumoto Y, Furusawa Y and Noda K 2010 *Phys. Med. Biol.* **55** 6721-37
[20] Hishikawa Y, Oda Y, Mayahara H, Kawaguchi A, Kagawa K, Murakami M and Abe M 2004 *Radioth. Oncol.* **73** S38-40
[21] Ohno T *et al* 2011 *Cancers* **3** 4046-60
[22] Kudo S *et al* 2016 *Int. J. Particle Therapy* **2** 464-7
[23] Nakayama Y *et al* 2016 *Int. J. Particle Therapy* **2** 478-80
[24] Noda K *et al* 2014 *Proc. Of IPAC2014* 2812-4
[25] Mobaraki A, Ohno T, Yamada S, Sakurai H and Nakano T 2010 *Cancer Sci.* **101** 1834-9
[26] Kamada T *et al* 2015 *Lancet Oncol.* **16** e93-100

IOP Conf. Series: Journal of Physics: Conf. Series **860** (2017) 012028 doi:10.1088/1742-6596/860/1/012028

Fiber optically coupled radioluminescence detectors: A short review of key strengths and weaknesses of BCF-60 and Al_2O_3:C scintillating-material based systems in radiotherapy dosimetry applications.

S Buranurak[1,3] and C E Andersen[2]

[1] Department of Physics, Khon Kaen University, 40002-Khon Kaen, Thailand
[2] Center for Nuclear Technologies, Technical University of Denmark, 4000-Roskilde, Denmark

[3] Corresponder's e-mail: bsirit@kku.ac.th

Abstract. Radiotherapy technologies have improved for several decades aiming to effectively destroy cancerous tissues without overdosing surrounding healthy tissues. In order to fulfil this requirement, accurate and precise dosimetry systems play an important role. Throughout the years, ionization chambers have been used as a standard detector for basic linear accelerator calibrations and reference dosimetry in hospitals. However, they are not ideal for all treatment modalities: and limitations and difficulties have been reported in case of (i) small treatment fields, (ii) strong magnetic field used in the new hybrid MRI LINAC /cobalt systems, and (iii) in vivo measurements due to safety-issues related to the high operating voltage. Fiber optically coupled luminescence detectors provide a promising supplement to ionization chambers by offering the capability of real-time in vivo dose monitoring with high time resolution. In particular, the all-optical nature of these detectors is an advantage for in vivo measurements due to the absence of high voltage supply or electrical wire that could cause harm to the patient or disturb the treatment. Basically, fiber-coupled luminescence detector systems function by radiation-induced generation of radioluminescence from a sub-mm size organic/inorganic phosphor. A thin optical fiber cable is used for guiding the radioluminescence to a photomultiplier tube or similar sensitive light detection systems. The measured light intensity is proportional to dose rate. Throughout the years, developments and research of the fiber detector systems have undergone in several groups worldwide. In this article, the in-house developed fiber detector systems based on two luminescence phosphors of (i) BCF-60 polystyrene-based organic plastic scintillator and (ii) carbon-doped aluminum oxide crystal (Al_2O_3:C) are reviewed with comparison to the same material-based systems reported in the literature. The potential use of these detectors for reference-class dosimetry in radiotherapy will be discussed with a particular emphasis on uses in small and large MV photon fields.

1. Background and introduction

Since the discovery of x-ray in 1895, the use of radiation for medical treatment has been revolutionized. Advanced radiotherapy technologies have been continuously improved and refined.

Nowadays, it has been accepted as an import part of cancer treatments that is offered to about 50% of all cancer patients [1]. Basically, radiotherapy can be separated into two main types: (i) external beam radiotherapy where high energy of ionizing radiations (e.g. x-rays, electrons, and heavy charged particles) generated by a linear accelerator, and (ii) brachytherapy where a small radioactive source (e.g. Iridium 192) is inserted inside a catheter or a slender tube and controlled using the remote after loading system to let it dwell in or close to the tumor volume inside a body cavity. Careful treatment planning and optimization are mandatorily required for both types of cancer treatment aiming to deliver dose to the tumor region effectively, while minimizing any harmful side effects on surrounding healthy tissues. Treatment errors larger than 5% are considered unacceptable [2]. Therefore, accurate and precise dosimetry plays a key role to achieve this goal.

Ionization chambers are routinely used in hospitals due to their special characteristics of high reproducibility and long-term stability. Although, they are excellent for basic linear accelerator calibrations and reference dosimetry, they are not universally ideal detectors for all new treatment modalities–some limitations and difficulties have been observed in small treatment fields as found in the Intensity Modulated Radiation Therapy (IMRT), the Volumetric Modulated Arc Therapy (VMAT) and in strong magnetic fields of the new hybrid MRI LINAC /cobalt systems. Das et al. [3] reported a variation in depth dose profiles of using a 6 MV beam delivery obtained with various types of detectors for small (1×1 cm^2), reference (10×10 cm^2), and large (40×40 cm^2) fields. The study provided unacceptable variation for small and large fields. For small fields, the main problems relate to the lack of lateral charge equilibrium of secondary electrons [4]. Whereas for large fields, the main problem concerns the increased induction of leakage noise current (i.e. large field-size projection compared to the small size of detectors causes more scatter of low energy electrons which contribute to the detector response).

Additionally, ionization chambers are generally constructed out of non-water equivalent materials in term of atomic composition and density. This causes significant perturbation effects that influence the accuracy of absorbed dose determination [5–7]. The studies by Scott et al. [8, 9] based on the Monte Carlo simulation provided important insights into the influence of detector density on detector response in non-equilibrium small photon fields. Three types of detectors were studied: a PinPoint ionization chamber, a diamond detector, and a diode. The study found that the modeled PinPoint detector under responded relative to an ideal, fully water-equivalent detector. In contrast, the high-density material solid-state detectors (silicon and diamond) over responded.

A promising supplement to ionization chambers is fiber optically coupled luminescence detectors. The system basically functions by using the optical sensing technique where the dosimeter probe consists of a small piece of a sub-mm sized phosphor sample attached to a thin optical fiber cable (typically Polymethylmethacrylate (PMMA), Polystyrene (PS) and Polyvinyltoluene (PVT)). Radiation-induced generation of radioluminescence (RL) from the phosphor sample is transmitted through the fiber cable to be recorded by a sensitive light detector. The measured light intensity is proportional to dose rate. At present, the prime phosphors can be separated into two types: (i) organic plastic scintillators (e.g. BCF-12, BCF-60, and BC-400), and (ii) inorganic crystalline phosphors (e.g. Al_2O_3:C, LiF:W, SiO_2:Cu, and KBr:Eu) [10].

For several years, fiber-coupled luminescence dosimetry has been subject of research and development in several groups worldwide (including, for example, University of Sydney NSW, the University of Texas MD Anderson Cancer Center, the Konkuk University, and Technical University of Denmark). The fiber detectors offer several attractive features over ionization chambers including: (i) real-time dose monitoring with high time resolution— this property is particularly used in pulse-to-pulse measurement of doses delivered by pulsed radiation sources like LINACs, (ii) negligible treatment perturbation due to sub-mm detector size, (iii) a wide dynamic range of measured dose from a mGy to several Gy, and (iv) no electrical wire or other electronic devices in the dosimeter probe head to disturb the treatment or cause harm to patient [10–14]. However, the major problem of these detectors is due to the fiber cable itself; radiation also induces generation of fluorescence and Cerenkov light from the fiber material during exposure, so-call stem signal. This contribution signal lead to inaccuracy of the measured doses.

This article presents an overview of the use of fiber-coupled luminescence detectors that focuses on two luminescence phosphors of (i) organic plastic scintillator (BCF-60) and (ii) carbon-doped aluminium oxide crystal (Al_2O_3:C). In the following sections, the basics of luminescence physics and the luminescence detection instruments between two detector systems are introduced, and then the important physical parameters that are necessary to be considered as sources of variability in radiotherapy dosimetry are identified. Finally, the potential use of both detector systems in small and large MV photon fields is briefly discussed.

2. Luminescence mechanisms and detection instruments

2.1. Detector probes

In this article, two in-house fiber-coupled luminescence detectors developed by the Center for Nuclear Technologies, Technical University of Denmark, are reviewed. Both detector probes look physically identical, except their sensitive parts. One was made by a piece of BCF-60 (i.e. 1 mm diameter and 2 mm long from Saint-Gobain Ceramic & Plastics Inc., France) [15], and another was made by a Al_2O_3:C crystal (i.e. 1×1×2 mm^3 from Landauer Inc., USA) [6]. Each of them was polished and mechanically coupled with a polished PMMA optical fiber cable (i.e. 1 mm diameter and 10–15 cm long from Mitsubishi Rayon Co., Japan) using UV-curing glue (NOA68) with a black epoxy cement was coated for light-tighten and bond protection (Fig.1).

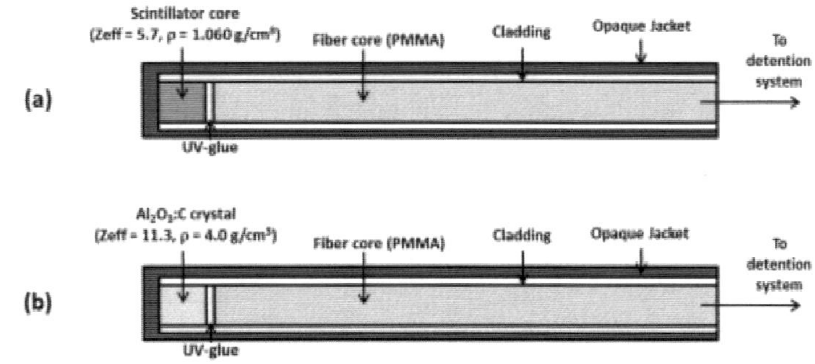

Fig.1. Sketch of the fundamental composition of (a) a fiber-coupled BCF-60 organic plastic scintillator probe, and (b) a fiber-coupled Al_2O_3:C crystal probe. Redraw from Buranurak [16]

2.2. Luminescence mechanisms

In principle, the luminescence mechanism in the organic plastic scintillators is completely different from the crystalline phosphors. When the BCF-60 is exposed, molecules of polystyrene base-substrate doped with perylene are excited, and de-excited after a short while (i.e. in range nano seconds or shorter) with RL emission. Perylene is an organic fluorescence dye used as a wavelength shifter to absorb shorter wavelength of UV released from the relaxation procedures of polystyrene molecules and emit longer wavelength of visible green instead (i.e. spectrum range: ~ (490–650 nm), emission peak: 530 nm (www.crystals.saint-gobain.com)). In case of Al_2O_3:C crystal, it is a perfect insulator with a wide energy band-gap of ~9.5 eV. When the crystal is subject to a beam of ionizing radiation, radiation-induced creation of electron-hole pairs migrate through the crystal via the de-localized conduction and valence bands. Some electrons are trapped in shallow, dosimetry and deep traps caused by dopant materials, impurities and other imperfections in the crystal structure. Other electrons can pass through to recombine with the holes at the recombination centers, resulting in the RL emission (i.e. the main emission from Al_2O_3:C is associated with the F-center luminescence which has the range of ~ (300–520 nm) with the centered at 420 nm and a ~ 35 ms lifetime [10, 17]. Additionally, all electrons trapped during exposure can be released to the conduction band using a stimulating light source. These electrons may recombine with the holes at the recombination centers resulting in the luminescence emission that is so-called the optically stimulated luminescence (OSL) [18, 19].

2.3. Luminescence detection instruments

As previously mentioned, the fiber-coupled luminescence detectors function by radiation-induced RL emissions from the luminescence phosphors which are guided through an optical fiber cable to be detected by a light detector. In principle, the measured RL signal should be proportional to the dose delivery rate. However, the contribution of stem signal, which is caused by radiation-induced generation of fluorescence and Cerenkov radiation in the fiber cable itself, leads super-imposed to the measured light signals that results in an inaccurate dose estimation. Over the years, several methods have been proposed to eliminate the stem effect, for example: background subtraction using dual fibers [12, 13], spectral filtration [20], temporal gate system [21], hollow-core fiber light guidance [22], and chromatic removal [23, 24]. Each method has both advantages and disadvantages depending on luminescence materials used and dosimetry applications.

The in-house developed ME30 luminescence reader is used for measuring RL signal and discriminating the unwanted stem signal from the fiber-coupled Al_2O_3:C probes. The system consists of four photomultiplier tubes (PMTs) (MP982, PerkinElmer Inc., USA) attached with 395−440 nm optical filters. The PMTs are operated in pulsed mode with a high-speed pre-scaler (1:8). The system basically functions by using the temporal gating circuit (compact-RIO 9074 controller, National Instruments Co. Ltd., USA) which is driven by the linear accelerator synchronization signal to block the counting events for an initial 25 μs after each synchronization pulse. Using this protocol, the main part of RL signal generated from the Al_2O_3:C crystal which has a relaxation time of ~ 35 ms can be measured, whereas the stem signal presented during ~5 μs of each accelerator gun pulse can be blocked. However, the temporal gating system cannot remove the stem signal from the fiber-coupled organic plastic scintillator probes because of the short luminescence lifetime for these materials (typically a few ns as reported by www.crystals.saint-gobain.com). The temporal stem removal technique only works in pulsed beams from linear accelerators when the luminescence time is longer than the pulse width.

The ME40 luminescence reader developed at the Technical University of Denmark is tailored for accelerator dosimetry with the fiber-coupled BCF-60 probes. The system basically functions using the chromatic removal method to separate the scintillation light from the stem signal. The system consists of two optical components of 45° dichroic mirrors (Edmund Optics Ltd., United Kingdom) using to split the total collected light into the blue and the green components. The collected light is firstly entered to the blue component where the stem signal dominated in the visible blue and soft UV light is reflected by a magenta/blue dichroic mirror and further to be deposited in the first PMT. While the scintillation light of BCF-60 dominated in the visible green can pass through this mirror to be reflected by a yellow/green dichroic mirror in the green component and eventually deposited in the second PMT. Figure 2 shows the simple sketches of (a) the fiber-coupled Al_2O_3:C probe equipped with the ME30 luminescence reader, and (b) the fiber-coupled BCF-60 probe attached to the ME40 RL detection instrument.

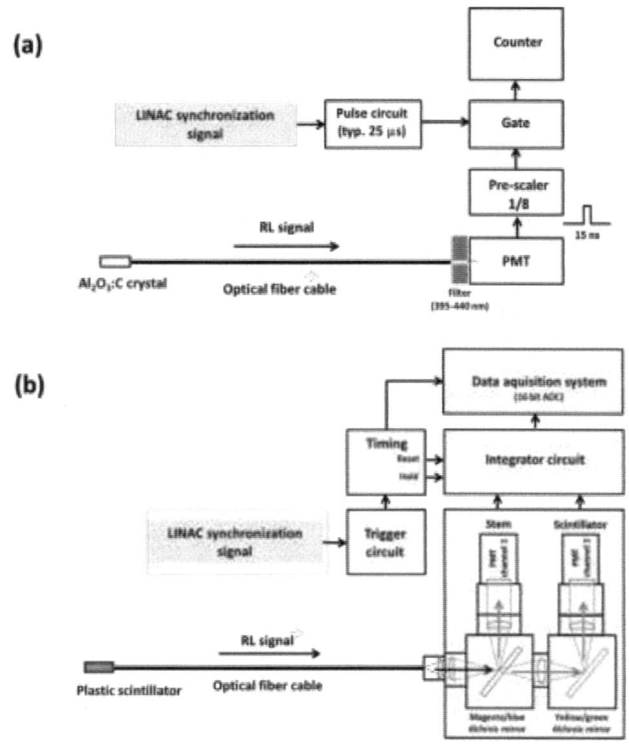

Fig.2. Schematic view of (a) the fiber-coupled Al_2O_3:C detector system applied with the temporal gate pulse circuit in ME30 instrument for stem signal subtraction, and (b) the fiber-coupled BCF-60 organic scintillator system equipped with the optical component of chromatic stem removal system and the data acquisition electronic circuits in ME40 instrument. Source: Buranurak [16].

3. Important dosimetric characteristics of the fiber-coupled luminescence detectors

In order to evaluate the performance of detector systems used in radiotherapy, these following dosimetric characteristics are necessary to be concerned.

3.1. Reproducibility

Measurements are always subject to various sources of uncertainty that will cause the measured result to differ from the true value [25]. In practical, such deviations are therefore required to be within an acceptable limit. Reproducibility is defined as the variation in measurements made on a subject under changing conditions (i.e. differences of measurement methods, instruments, observers, or observed times) [25, 26]. Therefore, reproducibility is a measure of the stability of measurement system.

Reproducibility of the fiber-coupled Al_2O_3:C detectors carried out using a 50 kV x-rays was found to be within 0.2% (1 SD) [11]. Additionally, the study by Aznar et al. (2004) reported that the reproducibility of the fiber-Al_2O_3:C system performed using the clinical photon beams of 6 MV and 18 MV (Varian Clinac 2300EX, Varian Medical System, Palo alto, USA) under reference conditions was found to be within 0.3% (1 SD) and 0.5% (1 SD), respectively. However, the RL/OSL signals from Al_2O_3:C fiber system was observed to have the day-to-day variability of the gain coefficient up to 2%. This suggests that the fiber-Al_2O_3:C system using the combined RL/OSL readout protocol requires frequent re-calibrations.

In case of the fiber-coupled BCF-60 organic scintillation detectors, the recent study by Beierholm et al. [14] reported that the reproducibility of this system carried out using a Varian TrueBeam™ in different beam quantities under the reference conditions was found to be within 0.4% (1 SD). While the day-to-day variability of the gain coefficient was found up to 2.5% (1 SD).

3.2. Detector density

The impact of detector density on the response of various detectors in small-field radiotherapy dosimetry was studied by Scott et al [8, 9]. It was found that a density correction factor is required to improve the accuracy of dose calculation. The density correction factor is defined as the ratio of Monte-Carlo calculated dose delivered to water and detector voxels at the same reference condition. As briefly mentioned in the introduction part, the studies observed under response for the gas-filled PinPoint detector (i.e. a low-density detector material), whereas over response was observed in the high-density solid-state silicon diode and diamond detectors. Based on Scott's concept, Monte-Carlo computations of density correction factors were carried out by Buranurak [16] without detailed modelling of the accelerator head. For the sensitive volume of the fiber-coupled BCF-60 detectors, a voxel of polystyrene-based organic scintillator was modeled. Additionally, a voxel of Al_2O_3:C crystal was also modeled to represent the sensitive volume of the fiber-coupled Al_2O_3:C detectors. The study found an overdose response up to 27% for the Al_2O_3:C voxel at the smallest field size of 0.25×0.25 cm^2. In contrast, no change was observed for the polystyrene voxel at the same field size. Therefore, the impact of detector density on the dose-response was found to decrease with increasing field size. For field size larger than 2×2 cm^2, it was found to be negligible. For the modeled voxels of air and silicon, the profiles of the density correction factors with field sizes were found to be consistent with the Scott's study. Interestingly, the profiles of the density correction factors with field sizes of the polystyrene-based modified-density materials such as polystyrene-based Al_2O_3:C, polystyrene-based silicon, and polystyrene-based air were found to be correlated as mirrored to their originals, with deviation ~13% was observed. This study suggests that the density effect was mostly related to field-size dependent changes in electron fluence spectra created in the water phantom. In other words, the density effect reported by Scott et al. [9] was reproduced on the basis of a simplified model for simulation without any changes of the primary photon fluence with field size. This study indeed supports that the polystyrene-based organic plastic scintillators can be used for small field dosimetry without any requirement of the density correction factors. This is also supported by Monte-Carlo studies [27, 28] and experimental studies [29, 30].

3.3. Energy dependence

In principle, changing in energy spectrum of ionizing radiation influences on the dosimetric response of a detector. Energy dependence therefore is an important physical parameter in dosimetry whenever the detector is used under conditions that are not identical to the conditions under which the detector was calibrated. For the fiber-coupled Al_2O_3:C detectors, the variation of output signals was observed to be 0.6% (1 SD) for both RL and OSL signals, when the detectors were exposed using 6 and 18 MV photon beams with a 2 Gy dose delivery at a 10 cm depth in water [11].

For the fiber-coupled polystyrene-based organic scintillation detectors, the detector sensitivity was observed to be energy independent for both photon and electron beams [13, 24]. This was confirmed by the Monte-Carlo studies [31]. However, in practical, the magnitude of the stem effect is of more concern because small systematic errors may lead to a large effect on the detector reading. For this reason, stem effect baseline correction is necessary to be taken into account. The study by Beierholm et al. [14] observed the energy dependence in the in-house fiber-coupled BCF-60 detectors within 0.4% (1 SD) for energies ranged from 6 MV to 15 MV. This was found to be consistent with the commercial fiber-coupled polystyrene organic scintillation detectors called Exradin W1 [32, 33].

3.4. Temperature dependence

Temperature effect in the fiber-coupled organic plastic scintillation detectors was firstly studied by Beddar et al. [13]. The study was performed using polyvinyltoluene-based organic scintillators which were exposed to 4 MV x-ray beam under the controlled temperature of 0 – 50 °C, no significant difference of the overall measurement uncertainty was observed. However, temperature effect in the fiber-coupled organic scintillators was revisited and studied again by Beddar [34] and Buranurak et al. [35]. According to the Buranurak's study, temperature effects in four polystyrene-based organic scintillators were investigated. All tests were carried out using a 50 kV x-ray source where Cerenkov light generation in the fiber cable would not influence the results. The temperature was controlled in the clinical range of 15 – 40 °C. Different temperature coefficients were observed for the four-tested scintillators with the maximum of –(0.55 ± 0.04)%/K for BCF-60. This was found to be consistent with the study by Wootton and Beddar [36].

Investigation of temperature dependence in the fiber-coupled Al_2O_3:C detectors was reported by Andersen et al [37]. All measurements were carried out using a 50 kV x-ray source with 10 – 40 °C temperature controlled. The study was found a temperature coefficient of –(0.21 ± 0.01)%/°C.

3.5. Dose-rate dependence

Ideally, the detector response for overall dose should remain constant regardless with the accelerator dose delivery rate applied [38]. Dose-rate effect is interesting for radiotherapy dosimetry. This is due to the fact that dose rate applied for detector calibration generally uses 600 MU/min but only low dose rate is practically used for preventing or reducing late side effects of radiotherapy [39, 40].

Dose-rate effect in fiber-coupled organic scintillation detectors was firstly investigated by Beddar et al. [13]. The effect of dose-rate was found to be nearly independent; 0.05% deviation of the detector response was observed when the delivered dose rate was changed from 100 – 400 cGy/min.

For the fiber-coupled Al_2O_3:C detectors, a small, but highly significant dose rate effect was found using the saturated RL readout protocol. The study by Andersen et al [37] reported a change in RL signal of –(0.22 ± 0.01) % when the delivered dose rate was increased from 100 to 600 MU/min. However, there was no clear explanation whether this effect was a feature of the luminescence mechanisms inside the crystal itself or the specific instrumentation used. The problem of dose-rate effect in the Al_2O_3:C fiber detectors was therefore further investigated by Buranurak and Andersen [41]. An iris collimated to 2 mm diameter was used to limit the intensity of RL transmitted to the PMT. All measurements were carried out using 6 MV x-ray to deliver dose of 300 MU with dose rate was varied from 100 to 600 MU/min, and source-to-surface distance (SSD) was varied from 70 to 140 cm. Measured RL counts between with and without iris were compared. The study found that the effect of dose-rate could be resolved using an iris to reduce RL signal which lead to the reduction of count-rate. Additional measurements indicated that the dose-rate slope was found to be decreased with increasing SSD; because of count-rate decreased. This study provided a clear evidence that the dose rate effect was mainly caused by a dead-time problem in the PMT rather than the RL kinetics of Al_2O_3:C crystal. Therefore, a simple dead-time correction factor can be used to provide dosimetry also under a wide range for clinically relevant dose rates.

4. Potential use of fiber-coupled luminescence detectors in small and large MV photon fields

Throughout the years, the use of fiber-coupled luminescence detectors with both organic plastic scintillators and Al_2O_3:C crystals have been applied in variety dosimetry applications. Some of these applications have been reported in literature, for example: dosimetry applications in external beam radiotherapy [37, 42, 43], brachytherapy [43–48], particle therapy [49–52], and total body irradiation [41]. In this part, benefits and drawbacks of both detector systems applied in the external beam radiotherapy with comparison of small and large MV photon fields are discussed.

4.1. Small-field dosimetry

Current and future treatment modalities such as intensity modulated radiation therapy (IMRT), Stereotactic radiotherapy and volumetric modulated arc therapy (VMAT) tend to employ smaller treatment fields to ensure optimal dose conformity over a small target volume while minimizing damage to surrounding healthy tissue. However, accurate dose measurement in small treatment fields is a well-known challenge for many detectors due to a large degree of uncertainty was observed [3, 53, 54]. Basically, the uncertainty is mainly caused by (i) partial occlusion of the primary beam by collimating devices, (ii) lack of lateral charge-particle equilibrium, (iii) inhomogeneity of dose-averaging over the sensitive volume of detectors, and (iv) density differences between detectors and their surrounding media [4, 55]. Therefore, a detector-specific correction factor in small fields is necessary to take into account [29, 54].

Fiber-coupled organic plastic scintillation detectors have been hypothesized to have a minimal perturbation to the treatment due to a sub-mm size diameter. Additionally, they provide many favorable dosimetric characteristics of nearly water-equivalent, independent of energy, angular, and dose delivery rate, dose linearity and resistance to radiation damage [13, 14, 28, 56]. Based on these benefits, the organic scintillation detectors look promising for reference measurements in order to determine the small-field correction factors for ionization chambers and other commercial detectors. An interesting study was conducted by Beierholm [29] to assess the current status of reference dosimetry and small-field dosimetry in clinical practice. Experimental study was carried out at six Danish clinics. Small-field correction factors for a PinPoint chamber and a diamond detector were estimated using the output factors measured from both detectors relative to the in-house fiber-coupled BCF-60 detectors used as reference. The small-field correction factors were also corrected for volume averaging. Deviation within 2% (1 SD) was observed at 1×1 cm^2 field size.

For the fiber-coupled Al_2O_3:C detectors, the experimental study related to the use of this detectors for small treatment field dosimetry and investigation of small-field correction factors has never been reported in literature. However, Monte-Carlo simulation indicated that a discrepancy in the output factors between the Al_2O_3:C and polystyrene-based organic scintillator probes of up to 4.3% [57].

4.2. Large field dosimetry

Total body irradiation (TBI) is classified as a large field radiotherapy treatment technique where a homogeneous dose is delivered to treat the entire body of patient. In practical, TBI is used for treatment of specific hematological diseases [58]. Additionally, it can be applied to be used as part of conditioning regimen for bone marrow or hematopoietic stem cell transplantation [59–62]. Throughout the years, the fiber-coupled luminescence detectors with both organic scintillators and Al_2O_3:C crystal offer the capability to be used in a wide range of dosimetry applications, and may be used as a reference dosimeter in small-field radiotherapy. However, the use of fiber detectors either coupled with organic scintillators or crystalline phosphors applied for large-field dosimetry as TBI treatment has not been reported previously. In practice, large-field dosimetry is challenge for the fiber systems due to the fact that the influence of the stem effect on the measured RL signals is increased proportionally with the increasing length of fiber cable subject to the primary beam. The recent study by Buranurak and Andersen [41] demonstrated the efficiency of stem signal subtraction comparing between the temporal gating circuit in the ME30 instrument used for the fiber-coupled Al_2O_3:C detectors and the chromatic stem removal system in the ME40 instrument used for the fiber-coupled BCF-60 organic scintillation detectors. Measured doses with field size variation from both detector systems were compared between two configurations: (i) minimum configuration where the detector probe was arranged in the solid phantom as the same calibration setup, and (ii) extra-fiber configuration where a ~ 2.2 m fiber cable was looped and placed on top of the phantom. The extra-fiber configuration was used to simulate the situation as TBI treatment where an approximately 2 m of the cable length was subject to the treatment beam. The study demonstrated failure of the chromatic system to remove the substantial amount of stem signal generated in the extra-fiber configuration. The inadequacy of the chromatic stem removal technique in large fields could be due to violations of the basic assumptions behind the technique: (i) that the spectrum of the stem signal remains constant regardless of the irradiation conditions, and (ii) linearity of the detector over the range of the light levels experienced by the system. In contrast, the study found the temporal gating system to essentially remove stem signal perfectly, no matter how much of the cable length is exposed.

5. Conclusions and future perspectives

Nowadays, the tendency in radiotherapy is going toward the development of precise beam-delivery machines and more conformal treatments to a small and well-defined tumor. To achieve this goal, the extensive QA programs together with accurate and precise dosimetry system play a key role in this work. Therefore, the development of improved detectors that potentially can be used as reference measurements in various applications remains relevant. Fiber-coupled luminescence detectors have been developed over two decades to fulfill this demand. The systems provide several attractive features for radiotherapy dosimetry including real-time in vivo dose monitoring with high time resolution, minimal perturbation to the treatment, high dynamic range of dose measurements, reliability and robustness. Especially, the fiber detectors coupled with organic scintillators offer favorable dosimetric characteristics of nearly water equivalent, high reproducibility, independency in energy, angular, and dose delivery rate, and negligible density effect. However, as general solid-state detectors, thermal effect has been observed in this fiber system. The correction factor for the influence of temperature variations is required for in vivo dosimetry.

Regarding these impressive characteristics, up to now, the fiber system based on organic plastic scintillators looks promising as the best candidate to be used for reference measurements in small treatment fields, rather than other real-time detectors. However, the problem of parasitic stem signal contribution is necessary to be concerned, and the stem effect is more significant relative to the increased length of fiber cable in the treatment beam. The chromatic stem removal technique used for the fiber-coupled organic scintillators is inefficient for large-field dosimetry as TBI treatment. The reason is possible due to the failure of fundamental assumptions in the chromatic technique–either the linearity in dose response when a huge amount of stem signal is generated, or the requirement of constant spectra of the stem signal. Background subtraction using dual fibers is a conventional technique to remove stem signal from the fiber-coupled organic scintillators alternative to the chromatic removal technique. This technique looks promising to remove the huge stem signal generated during TBI treatment. However, the practical use of this technique for large-field dosimetry needs to investigate further.

In contrast, the temporal gating circuit used for the fiber-coupled Al_2O_3:C crystal offers the perfect stem signal subtraction; even if the huge stem signal is generated. However, this detector system provides unimpressive dosimetric characteristics, thus many correction factors are required for more accurate dose estimation.

It is the opinion of the authors that the fiber-coupled luminescence detectors with both organic scintillators and Al_2O_3:C crystal provide good potential to be used for radiotherapy dosimetry, but in different circumstances. The organic scintillator system is suitable for small-field dosimetry due to nearly perturbation-free in small treatment fields, whereas the crystal system can be used in large-field dosimetry of pulsed accelerator beams based on the near-perfect stem signal subtraction of the temporal gating technique.

References

[1] Urbański B 2012 The future of Radiation Oncology: Considerations of Young Medical Doctor *Reports Pract. Oncol. Radiother.* **17**(5) 288–93
[2] Shalek R 1977 Determination of Absorbed Dose in a Patient Irradiated by Beams of X or Gamma Rays in Radiotherapy Procedures *J. Med. Phys.* **4**(5) 461
[3] Das IJ, Ding GX, and Ahnesjö A 2008 Small fields: nonequilibrium radiation dosimetry *J. Med. Phys.* **35**(1) 206–15
[4] Aspradakis MM, Byrne JP, Palmans H, Duane S, Conway J, Warrington AP, and Rosser K 2010 IPEM report 103: Small field MV photon dosimetry (IAEA-CN--182)
[5] Nilsson B, Montelius A and Andreo P 1996 Wall effects in plane-parallel ionization chambers *J. Phys. Med. Biol.* **41**(4) 609–23
[6] Nahum AE 1996 Perturbation effects in dosimetry: Part I. Kilovoltage x-rays and electrons *J. Phys. Med. Biol.* **41**(9) 1531–80
[7] Sauer O and Wilbert J 2007 Measurement of output factors for small photon beams *J. Med. Phys.* **34**(6) 1983–88
[8] Scott AJD, Nahum AE, and Fenwick JD 2008 Using a Monte Carlo model to predict dosimetric properties of small radiotherapy photon fields *J. Med. Phys.* **35**(10) 4671–84
[9] Scott AJD, Kumar S, Nahum AE, and Fenwick JD 2012 Characterizing the influence of detector density on dosimeter response in non-equilibrium small photon fields *J. Phys. Med. Biol.* **57**(14) 4461–76
[10] Andersen CE 2011 Fiber-coupled Luminescence Dosimetry in Therapeutic and Diagnostic Radiology *Concepts Trends Med. Radiat. Dosim. AIP Conf. Proc.* **1345** 100–19
[11] Aznar MC, Andersen CE, Botter-Jensen L, Back SA, Mattsson S, Kjær-Kristoffersen F, and Medin J 2004 Real-time optical-fibre luminescence dosimetry for radiotherapy: physical characteristics and applications in photon beams. *J. Phys. Med. Biol.* **49**(9) 1655–69

IOP Conf. Series: Journal of Physics: Conf. Series **860** (2017) 012028 doi:10.1088/1742-6596/860/1/012028

[12] Beddar S, Mackie TR, and Attix FH 1992 Water-equivalent plastic scintillation detectors for high- energy beam dosimetry: I Physical characteristics and theoretical considerations *J. Phys. Med. Biol.* **37**(10) 1883–900

[13] Beddar S, Mackie TR, and Attix FH 1992 Water-equivalent plastic scintillation detectors for high- energy beam dosimetry: II Properties and measurements *J. Phys. Med. Biol.*. **37**(10) 1901–13

[14] Beierholm AR, Behrens CF, and Andersen CE 2014 Dosimetric characterization of the Exradin W1 plastic scintillator detector through comparison with an in-house developed scintillator system *J. Radiat. Meas.* **69** 50–6

[15] Beierholm AR, Ottosson RO, Lindvold LR, Behrens CF, and Andersen CE 2011 Characterizing a pulse-resolved dosimetry system for complex radiotherapy beams using organic scintillators *J. Phys. Med. Biol.* **56**(10) 3033-45

[16] Buranurak B 2016 Toward reference-class fiber-coupled luminescence dosimetry for radiotherapy *PhD thesis* (Denmark: Technical University of Denmark)

[17] Chithambo ML, Nyirenda AN, Finch AA, and Rawat NS 2015 Time-resolved optically stimulated luminescence and spectral emission features of α-Al2O3:C *J. Phys. B Condens. Matter.* **473** 62–71

[18] Polf JC, Yukihara EG, Akselrod MS, and McKeever SWS 2004 Real-time luminescence from Al2O3 fiber dosimeters *J. Radiat. Meas.* **38**(2) 227–40

[19] Yukihara EG, McKeever SWS, and Akselrod MS 2014 State of art: Optically stimulated luminescence dosimetry – Frontiers of future research *J. Radiat. Meas.* **71** 15–24

[20] De Boer SF, Beddar S, and Rawlinson JA 1999 Optical filtering and spectral measurements of radiation-induced light in plastic scintillation dosimetry *J. Med. Phys.* **38**(7) 945–58

[21] Clift MA, Sutton RA, and Webb DV 2000 Water equivalence of plastic organic scintillators in megavoltage radiotherapy bremsstrahlung beams *J. Phys. Med. Biol.* **45**(7) 1885–95

[22] Lambert J, Yin Y, McKenzie DR, Law S, and Suchowerska N 2008 Cerenkov-free scintillation dosimetry in external beam radiotherapy with an air core light guide *J. Phys. Med. Biol.* **53**(11) 3071–80

[23] Fontbonne JM, Iltis G, Ban G, Battala A, Vernhes JC, Tillier J, Bellaize N, Le Brun C, Tamain B, Mercier K, and Motin JC 2002 Scintillating fiber dosimeter for radiation therapy accelerator *J. IEEE Trans. Nucl. Sci.* **49**(5). 2223–27

[24] Frelin AM, Fontbonne JM, Ban G, Colin J, Labalme M, Batalla A, Isambert A, Vela A, Leroux T 2005 Spectral discrimination of Cerenkov radiation in scintillating dosimeters *J. Med. Phys.* **32**(2005) 3000–6

[25] Bartlett JW and Frost C 2008 Reliability, repeatability and reproducibility: analysis of measurement errors in continuous variables *Ultrasound in Obstetrics and Gynecology* (New York: Wiley & Sons) **31**(4) 466–75

[26] Taylor BN and Kuyatt CE 1994 Guidelines for Evaluating and Expressing the Uncertainty of NIST Measurement Results. *NIST Technical Note 1297* (US: International Institute of Standards and Technology)

[27] Wang LLW, Klein D, and Beddar S 2010 Monte Carlo study of the energy and angular dependence of the response of plastic scintillation detectors in photon beams *J. Med. Phys.* **37**(10) 5279–86

[28] Wang LLW and Beddar S 2011 Study of the response of plastic scintillation detectors in small-field 6 MV photon beams by Monte Carlo simulations *J. Med. Phys.* **38**(3) 1596–99

[29] Beierholm AR, Behrens CF, Patrik S, Rønde HS, Biancardo S.BN, Aznar MC, Thomsen JB, Præstegaard L, Nyvang L, Riis HL, Pedersen KM, Helt-Hansen J, and Andersen CE 2014 Reference dosimetry and small-field dosimetry in external beam radiotherapy: Results from a Danish intercomparison study *Annual Report: DTU Nutech No. 8(EN)* p.29 (Denmark: Technical University of Denmark)

[30] Morin J, Beliveau-Nadeau D, Chung E, Seuntjens J, Theriault D, Archambault L, Beddar S, and Beaulieu L 2013 A comparative study of small field total scatter factors and dose profiles using plastic scintillation detectors and other stereotactic dosimeters: the case of the

CyberKnife *J. Med. Phys.* **40**(1) 11719

[31] Beddar S, Briere TM, Mourtada FA, Vassiliev ON, Liu HH, and Mohan R 2005 Monte Carlo calculations of the absorbed dose and energy dependence of plastic scintillators *J. Med. Phys.* **32**(5) 1265–69

[32] Beierholm AR, Behrens CF, and Andersen CE 2016 Comment on 'Characterization of the Exradin W1 scintillator for use in radiotherapy' [J. Med. Phys. 42 297–304 (2015)] Correspondence Comment on 'Characterization of the Exradin W1 scintillator' *J. Med. Phys.* **304**(2015) 1–4

[33] Carrasco P, Jornet N, Jordi O, Lizondo M, Latorre-Musoll A, Eudalda T, Ruiz A, and Ribas M 2015 Characterization of the Exradin W1 scintillator for use in radiotherapy Characterization of the Exradin W1 scintillator for use in radiotherapy *J. Med. Phys.* **42**(297) 297–304

[34] Beddar S 2012 On possible temperature dependence of plastic scintillator response *J. Med. Phys.* **39** 6522

[35] Buranurak S, Andersen CE, Beierholm AR, Lindvold LR 2014 Temperature variations as a source of uncertainty in medical fiber-coupled organic plastic scintillator dosimetry *J. Med. Phys.* **56**(2013) 307-311

[36] Wootton L and Beddar S 2014 Temperature dependence of BCF plastic scintillation detectors *J. Phys. Med. Biol.* **58**(9) 2955–67

[37] Andersen CE, Damkjær SMS, Kertzscher G, Greilich S, and Aznar MC 2011 Fiber-coupled radioluminescence dosimetry with saturated Al 2O3:C crystals: Characterization in 6 and 18 MV photon beams *J. Radiat. Meas.* **46**(10) 1090–98

[38] O'Keeffe S, McCarthy D, Woulfe P, Grattan MW, Hounsell AR, Sporea D, Vata I, Leen G, and Lewis E 2015 A review of recent advances in optical fibre sensors for in vivo dosimetry during radiotherapy *Br. J. Radiol.* **88**(1050) 20140702

[39] Bentzen SM 2006 Preventing or reducing late side effects of radiation therapy: radiobiology meets molecular pathology *J. Nat. Rev. Cancer.* **6**(9) 702–13

[40] Ling CC, Gerweck LE, Zaider M, and Yorke E 2010 Dose-rate effects in external beam radiotherapy redux *J. Radiother. Oncol.* **95**(3) 261–68

[41] Buranurak S and Andersen CE 2016 Fiber-coupled Al2O3:C radioluminescence dosimetry for total body irradiations *J. Radiat. Meas.* **93** 46–54

[42] Andersen CE, Marchmann CJ, Aznar MC, Botter-Jensen L, Kjær-Kristoffersen F, and Medin J 2006 An algorithm for real-time dosimetry in intensity-modulated radiation therapy using the radioluminescence signal from Al2O3:C *J. Radiat. Prot. Dosim.* **120**(1–4) 7–13

[43] Archambault L, Briere TM, Pönisch F, Beaulieu L, Kuban DA, Lee A, and Beddar S 2010 Toward a Real-Time In Vivo Dosimetry System Using Plastic Scintillation Detectors *Int. J. Radiat. Oncol.***78**(1) 280–87

[44] Andersen CE, Nielsen SK, Greilich S, Helt-Hansen J, Lindegaard JC, and Tanderup K 2009 Characterization of a fiber-coupled Al2O3:C luminescence dosimetry system for online in vivo dose verification during 192Ir brachytherapy. *J. Med. Phys.***36**(3) 708–18

[45] Kertzscher G, Andersen CE, Edmund JM, and Tanderup K 2011 Stem signal suppression in fiber-coupled Al2O3:C dosimetry for 192Ir brachytherapy *J. Radiat. Meas.* **46**(12) 2020–24

[46] Klawikowski SJ, Zeringue C, Wootton LS, Ibbott GS, and Beddar S 2014 Preliminary evaluation of the dosimetric accuracy of the in vivo plastic scintillation detector OARtrac system for prostate cancer treatments *J. Phys. Med. Biol.* **59**(9) 27–36

[47] Suchowerska N, Jackson M, Lambert J, Yin YB, Hruby G, and McKenzie DR 2011 Clinical trials of a urethral dose measurement system in brachytherapy using scintillation detectors *Int. J. Radiat. Oncol. Biol. Phys.***79**(2) 609–15

[48] Lambert J, McKenzie DR, Law S, and Suchowerska N 2006 A plastic scintillation dosimeter for high dose rate brachytherapy *J. Phys. Med. Biol.* **51**(21) 5505–16

[49] Greilich S, Edmund JM, Jain M, and Andersen CE 2008 A coupled RL and transport model for mixed-field proton irradiation of Al2O3:C *J. Radiat. Meas.***43**(2–6) 1049–53

[50] Edmund JM, Andersen CE, Marckmann CJ, Aznar MC, Akselrod MS, and Bøtter-Jensen 2006 CW-OSL measurement protocols using optical fibre Al2 O3:C dosemeters *J. Radiat. Prot.*

Dosim. **119**(1–4) 368–74

[51] Ghergherehchi M, Afarideh H, Ghannadi M,. Mohammadzadeh A, Aslani GR, and Boghrati B 2010 Proton Beam Dosimetry: a Comparison between a Plastic Scintillator, Ionization Chamber and Faraday Cup *J. Radiat. Res.***51**(4) 423–30

[52] Jang KW, Yoo WJ, Shin SH, Shin D, and Lee B 2012 Fiber-optic Cerenkov radiation sensor for proton therapy dosimetry. *J. Opt. Express*. **20**(13) 13907–14

[53] Verhaegen F 1998 Monte Carlo dosimetry study of a 6 MV stereotactic radiosurgery unit *J. Phys. Med. Biol.* **43**(10) 2755-68

[54] Alfonso R, Andreo P, Capote R, Huq MS, Kilby W, Kjäll P, Mackie TR, Palmans H, Rosser K, and Seuntjens J 2008 A new formalism for reference dosimetry of small and nonstandard fields *J. Med. Phys.* **35**(11) 5179–86

[55] Beaulieu L, Goulet M, Archambault L, and Beddar S 2013 Current status of scintillation dosimetry for megavoltage beams *J. Phys. Conf. Ser.* **444**(1) 12013

[56] Mijnheer B, Beddar S, Izewska J, and Reft C 2013 In vivo dosimetry in external beam radiotherapy *J. Med. Phys.* **40**(7) 70903

[57] Beierholm AR 2011 Pulse-resolved radiotherapy dosimetry using fiber-coupled organic scintillators *PhD thesis* (Denmark: Technical University of Denmark)

[58] Khan FM 2003 *The physics of radiation therapy:3rd ed.* (US: Lippincott Williams & Wilkins)

[59] Quast U 1987 Total body irradiation- review of treatment techniques in Europe *J. Radiother. Oncol.* **9**(2) 91–106

[60] Mangili P, Fiorino C, Rosso A, Cattaneo GM, Parisi R, Villa E, and Calandrino R 1999 In-vivo dosimetry by diode semiconductors in combination with portal films during TBI: Reporting a 5-year clinical experience *J. Radiother. Oncol.* **52**(3) 269–76

[61] Aydogan B, Mundt AJ, and Roeske JC 2006 Linac-based intensity modulated total marrow irradiation (IM-TMI). *J. Technol. Cancer Res. Treat.* **5**(5) 513–19

[62] Chen HH, Wu J, Chuang KS, Lin JF, Lee JC, and Lin JC 2013 Total body irradiation with step translation and dynamic field matching *J. Biomed Res. Int.* **2013** 216034

Acknowledgments

On behalf of co-author; Claus E. Andersen, and myself, I would like to thank Lars R. Lindvold, senior scientist at DTU Nutech for very kind suggestion and support. I would also like to thank all staff of the International Nuclear Science and Technology Conference (INST 2016) for organizing a good conference and giving me an opportunity to present this work with a registration fee waiver awarded.

Cyclotron production of ^{48}V via ^{nat}Ti(d,x)^{48}V nuclear reaction; a promising radionuclide

A R Usman[1], M U Khandaker[1,3] and H Haba[2]

[1] Department of Physics, University of Malaya, 50603 Kuala Lumpur, Malaysia
[2] Nishina Center for Accelerator-Based Science, RIKEN, Wako, Saitama 351-0198, Japan

[3] Corresponding author e-mail: mu_khandaker@yahoo.com

Abstract. In this experimental work, we studied the excitation function of ^{nat}Ti(d,x)^{48}V nuclear reactions from 24 MeV down to threshold energy. Natural titanium foils were arranged in the popular stacked-foil method and activated with deuteron beam generated from an AVF cyclotron at RIKEN, Wako, Japan. The emitted γ activities from the activated foils were measured using an offline γ-ray spectrometry. The present results were analyzed, compared with earlier published experimental data and also with the evaluated data of Talys code. Our new measured data agree with some of the earlier reported experimental data while a partial agreement is found with the evaluated theoretical data. In addition to the use of ^{48}V as a beam intensity monitor, recent studies indicate its potentials as calibrating source in PET cameras and also as a (radioactive) label for medical applications. The results are also expected to further enrich the experimental database and also to play an important role in nuclear reactions model codes design.

1. Introduction

The use of short-lived positron emitting radionuclides in medical diagnostic procedures is nowadays a very well developed technique, most especially as imaging modality in positron emission tomography. In such procedures, the short-lived radionuclides, like ^{15}O ($T_{1/2}$= 2 min), ^{11}C ($T_{1/2}$ = 20.4 min) or ^{18}F ($T_{1/2}$ = 110 min), are usually labeled with pharmaceuticals of biological origin. By using these formed radiopharmaceuticals (e.g $^{11}CH_3I$, ^{18}FDG) with PET cameras, excellent images of heart, brain or tumors can be obtained for different kind of medical examinations. However, the short-lived nature of these positron emitters is usually a disadvantage when performing studies of slow biological processes [1]. The longer lived positron emitters (also known as non-standard positron emitters or Innovative PET radionuclides) have been reported as more suited to the study of slow metabolic activities and to labelling of organic compounds [2]. The decay properties of the longer-lived radionuclides play an important role in obtaining quality and volume of images as it allows ample time for image measurement during biological studies.

The usage of metallic titanium (Ti) in medical field is wide ranging. Since it has not been known to react with human body, titanium metal is widely used in several biological implants such as artificial hips and formation of pins for setting bones [3, 4]. On the other hand, charged particle bombardment of natural or enriched titanium isotopes produce several radionuclides useful in medical and biological fields. The production of ^{48}V via proton or deuteron has been reported by several studies. Vanadium-48 ($T_{1/2}$=15.9735 d), a positron emitting radionuclide, which decays to stable ^{48}Ti by EC (50.1%) and β^+(49.9%), find its acceptance as a monitor radioisotope mainly due to its decay characteristics and its relatively long half-life. The investigated radionuclide (^{48}V) is recently finding an increasing

IOP Conf. Series: Journal of Physics: Conf. Series **860** (2017) 012029 doi:10.1088/1742-6596/860/1/012029

potentials in different fields such as, from being radioactive tracer in biological processes [5] and material science [6] studies to its applicability in renal artery brachytherapy procedure [7]. Furthermore, ^{48}V was associated with some important biochemical characteristics such as anti-carcinogenic effect and thus tipped to be labeled with some compounds for *in vivo* studies [8]. In addition, a conducted study on 'alternative radionuclide' to the conventional use of ^{68}Ge ($T_{1/2}$ =271 d) in PET to further improve the quality of images has suggested the use of ^{48}V as a transmission source for correction of possible attenuation in the positron emission tomography technique [9]. Therefore, the need of precise measurement of excitation function for ^{48}V is very important.

An extensive survey on the reported experimental data on production of the beam monitoring and medically important ^{48}V via deuteron bombardment route indicated several studies are available. However, there exist discrepancies among the experimental literature, leading to the need of more measurements. In beam monitoring purpose and medical applications, high accuracy of radionuclide cross-sections are especially important and therefore the need for additional measurements are vital. Present study reports new experimental cross-sections of ^{48}V via deuteron bombardment on natural titanium foils.

2. Experimental setup and Analysis

Present study was performed using similar main experimental procedures to some of our previous works [10-14]. The irradiation of target foils was performed using the well-established stacked foil activation technique while measurements of activities of the ^{48}V was achieved via HPGe γ-ray spectrometry. The excitation function of ^{48}V radionuclide has been reported in the energy region of 2.5 to 23.8 MeV. Additional details relevant to present study are explained under the sub-sections below.

2.1 Targets, bombardment and activity measurements

Metallic titanium (20.32-μm thick; 99.99% purity; Goodfellow, UK) foil of natural isotopic abundance (^{46}Ti: 8.25 %; ^{47}Ti: 7.44 %; ^{48}Ti: 73.72 %; ^{49}Ti: 5.41 %; ^{50}Ti: 5.18 %) [15] was used as the primary target material. For beam monitoring and additional excitation functions measurements, several high purity (99.99% purity; Goodfellow, UK) natural metallic foils of nickel, and platinum were interleaved between the titanium foils in the stack. The target foils were weighed with high precision electronic balance for precise determination of foil thickness. A stack was prepared with uniform foil dimensions of 15 × 15 mm^2 following the size of the target holder to ensure proper focusing of the incident beam to center of the stack. The prepared stack was then held in a water-cooled target holder, which serves as a Faraday cup and then bombarded for 2.0 h with 24-MeV deuteron beam of characteristic average beam current of about 200 nA, from the AVF cyclotron of RIKEN RI Beam Factory. Through a tantalum slit in the particle exit channel of the cyclotron, the beam was collimated to 9-mm diameter onto the target foils. The whole processes of stack preparation, irradiation and activity measurements were carried out at the Nishina Center for Accelerator-Based Science, RIKEN, Wako, Saitama, Japan. The emitted γ-rays were measured without any chemical separation using a high resolution HPGe γ-ray spectrometer (ORTEC; GEM-25185P; 55.1-mm crystal diameter and 52.0-mm thickness; operating voltage: +2000 V; relative efficiency: 25%) which was coupled to a 4096 multi-channel analyser and other associated electronics. A cooling period of 9 to 10 days were allowed before the measurements of this radionuclide based on its half-life so as to ensure interference free cross-sections. The measurements were repeated several times at different source-to-detector distances. Similarly, all measurements were done with consideration of dead time less than 10 % by adjusting the source to detector distances. The gamma analysis of the spectra was done using the Maestro (Ver. 7.01; ORTEC) gamma vision program.

The efficiencies of the detector used at various source-to-detector distances were evaluated using multi-nuclide γ-ray standard source obtained from DBA Isotopes Products Laboratories (USA). Further details of major procedure involved in the calculations of the efficiencies at various source-to-distances were earlier reported in our similar studies [14].

2.2 Determination of beam intensity, foil energies and cross-sections

The beam intensity was determined from measured activity of $^{nat}Ni(d,x)^{61}Cu$ from the front foil facing the beam, using the IAEA [16] $^{nat}Ni(d,x)^{61}Cu$ recommended monitor reaction cross-section. The intensity was considered as a constant in the stack and was used to deduce cross-sections for each foil in the stack. The uniformity of the deuteron beam intensity along the irradiated foils was further ascertained by calculating cross-sections of ^{61}Cu from the other Ti foils in the stack, and comparing with the IAEA [16] recommended values. The presence of other metallic foils of different densities and thicknesses helped in slowing down the incident deuteron beam along the stack. The degradation of the initial 24 MeV deuteron beam energy along the stacked foils was calculated using a computer program, SRIM-2003 software [17]. The average deuteron-energy on each foil was solely dependent on the foil position in the stack.

Taking into account some of the target parameters such as density, thickness and the incident bombarding particles, the cross-sections, σ (E), of the assessed radionuclide were computed using the well-known activation formula [10, 14].

Similarly, we adopted decay data of the investigated reaction products from the ENSDF library [18] retrieved via the interface of Live Chart of nuclides [19]. The Q-values and threshold energies were calculated based on the AME mass evaluation [20], extracted through the Q-tool system [21] and they are as well presented in the Table 1.

Table 1. Decay data of the assessed ^{48}V.

Nuclide	Half-life	Decay mode (%)	E_γ (keV)	I_γ (%)	Contributing reactions	Q-Value (MeV)	Threshold (MeV)
^{48}V	15.9735 d	$\varepsilon+\beta^+$:100	944.130	7.870 7	$^{47}Ti(d,n)^{48}V$	4.6	0.0
			983.525	99.98 4	$^{48}Ti(d,2n)^{48}V$	-7.0	7.3
			1312.106	98.2 3	$^{49}Ti(d,3n)^{48}V$	-15.2	15.8

The major sources of uncertainties in our experiment were considered during the cross-section calculation and are summarized in Table 2. An overall uncertainty in present cross-sections in the range of 0.01 - 22.81% was obtained by quadratic summing of the individual uncertainties listed in the Table 2.

Table 2: Evaluated uncertainties on cross-sections calculations.

Nuclide	Uncertainties (%)					
	γ-ray intensity (ΔIγ/ Iγ)	beam intensity	detector efficiency	thickness of target	γ-ray counting statistics	total uncertainty
^{48}V	0.040	5	4	2	0.3 - 5.9	0.01 - 22.81

3. Results and discussion

The production cross-sections of the ^{48}V radionuclide are tabulated in Table 1 while its excitation function is plotted in Fig. 1 together with earlier reported experimental measurements obtained from the EXFOR library [22] and the evaluated data in the TENDL-2014 library [23] which provides the output of the TALYS code [24]. In some cases, isotopic cross-sections reported in some earlier measurements [25-27] were normalized by multiplying with their natural isotopic abundance for comparison with our elemental cross-sections on the Figure. Similarly, the reported work in [28] was normalized based on the recommendation of the authors in their latter publication [29].

3.1. Production of independent ^{48}V

In addition to ^{48}V, there is also ^{48}Sc which was simultaneously produced from the deuteron irradiated Ti foils. Concurrently, both ^{48}V ($T_{1/2}$ = 15.9735 d) and the ^{48}Sc ($T_{1/2}$ =43.67 h,) decays to the same stable ^{48}Ti nuclide, and thus have some shared gamma lines as indicated in Table 2. There are three possible ways to assess independent ^{48}V production cross-sections. First, it is possible to use any of the interfering gamma line of ^{48}Sc with ^{48}V for the measurement of cross-sections of the ^{48}V but a computational technique is necessary in order to latter isolate the independent ^{48}V contributions in the calculated cross-sections. This is usually technical and was thus avoided. The second possibility is to use the independent gamma line of the ^{48}V (E_γ =944.130 keV), although it has a relatively low intensity (I_γ = 7.870 %). In the third method, since ^{48}V has longer half-life compared to ^{48}Sc, a long cooling period is usually allowed before the measurement of ^{48}V, sufficient enough until the interfering ^{48}Sc radionuclide completely decay. Thus, a cooling period of at least 10 half-lives of ^{48}Sc ($T_{1/2}$ =43.67 h) is usually needed for it to completely decay, that is about 450 hours from end of bombardment of the foil. This last procedure was followed in the present study since our measurements were made 9 to 10 days from the end of irradiation.

The major nuclear reactions contributing to the formation of ^{48}V are listed in Table 1. There are several experimental studies for comparison. Figure 1 compared the present studies with some of the literature experimental data [25-34] and also with IAEA recommended data. The Figure also compared our results with the extracted data of TENDL-2014 library, which was calculated based on TALYS code. Our data agree with the recommended data of IAEA but not with some few of the experimental literature data. On the other hand, the TENDL-2014 evaluated cross-sections have overestimated the present measurement.

Table 3: Measured cross-sections for ^{nat}Ti(d,x)^{48}V

Energy (MeV)	Cross-section (mb)
23.8 ± 0.5	224.8 ± 15.2
22.3 ± 0.5	261.6 ± 17.6
20.4 ± 0.5	308.0 ± 20.7
18.6 ± 0.5	336.4 ± 22.6
16.8 ± 0.6	339.0 ± 22.8
12.8 ± 0.6	271.1 ± 18.2
6.3 ± 0.8	21.6 ± 1.5
4.6 ± 0.9	16.7 ± 1.1
2.50 ± 1.50	0.80 ± 0.01

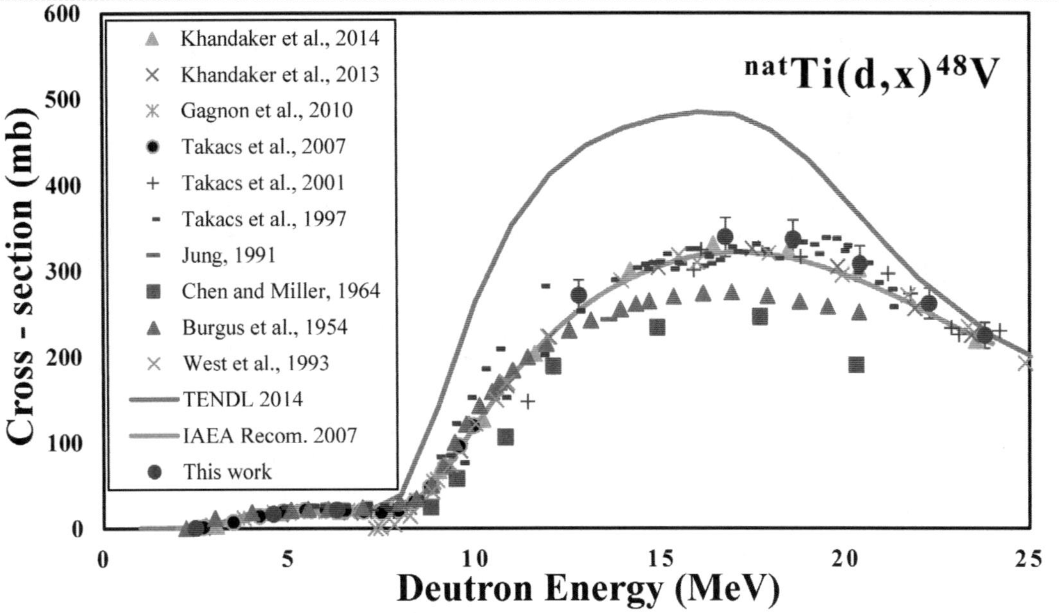

Figure 1. Excitation function of $^{nat}Ti(d,x)^{48}V$ reaction.

4. Conclusion

This work reports new experimental cross-sections for the $^{nat}Ti(d,x)^{48}V$ reaction in the energy region of 2.5 – 23.8 MeV. The measured excitation function was compared with the available experimental literature data and with the theoretical data extracted from the TENDL-2014 library. The new cross-sections confirm the previous available data in the EXFOR database with some additional insight. A large discrepancy was found between the present experiment cross-sections and the theoretical data of the TENDL-2014 library, which was based on Talys code. Present result could play an important role for improvements of prediction ability of the TALYS code.

Acknowledgments

Present work was performed at RI Beam Factory operated by RIKEN Nishina Center and CNS, University of Tokyo Japan. The work was partially supported by the University of Malaya Research Grants PG102-2014A and RP006D-13AFR.

References

[1] Amjed N, Hussain M, Aslam M N, Tárkányi F and Qaim S M 2016 Evaluation of nuclear reaction cross sections for optimization of production of the emerging diagnostic radionuclide ^{55}Co *Applied Radiation and Isotopes* **108** 38-48
[2] Qaim S M 2011 Development of novel positron emitters for medical applications: nuclear and radiochemical aspects *Radiochimica Acta International journal for chemical aspects of nuclear science and technology* **99** 611-25
[3] Arifvianto B, Leeflang M A and Zhou J 2015 The compression behaviors of titanium/carbamide powder mixtures in the preparation of biomedical titanium scaffolds with the space holder method *Powder Technology* **284** 112-21
[4] Singh R, Lee P D, Jones J R, Poologasundarampillai G, Post T, Lindley T C and Dashwood R J 2010 Hierarchically structured titanium foams for tissue scaffold applications *Acta Biomaterialia* **6** 4596-604
[5] Xuan Tham L, Nagasawa N, Matsuhashi S, Ishioka N S, Ito T and Kume T 2001 Effect of

radiation-degraded chitosan on plants stressed with vanadium *Radiation Physics and Chemistry* **61** 171-5

[6] Rorat E, Petelenz B, Marczewska B and Ochab E 2005 Thermoluminescence dosimetry of model line sources containing vanadium-48 *Radiation Measurements* **39** 495-501

[7] Arbabi A, Sadeghi M and Joharifard M 2009 Irradiation and dosimetry of Nitinol stent for renal artery brachytherapy *Applied Radiation and Isotopes* **67** 129-32

[8] De Cremer K, Cornelis R, Strijckmans K, Dams R, Lameire N and Vanholder R 2002 Behaviour of vanadate and vanadium–transferrin complex on different anion-exchange columns. Application to in vivo 48V-labelled rat serum *Journal of Chromatography B* **775** 143-52

[9] Hichwa R D, Kadrmas D, Watkins G L, Wollenweber S D, Maniam S, Boles Ponto L L, Richmond J C W and Koeppel J A 1995 Vanadium-48: A renewable source for transmission scanning with PET *Nuclear Instruments and Methods in Physics Research Section B: Beam Interactions with Materials and Atoms* **99** 804-6

[10] Khandaker M U, Kim K, Lee M, Kim K S and Kim G 2011 Excitation functions of (p,x) reactions on natural nickel up to 40 MeV *Nuclear Instruments and Methods in Physics Research Section B: Beam Interactions with Materials and Atoms* **269** 1140-9

[11] Khandaker M U, Haba H, Otuka N and Usman A R 2014 Investigation of (d,x) nuclear reactions on natural ytterbium up to 24 MeV *Nuclear Instruments and Methods in Physics Research Section B: Beam Interactions with Materials and Atoms* **335** 8-18

[12] Khandaker M U, Haba H, Murakami M and Otuka N 2015 Production cross-sections of long-lived radionuclides in deuteron-induced reactions on natural zinc up to 23 MeV *Nuclear Instruments and Methods in Physics Research Section B: Beam Interactions with Materials and Atoms* **346** 8-16

[13] Usman A R, Khandaker M U, Haba H, Murakami M and Otuka N 2016 Measurements of deuteron-induced reaction cross-sections on natural nickel up to 24 MeV *Nuclear Instruments and Methods in Physics Research Section B: Beam Interactions with Materials and Atoms* **368** 112-9

[14] Usman A R, Khandaker M U, Haba H, Otuka N, Murakami M and Komori Y 2016 Production cross-sections of radionuclides from α-induced reactions on natural copper up to 50MeV *Applied Radiation and Isotopes* **114** 104-13

[15] Berglund M and Wieser M E 2011 Isotopic compositions of the elements 2009 (IUPAC Technical Report) *Pure and Applied Chemistry* **83** 397-410

[16] Qaim S M, Tárkányi F T, Obložinský P, Gul K, Hermanne A, Mustafa M, Nortier F, Scholten B, Shubin Y N and Takács S 2002 Charged-particle cross section database for medical radioisotope production *Journal of Nuclear Science and Technology* **39** 1282-5; https://www-nds.iaea.org/medical/monitor_reactions.html (updated in April 2007).

[17] Ziegler J F 2004 SRIM-2003 *Nuclear Instruments and Methods in Physics Research Section B: Beam Interactions with Materials and Atoms* **219–220** 1027-36

[18] Burrows T W 2006 Nuclear Data Sheets for A = 48 *Nuclear Data Sheets* **107** 1747-922

[19] IAEA 2009 LiveChart of Nuclides, Interractive Chart of Nuclides *International Atomic Anergy. Available at https://www-nds.iaea.org/medical/monitor_reactions.html*

[20] Audi G, Wapstra A H and Thibault C 2003 The AME2003 atomic mass evaluation: (II). Tables, graphs and references *Nuclear Physics A* **729** 337-676

[21] Qtool 2011 Calculation of Reaction Q-values and thresholds, Los Alamos National Laboratory, T-2 Nuclear Information Service, Available from: < http://t2.lanl.gov/nis/data/qtool.html>.

[22] N. Otuka et al. 2014 Towards a more complete and accurate experimental nuclear reaction data library (EXFOR): International collaboration between Nuclear Reaction Data Centres (NRDC) *Nucl. Data Sheets* **120** 272-6

[23] Koning A J, Rochman D, S.C van der Marck1, Kopecky J, Sublet J C, Pomp S, Sjostrand H, Forrest R and al. e 2014 Talys Evaluated Nuclear Data Library(TENDL-2014). Nuclear Research and Consultancy Group (NRG) Petten, The Netherlands. Available at ftp://ftp.nrg.eu/pub/www/talys/tendl2014/tendl2014.html)

[24] Koning A J and Rochman D 2012 Modern nuclear data evaluation with the TALYS code system *Nucl. Data Sheets* **113** 2841–934

[25] West Jr H, Lanier R and Mustafa M 1993 3. Excitation Functions for the Nuclear Reactions on Titanium Leading to the Production of ^{48}V, ^{44}Sc, and ^{47}Sc, by Proton, Deuteron and Triton Irradiations at 0-35 MeV _ *MASI£ R* 3

[26] Burgus W H, Cowan G A, Hadley J W, Hess W, Shull T, Stevenson M L and York H F 1954 Cross sections for the reactions ^{48}Ti(d,2n)^{48}V, ^{52}Cr(d,2n)^{52}Mn and ^{56}Fe(d,2n)^{56}Co *Physical Review* **95** 750-1

[27] Chen K and Miller J 1964 Comparison between reactions of alpha particles with scandium-45 and deuterons with titanium-47 *Physical Review* **134** B1269

[28] Takácss S, Sonck M, Scholten B, Hermanne A and Tárkányi F 1997 Excitation functions of deuteron induced nuclear reactions on ^{nat}Ti up to 20 MeV for monitoring deuteron beams *Applied Radiation and Isotopes* **48** 657-65

[29] Takács S, Király B, Tárkányi F and Hermanne A 2007 Evaluated activation cross sections of longer-lived radionuclides produced by deuteron induced reactions on natural titanium *Nuclear Instruments and Methods in Physics Research Section B: Beam Interactions with Materials and Atoms* **262** 7-12

[30] Takács S, Szelecsényi F, Tárkányi F, Sonck M, Hermanne A, Shubin Y, Dityuk A, Mustafa M G and Youxiang Z 2001 New cross-sections and intercomparison of deuteron monitor reactions on Al, Ti, Fe, Ni and Cu *Nuclear Instruments and Methods in Physics Research Section B: Beam Interactions with Materials and Atoms* **174** 235-58

[31] Khandaker M U, Haba H, Kanaya J and Otuka N 2013 Excitation functions of (d,x) nuclear reactions on natural titanium up to 24 MeV *Nuclear Instruments and Methods in Physics Research Section B: Beam Interactions with Materials and Atoms* **296** 14-21

[32] Jung P 1992 *Nuclear Data for Science and Technology,* ed S Qaim: Springer Berlin Heidelberg) pp 352-4

[33] Gagnon K, Avila-Rodriguez M A, Wilson J and McQuarrie S A 2010 Experimental deuteron cross section measurements using single natural titanium foils from 3 to 9 MeV with special reference to the production of ^{47}V and ^{51}Ti *Nuclear Instruments and Methods in Physics Research Section B: Beam Interactions with Materials and Atoms* **268** 1392-8

[34] Khandaker M U, Haba H, Kanaya J, Otuka N and Kassim H A 2014 Activation Cross-sections of Deuteron-induced Nuclear Reactions on Natural Titanium *Nuclear Data Sheets* **119** 252-4

IOP Conf. Series: Journal of Physics: Conf. Series **860** (2017) 012030 doi:10.1088/1742-6596/860/1/012030

Innovative research of plasma physics for life sciences

D Boonyawan

Plasma and Beam Physics Research Facility, Department of Physics & Materials Science, Faculty of Science, Chiang Mai University, Chiang Mai 50200, Thailand

Corresponding author e-mail: dheerawan.b@cmu.ac.th

Abstract. In medicine, cold atmospheric plasma (CAP) for the medical treatment is a new field in plasma application, called plasma medicine. CAP contains mix of excited atoms and molecules, UV photons, charged particles as well as reactive oxygen species (ROS) and reactive nitrogen species (RNS). Typical species in air-CAPs are O_3, OH, N_xO_x, and HNO_x. The current developments in this field have fuelled the hope that CAP could be an interesting new therapeutic approach in the treatment of cancer. CAP apparently demonstrated effect on cancer cell apoptosis which did not induce cell necrosis or disruption. Moreover, CAP seemed to be selective for cancer cells since it was more effective in tumor cells than in normal non-neoplastic cells. In bioscience, dentistry and veterinary medicine : Since CAP, is delivered at room temperature, which results in less damaging effects on living tissue, while still has the efficiency in disinfection and sterilization. Recent studies proved that it is able to inactivate gram-negative and gram-positive bacteria, fungi, virus, spore, various parasites, and foreign organisms or pathogens without harming tissue. Moreover, cold plasma has been used effectively in medical field such as dental use, inducing apoptosis of malignant cells, stopping bleeding, promoting wound healing and tissue regeneration. Sericin hydrolysates, originating from silkworm is found support cell proliferation, expand cell adhesion and increase cell yield. The covalent linkage between a bioactive protein molecule and polystyrene dish surface via a carbon intermediate layer can slow down the release rate of protein compound into the phosphate buffer saline (PBS) solution. We found that a-C films and a-C:N_2 films show good attachment of human bone marrow-derived mesenchymal stem cells (hBM-MSCs). All of carbon modified-Polystyrene(PS) dishes revealed the less release rate of sericin molecules into PBS solution than PS control.

1. Introduction

In term of physics and chemistry, plasma is a completely or partially ionized gas with ions, electrons and uncharged particles such as atoms, molecules, and radicals, regarded as fourth state of matter. The other states of matter are liquid, gas, and solid. Plasma does not have a definite shape or volume as gas; however, it will form structures when put under a magnetic field, which is not seen in gas. From a macroscopic point of view, plasma is electrically neutral. However, plasma is electrically conductive and a lot of free charge carriers are contained in it.

Depending on amounts of energy transferred, the plasma properties change in terms of electron density and electron temperature. These two parameters define plasmas into two categories: thermal equilibrium plasma or thermal plasma and non-thermal equilibrium plasma or cold plasma.

• Thermal plasma, transitions and chemical reactions are mainly controlled by plasma particles collisions, not by radiative processes. Moreover, collision phenomena are micro-reversible in thermal plasma, suggesting that each kind of collision is balanced by its inverse such as excitation/de-excitation; ionization/recombination; kinetic balance [1]. Therefore, in thermal plasma the plasma temperature is equal to the gas temperature.

• Cold plasma, described by two temperatures: electron temperature (T_e) and ion temperature (T_i). Because of the large mass difference between electrons and ions, the plasma temperature or gas temperature is determined by ion temperature. On the other hand, the electron-induced de-excitation rate of the atom is generally lower than the corresponding electron-induced excitation rate because of a significant radiative de-excitation rate. Therefore, the density distribution of excited atoms in cold plasma is possible to depart from Boltzmann distribution, saying that the plasma temperature is much lower than the electron temperature [2-5].

2. Plasma for application in life sciences

2.1. The cold atmospheric plasma (CAP)

There are many different types of cold plasma sources for bio- or life-science applications. However, dielectric barrier discharge (DBD) atmospheric pressure, room air plasma has great potential for aesthetic medicine. In his report, Fridman et al. has confidently concluded that the direct plasma or filamentary DBD is much more potent in bacterial eradication than the indirect plasma or glow DBD type [6-8]. Nevertheless, the direct plasma source using the DBD technique can work only at very small millimeters distance, the gap between the powered or floating electrode and skin as second or ground electrode. The problem with using a DBD electrode on a non uniform skin surface is that the geometry and variation of skin electrical properties make it difficult to apply the flat planar floating electrode clinically. It is difficult to maintain and fix small-millimeter gap distances to create a uniform and stable homogeneous plasma beam without uncontrollable streamer discharge. A novel CAP device operated in mixed filamentary and glow discharge using the FE-DBD configuration has been developed by Photo Bio Care, Thailand as shown schematically in Fig. 1. The device can create a low-temperature, sustained, homogeneous micro-filament beam (23) that emits from the, round shape designed, dielectric floating electrode (22) when it comes in direct contact with a dry or mildly moist skin surface; e.g. wounds (24). The round shape of the dielectric electrode tip can be varied from a few millimeters up to 25 mm. The electrode used in this study is 18 mm in diameter. Such geometry creates regular, small gap distances once the electrode is mildly pressed to the skin surface and is moved constantly.

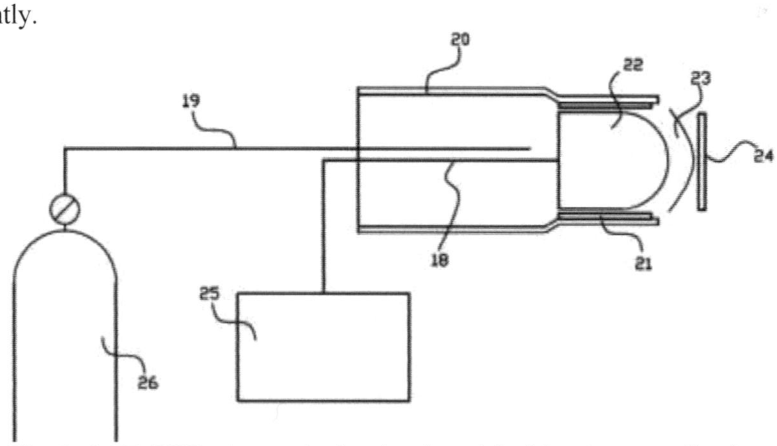

Fig. 1. A FE-DBD plasma device developed for bioscience applications.

The power generator (25) drives high-frequency pulse waves in the range of 15–20 KHz. Peak-to-peak voltage is ~6–7 kV, delivered as pulsing and adjustable rate from 10 to 110 Hz. The maximum in-put power from a normal household socket is 40 watts. Power intensity is adjustable from 1 to 10. The plasma power output is in the range of 0.2 – 1 watt, adjustable according to intensity and repetitive pulse setting. The automatic preset is at level 05 intensity at 50 Hz, with average CAP power of 0.62

watts. The system is set at a duration limit of 20 minutes per treatment. Also, it is equipped with Argon gas (26) to perform a hybrid mode CAP operation.

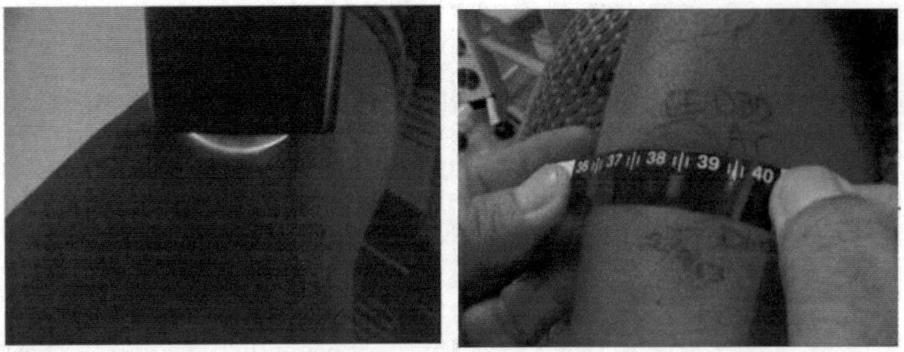

Fig. 2. CAP was applied to arm skin and temperature strip was measured afterward, from left to right. The setting was 5/50 at 24°C 50% RH [9].

Table 1. Skin temperature and sense of power of volunteers with CAP application time [9].

Application time (s)	-	2	3	4	5
Measured temp (°C)	26.5	29.5	32.0	35.5	39.0
Sense of volunteer	-	feel warm	warm	Quite warm	hot

2.1.1. Radicals in the CAP. In CAP, feed or air gas molecules are excited and ionized through the collisions with energetic electrons. And most of them are responding by electron-induced excitation mode in the non-equilibrium plasma. Therefore, the active species such as excited molecules, atoms, and ions in CAP emits light through a significant radiative de-excitation. Optical emission spectroscopy (OES) is well known as a non-invasive and non-disturbing technique for general plasma diagnostics as shown in Fig. 3.

To determine number of radicals particularly in this CAP, the spectroscopy can be rearranged from OES to UV-absorption spectroscopy [10]. And the density of ground state hydroxyl (OH) radicals, could be carried out by UV absorption spectroscopy using the Lambert-Beer's law,

$$[\mathrm{OH}] = -\frac{1}{\sigma \cdot l}\ln\left(\frac{\mathrm{I}^{t+p}-\mathrm{I}^{p}}{\mathrm{I}^{s}}\right),$$

where σ is the attenuation cross section of for absorbing species of OH radical, and l is the plasma spatial depth which is the UV beam passed through.

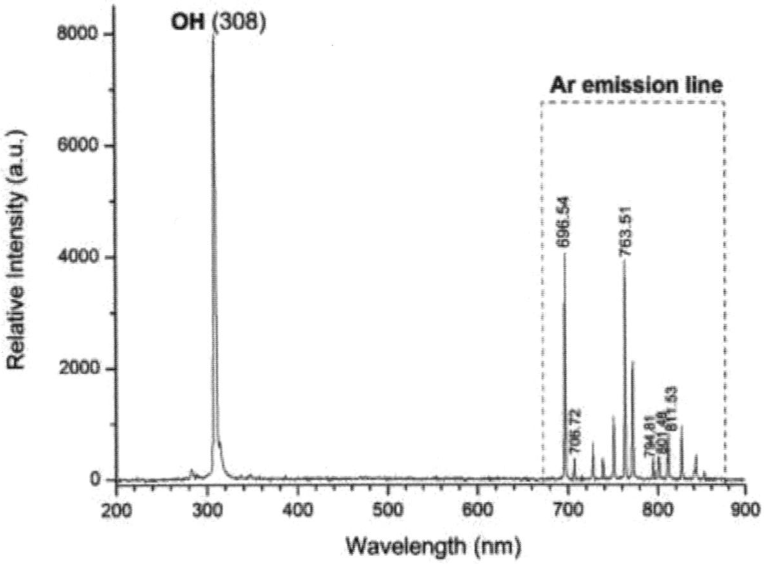

Fig. 3. A typical optical emission spectroscopy of CAP using Argon as a working gas [10].

Whereas I^{t+p} is the total light intensity or the sum of the transmitted light and plasma light measured with UV lamp and plasma on, I^{p} is the plasma intensity with the source off and I^{s} is the intensity of the source with the plasma off. The numbers of OH radical based on the absorbed band head at 306.4 nm from electronic transition $A^2\Sigma \rightarrow X^2\Pi$ has been investigated versus Argon flow rate with some settings of the hybrid CAP and summarized in Table 2.

Table 2. Numbers of OH radical (x10^{15} cm^{-3}) from the hybrid CAP at some settings [10].

Flow / Setting	5/10	5/50	5/110
@ Ar 4.5 lpm	.74	1.8	2.2
@ Ar 5 lpm	.68	1.7	2.0
@ Ar 6 lpm	.42	.76	.95

In a Ar-CAP, the OH radical is primarily ascribed by electrons and Ar metastables (Ar_m) induced by dissociation of water molecules in ambient air. Two dominant mechanisms are: $e^- + H_2O \rightarrow OH + H + e^-$, $k_1 = 2 \times 10^{-18\sim-16}$ m^3 s^{-1} and $Ar_m + H_2O \rightarrow OH + H + Ar$, $k_2 = 4.5 \times 10^{-16}$ m^3 s^{-1} where k_1 is estimated at typical conditions for low temperature plasma, i.e. low ionization degree ($\sim10^{-4}$), an electron temperature of 1–2 eV.

2.1.2. The efficacy of CAP in bioscience. Through the oxidation process, chemically reactive oxygen species (ROS) are known to be effective in the disruption of the bacterial cell wall [11-13]. The role of oxygen radicals in plasma sterilization has been also demonstrated: the rate of bacterial inactivation in the oxygen plasma group was greater than that in the helium plasma group. This is due to the presence of oxygen-based active species, such as atomic oxygen, the metastable singlet state of oxygen, and ozone (O_3). Also, in the presence of moisture, the OH radical, is expected to play a significant role by chemically attacking the outer structures of bacterial cells. This role also affirmed by the ATR-FTIR spectra of Methicillin-resistant Staphylococcus aureus (MRSA) specimens in Fig. 4. In gram-negative bacteria like E.coli main mechanism comes from OH radicals form carboxylic acid. In this case OH radical is strong and affect in disrupting an integrity of phospholipids layer (P=O: phosphine oxide) of cell membrane. The effect is clearer when <0.5% oxygen was added and oxygen radicals (O-) enhanced damage via increment of alcohol (C-O).

Scanning electron micrographs in Fig.5 is the instant effect of CAP treatment on *C. albicans* and *E. faecalis* biofilms [9]. It can clearly see that cells wall show ruptures or shrinkage for all over. These un-healthy cells in biofilm are unable to proliferate.

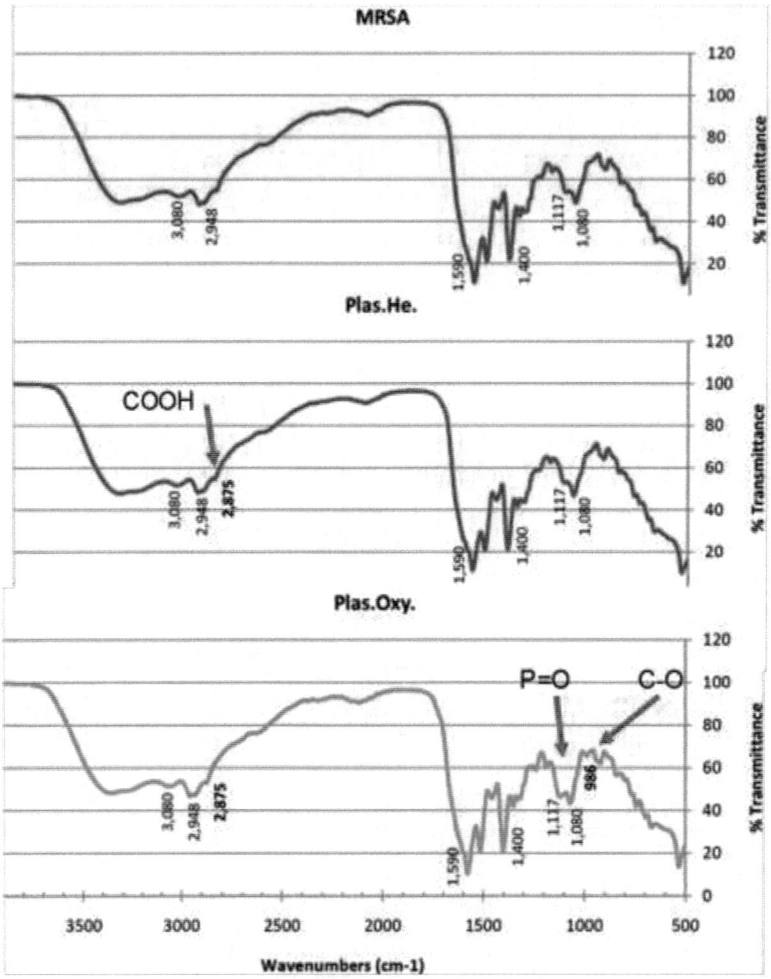

Fig. 4. ATR-FTIR spectra of MRSA specimens revealed reactive species in He/O_2-CAP interact with phospholipids layer.

IOP Conf. Series: Journal of Physics: Conf. Series **860** (2017) 012030 doi:10.1088/1742-6596/860/1/012030

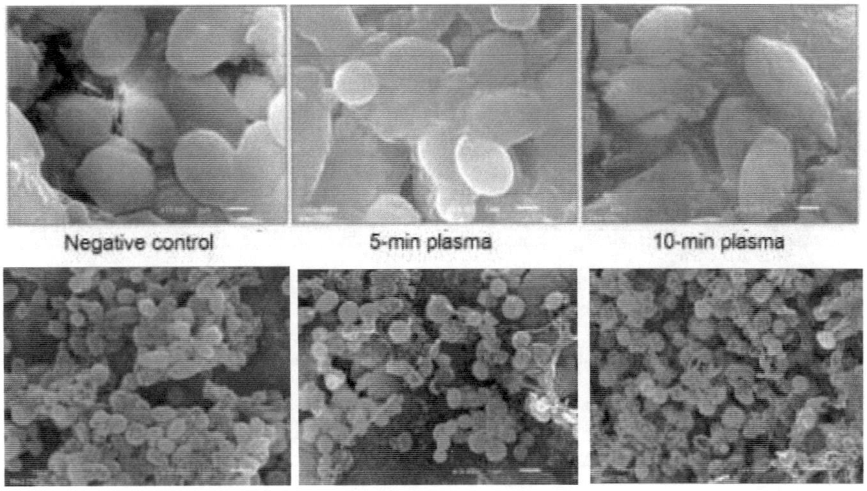

Fig. 5. Instant effect of the CAP treatment on *C.albicans* (top) and *E.faecalis* (bottom) biofilms [9].

2.1.3. Case study: the wound healing. A chronic wound such as bed sore or pressure sore is one of the illness that create a heavy and chronic burden to patient and health care service system. Pressure sores are soft-tissue injuries resulting from an unrelieved pressure of soft tissue over a bony prominence. Most of them will turn to a chronic wound stage with multiple drug resistant bacterial superimposed infections. Such wound usually becomes heavy exudative forming biofilm and necrotic tissue debris. In this study [14] 42 patients were invited to participate if they had chronic pressure ulcer grade 3 or 4. They were divided into 2 groups using block randomized technique. Fig. 6 shows the significantly reduction of bacterial load from wound after the first week by Ar-hybrid CAP treatment compare with standard wound care as control group ($p = 0.001$ for week 1 and 4, $p = 0.002$ for week 8). The plasma treatment was carried out once weekly with 1 min/cm^2 plasma dose.

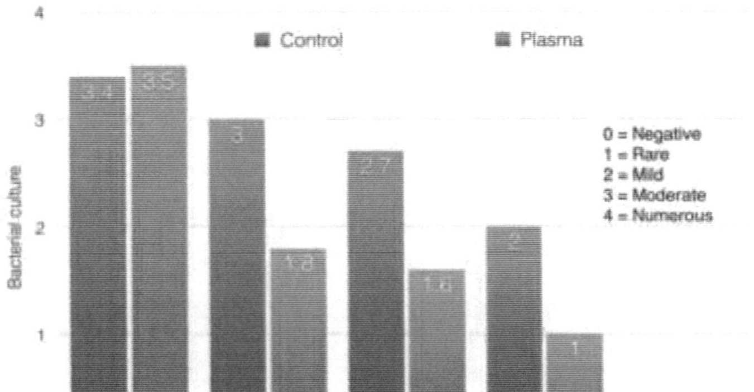

Fig. 6. Bacterial culture reduction between control and CAP treatment group. ($p = 0.001$ for week 1 and 4, $p = 0.002$ for week 8)

Fig. 7 shows the follow-up of wound healing by the plasma treatment. The number of wounds that reduce in size compared to the initial assessment were statistically different at second week after 2 treatments. The number of wounds that reduce in amount of exudate compared to the initial assessment were statistically different at first week after one treatment. The number of wounds that improve tissue type of wound base compared to the initial assessment were statistically different at second week after 2 treatments. The number of wounds that improve in overall wound healing process compared to the initial assessment were statistically different at first week after one treatment. This in-vivo clinical results may allow the conclusion that CAP is a promising new option for the treatment of chronic wounds.

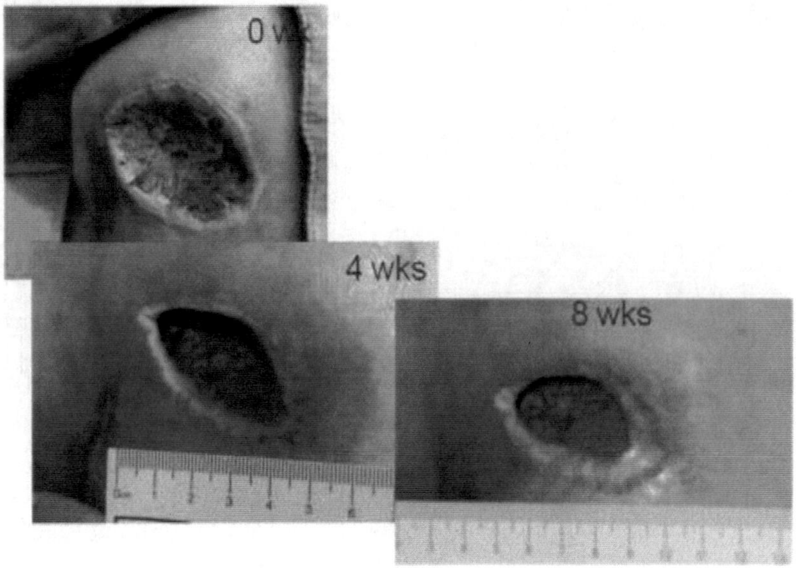

Fig. 7. Photographs of an infected wound for 8 weeks of Ar-hybrid CAP treatment.

2.2. The bioactive molecule layer for biomaterials [15]

2.2.1. We studied the effect of a covalent linkage between a bioactive protein molecule and polystyrene dish surface via a carbon intermediate layer can slow down the release rate of protein compound into the phosphate buffer saline (PBS) solution. Films of amorphous carbon (a-C) and functionalized carbon were deposited on polystyrene (PS) culture dish surfaces using the RF plasma-enhanced chemical vapor deposition (PECVD) system as shown in Fig. 8. Cell attachment and proliferation assay were quantitatively evaluated. Cells were plated at a density of 1×10^4 *cells* per dish with 150 μl of WST-1 reagent (Roche, USA) and incubated at 37°C for 4 *h*. The proliferation assay was analyzed using the OD450 value at day 1 and day 7 in comparison to the control at day 1, which was presented as percentage of attached and proliferated cells.

Fig. 8. RF PECVD system for a-C film synthesis on PS dishes.

The synthesized a-C based-films can increase the hydrophlilicity and biocompatibility of PS dish, especially a-C films and a-C:N_2 films which show the good attachment of human bone marrow-derived mesenchymal stem cells (hBM-MSCs). The carbon-coated PS dishes grafting with sericin protein were used in serum-free condition. We also found that hBM-MSCs have higher % of proliferated cells at day 7 for the modified dishes with carbon films and coated with sericin than the PS control coated with sericin only.

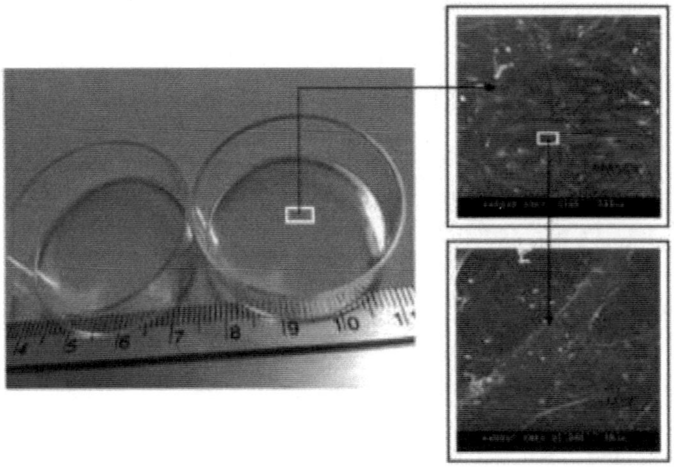

Fig. 9. The synthesized a-C based-films with sericin (right) can support cell proliferation.

2.2.2. A wound dressing surface modification method was developed by using PECVD of mixed ammonia (NH_3) and acetylene (C_2H_2). Plasma active species such as carbon and nitrogen radicals form film with active sites to create a covalent bond between Ag-SD compound and the cotton gauze surface. This novel method was targeted at controlling the delivery of the Ag-SD release while covering and protecting the wound, reducing the drug dosage and dosing times as well as keeping the wound dry by additional water repellant property.

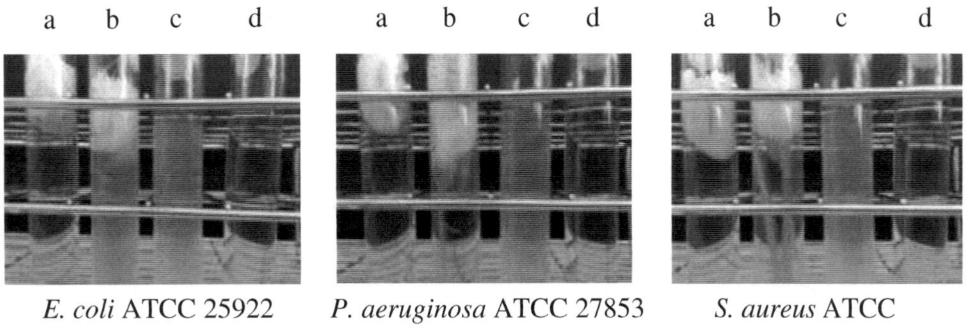

Fig. 10. The antimicrobial activity of modified gauze dressing (a) un-modify (b) control (c&d).

Fig. 10 shows the testing result on the effectiveness of the cotton wound dressing on antibacterial after exposure to *E. coli*, *P. aeruginosa*, and *S. aureus*, respectively. The plasma-treated gauze dressing plus Ag-SD exhibited clear appearance of the Fluid Thioglycollate Medium (FTM) as shown in tube a, indicating no bacterial growth, when compared with turbid appearance which indicated bacterial growth in gauze dressing standard and the control. This result demonstrates that plasma-treated gauze dressing performs well effectively to inhibit growth of tested bacterial strains which are commonly found on wound.

IOP Conf. Series: Journal of Physics: Conf. Series **860** (2017) 012030 doi:10.1088/1742-6596/860/1/012030

Acknowledgements

Author would like to express all sincere thanks to Y. Thana, S. Klinhom, S. Chinsawananon, S. Khemaleelakul, S. Tunma, C. Chutsirimongkol M.D. and Prof. A. Chuangsuwanich M.D. for their devoted efforts and all contributions in this article. DB also thanks Chiang Mai University, Thailand Center of Excellence in Physics (ThEP) and Thailand Center of Excellence for Life Sciences (TCELS) for their supports to the plasma bio/life sciences project.

References

[1] M. Moisan, M. D. Calzada, A. Gamero, A. Sola, J. Appl. Phys. 1996, 80, 46.
[2] R. H. Huddlestone, S. L. Leonard, Plasma Diagnostic Techniques, Academic Press, New York, 1965.
[3] H. R. Griem, Plasma Spectroscopy, McGraw-Hill, New York, 1964.
[4] W. Lochte-Holtgreven, Plasma Diagnostics, North-Holland, Amsterdam, 1968.
[5] M. Mitchner, C. H. Kruger, Partially Ionized Gases, Wiley, New York, 1973.
[6] Kalghatgi S, Kelly CM, Cerchar E, Torabi B, Alekseev O, Fridman A, et al. Effects of non-thermal plasma on mammalian cells. PLoS One. 2011 Jan 21;6(1):e16270.
[7] Kalghatgi S, Friedman G, Fridman A, Clyne AM. Endothelial cell proliferation is enhanced by low dose non-thermal plasma through fibroblast growth factor-2 release. Ann Biomed Eng. 2010 Mar;38(3):748–57.
[8] Kalghatgi SU, Fridman A, Friedman G, Clyne AM. Cell proliferation following non-thermal plasma is related to reactive oxygen species induced fibroblast growth factor-2 release. Conf Proc IEEE Eng Med Biol Soc. 2009;2009:6030–3.
[9] D. Boonyawan, W. Kerdwichai, S. Khemaleelakul NRCT Reserch Project Report 2015.
[10] Y.Thana, A.Ngamjarurojana, D.Boonyawan Analysis of cold atmospheric-pressure bio-medicine plasmas by using UV absorption spectroscopy 2016 doi:10.1016/j.surfcoat.2016.05.025
[11] Kuo SP. Plasma Assisted Decontamination of Bacterail Spores. The Open Biomedical Engineering Journal. 2008;2:36-42.
[12] Stoffels E, Sakiyama Y, Graves DB. Cold Atmospheric Plasma: Charged Species and Their Interactions With Cells and Tissues. IEEE Transactions on plasma science 2009;36(4):1441-57.
[13] Tolba O, Loughrey A, Goldsmith CE, Millar BC, Rooney PJ, Moore JE. Survival of epidemic strains of nosocomial- and community-acquired methicillin-resistant Staphylococcus aureus on coins. Am J Infect Control. 2007 Jun;35(5):342-6.
[14] A. Chuangsuwanich, T. Assadamongkol, C. Chutsirimongkol, D. Boonyawan The Healing Effect of Low Temperature Atmospheric Pressure Plasma in Pressure sore: A Randomized Controlled Trial International Journal of Lower Extremity Wounds 2016 "in press".
[15] Tunma, S., Song, D.H., Kim, S.E., Kim, K.N., Han, J.G. and Boonyawan, D., Immobilization of sericin molecules via amorphous carbon plasma modified-polystyrene dish for serum-free culture, Applied Surface Science 2013 (283) pp 930–940.

IOP Conf. Series: Journal of Physics: Conf. Series **860** (2017) 012031 doi:10.1088/1742-6596/860/1/012031

Improvement of water quality using dielectric barrier discharge plasma

N T Quyen[1], T Traikool[2], R Nitisoravut[3] and T Onjun[1,4]

[1] School of Common and Graduate Students, Sirindhorn International Institute of Technology, Thammasat University, Pathum Thani, 12120 Thailand
[2] School of Manufacturing Systems and Mechanical Engineering, Sirindhorn International Institute of Technology, Thammasat University, Pathum Thani, 12120 Thailand
[3] School of Bio-Chemical Engineering and Technology, Sirindhorn International Institute of Technology, Thammasat University, Pathum Thani, 12120 Thailand

[4]Author to whom any correspondence should be addressed: thawatchai@siit.tu.ac.th

Abstract. The improvement of water quality using by atmospheric plasma produced from a dielectric barrier discharge (DBD) was studied. An experiment was set-up with a 4 mm diameter pipe, which contains 2 electrodes and has an air flow with the rate of 15 liters per minute. Surface water, domestic wastewater and DI water were treated with the DBD plasma for some period of time. Electricity was supplied at 3.5 kV with the frequency of 5.5 kHz. Some key parameters of water quality includes the level of chemical oxygen demand (COD), total suspended solid (TSS), color, and odor are measured before and after. The result showed that strong acid with pH below 2 was observed after 60 minutes plasma treatment for the DI water, while the surface water and wastewater needs about 120 minutes to pH below 2 even though the pH value are about the same at the beginning. Moreover, It was formed that the COD, TSS microorganism was noticeably decreased, therefore the increasing of transparency level. This result confirms that atmospheric DBD plasma generated acidity in water as reduce amount of organic and suspended solid in water.

1. Introduction

Waste water treatment is one of crucial issues in many cities around the world, required a balance between economic development and environmental protection. Traditional sewage and water treatment technologies can be specifically classified into three categories: biological, chemical and physical processes. Biological methods often use to treat both municipal and industrial waste waters, but they are generally not successful at degrading many toxic organic compounds. Chemical processes typically involve adding a chemical substance to a system to initiate a transformation. Some commonly added chemicals, such as chlorines, can cause negative effects on both human and environmental health. Physical processes are generally served as removing contaminants from the bulk fluid phase by concentrating them in a liquid or sludge phase. Many drinking water systems are challenged with difficult to treat target compounds such as organics (e.g., NDMA) and pathogens (e.g., Cryptosporidium, viruses, etc.)[6]. In fact, due to some properties of atmospheric plasma, it has been used in many applications, such as cleaning or modifying material surface. It is interesting to know the effectiveness of atmospheric plasma to improve the water quality, especially for waste water.

IOP Conf. Series: Journal of Physics: Conf. Series **860** (2017) 012031 doi:10.1088/1742-6596/860/1/012031

In order to adapt to sustainable development, the improving water resource management typically refers to recycling wastewater. This has been set as the first priority in many countries around the world. Plasma technology was suggested as the most promising technique to solve this task. The use of plasma processes was known as environmental friendly technique, safety and cost- effective alternatives. Recent studies about plasma technology in water treatment almost focus on direct plasma and thermal plasma because they require complex equipment as plasma reactor and small scale or apply plasma as additional method such as using UV, Ozone to disinfection [6 -7, 10-12]. This study is presenting a simple, comprehensive and large- scale method using indirect atmospheric dielectric barrier discharge (DBD) plasma.

Plasma technology based on applying electrical current, ions, high-reactive short-lived radicals are generated in plasma from ambient air or atmospheric oxygen such as oxygen and nitrogen reactive radicals, ozone, H+, etc. Hence, reactive particles are generated during discharge and rapidly react with dissolved pollutant in water due to their high reactivity. The oxidation processes of ozone and oxygen radicals are major factors to reduced pollutant in water and they are commonly using techniques. Alternatively, they were called as advantaged oxidation processes (AOP). However, the reaction between reactive oxygen species, nitrogen species and water were strongly supposed to acidify in water. The acidification of deionized water under direct exposure to dielectric barrier discharge (DBD) plasma was demonstrated in Natalie et.al. and Oehmingen et.al. Natalie's research shown that deionized water has pH≈ 2 after 90 minutes treat with DBD plasma [1-2]. So, the control acidity in water during plasma treatment is important in using plasma technology to water treatment.

This study observed indirect DBD plasma system working as a comprehensive water treatment. The sewage, surface water and DI water were used in the experiment. Some key parameters were observed, especially total suspended solid (TSS), pH, chemical oxygen demand (COD), and microorganism. Treatment mechanism of system based on dynamic gas moving in water tank and the advantage of oxidation processes. In plasma phase, there are the strong energy electrons, ions, UV rays and advanced oxidation such as HO*, O*, H*, O3, H2O2 (hydrogen peroxide). When water contacted to plasma, the ionic bonds, covalent linking of organic, inorganic and bacteria are broken down and killed by advanced oxidation. Moreover, acidification in water has strong effect on inactive bacteria. Therefore, the quality of wastewater is improved and then it might be pumped out.

2. Experimental

The experiment setup is shown in Figure 1. The power supply is a laboratory-made free running oscillator which was set at frequency of 5.5.kHz, and Voltage was set at 3.5 kV. This system works at the atmosphere pressure and using air as carrier gas. A glass tube (2) with one inner copper electrode and aluminum external electrode wrap around will be used as dielectric barrier. The distance between two electrodes is approximately 1 mm. This tube also connected to a gas pump with an attached valve control (4). Gas speed was supplied at 15 litter/ minutes. The high-voltage was supplied to two electrodes and created a plasma, the gas passed through a glass tube and get energy from high voltage electricity and then it went down into a water tank (5). When the air passed the glass tube, they are excited and jumped into water. Reactive oxygen species (ROS) and reactive nitrogen species (RNS) are usually found in exhaust stream. Represents for these species are O_2^*, O^*, O_3, OH^*, NO^*, NO_2^*, etc.

IOP Conf. Series: Journal of Physics: Conf. Series **860** (2017) 012031 doi:10.1088/1742-6596/860/1/012031

[15, 16, 17]. These compounds can react with water in the tank.

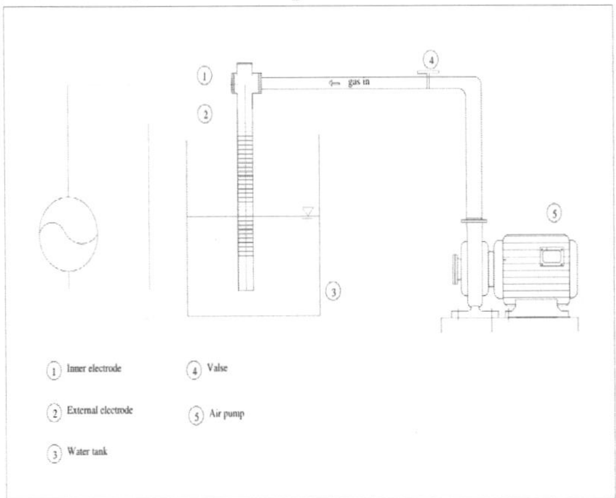

Figure 1. Experiment schematic

In this experiment, each 100 ml of deionized (DI) water, domestic wastewater, surface water was treated in 120 minutes and kept in 168 hours without plasma. All of those samples of water were collected in SIIT, Thammasat University. Some parameters were employed including pH values measurement by pH paper in every 15 minutes during plasma treated and every 8 hours after that. The temperature measured by thermometer in every 30 minutes. Chemical Oxygen demand (COD) of wastewater and surface water were recorded after 1 hour and 2 hours treated. The organoleptic quality was observed during the experiment with all samples.

3. Results and discussion

The first observation of water after the plasma treatment is the improvement of organoleptic quality. They have almost non-color and odorless. It indicates that a large amount of organic matter has decomposed. Total suspended solid of water has noticeably reduced.

Total suspended solids (TSS) is the most visible indicator of water quality. TSS are particles which sizes are greater than 2 microns found in the water column. In another hand, TSS are total quantity measurements of solid material per volume of water though most suspended solids. These suspended solids are made up of inorganic materials, bacteria and algae can also contribute to the total solids concentration. TSS affect to water turbidity and water clarity also. TSS can be used to calculate sedimentation rates.

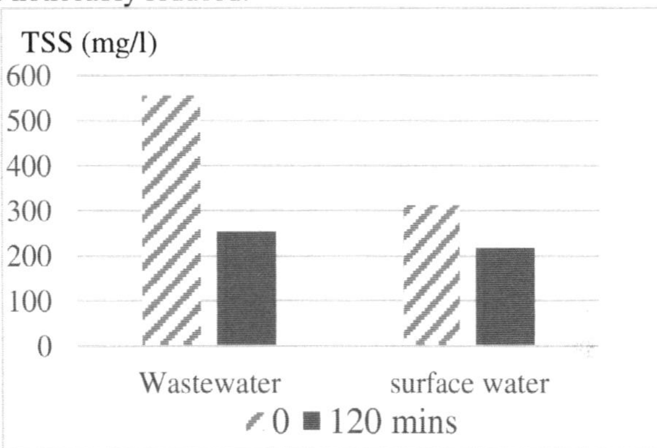

Figure2. Total suspended solid of wastewater and surface water before and after plasma treated.

In our study, the values of TSS were recorded in both wastewater and surface water (Figure 2). TSS of wastewater and surface water reduced to 54.32% and 30.12%, respectively. These superoxide radicals generated in plasma phase play an important role to removed organic compound in wastewater (normally >50%), which is main reason cause the color of water [7, 18, 19]. Moreover, visual inspection recorded the reduced of large-size solids. It seems treatment process causes a breakdown in large-size solids without any additional physical methods. However, the mechanism should be detailed explain.

IOP Conf. Series: Journal of Physics: Conf. Series **860** (2017) 012031 doi:10.1088/1742-6596/860/1/012031

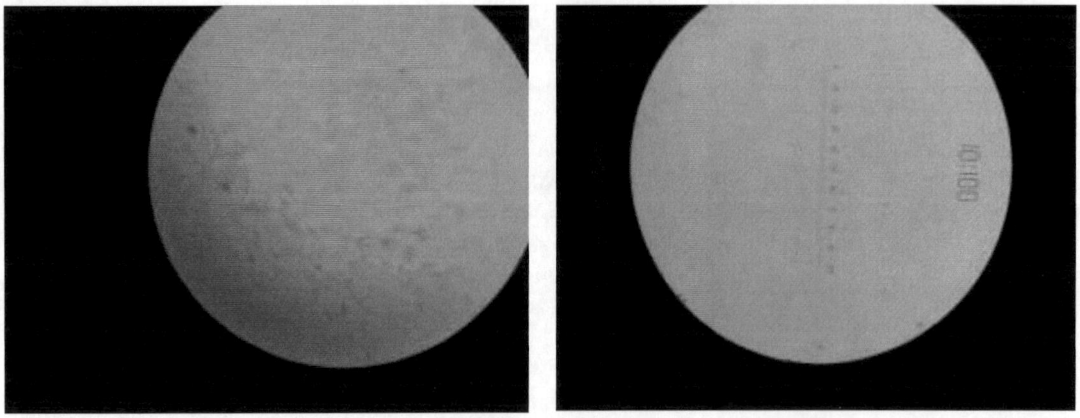

(a) (b)

Figure 3. Microscope picture of surface water before(a) and after treated (b)

Figure 3 shows the decreasing of microorganism in surface water that was observed in a microscope. Wastewater has not made a clear difference. Before plasma treatment, surface water samples have many algae and planktons, and most of them were killed during plasma treatment. Algae and bacteria, plankton in water might be killed by oxidation and acidity in water. When oxidants such as ozone contacts with bacteria cell, which keep bacteria in shape and protect bacteria, a reaction called an oxidative burst occurs which literally creates a tiny hole in the cell wall. A newly created hole in the cell wall has injured the bacterium, after thousands of ozone collisions in a few seconds, the bacterial wall can no longer maintain its shape and the cell got killed. Moreover, the acidification in water during working of plasma system also killed bacteria by reaction of reactive oxygen species. Superoxide and hydrogen peroxide (H2O2) are produced while generating plasma. Both O_2^- and H_2O_2 can damage a variety of biomolecules and enzyme. These types of damage result in metabolic defects, including auxotrophy for aromatic, branched-chain, and sulfur-containing amino acids. Superoxide cannot damage membranes in bacteria but they destroyed bacteria through damage DNA and enzyme of bacteria [1,2,14]. Results of increase acidity in water also support this opinion.

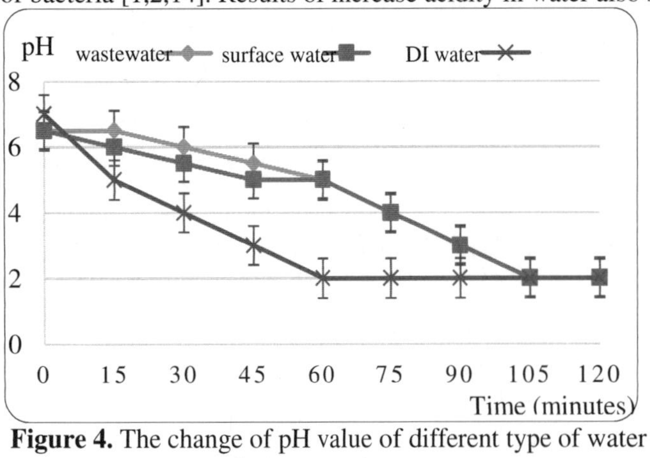

Figure 4. The change of pH value of different type of water during experiment.

In this experiment, the system is indirect system, which has the same principle with Oeminhem *et al.* experiment. Indirect plasma technologies generate plasma in a location away or near to but not direct within sample to be treated. In this case, the main source of the pH change in the liquid is a chemical change – HNO_3, HNO_2, O_2 radical and O_3 that are created by the plasma in the air are reaching the liquid and becoming dissolved there.

The detected of nitrate/ nitrous acid was confirmed in both Natalie *et al.* and Oeminhem *et al.* study. Atmospheric plasma causes some chemical reactions in the air, typical creation ozone (O_3) and oxidize nitrogen, then dissolve in water and cause chemical change. Although the formation of nitrate/nitrous acid in water was confirmed, they still do not have strong acidity, which indicated by pH≈ 2. The other important things are superoxide radicals, which were the result of reaction between ozone and water. This result and mechanism also explained in Natalie *et al.* and others study [1,3]. The acid, which from superoxide O_2, and H^+ radical was called plasma acid. Figure 4 shows pH dependence on time of four types of water: deionized (DI) water, surface water, and wastewater. It clearly shows a rapid increase in the acidity of all types of water. Although the time to achieved strong acid is different, the strong acid which has pH≈ 2 was detected in all types of water. In other words, DBD plasma created acidity in all types of water. The similar results with DI water were reported in Natalie *et al.* and Oeminhem *et al.* [1, 2]. DI water achieved pH≈ 2 after 60 minutes treated and keep it stable in a long time later while wastewater and surface water need 135 minutes to gain pH ≈ 2.

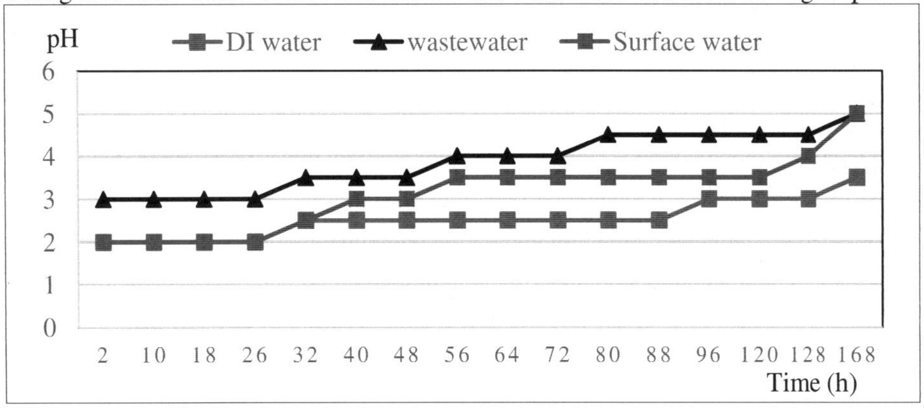

Figure 5. The increases of pH values of DI water, wastewater and surface water after treatment without plasma exposure.

Figure 5 show the sustainability of water when leaving them without addition of plasma. Plasma acid is stable in a long time after plasma treatment, water sample was kept in 7 days without additional plasma. The pH values of water show that plasma acid, which was created in treating time stable for 26 hours and slowly increase after that. After 7 days without addition plasma treatment, pH value of wastewater and surface water are 5 while DI water has pH≈ 3.5. Although pH≈ 5 is not recommended, the output water is not dangerous affect to human. These results suggest that plasma can use as chemical method to kill bacteria line chlorine. However, it also suggests that keeping treated water in tank or raceway system at least 7 days is necessary when we use plasma treatment as wide method.

The slow increase of acidity in wastewater and surface water suggested that the influence of other compounds in water to increasing of acidity. The most suspicious element is organic compounds, which has a large part in both wastewater and surface water and be easily oxidized by ozone and reactive oxygen radicals. Chemical oxygen demand (COD) of wastewater and surface water were measured before and after treatment to observe the effect of plasma system to dissolve pollutants in water.

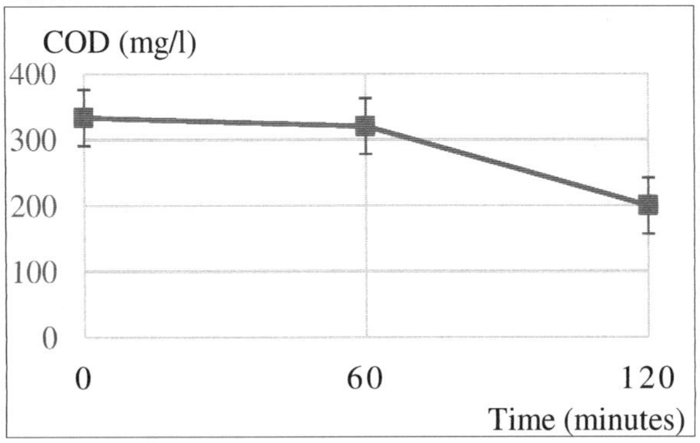

Figure 6. COD of waste water during treated.

Figure 6 shows the reducing of organic compounds in wastewater during experiment. COD of wastewater decreased 40% after 120 minutes treated. Organic compounds is a major component to be treated in wastewater treatment. It also caused color and odor in wastewater due to the anaerobic decomposition of organic compounds. In other words, due to the shortage of oxygen in the water.

The impression results in terms of sensory were also recorded in this experiment, a large part of the sewage smell and the color was disappeared. Wastewater after exposure to DBD plasma becomes transparent and odorless. This results suggested that DBD plasma has important contribute to increasing of dissolve oxygen in wastewater and a part of ozone was used to oxidize organic matter in sewage. Because a large amount of ozone was generated by DBD plasma in gas phase, so oxidized by ozone become the most likely factor. The general chemical reacts follow [4,5]:

$$C + 2O_3 \rightarrow CO_2 + 2O_2$$

The decreasing of COD of surface water (Figure 7) was also recorded in this experiment. After 120 minutes exposure to DBD plasma, 33% organic compounds in surface water were decomposed. The result with surface water is approximately identical with wastewater. Organoleptic quality also improved by DBD plasma. Although ozone is a strong oxidizer, DBD plasma cannot oxidize all of organic compounds in water, there was a part of organic compounds still remain in water after treatment. The remaining organic matters in both wastewater and surface water may in the form of insoluble suspended solids or non-oxidized by ozone and oxygen radicals.

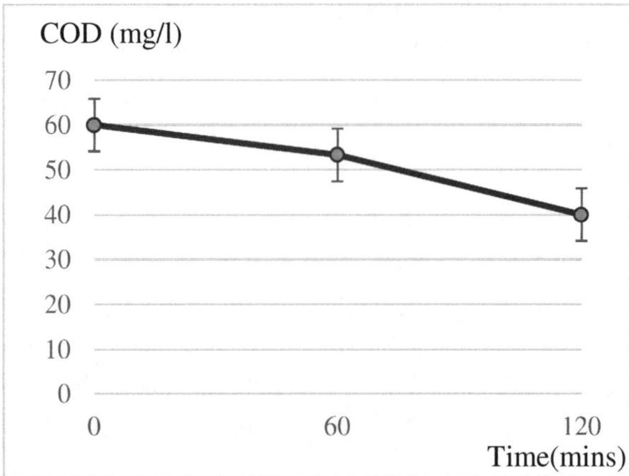

Figure 7. COD of surface water during treated.

These results suggest that plasma treatment is also working as a microbiological method to reduce organic compounds though it has lower efficient than microbiological method. So, if we want to use plasma as a comprehensive system, the addition microbiological is suggested.

4. Conclusion

In this experiment, the atmospheric DBD plasma has potential to apply in water treatment as a comprehensive system without addition of physical or chemical methods. After treated by DBD plasma, both waste water and surface water quality were improved. TSS of both types of water has significantly reduced and also help improve visual water quality. The number of organic compounds in wastewater and surface water have decreased after plasma treated. Chemical oxygen demand of wastewater and surface water reduced 40% and 33% after 120 minutes treating, respectively. Atmospheric DBD also plasma caused acidify in all types of water includes DI water, surface water and wastewater. Strongly acid with pH $\approx$ 2 was detected with all types of water. This acid condition is stable for 26 hours.

References

1. Natalie Shainsky, D. Dobrynin, U. Ercan, S. G. Joshi, H.JI, A.Brooks, G.Fridman, Y. Cho, A.Fridman, G. Friedman, *Plasma processess and polymers*, 2012.
2. K. Oehmigen, M.Hahnel, R. Brandenburg, Ch. Wilke, K.-D. Weltmann, Th.von Woedtke, Plasma processess and polymers, Vol 7. Pp 250-257, 2010.
3. K. Takahashi, K. Satoh, H. Itoh, H. Kawaguchi, I. Timoshkin, M. Given, and S. MacGregor, Japanese Journal of Applied Physics, Vol. 55, 2016, http://doi.org/10.7567/JJAP.55.07LF01
4. Paul Westerhoff, G. Aiken, G. Amy and J. Debroux, Water research, Vol. 33, 1999, P 2265-2276.
5. J. Świetlik, A. Dąbrowska, U. Raczyk-Stanisławiak, J. Nawrocki, Water research, Vol 38, 2004, P 547-558.
6. Jen-Shi Chang, International journal of Plasma Environmental Science and Technology, Vol 3, No.2, September 2009.
7. T.N. Dam, N. Harada, IEEE 2014 Region 10 conference, 2014, 2159-3442.
8. Malik MA, Ghaffar A, Malik SA (2001) Plasma Sources Sci Techno 10:82
9. Locke BR, Sato M, Sunka P, Hoffmann MR, Chang JS (2006) Ind Eng Chem Res 45:882
10. Shi J, Bian W, Yi X (2009) J Hazard Mater 171:924.

11. B.R Locke, M. Sato, P. Sunka, M.R. Hoffmann and J.S. Chang, Industrial & Engineering Chemistry Research, Vol. 45, pp 882-905, 2005.
12. Shinfuku Nomura, Toyota Hiromichi, United states patent of publication, US 2005/0227115 A1, 2005.
13. Bernier et al. United states patent, US 6,187,206 B1, Feb. 13, 200.
14. Jame M. Slauch, Mol Microbiol. 2011 May; 80(3): 580–583.
15. A. A. H. Mohamed, S. Suddala, M.A. Malik, K.H. Schoenbach, Plasma Science, ICOPS 2004. IEEE Conference Record, 2004.
16. P. Attri, Y. H. Kim, D. H. Park. J. H. Park, Y. J. Hong , H. S. Uhm; K. N. Kim, A. Fridman, E. H. Choi, Science report, Vol. 5, No. 9332, 2015, Doi. 10.1038/scep09332.
17. S. J. Kim, T.H. Chung, Science report, Vol. 6, No. 20332, 2016, Doi. 10.1038/scep20332.
18. M. Makowska, M. Spychała, Pol. J. Environ. Stud., Vol. 23, No. 1 (2014), 131-137.
19. M. G. Antoniou, C. Zhao, Kevin E. O'Shea, G. Zhang, D. D. Dionysiou, C. Zhao, C. Han, M. N. Nadagouda, H. Choi, T. Fotiou,T. M. Triantis and A. Hiskia, Photocatalysis: Applications, 2016, pp. 1-34 DOI: 10.1039/9781782627104-00001, eISBN:978-1-78262-710-4.

Acknowledgments
The authors gratefully acknowledge the financial support provided by Thammasat University, Thailand under the TU Research Scholar, Contract No. 1/2559.

IOP Conf. Series: Journal of Physics: Conf. Series **860** (2017) 012032 doi:10.1088/1742-6596/860/1/012032

Probability of detection model for the non-destructive inspection of steam generator tubes of PWRs

N Yusa

Department of Quantum Science and Energy Engineering, Graduate School of Engineering, Sendai, Miyagi, Japan

E-mail: noritaka.yusa@qse.tohoku.ac.jp

Abstract. This study proposes a probability of detection (POD) model to discuss the capability of non-destructive testing methods for the detection of stress corrosion cracks appearing in the steam generator tubes of pressurized water reactors. Three-dimensional finite element simulations were conducted to evaluate eddy current signals due to stress corrosion cracks. The simulations consider an absolute type pancake probe and model a stress corrosion crack as a region with a certain electrical conductivity inside to account for eddy currents flowing across a flaw. The probabilistic nature of a non-destructive test is simulated by varying the electrical conductivity of the modelled stress corrosion cracking. A two-dimensional POD model, which provides the POD as a function of the depth and length of a flaw, is presented together with a conventional POD model characterizing a flaw using a single parameter. The effect of the number of the samples on the PODs is also discussed.

1. Introduction

At the present, the emergence of flaws accompanying plant ageing is one of the most important issues in the nuclear industry. Harmful flaws must be detected in their early stages using a periodical non-destructive inspection in order to take suitable actions following proper rules, such as ASME Sec. XI. In contrast, not all flaws are actually harmful. For example, a tiny flaw in a thick structure would not affect the integrity of the structure at all. This indicates the importance of quantitatively evaluating the capability of non-destructive testing methods not on the basis of a minimum detectable flaw size but depending on whether or not they can surely detect flaws that must be detected. Aerospace industries have demonstrated the effectiveness of a concept of probability of detection (POD) to evaluate the capability of non-destructive testing methods from this point of view[1,2], and recent studies have pointed out that the concept of POD would be effective in nuclear industries[3]. However, a quite limited number of studies have actually evaluated non-destructive testing methods for nuclear industries using POD[4].

One of the largest reasons for this is conventional POD characterized a flaw using a single parameter. Conventional PODs used in aerospace industries have targeted mainly fatigue cracks appearing at the rivet holes of airplanes. It would be reasonable to characterize the fatigue cracks using a single parameter because they would have almost the same aspect ratio. In contrast, various flaws would appear in various components in nuclear power plants, which indicates the difficulty in characterizing a flaw using a single parameter. To address this issue, earlier studies of the authors have proposed how to construct POD with multiple flaw parameters[5,6] without postulating a closed-form of signal response or a constant variance[7,8]. The studies demonstrated its effectiveness by evaluating the POD from eddy current signals. However, what they considered was a relatively simple problem of detecting artificial slits introduced in austenitic stainless steel plates.

This study discusses the application of the POD to more practical inspection in nuclear industries, more specifically eddy current inspection of stress corrosion cracks appearing at the outer surface of steam generator tubes of pressurized water reactors. Signals due to the stress corrosion cracks are obtained by three-dimensional finite element simulations on the basis of earlier studies discussing how to model stress corrosion cracks from the viewpoint of electromagnetic non-destructive testing. This study also describes a conventional POD model characterizing a flaw using a single parameter to discuss the advantage of a POD model with multiple parameters. The number of samples necessary to construct a reliable POD is also evaluated.

2. Eddy current signals

Figure 1 illustrates the configuration of the numerical simulations this study conducted to obtain eddy current signals. An absolute pancake probe scans the inner surface of a steam generator tube to gather signals due to an axial flaw appearing on the inner surface of the tube. The probe has a height, inner and outer diameters of 0.8, 1.0 and 3.2 mm, respectively. The exciting frequency and lift-off are 100 kHz and 1 mm, respectively. The simulations considered 1/4 of the tube. The thickness, conductivity and relative permeability of the tube are 1.27 mm, 1 MS/m. and 1, respectively.

The simulations calculated signals due to 96 rectangular slits and those due to 96 stress corrosion cracks, which would be sufficient to discuss the dependency of the POD model on the number of flaws because conventional POD models requires 40 or 60 flaws to obtain reliable POD. Parameters used in the simulations are listed in Table 1. The rectangular slits were modeled as air region. In contrast, the stress corrosion cracks were modeled as a rectangular region with a constant width and uniform conductivity inside. The conductivity was randomly set so that it followed a normal distribution with a mean of 0.1 MS/m and standard deviation of 0.03 MS/m on the basis of earlier studies discussing how to model stress corrosion cracks in finite element simulations [9,10].

The simulations were conducted using finite element program on the basis of **A**-ϕ formulation and FEM-BEM coupling discretization [11]. All signals were calibrated so that the maximum signal due to an artificial slit with a length of 20 mm and a depth of 20% tube wall thickness becomes 1.

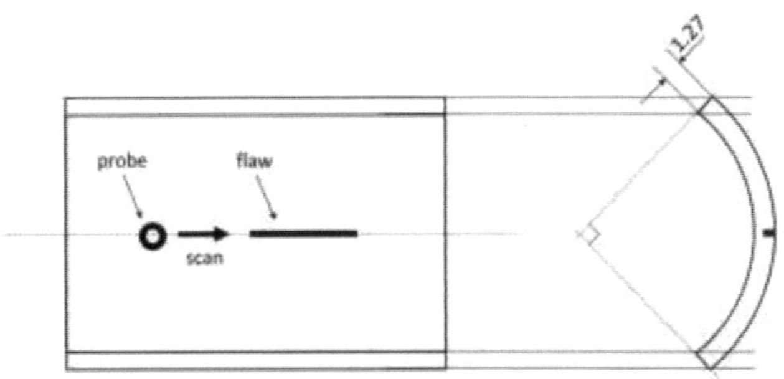

unit: mm

Figure 1. Configuration of numerical simulation

Table 1. Parameters of flaws considered in the numerical simulations

Parameters	Value
Width	0.2 [mm]
Length	2, 4, 6, 8, 10, 12, 16, 20 [mm]
Depth	5, 10, 15, 20, 25, 30, 35, 40, 50, 60, 70, 80 [%wt*]

*wt: tube wall thickness (=1.27 mm)

3. POD models

This study evaluates POD using two approaches: a conventional $\hat{a}$-a approach and two-dimensional POD approach. Both the approaches do not censor any signals; they adopted the same decision threshold, $\hat{a}_{th}$=1.0, to compare their results clearly. When the number of stress corrosion cracks used to construct POD is less than 96, signals used were randomly selected among the 96 simulated ones. All simulations to construct POD were conducted on R using software developed by the author.

3.1. Conventional POD model [1,2]

This model assumes that the maximum signal due to a flaw, $\hat{a}$, is represented as

$$\ln(\hat{a})=\beta_0+\beta_1\ln(a)+N(0,\sigma^2), \quad (1)$$

where a and $N(0, \sigma^2)$ denote the depth of the flaw and a normal distribution with a mean of 0 and standard deviation of σ. After β_0, β_1 and σ are estimated using a maximum likelihood analysis, POD(a) is calculated as the probability that $\hat{a}$ exceeds $\hat{a}_{th}$ as

$$\mathrm{POD}(a)=\Phi((\hat{a}-\hat{a}_{th})/\sigma) \quad (2)$$

where Φ stands for the cumulative distribution function of the standard normal distribution. Confidence bounds of POD(a) are calculated using a bootstrap analysis [12]. Note that only signals due to stress corrosion cracks are used in this conventional POD analysis, and those due to artificial slits do not contribute to the construction of the POD.

3.2. Two-dimensional POD model [5,6]

A two-dimensional POD model evaluates POD as a function of the depth and the length of a flaw. Specifically, this model assumes that the maximum signal due to a flaw, $\hat{a}$, is represented as

$$\hat{a}=N(\mu_1,\sigma_1^2)a_{sim}(d,l)+N(\mu_2,\sigma_2^2), \quad (3)$$

where $a_{sim}(d,l)$ stands for the maximum signal due to an artificial slit with a depth of d and a length of l obtained by numerical simulations. The two normal distributions represent the variation of signals relevant and irrelevant to the amplitude of the signals. The four unknown parameters, μ_1, σ_1, μ_2, and σ_2, are estimated on the basis of signals due to stress corrosion cracks using a maximum likelihood analysis. The probability that a flaw with a depth of d and a length of l is detected, POD(d,l), is calculated as

$$\mathrm{POD}(d,l)=\Phi(((\mu_1 a_{sim}(d,l)+\mu_2)-a_{th})/(a_{sim}(d,l)^2\sigma_1^2+\sigma_2^2)^{1/2}). \quad (4)$$

The confidence bounds are calculated using a bootstrap analysis.

4. Results and discussion

Figure 2 shows the POD curves obtained using the conventional POD model. The figure demonstrates that the number of samples, denoted as N_{data}, has a large effect on the POD curve. When the number of samples is less than 40, $a_{90/95}$ becomes more than 80% wall thickness. This indicates that either the detectability is very poor or the POD does not represent the detectability reasonably. Figure 3 presents the result of the maximum likelihood analysis to correlate $\hat{a}$ with a when N_{data}=96 and 10. The scatter of the signals is quite large, which clearly indicates the limitation in characterizing flaws using only one parameter.

Figures 4 and 5 present the results of the analyses using the two-dimensional POD model. Because of the difficulty in presenting a three-dimensional surface quantitatively, the figures show the contour line of the POD surface on a two-dimensional plane. The figures illustrate that the POD depends both on the depth and length of a flaw. Specifically, it is difficult to detect a short, deep flaw, and the length of a flaw has little effect on signals when a flaw is longer than a certain value. These results agree well with the general characteristics of eddy current testing. The figures also reveal that the two-dimensional POD model requires a much lower number of samples compared with the conventional

IOP Conf. Series: Journal of Physics: Conf. Series **860** (2017) 012032 doi:10.1088/1742-6596/860/1/012032

POD model. The results of the maximum likelihood analyses when N_{data}=96 and 10 are shown in figures 6 and 7, respectively. The figures do not differ significantly in spite of the large difference in the number of the samples used, which supports that the two-dimensional POD model does not require so large number of samples.

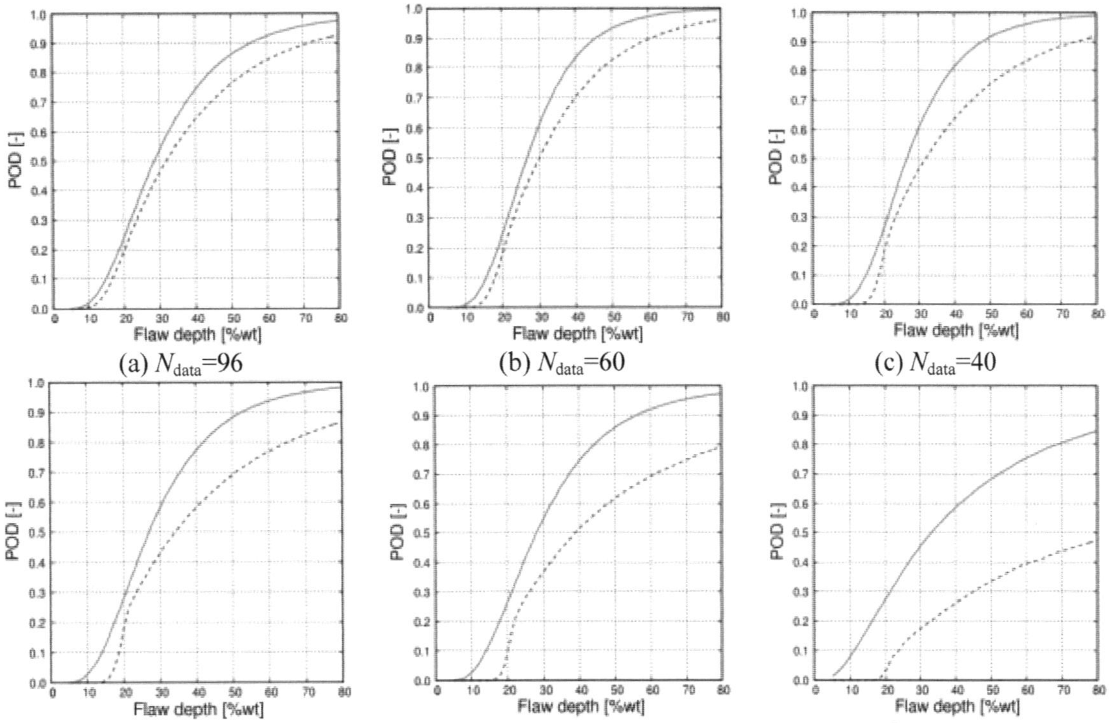

(a) N_{data}=96 (b) N_{data}=60 (c) N_{data}=40

(d) N_{data}=30 (e) N_{data}=20 (f) N_{data}=10

Figure 2. POD curve (solid line) with lower 95% confidence bound (broken line) obtained using the conventional POD model.

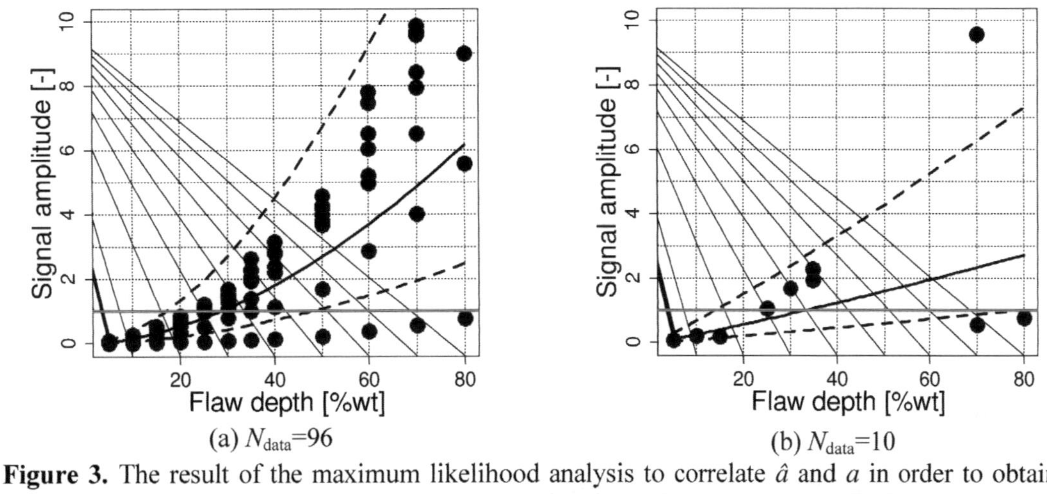

(a) N_{data}=96 (b) N_{data}=10

Figure 3. The result of the maximum likelihood analysis to correlate $\hat{a}$ and a in order to obtain the result shown in figure 2. The circles indicate the signals used to construct the POD curves. The black solid and broken lines represent the mean and mean$\pm$standard deviation of a. The horizontal line shows a=a_{th}=1.

IOP Conf. Series: Journal of Physics: Conf. Series **860** (2017) 012032 doi:10.1088/1742-6596/860/1/012032

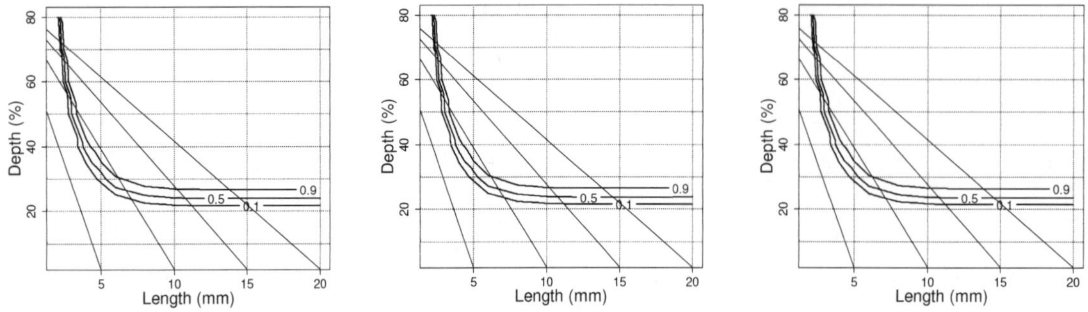

(a) N_{data}=96 (b) N_{data}=60 (c) N_{data}=40

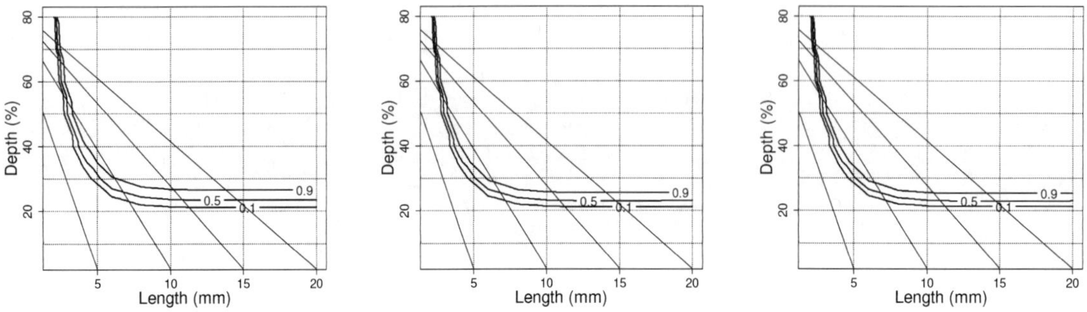

(d) N_{data}=30 (e) N_{data}=20 (f) N_{data}=10

Figure 4. Contour lines of POD=0.1, 0.5 and 0.9 obtained using the two-dimensional POD model with different number of samples.

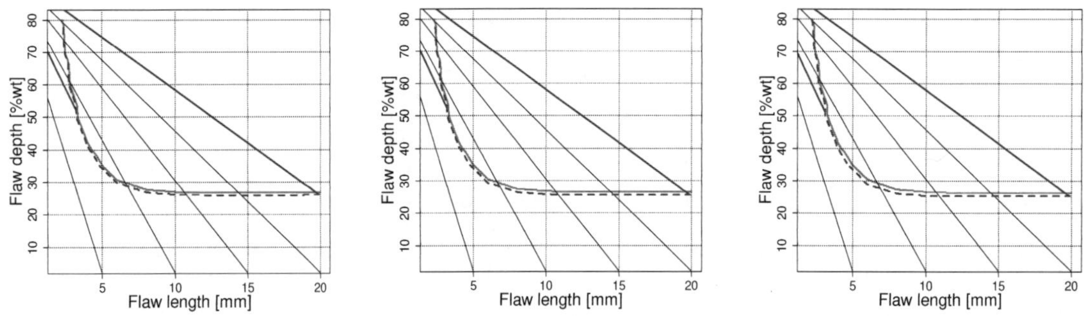

(a) N_{data}=96 (b) N_{data}=60 (c) N_{data}=40

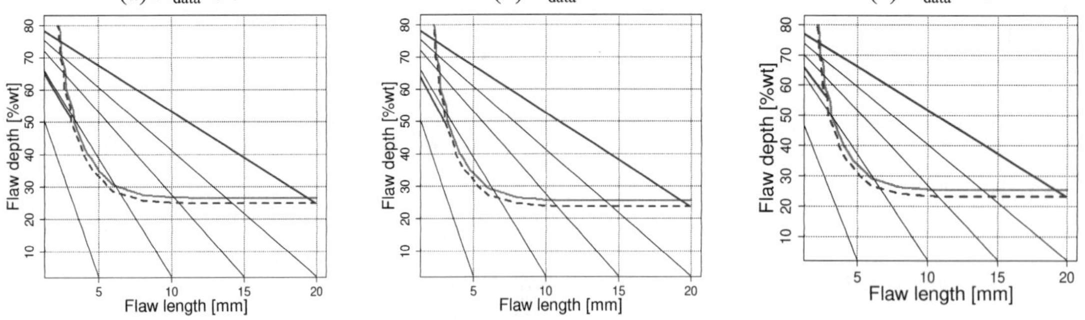

(d) N_{data}=30 (e) N_{data}=20 (f) N_{data}=10

Figure 5. POD=0.9 (solid line) with their lower 95% confidence bounds (broken line) obtained using the two-dimensional POD model with different number of samples.

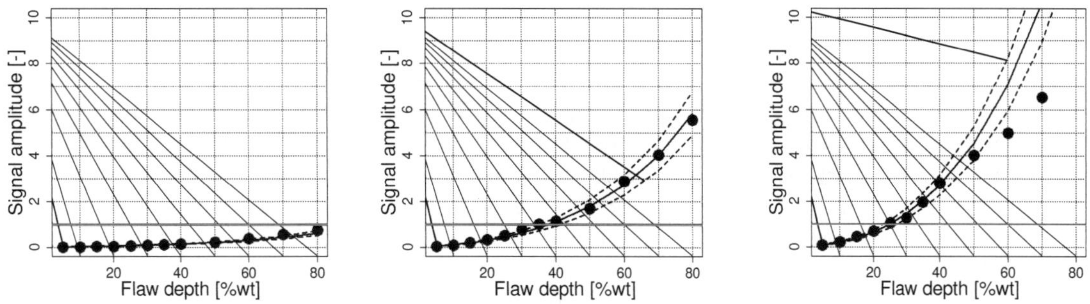

(a) Flaw length = 2 mm (b) Flaw length = 4 mm (c) Flaw length = 8 mm

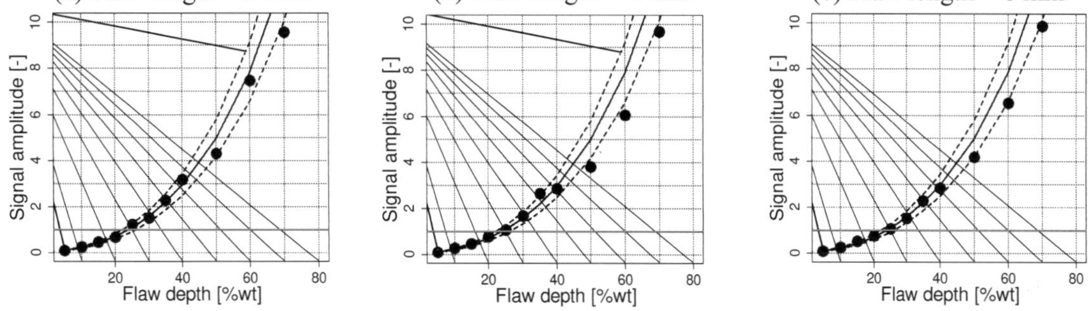

(d) Flaw length = 12 mm (e) Flaw length = 16 mm (f) Flaw length = 20 mm

Figure 6. The results of the maximum likelihood analyses to correlate eddy current signals with flaw parameters in two-dimensional POD model with N_{data}=96. Results concerning flaw lengths of 6 and 10 mm are excluded to avoid redundancy. The horizontal line shows a=a_{th}=1.

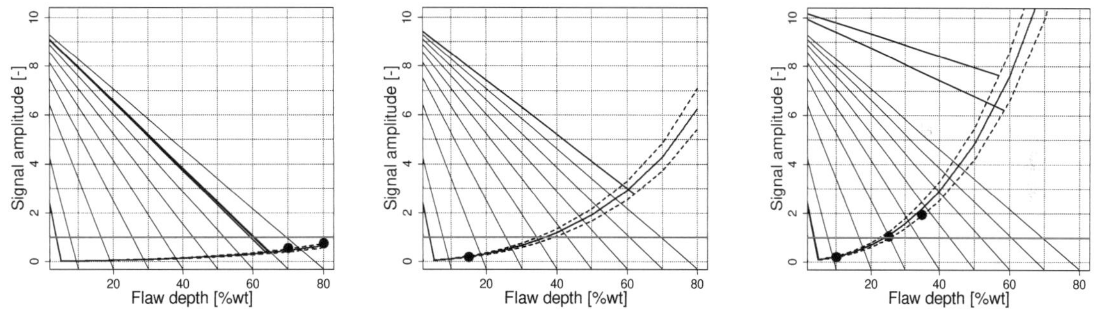

(a) Flaw length = 2 mm (b) Flaw length = 4 mm (c) Flaw length = 8 mm

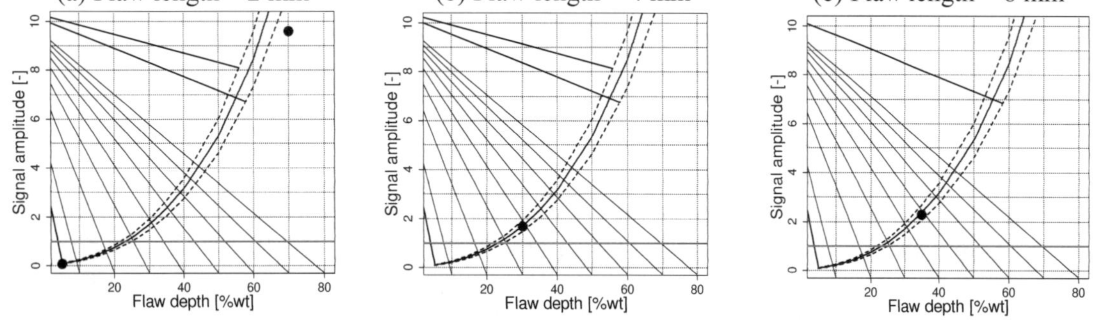

(d) Flaw length = 12 mm (e) Flaw length = 16 mm (f) Flaw length = 20 mm

Figure 7. The results of the maximum likelihood analyses to correlate eddy current signals with flaw parameters in two-dimensional POD model with N_{data}=10. The horizontal line shows a=a_{th}=1.

5. Conclusion

This study evaluated the POD of the eddy current inspection of stress corrosion cracks appearing on the outer surface of steam generator tubes of PWRs. Finite element simulations were conducted to evaluate signals due to cracks; accounting for both the depth and length of a crack explicitly to evaluate POD quantitatively. Furthermore, combinational use of numerical simulations to evaluate the general dependency of signals on flaw parameters enables the construction of a reliable POD using fewer samples compared with a conventional POD analysis using only experimental data [13,14]. Whereas this study considered the eddy current inspection of steam generator tubes of PWR, the two-dimensional POD model used in this study should be applicable to discuss POD of any non-destructive testing inspections, regardless of the target is either PWR or BWR, so long as signals obtained by the inspections are given as a quantitative one and numerical simulations can calculate signals due to a flaw with a known profile.

References

[1] Berens AP. NDE Reliability data analysis, ASM Handbook, Volume 17: Nondestructive Evaluation and Quality Control: ASM International, 1989;689-670

[2] Department of Defense Handbook: Nondestructive Evaluation System Reliability Assessment: USDOD;2009, MIL-HDBK-1823A

[3] Gandossi L, Simola K and Shepherd B 2009 The link between risk-informed in-service inspection and inspection qualification *Insight* **51** 16-20

[4] Gandossi L and Simola K 2010 Derivation and use of probability of detection curves in the nuclear industry *Insight* **52** 657-63

[5] Yusa N and Knopp J 2016 Evaluation of probability of detection (POD) studies with multiple explanatory variables *J. Nucl. Sci. Tech.* **53** 574-9

[6] Yusa N, Chen W and Hashizume H 2016 Demonstration of probability of detection taking consideration of both the length and the depth of a flaw explicitly *NDT&E Int.* **81** 1-8

[7] Pavlovic M, Takahashi K and Muller C 2012 Probability of detection as a function of multiple influencing parameters *Insight* **54** 606–11

[8] Kanzler D, Muller C and Pitkanen J 2014 Probability of detection for surface-breaking holes with low-frequency eddy current testing – a non-linear multi-parametric approach *Insight* **56** 664-8

[9] Yusa N and Hashizume H 2009 Evaluation of stress corrosion cracking as a function of its resistance to eddy currents *Nucl. Eng. Des.* **239** 2713-8

[10] Yusa N and Hashizume H 2011 Four-terminal measurement of the distribution of electrical resistance across stress corrosion cracking *NDT&E Int.* **44** 544-6

[11] Matsuoka F and Kameari 1998 A Calculation of three dimensional eddy current by FEM-BEM coupling method *IEEE Trans. Magn* **24** 182-5

[12] Goursolle T, Fauret T and Juliac E 2016 Effect of data amount on probability of detection estimation: application to eddy current testing *19th World Conference on Non-destructive Testing* 2016/06/13-17 Munich, Germany

[13] Joo K and Hong J 2015 Development of probability of detection funcitons for evaluating integrity of axial outside diameter stress corrosion cracks in steam generator *J. Nucl. Sci. Tech.* **51** 63-9

[14] Lee J, Park J, Kim H and Chung H 2010 Evaluation of ECT reliability for axial ODSCC in steam generator tubes *Int. J. Pressure Vessels and Piping* **87** 46-51

IOP Conf. Series: Journal of Physics: Conf. Series **860** (2017) 012033 doi:10.1088/1742-6596/860/1/012033

Fuel burnup analysis for Thai research reactor by using MCNPX computer code

S Sangkaew[1], T Angwongtrakool and B Srimok

Bureau of Nuclear Safety Regulation, Office of Atoms for Peace,
16 Vibhawadee-Rangsit Road, LadYao, Chatuchak, Bangkok, 10900, Thailand

[1] Corresponding author e-mail: Suksit.s@oap.go.th

Abstract. This paper presents the fuel burnup analysis of the Thai research reactor (TRR-1/M1), TRIGA Mark-III, operated by Thailand Institute of Nuclear Technology (TINT) in Bangkok, Thailand. The modelling software used in this analysis is MCNPX (MCNP eXtended) version 2.6.0, a Fortran90 Monte Carlo radiation transport computer code. The analysis results will cover the core excess reactivity, neutron fluxes at the irradiation positions and neutron detector tubes, power distribution, fuel burnup, and fission products based on fuel cycle of first reactor core arrangement.

1. Introduction

One piece of crucial information for regulatory purpose is the burnup data of the regulated reactor. The burnup data of each fuel element will reflect reactivity, neutron flux, power distribution, and fission products. Each fuel reshuffling will be permitted by the regulator only when such reshuffling will abide the operating limit conditions, in particular excess reactivity, thermal power, and fuel temperature. In addition, the fission products generated present some concern regarding activity level when moving fuel elements from one place to another within the reactor pool or transferring fuel elements to a storage place. Besides safety concern, the regulator also focuses on the remaining quantity of U-235 and generated quantity of plutonium for the safeguards purpose.

Several tools exist in calculating fuel burnup. One popular and reliable software tool is MCNPX (Monte Carlo N-Particle eXtended Transport Code) developed by Los Alamos National Laboratory [1]. It is a Fortran90 Monte Carlo radiation transport computer code that models the interaction of radiation with matter, capable of burnup analysis.

This paper aims to present the burnup analysis of the Thai research reactor (TRR-1/M1), TRIGA Mark-III, at Thailand Institute of Nuclear Technology (TINT) in Bangkok, Thailand. Specifically, only the first core arrangement of fuel elements is analyzed. The results will further serve as an input for the calculation of the subsequent core arrangements to benchmark with the actual operating data and to evaluate radiation safety as well as to assess nuclear security and safeguards.

2. The TRR-1/M1 Reactor

2.1. The Description of TRR-1/M1 Reactor

The TRR-1/M1 reactor is an open pool type TRIGA Mark-III research reactor, moderated and cooled by light water. The TRR-1/M1 reactor was converted from an MTR-type reactor (TRR-1) into rod types in 1975 and it has been operated for the modified type reactor (TRIGA Mark-III) since 1977.

The TRR-1/M1 reactor was designed for operation at a maximum steady state thermal power of 2000 kW. However, in 2012, the maximum thermal power is limited at 1300 kW for operation. The perspective view of the TRR-1/M1 reactor is shown in Figure 1.

The fuel for the TRR-1/M1 reactor in the first core loading is the solid, homogeneous mixture of uranium-zirconium hydride (UZrH) alloy with Uranium-Zirconium atomic ratio of 1.6 to 1.7. The power level of the reactor is controlled with five control rods; a safety rod, a regulating rod, two shim rods, and a transient rod. The safety rod, regulating rod, and shim rods are fuel follower control rods. The neutron absorber for all control rods is boron carbide in a solid form. The cross-section view of the reactor core configuration is shown in Figure 2.

The reactor core has a cylindrical configuration. There are 121 locations in the core that can be filled by fuel elements, control rods, neutron source, irradiation tubes, or neutron detectors. Elements are arranged in seven concentric rings in hexagonal geometry and spaces between the rods are filled with light water as coolant and moderator. The number of fuel elements in the core varies for each core loading. As mentioned above, each fuel element in the first core loading (core number 1) has about 38 grams of uranium (about 8.5 weight percent) with 20% U-235 enrichment. However, in the subsequent core loading, fuel elements made of uranium-erbium-zirconium-hydride (UErZrH) alloy are added. Each newly added fuel element has about 98 grams of uranium (about 20 weight percent) with 20% U-235 enrichment.

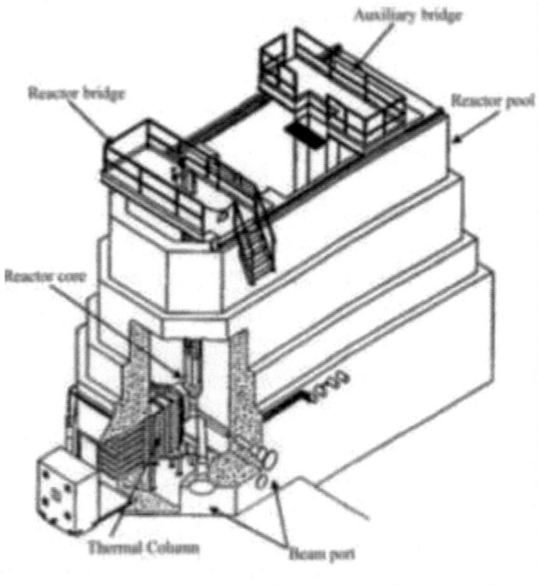

Figure 1. Perspective view of the TRR-1/M1 reactor.

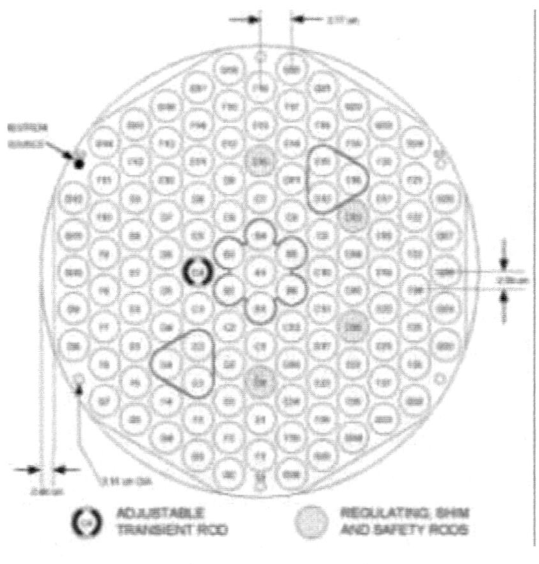

Figure 2. Cross-section view of the TRR-1/M1 reactor core.

2.2. Modelling of TRR-1/M1

The reactor was modeled in 3D with non-repeated structures in order to separate burnup calculation for each fuel element. The core number 1 model consisted of 97 fuel elements, four fuel follower control rods, a transient control rod, a central thimble irradiation tube, and three neutron detector tubes. The fuel elements were explicitly modeled non-homogeneously with the sections of fuel (UZrH), zirconium rod through the center of active fuel section, top and bottom graphite reflectors, molybdenum disk, top and bottom stainless steel end fittings, and stainless steel cladding. The control rods were modeled with the sections of boron carbide, fuel follower with zirconium rod through the center of active fuel section and void region. In the model, the central thimble was filled with water, and the three neutron detector tubes were void.

In the model, the UZrH density was intentionally varied in each fuel element in order to reflect the actual fuel measurement data from the manufacturer. All the geometric and material data were taken from the safety analysis report (SAR) prepared by Thailand Institute of Nuclear Technology in 2012 [2]. The physical properties of fuel and fuel follower elements are shown in Table 1.

Table 1. Physical properties of fuel and fuel follower elements of TRR-1/M1.

	Fuel element	Fuel follower control rod
Uranium content (wt%)	8.5, 20	8.5, 20
Hydrogen/zirconium ratio	1.6	1.6
Erbium content (wt%)	0.0, 0.5	0.0, 0.5
Enrichment of U235 (wt%)	20	20
Outer diameter (cm)	3.734	3.429
Fuel diameter (cm)	3.632	3.327
Fuel length (cm)	38.1	38.1
Diameter of zirconium rod (cm)	0.635	0.635
Cladding thickness (cm)	0.051	0.051
Absorber diameter (cm)	-	3.327
Absorber length (cm)	-	38.1

In the model, the burnup calculation for the core number 1 was based on the reactor operation on weekdays at 1 MW, the average reactor power at steady state. Also in the model, the reactor was assumed to be operated continuously 24 hours for 5 days (1 MWD for each operating day) and to be shut down for 2 days for each week. Since the control rod position could not be adjusted real time in the MCNPX model, the control rod was simply fixed at 29.53 cm out of 38.1 cm (77.5% of fully withdrawn position), which was the maximum position used in the actual reactor operation for core number 1, taken from Thai research reactor (TRR-1/M1) operation log book volumes 25-27 (1977-1979). The end of cycle for the core number 1 had the accumulated burnup of 61.22 MWD before shutting down for maintenance and fuel reshuffling for the core number 2.

In the model using MCNPX, the average UZrH fuel temperature was assumed to be at 600 K and the ENDF/B-VI.2 cross section libraries were used. The temperature limitation in MCNPX data library for the structural materials such as graphite reflector and stainless steel are only available at 293.6 K, 600 K, 900 K, 1200 K, and 2500 K. The structural materials were then assumed to be at 293.6 K because the temperatures of those structures must be below the fuel temperature. Zirconium and hydrogen in the ZrH fuel were assumed to be at 600 K and the $S(\alpha,\beta)$ thermal scattering cross section in the ENDF5 libraries was used. However, hydrogen in H_2O and carbon in graphite were assumed to be at 300 K and the same ENDF5 cross section libraries were used. In the model using MCNPX, the calculations were performed to find effective multiplication factor (k_{eff}), radial power peaking distribution, axial neutron flux in central thimble, axial neutron flux in neutron detector tubes, uranium depletion for each fuel elements and fission products. The axial and radial views of MCNPX model of the TRR-1/M1 of TINT are shown in Figure 3 and Figure 4 respectively.

The computing tool used in this calculation is a personal computer with Intel® core™ 2 Quad CPU Q8300 2.50 GHz, 4GB RAM under WINDOWS server® 2008 enterprise operating system. The calculation was performed with 115 cycles on a nominal source size of 10,000 particles per cycle. To avoid data fluctuations during the initial cycles, the first 15 cycles were skipped.

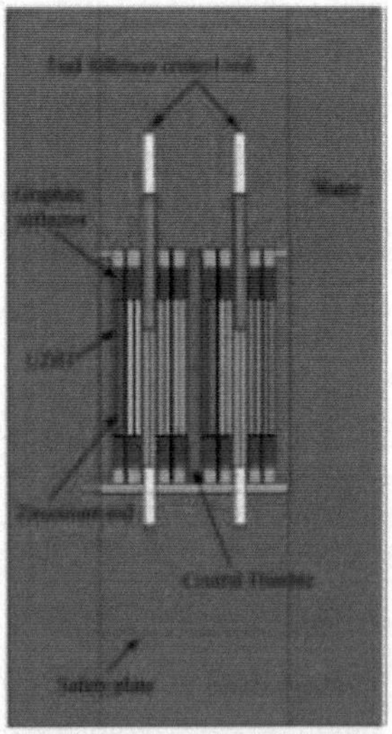

Figure 3. The axial view of MCNPX model of the TRR-1/M1

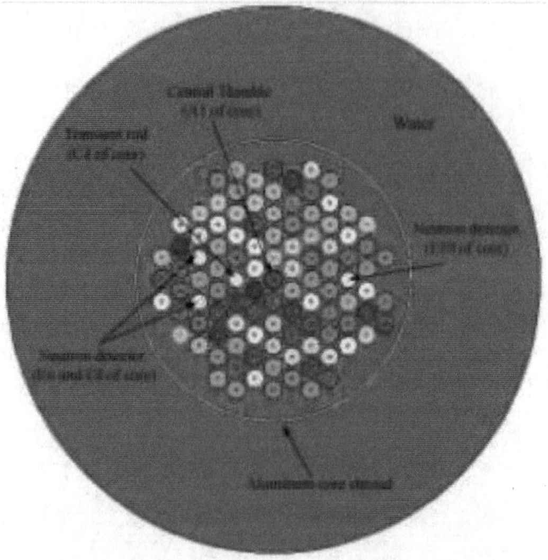

Figure 4. The radial view of MCNPX model of the TRR-1/M1

3. Result and discussion

3.1. Reactivity and material concentration

The core excess reactivity at the beginning of cycle (BOC) using MCNPX is calculated to be \$6.79 (k_{eff} = 1.04989 ±0.00068). The comparisons for the core excess reactivity at the beginning of cycle (BOC) are shown in Table 2. Table 2 also compares the result from an unpublished work using SCALE [3]. After a few days, a sharp reactivity loss of \$4.31 (k_{eff}= 1.03111±0.00075) appears due to the buildup of neutron poison, mainly xenon-135. Xenon-135 reaches equilibrium after seven operating days.

Table 2. The core excess reactivity at the beginning of cycle (BOC)

	Actual data	MCNPX	SCALE
Reactivity (\$)	\$7.43	\$6.79	\$7.17
k_{eff}	1.05486	1.04989	1.05287
σ	-	0.00068	0.00061

At the end of cycle (EOC) with the burnup of 61.22 MWD, the core excess reactivity is calculated to be about \$2.50 (k_{eff} = 1.01781 ±0.00074). The core excess reactivity as a function of operating time is shown in Figure 5 and the Xenon buildup as a function of operating time is shown in Figure 6. From the BOC to the EOC, uranium-235 is burned while Plutonium-239 is steadily produced by capture reaction on uranium-238. The uranium, plutonium, caesium and samarium concentrations as a function of operating time are presented in Figure 7 to Figure 10, respectively.

IOP Conf. Series: Journal of Physics: Conf. Series **860** (2017) 012033 doi:10.1088/1742-6596/860/1/012033

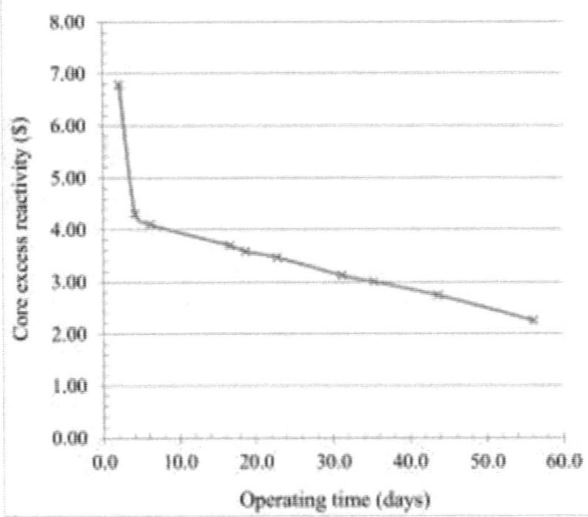

Figure 5. Core excess reactivity as a function of operating time.

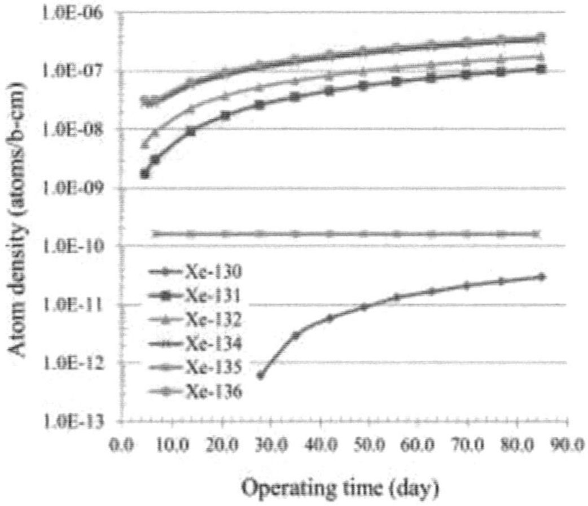

Figure 6. Xenon build-up as a function of operating time.

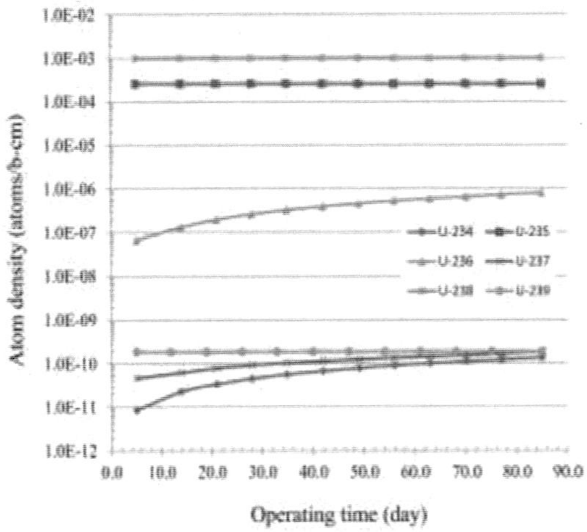

Figure 7. Uranium concentration as a function of operating time.

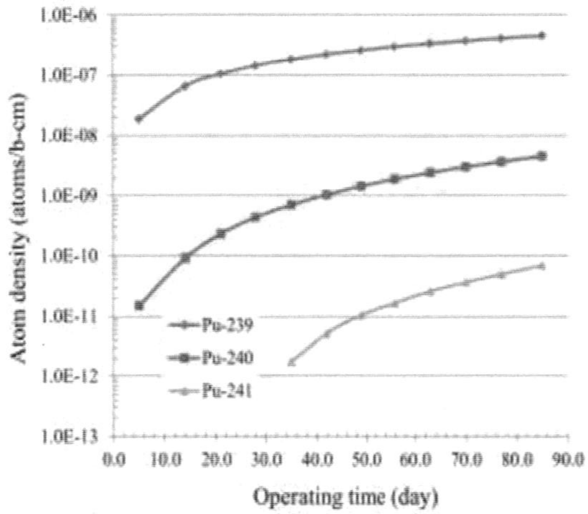

Figure 8. Plutonium concentration as a function of operating time.

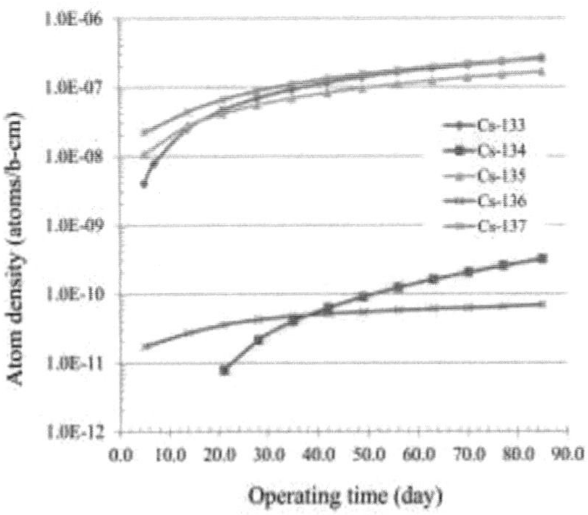

Figure 9. Caesium concentration as a function of operating time.

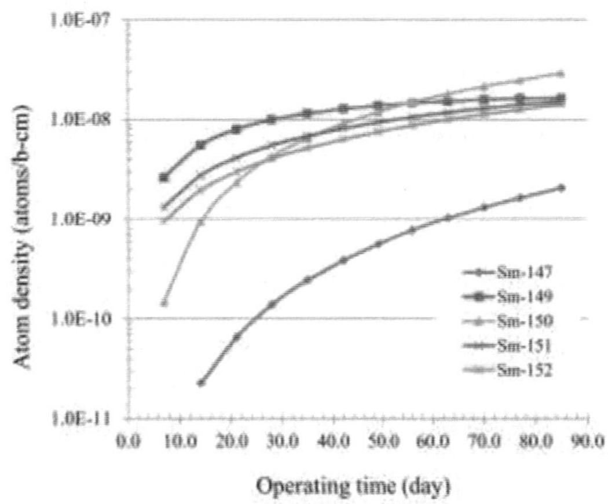

Figure 10. Samarium concentration as a function of operating time.

3.2. Power distribution and U-235 depletion

For both the beginning of cycle (BOC) and end of cycle (EOC), the maximum power factors of fuel are found to be in the B ring due to higher thermalization of neutrons in central region of the reactor core. At the BOC, the maximum power factor is 1.70 at the B2 position and the minimum is 0.54 at the G30 position. At the EOC, the maximum power factor is 1.66 at the B5 position and the minimum is 0.55 at the G30 position, which is the same position as in the BOC. The difference between the maximum and minimum power factor positions is about 70% for both the BOC and the EOC. The maximum power factors of the EOC and BOC differ by about 2%. The radial power factor distribution of fuel and fuel follower elements in the reactor core at the BOC is shown in Figure 11.

The individual %U-235 depletion of fuel and fuel flower elements in the reactor core at EOC (61.22 MWD) is shown in Figure 12. The maximum fuel depletion of around 3.27% (1.24 grams of U-235) occurs in the B ring. From the centre to the rim of the reactor core, the U-235 depletion decreases and reaches the minimum value of 1.07% (0.41 grams of U-235) at the G30 position. The average burnup of the core is about 2.00% or roughly 76 grams of U-235

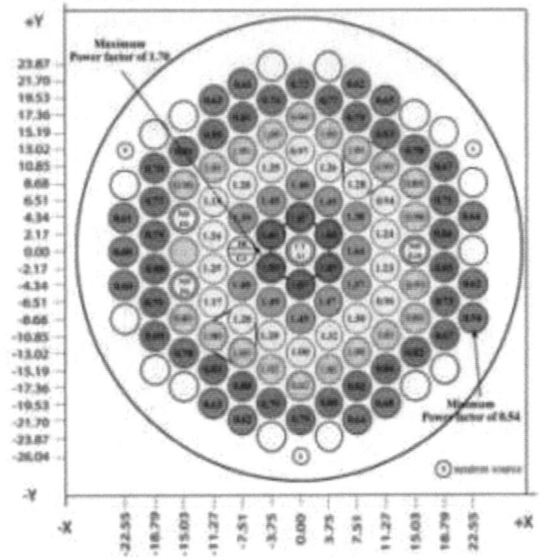

Figure 11. Radial power factor distribution at BOC of reactor core

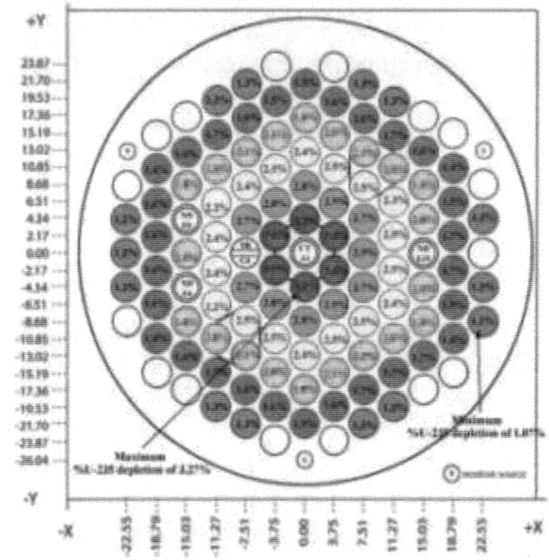

Figure 12. %U-235 depletion of fuel and fuel flower elements at EOC of reactor core

3.3. Neutron flux

The axial neutron fluxes distributions in the reactor core at the central thimble (CT) and at the three neutron detector tubes (at the E06, E08, and E19 positions) are shown in Figure 13 and Figure 14, respectively.

For the central thimble (CT) at the A1 position of the reactor core, the neutron flux distributions at the BOC and EOC are in parabolic shape. The maximum neutron fluxes at both the BOC and EOC occur at the middle of the CT due to high neutron density in the active region of the reactor core. At the middle of the CT, neutron fluxes at the BOC are little higher than neutron fluxes at the EOC by about 2% to 5%. Thermal neutron fluxes at the BOC and EOC are 7.01×10^{12} n/cm^2-s and 6.86×10^{12} n/cm^2-s, respectively. Epithermal neutron fluxes at the BOC and EOC are 3.09×10^{13} n/cm^2-s and 2.99×10^{13} n/cm^2-s, respectively. And fast neutron fluxes at the BOC and EOC are 1.12×10^{13} n/cm^2-s and 1.06×10^{13} n/cm^2-s, respectively.

The neutron fluxes of the three neutron detector tubes (ND) at the E06, E08, and E19 positions of the reactor core represent the total energy range (0.0-15.0 MeV) unlike the neutron fluxes at the CT, which represent the thermal, epithermal and fast neutron energy ranges. This is due to the nature of a neutron detector that measures neutrons for the whole energy range. The neutron flux distributions of neutron detector tubes at the BOC and EOC are also in similar parabolic shape as the neutron flux distributions of the CT. The maximum neutron flux at the EOC is nearly the same as the maximum neutron flux at the BOC, differing by about 1%. The average neutron flux at the E19 position appears higher than the average neutron fluxes at the E06 and E08 positions by about 12%. This is mainly because the E19 position is surrounded by more fuel elements. At the EOC, the maximum neutron fluxes at the E06, E08 and E19 positions are 4.09 x 10^{13}, 4.08 x 10^{13}, 4.60 x 10^{13} n/cm^2-s, respectively.

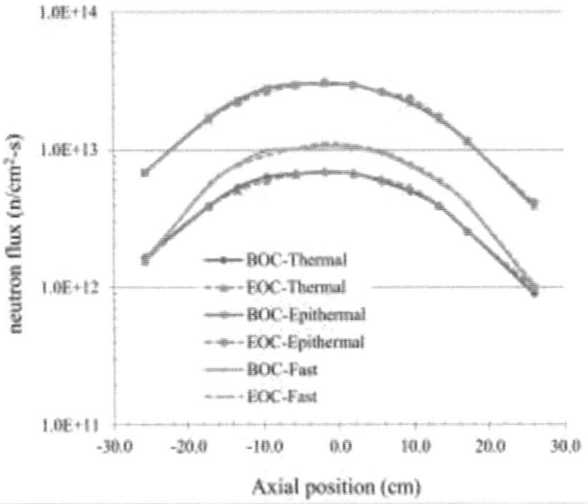

Figure 13. Thermal, epithermal and fast neutron fluxes distribution at central thimble (CT)

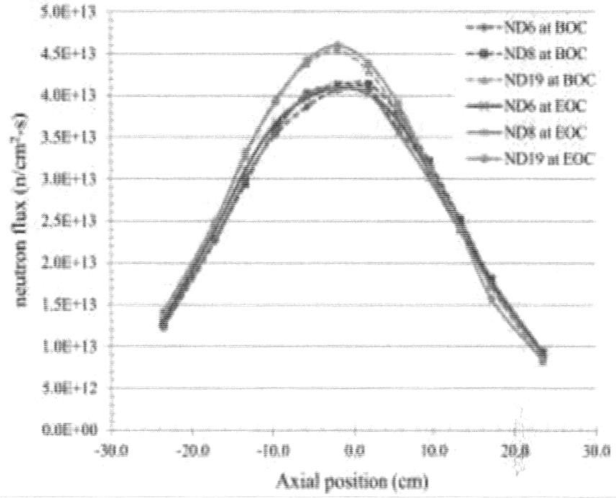

Figure 14. Total neutron fluxes distribution at neutron detector tubes (ND)

4. Conclusion

This paper presents a preliminary analysis of fuel burnup of the TRR-1/M1 Thai research reactor. The analysis covers the calculation of core excess reactivity, neutron flux, power distribution and fission products. The results show that even though reactivity decreases greatly at the EOC, the power and neutron flux distributions at the EOC are less than at the BOC by only about 2%. Also, the neutron flux distributions of the CT and ND at the BOC and EOC differ by only about 1-5 %. This means the reactor core behaves almost the same throughout the operation from the BOC to the EOC. The analysis results such as depletion of uranium and production of fission products will be considered and served as the input for the burnup analysis of the next core loading. In addition, the results from this paper will be benchmarked in the future with the actual experiment results or the calculation results from other calculation software.

References

[1] Hendricks J S, McKinney G W, Fensin M L, James M R, Johns R C, Durkee J W, Finch J P, Pelowitz D B, Waters L S and Johnson M W 2008 *MCNPX 2.6.0 Extensions* (Tennessee: Oak Ridge National Laboratory)

[2] Thailand Institute of Nuclear Technology 2012 *Safety Analysis Report for Thai Research Reactor-1/Modification 1 (TRR-1/M1) Rev* 1 (Bangkok: Thailand Institute of Nuclear Technology) chapter 5 p 10

[3] Rearden B T and Jessee M A 2016 *SCALE Code System Version 6.2.1* (Tennessee: Oak Ridge National Laboratory)

[4] Chiesa D, Clemenza M, Pozzi S, Monica S, Alloni D, Magrotti G, Manera S, Prata M, Salvini A, Cammi A, Zanetti M and Sartori A 2016 *Annals of nuclear Energy* **96** 270

[5] Amina E A, Bashterb I I, Hassanb N M and Mustafab S S 2016 *Applied Radiation and Isotopes* **113** 70

[6] Türkmena M and Çolakb Ü 2014 *Progress in Nuclear Energy* **77** 152
[7] Abdelghafar Galahom A, Bashterb I I and Azizc M 2014 *Progress in Nuclear Energy* **77** 65
[8] El Bakkri B, El Bardouni T, Nacir B, El Younoussi C, Boulaich Y, Boukhal H and Zoubair M 2013 *Annals of Nuclear Energy* **51** 112
[9] El Bakkari B, El Bardouni T, Nacir B, El Younoussi C, Boulaich Y and Boukhal H 2013 Annals of Nuclear Energy **60** 242
[10] Lyric Z Idris,Mahmood M Sayem, Motalab M Abdul and Khan J Haque 2013 *Annals of Nuclear Energy* **55** 225
[11] Hudaa M Q, Bhuiyana S I and Obarab T 2008 *Annals of nuclear Energy* **35** 147
[12] Zagar T, Ravnik M and Jeraj R 2000 *Proc. Int. Conf. on Nuclear Energy in Central Europe 2000 (Bled)* (Ljubljana: Nuclear Society of Slovenia)
[13] Persic A, Ravnik M and Zagar T 1998 *Proc. Int. Conf. on Nuclear Energy in Central Europe 1998* (Terme Catez) (Ljubljana: Nuclear Society of Slovenia) p 35
[14] Hudaa M Q, Bhuiyana S I and Obarab T 2008 *Annals of nuclear Energy* **35** 147

IOP Conf. Series: Journal of Physics: Conf. Series **860** (2017) 012034 doi:10.1088/1742-6596/860/1/012034

Impurity accumulation and performance of ITER and DEMO plasmas in the presence of transport barriers

B Chatthong[1,4], J Promping[2] and T Onjun[3]

[1] Department of Physics, Faculty of Science, Prince of Songkla University, Hat Yai, Songkla, Thailand
[2] Thailand Institute of Nuclear Technology, Bangkok, Thailand
[3] School of Manufacturing Systems and Mechanical Engineering, Sirindhorn International Institute of Technology, Thammasat University, Pathum Thani, Thailand

[4] Corresponding author e-mail: boonyarit.ch@psu.ac.th

Abstract. In this work, the impurity accumulations and their performance in the presence of both ITB and ETB in ITER and DEMO plasmas are investigated using a BALDUR integrated predictive modelling code. In these simulations, a combination of a neoclassical transport model NCLASS and an anomalous transport model Mixed Bohm/gyro-Bohm is used. The boundary condition is described at the top of the pedestal, which is calculated theoretically based on a combination of magnetic and flow shear stabilization pedestal width scaling and an infinite-n ballooning pressure gradient model. The toroidal flow is calculated based on the NTV (neoclassical toroidal viscosity) toroidal velocity model. The time evolution of plasma temperature and density profiles of ITER and DEMO (Korean K-DEMO and Japanese DEMO models A, B and C) plasmas are simulated in *H*-mode scenario with and without ITB formation. It is found that Japanese DEMO model C yields highest plasma temperature; while Korean DEMO yields the best plasma performance among those designs considered. Impurity accumulation is found to be highest in Japanese DEMO model B.

1. Introduction

Generally, the fusion reactions in tokamak experiments can be improved with the increase of plasma density, temperature and confinement time. It was discovered that when the plasma made a transition to a high confinement mode (*H*-mode), resulting from formation of an edge transport barrier (ETB), the plasma performance is significantly increased [1, 2]. The performance of the plasma can be further enhanced with the formation of an internal transport barrier (ITB) in the plasma core [3]. The formation of transport barriers is very important that the burning plasma experiment like ITER project [4] is designed to perform in *H*-mode operational regime. Furthermore, in the next step of fusion experiments, the DEMO experiments must operate based on advance scenario. In other words, both ITB and ETB must be simultaneously formed and sustained, so the plasmas are able to operate in a steady state fashion.

Density control is one of the essential parts for burning plasma in a fusion reactor. Impurity particles are composed of two different sources: intrinsic helium which is a product of deuterium-tritium reaction and impurities that come from plasma-wall interaction process such as beryllium, carbon, tungsten etc. Impurity accumulation can severely detriment fusion performance because these particles dilute the plasma and radiate energy away from the plasma core [5]. Experimentally, it was found that formation of ITB lead to impurity accumulation in the plasma core. For example, in JT-60U reverse shear experiments, strong peak impurity (argon and carbon) density profile was observed [6].

IOP Conf. Series: Journal of Physics: Conf. Series **860** (2017) 012034 doi:10.1088/1742-6596/860/1/012034

The helium density was not observed to be strongly peaked because it has low Z value [7]. Hence, the helium diffusivity and convection velocity found by JT-60U team illustrated that there was sufficient exhaust even in the plasma with ITB formation [8].

In this paper, a 1.5-dimensional BALDUR integrated predictive modelling code is used to simulate the time evolution profiles of electron and ion temperatures and electron, hydrogenic and impurity densities for ITER- and DEMO-like plasmas. ITB formation is simulated through a semi-empirical transport model called Mixed Bohm/gyro-Bohm. The model assumes that ITB formation occurs when anomalous transport is quenched due to flow shear and magnetic shear stabilization [9]. The flow shear is calculated from the radial electric field through the force balance equation, where toroidal rotation term is needed. In this work, a theoretically based toroidal velocity model is calculated based on NTV concept [10]. ETB formation is modelled based on the works in Ref. [11], where ETB is calculated using a pedestal model. The pedestal temperature is determined using a pedestal width model with pressure gradient limits ballooning mode instability. The simulation results will examine plasma performance, i.e. electron and ion temperature and impurities density.

The paper is organized as follows: the simulation setup as well as the important models used are introduced in section 2; simulations results of ITER and DEMOs and discussion are given in section 3; and the conclusion is presented in section 4.

2. BALDUR and Simulations Setup

All simulations in this work are carried out using an 1.5D BALDUR integrated predictive modelling code[12]. BALDUR is a time-dependent transport modelling code, which can compute tokamak plasma quantities such as electron and ion temperatures, electron, hydrogenic and impurities densities, safety factor q and toroidal flow [13, 14]. The code self-consistently computes those physical quantities by combining several physical processes together in form of modules. Important modules used by BALDUR are transport, plasma heating, particle flux, boundary conditions and sawtooth oscillations modules. This section discusses the three main modules used in the simulations: neoclassical transport module, toroidal velocity module and pedestal module. In short, BALDUR calculates the plasma profiles by solving the transport equations. The transport is composed of neoclassical transport, computed by NCLASS, and anomalous transport, where the diffusivities are calculated using Mixed Bohm-Gyro-Bohm model (section 2.1). The toroidal velocity is an ingredient needed to compute the anomalous transport; it is calculated based on neoclassical toroidal viscosity (NTV) physics (section 2.2). The boundary of the transport equations are set at the top of pedestal, where its value is calculated based on a pedestal model (section 2.3).

2.1. Mixed Bohm/Gyro-Bohm model

In this work, an anomalous transport at the core of the plasma is modelled using a semi-empirical transport model called mixed Bohm/gyro-Bohm (Mixed B/gB) [9]. This version of Mixed B/gB includes ITB formation modelling based on the well-known assumption that ITB can form when the anomalous transport is suppressed by the ω_{ExB} flow shear and magnetic shear [9]. ITB effect is implemented using a cut-off Heaviside function in the Bohm term. The model computes electron diffusivity χ_e, ion diffusivity χ_i, hydrogenic diffusivity D_H and impurity diffusivity D_Z as follows:

$$\chi_e = 1.0\chi_{gB} + 2.0\chi_B, \tag{1}$$

$$\chi_i = 0.5\chi_{gB} + 4.0\chi_B, \tag{2}$$

$$D_H = D_Z = (0.3 + 0.7\rho)\frac{\chi_e \chi_i}{\chi_e + \chi_i}, \tag{3}$$

where

$$\chi_{gB} = 5\times10^{-6}\sqrt{T_e}\left|\frac{\nabla T_e}{B_\phi^2}\right|, \tag{4}$$

$$\chi_B = \chi_{B_0} \times \Theta\left(-0.14 + s - \frac{1.47\omega_{E\times B}}{\gamma_{ITG}}\right), \tag{5}$$

$$\chi_{B_0} = 4x10^{-5} R \left|\frac{\nabla(n_e T_e)}{n_e B_\phi}\right| q^2 \left(\frac{T_e(0.8\rho_{max}) - T_e(\rho_{max})}{T_e(\rho_{max})}\right), \tag{6}$$

where χ_{gB} is the gyro-Bohm contribution, χ_B is the Bohm contribution, ρ is normalized minor radius, T_e is the local electron temperature in keV, B_ϕ is the toroidal magnetic field, s is the magnetic shear, ω_{ExB} is the shearing rate, γ_{ITG} is the linear growth rate, R is the major radius, and n_e is the local electron density. The linear growth rate γ_{ITG} can be calculated as v_{th}/qR, where v_{th} is the electron thermal velocity.

The ωExB shearing rate can be calculated as [15, 16]:

$$\omega_{E\times B} = \left|\frac{RB_\theta^2}{B_\phi}\frac{\partial(E_r/RB_\theta)}{\partial\psi}\right|, \tag{7}$$

where B_θ is the poloidal magnetic field, Ψ is the poloidal flux, and Er is the radial electric field, which can be calculated as follows:

$$E_r = \frac{1}{Zen_i}\frac{\partial p_i}{\partial r} - v_\theta B_\phi + v_\phi B_\theta \tag{8}$$

where p_i is plasma pressure, v_θ and v_ϕ are the poloidal and toroidal velocities, respectively, n_i is the ion density, and Z is the ion charge number. The poloidal flow is estimated using NCLASS module, while the toroidal flow is calculated as discussed in section 2.2.

2.2. Toroidal velocity model

According to equations (7) and (8), the toroidal velocity is one of the elements needed for the suppression of anomalous transport by the flow shear, leading to ITB formation. The model used in this work is based on NTV physics, where the offset toroidal rotation is caused by the neoclassical toroidal viscosity dissipation. This dissipation is a result of symmetry breaking generated by a non-axisymmetric field [17]. Additionally, the rotation is theoretically described to be intrinsic, hence its existence is crucial for enhancement operation of tokamak experiments. The development and implementation of the model into BALDUR code can be seen in Ref. [10], where a local and strong ITB is to be expected in ITER. In short, the model is in the form:

$$v_\phi \propto \frac{dT_i}{dr}, \tag{9}$$

implying that the rotation is driven by local ion temperature gradient.

2.3. Pedestal model

Throughout this work, the boundary condition for the transport equations is set to be at the top of the pedestal [18]. This is the location where the edge transport barrier (ETB) is observed. So all simulations are set to be in *H*-mode operation regime. The pedestal region is the edge area of the plasma with steepest gradient compared to other regions. The pedestal temperature T_{ped} is calculated as follows:

$$T_{ped} = \frac{1}{2kn_{ped}}\Delta\left|\frac{\partial p}{\partial r}\right| \tag{10}$$

where n_{ped} (m^{-3}) is pedestal density, k is the Boltzmann's constant, and Δ is the pedestal width. Details of the models can be seen in Ref. [11]. In essence, the pedestal pressure gradient and pedestal width are limited by the ballooning mode instability and the magnetic and flow shear stabilization [19, 20].

Note that the pedestal density is empirically assumed to be a fraction (0.71) of electron line average density [21].

3. Results and Discussions

3.1. ITER simulations

In this work, standard type I ELMy *H*-mode ITER simulations are investigated using BALDUR. The design parameters used in the simulations are: major radius *R* of 6.2 m, minor radius *a* of 2.0 m, plasma current I_p of 15.0 MA, toroidal field of 5.3 T, elongation κ of 1.7, triangularity δ of 0.33, RF (radio frequency) heating of 7.0 MW, NBI (neutral beam injection) heating of 33.0 MW and line average density n_l of 1.0x1020 m^{-3} [22]. Figure 1 illustrates simulation results of ITER-like plasma. The top left panel shows central ion temperature T_i (top) and central electron temperature T_e (bottom). It can be seen that initially the central temperatures are slowly increasing until around 100 seconds where the plasma has reach quasi-stationary state. This is because the plasma current has been set to slowly increase to the designated value. At quasi-stationary state, the temperatures are fluctuated around 10% of their average values. This is due to sawtooth instability, which occurs at the center of plasma. Quantitatively, the ion temperature fluctuates around 40 keV and electron temperature around 35 keV.

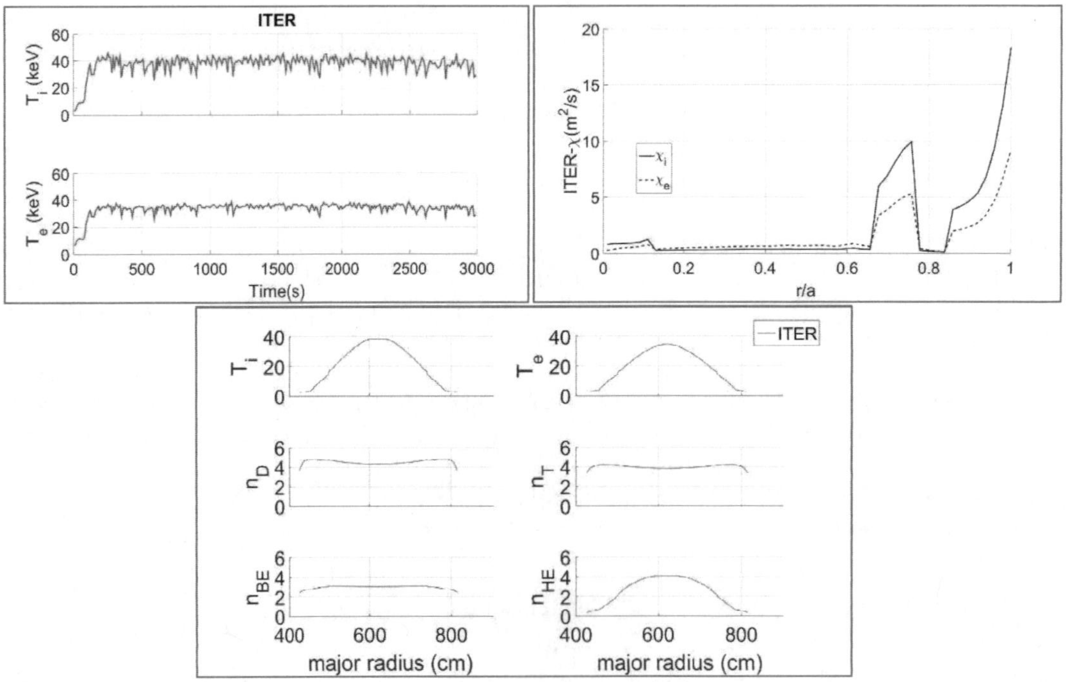

Figure 1. Time evolution of temperatures at plasma center (top left), transport diffusivities (top right) and plasma density and temperature profiles during quasi-stationary state (bottom) for ITER.

The bottom panels of figure 1 illustrate profiles of ion temperature, electron temperature, deuterium density (n_D), tritium density (n_T), beryllium density (n_{BE}), and helium density (n_{HE}) as a function of normalized minor radius *r/a* at arbitrary time during quasi-stationary state (2,900 seconds). Note that units of densities are 1x10^{19} m^{-3} for deuterium and tritium and 1x10^{18} m^{-3} for beryllium and helium. Obviously, the plasma temperatures are highest at plasma center and decrease toward the edge. It appears that a wide ITB are formed from *r/a* = 0.2 to 0.6. This agrees with the profiles of ion diffusivity χ_i and electron diffusivity χ_e as shown in the top right panel of figure 1. Moreover, deuterium and tritium densities have roughly the similar profiles, which is quite flat, just above 4x10^{19} m^{-3}, with strong gradient near the edge. Interestingly, beryllium is found to be able to penetrate into

IOP Conf. Series: Journal of Physics: Conf. Series **860** (2017) 012034 doi:10.1088/1742-6596/860/1/012034

the plasma core, though its profile is rather flat. Whereas, helium is found to accumulate in the plasma core so it is also affected by ITB formation.

3.2. DEMO simulations

Two different DEMO projects are investigated in this work: Korean DEMO (K-DEMO) and Japanese DEMO. The design parameters of K-DEMO used in simulations are R = 6.8 m, a = 2.1 m, I_p = 12 MA, toroidal field of 7.4 T, κ = 2.0, and δ = 0.625 [23]. For Japanese DEMO, three different models are investigated; model A, model B and model C. The design parameters used in simulations can be seen in table 1 [24]. In addition to the parameter used in table 1, it is calculated that tokamak built by model C will have the most weight of 23,900 ton, while the tokamak built by model C will have the lowest weight of 15,700 tons. Example of DEMO simulations are shown in figures 2 and 3, which are the simulation results from K-DEMO and Japanese DEMO model C, respectively.

Table 1. Parameters used for simulations of Japanese DEMOs [24].

parameter	model A	model B	model C
Major radius (m)	5.1	5.5	6.5
Minor radius (m)	2.1	2.1	2.1
Aspect ratio	2.5	2.6	3.1
Elongation	2.05	2.0	1.9
Triangularity	0.1	0.4	0.4
Magnetic field (T)	5.6	6.0	6.8
Plasma current (MA)	17.4	16.7	15.0
Current drive power (MW)	63	59	54

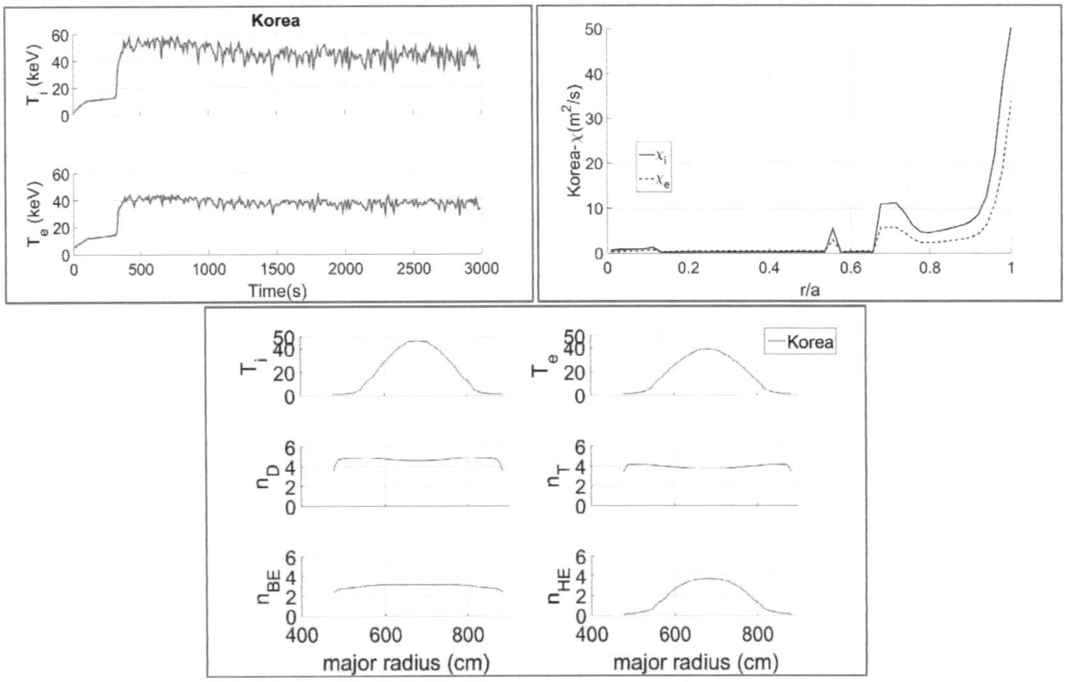

Figure 2: Time evolution of temperatures at plasma center (top left), transport diffusivities (top right) and plasma density and temperature profiles during quasi-stationary state (bottom) for K-DEMO.

The top left panel of figure 2 shows that the plasma has reached quasi-stationary state at around 400 seconds. The ion temperature fluctuates around 45 keV and electron temperature around 40 keV. Furthermore, in K-DEMO a wide ITB formation can be found from r/a = 0.1 to 0.5 as illustrated by the drop in diffusivities (top right panel of figure 2). It also appears that deuterium, tritium, and beryllium densities profiles are relatively flat, with a sharp gradient at plasma edge for deuterium and tritium. Similarly, helium is also accumulated inside the ITB.

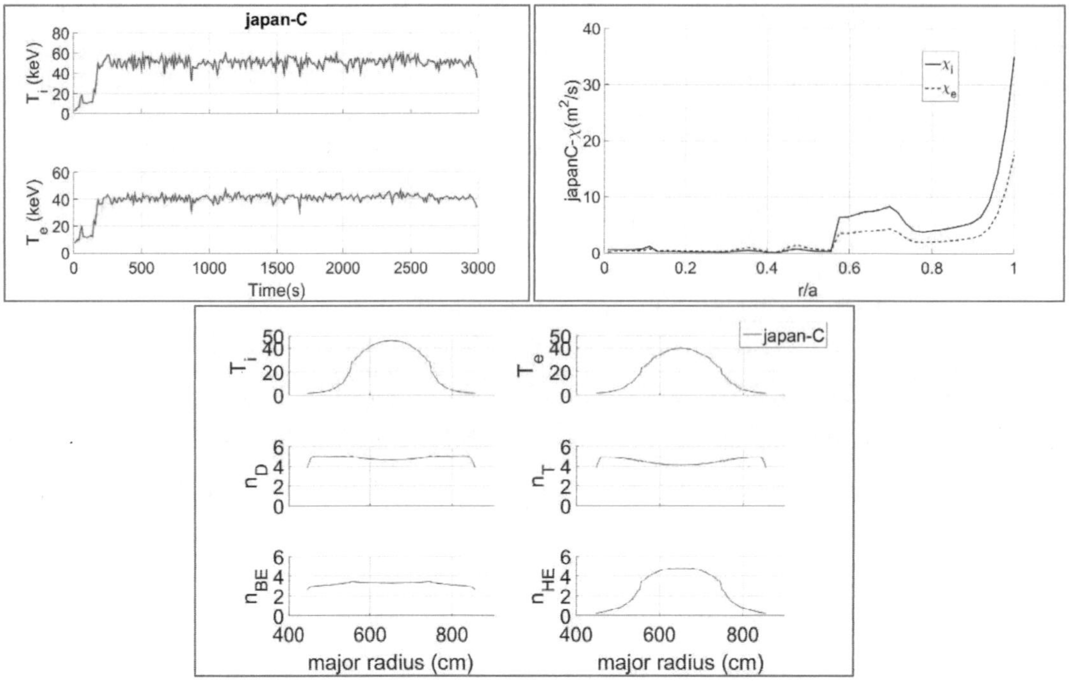

Figure 3: Time evolution of temperatures at plasma center (top left), transport diffusivities (top right) and plasma density and temperature profiles during quasi-stationary state (bottom) for Japanese DEMO model C.

Figure 3 illustrates simulation results of Japanese DEMO model C. Evidently, the plasma has reached a quasi-stationary state after 300 seconds. The ion temperature fluctuates around 50 keV and electron temperature around 40 keV. Furthermore, in Japanese DEMO model C, a wide ITB formation can be found from r/a = 0.1 to 0.5 as illustrated by the drop in diffusivities (top right panel of figure 3). It also appears that deuterium, tritium, and beryllium densities profiles are relatively flat, with a sharp gradient at plasma edge for deuterium and tritium. Helium peaking profile is found inside the ITB.

3.3. Performance comparison

Table 2 summarizes the performance of each tokamak simulated in this work. Note that W_{TOT} represents plasma total power yielded, P_α is alpha power, P_{aux} is auxiliary heating power and Fusion Q is the ratio of the output power to the input power, which is used to represent plasma performance. It can be calculated as:

$$Q = \frac{5 \times P_\alpha}{P_{aux}}. \tag{11}$$

Overall, Japanese DEMO model C yields the highest temperatures at plasma center, while ITER yields the lowest. The same machine also yields highest fusion power (W_{TOT}), which includes both power from alpha and neutron particles, but Japanese DEMO model B yields the most alpha heating power. On the other hand, Japanese DEMO model A yields lowest fusion power and ITER yields the least alpha heating power. Nevertheless, considering for fusion performance is best through the fusion Q. Apparently, K-DEMO yields highest value of Q. This is because it uses less heating input of 40 MW, comparing to 54 MW used in Japanese DEMO model C. On the other hand, Japanese DEMO model A is the worst in term of fusion performance comparison. For impurity accumulation, it is found from simulations that beryllium concentration at plasma center is highest in Japanese DEMO model B and lowest in ITER. Helium accumulation is highest in Japanese DEMO model B and lowest in K-DEMO.

Table 2. Comparison of simulation results from ITER and DEMOs (average values during quasi-stationary state).

Parameters	ITER	K-DEMO	J-DEMO A	J-DEMO B	J-DEMO C
$T_{i,0}$ (keV)	37.99	43.42	38.66	44.43	51.00
$T_{e,0}$ (keV)	34.15	37.27	34.45	37.86	41.35
W_{TOT} (MW)	426	441	404	461	534
P_{AUX} (MW)	40	40	63	59	54
P_{α}(MW)	133	153	151	187	176
Q	16.63	19.13	11.98	15.85	16.30
$n_{BE,0}$ (10^{18} m^{-3})	3.03	3.12	3.90	4.01	3.37
$n_{HE,0}$ (10^{18} m^{-3})	4.13	3.59	4.25	5.16	5.02

4. Conclusion

Simulations of ITER, Korean DEMO and Japanese DEMO (models A, B and C) in the presence of both ITB and ETB are carried out using BALDUR integrated predictive modelling code. The core transport is predicted using NCLASS and Mixed B/gB models with the toroidal velocity predicted using NTV offset rotation model and the boundary condition predicted using the magnetic and flow shear stabilization width scaling and an infinite-n ballooning pressure gradient limiting model. The ITB formation is found in all simulations, resulting an increase of fusion reactions and, consequently, helium accumulation and a decrease of beryllium accumulation in the core. In general, the plasma starts with slowly ramp-up phase until quasi-stationary state is achieved, where plasma temperatures fluctuate around 10% due to the sawtooth oscillation. It is found that among the designs considered, the best plasma performance, based on Q value, is observed in K-DEMO even though Japanese DEMO model C yields higher plasma temperatures and total fusion power. On the contrary, Japanese DEMO model A yields lowest plasma performance among the designs studied. For future work, this set of codes can be used to also simulate the performance of other DEMO designs such as European DEMO, Indian DEMO and Chinese DEMO. Based on all results, the performance optimization of the future DEMO can be predicted.

Acknowledgments

This work was supported by the National Research Council of Thailand (NRCT) under Contract No. 224184 and the research fund from Prince of Songkla University under Contract No. SCI59005N. This project is a part of a collaborative research under “Center of Plasma and Nuclear Fusion Technology (CPaF)”.

References

[1] ASDEX Team 1989 *Nucl. Fusion* **29** 1959
[2] Wagner F *et al* 1982 *Phys. Rev. Lett.* **49** 1408
[3] Connor J W, Fukuda T, Garbet X, Gormezano C, Mukhovatov V, Wakatani M *et al* 2004 *Nucl. Fusion* **44** R1
[4] Aymar R, Barabaschi P and Shimomura Y 2002 *Plasma Phys. Control. Fusion* **44** 519
[5] Wesson J 2004 *Tokamaks* (Clarendon Press)
[6] Takenaga H *et al* 2003 *Nucl. Fusion* **43** 1235
[7] Doyle E J *et al* 2007 *Nucl. Fusion* **47** S18
[8] Takenaga H and JT-60 Team 2001 *Phys. Plasmas* **8** 2217
[9] Tala T J J, Parail V V, Becoulet A, Corrigan G, Heading D J, Mantsinen M J, Strand P I *et al* 2002 *Plasma Phys. Control. Fusion* **44** A495
[10] Chatthong B and Onjun T 2013 *Nucl. Fusion* **53** 013007

IOP Conf. Series: Journal of Physics: Conf. Series **860** (2017) 012034 doi:10.1088/1742-6596/860/1/012034

[11] Onjun T, Bateman G, Kritz A H and Hammett G 2002 *Phys. Plasmas* **9** 5018-5030
[12] Singer C E *et al* 1988 *Comput. Phys. Commun.* **49** 275-398
[13] Chatthong B, Onjun T and Singhsomroje W 2010 *Nucl. Fusion* **50** 064009
[14] Onjun T and Pianroj Y 2009 *Nucl. Fusion* **49** 075003
[15] Zhu P, Bateman G, Kritz A H and Horton W 2000 *Phys. Plasmas* **7** 2898-2908
[16] Pankin A Y *et al* 2005 *Plasma Phys. Control. Fusion* **47** 483
[17] Zhu W *et al* 2006 *Phys. Rev. Lett.* **96** 225002
[18] Kritz A H *et al* 2004 *Comput. Phys. Commun.* **164** 108-113
[19] Connor J W 1998 *Plasma Phys. Control. Fusion* **40** 191
[20] Sugihara M, Igitkhanov Y, Janeschitz G, Hubbard A E, Kamada Y, Lingertat J, Osborne T H and Suttrop W 2000 *Nucl. Fusion* **40** 1743
[21] Bateman G, Onjun T and Kritz A H 2003 *Plasma Phys. Control. Fusion* **45** 1939
[22] Shimada M *et al* 2007 *Nucl. Fusion* **47** S1
[23] Kim K *et al* 2015 *Nucl. Fusion* **55** 053027
[24] Tobita K *et al* 2006 *Fusion Engineering and Design* **81** 1151-1158

IOP Conf. Series: Journal of Physics: Conf. Series **860** (2017) 012035 doi:10.1088/1742-6596/860/1/012035

Neutronics and thermal hydraulic analysis of TRIGA Mark II reactor using MCNPX and COOLOD-N2 computer code

K Tiyapun[1] and S Wetchagarun

Reactor Center, Thailand Institute of Nuclear Technology, 16 Vibhavadi Rangsit St., Bangkok, 10900, Thailand

[1] Corresponding author e-mail: kanokrat@tint.or.th

Abstract. The neutronic analysis of TRIGA Mark II reactor has been performed. A detailed model of the reactor core was conducted including standard fuel elements, fuel follower control rods, and irradiation devices. As the approach to safety nuclear design are based on determining the criticality (k_{eff}), reactivity worth, reactivity excess, hot rod power factor and power peaking of the reactor, the MCNPX code had been used to calculate the nuclear parameters for different core configuration designs. The thermal-hydraulic model has been developed using COOLOD-N2 for steady state, using the nuclear parameters and power distribution results from MCNPX calculation. The objective of the thermal-hydraulic model is to determine the thermal safety margin and to ensure that the fuel integrity is maintained during steady state as well as during abnormal condition at full power. The hot channel fuel centerline temperature, fuel surface temperature, cladding surface temperature, the departure from nucleate boiling (DNB) and DNB ratio were determined. The good agreement between experimental data and simulation concerning reactor criticality proves the reliability of the methodology of analysis from neutronic and thermal hydraulic perspective.

1. Introduction

The research reactor (TRIGA- Mark III) at Thailand Institute of Nuclear Technology (TINT) referred to as TRIGA Research Reactor Modified I (TRR-1/M1) is design for steady state and square wave operation up to a power level of 2 MW(th). The three-dimensional model of the TRR-1/M1 reactor was performed using the Monte Carlo code MCNPX.[1] The validation is an essential aspect of developing an accurate reactor physics model. Several studies in neutronic analysis, criticality experiment and MCNP simulation has been conducted for TRIGA Mark II reactor.[2, 3] The integral neutronics parameters, such as core excess reactivity, reactivity worth and integral reactivity curves of the control rods, and criticality value (k_{eff}), were calculated for the TRR-1/M1 and compared with the measurement. The objective of the thermal and hydrodynamic design is to safely remove the heat generated in the fuel without producing stream void formations, excessive fuel temperatures and without approaching the critical heat flux under steady-state operating conditions. The steady-state thermal hydraulic and safety analysis of TRR-1/M1 is performed to ensure that all important thermal hydraulic parameters are within the safety margin according to the operational limit and condition of the reactor. Boulaich Y. et al. provided calculations and the experiments related to the steady state thermal hydraulic analysis of Moroccan TRIGA Mark II reactor by using PARET/ANL and COOLOD-N2 codes.[4] The TRR-1/M1 fuel centerline temperature, cladding temperature, the coolant temperature, the departure from nucleate boiling (DNB) and DNB ratio (DNBR) were calculated with the thermal hydraulic codes COOLOD-N2[5] based on the axial power factor distributions obtained by Monte Carlo code MCNPX with the ENDF/B-VI nuclear data libraries. The validation of the calculation was compared with the TRR-1/M1 experimental and operational data for steady-state operations.

2. Computational Methods

2.1. MCNPX Computer Code and Experiments

The TRR-1/M1 lattice was modelled as a hexagonal prism, solids with eight faces (Figure 1). The fuel elements is homogeneous mixture of uranium-zirconium hydride alloy with the uranium-to-zirconium atom ratio of 1.6 to 1.7. The uranium in the uranium-zirconium hydride mixture is enriched to approximately 20% U-235. There are two types of fuel elements loaded in TRR-1/M1 core including 8.5% uranium by weight type and 20% uranium by weight type. The 20 wt.% fuel element is a mixture of uranium-erbium-zirconium-hydride (UErZrH) alloy containing approximately 0.5 wt.% erbium. In this study the core configuration consists of 20 wt.% fuel elements are loaded in the D ring, other locations in the core are loaded with 8.5 wt.% fuel elements, total of 105 fuel elements including fuel follower control rod (FFCRs), 3 locations of neutron detectors (position) and 12 locations of in-core irradiation facilities (including pneumatic transfer system) for in-core utilization

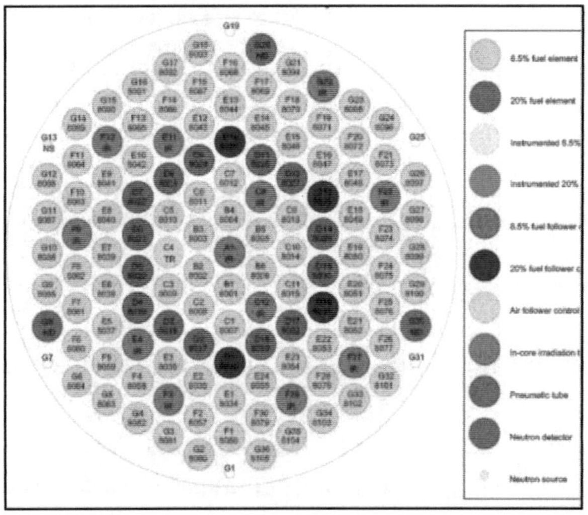

Figure 1. Fuels and FFCRs with 20 wt%. and 8.5wt%. fuel elements.

The fuel element is approximately 3.73 cm in diameter and 73.15 cm in overall length and the active part of the fuel element is 38.1 cm long. The power level of the reactor is controlled with five control rods, a safety rod, a regulating rod, two shim rods and a safety-transient rod. The regulating, shim, and safety rods are sealed 304 stainless steel tubes approximately 109 cm long by 3.43 cm in diameter in which the uppermost 16.5 cm section is an air void and the next 38.1 cm is the neutron absorber (boron carbide in solid form). The fuel elements figures are presented in Figure 2a and 2b.

IOP Conf. Series: Journal of Physics: Conf. Series **860** (2017) 012035 doi:10.1088/1742-6596/860/1/012035

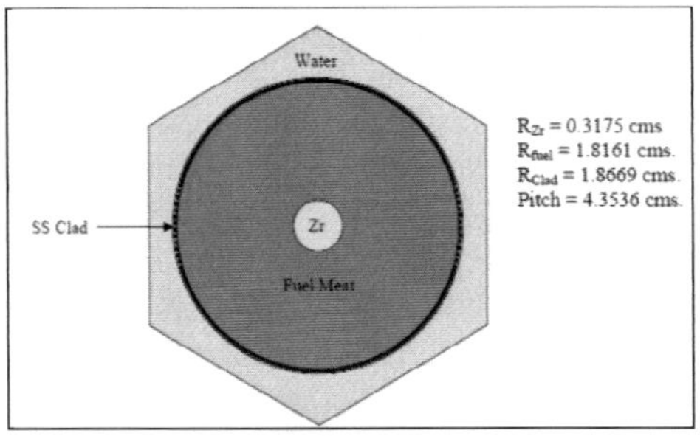

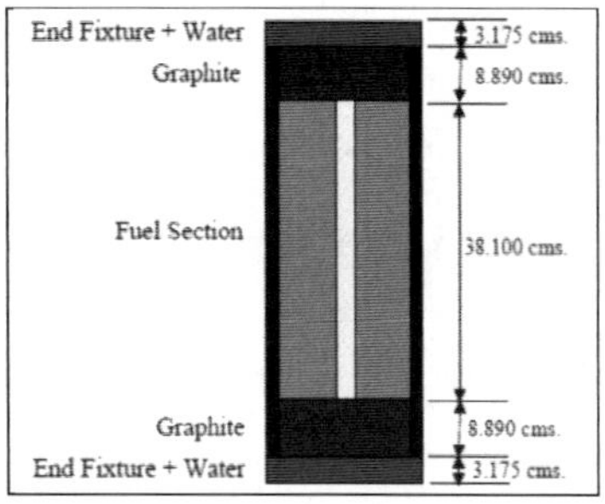

Figure 2a. Fuel element (Horizontal Cross Section) **Figure 2b**. Fuel element (Vertical Cross Section)

In the MCNPX criticality calculation, it simulates a number (N_0) of neutron from their birth at fission to their death by escape of absorption. After that, a number of (N_1) of fission neutrons are selected for simulation in the next cycle, where each particle is weighted by (N_0/N_1).[1] The multiplication factor can be expressed from the k-static Boltzmann equation.[1] The k_{eff} values can be converted into reactivity values using the following equation:

$$\rho = \frac{\frac{k_{eff} - 1}{k_{eff}}}{\beta_{eff}} \qquad (1)$$

Where ρ is the reactivity value in units of dollar ($) and β_{eff} is the fraction of effective delayed neutrons (β_{eff} = 0.007 for TRIGA fuel types).[5] The calculation of the control rods reactivity worth simulated explicitly the experiment, which was conducted by the positive period method.[6] The simulation with the control in the critical position calculating the k_{eff0} of the core. Then one of the control rods was withdrawn at a certain position, calculating the new k_{eff}. The control rod worth represented by reactivity ρ for that position was determine by comparing the k_{eff} and k_{eff0} as explained in Eq. 2.[7]

$$\rho = \rho_0 - \rho_1 = \left(1 - \frac{1}{k_{eff0}}\right) - \left(1 - \frac{1}{k_{eff}}\right) = \frac{1}{k_{eff}} - \frac{1}{k_{eff0}} \qquad (2)$$

The MCNPX model is used for the calculation of power peaking parameters including hot rod power peaking factor f_{HR}, axial power peaking factor f_Z and radial power peaking factor f_R. These three factors are important for steady state operation; they determine the maximum total power released by one fuel elements as well as its axial and radial peaking values, which are used as parameters in the thermal hydraulic calculation. The hot rod power peaking factor is defined as the ratio between the maximum power released by one fuel rod P_{rod} and the average power per element in the core $\overline{P}_{core}$

$$f_{HR} = \frac{(P_{Rod})_{max}}{\overline{P}\ Core} \quad (3) \quad and\ \overline{P}\ Core = \frac{P}{N_{EL}} \quad (4)$$

Where P is the total power, which is 1.2 MW for TRR-1/M1 and the N_{EL} is the number of fuel elements which is 105 fuel elements (100 fuel elements +5 fuel followers control rods).

2.2. Rod Reactivity Worth and the Criticality Experiments

The control rods reactivity worth experiment was measured by the positive period method.[6] The reactor is made critical at 15 Watts. The test rod is withdrawn a small distance so that the reactor is slightly supercritical and the power starts to increase. The reactor period is determined from the doubling time. The previous procedure is repeated until the rod test has been calibrated along its whole length. From the observed periods the corresponding reactivity are calculated using the inhour equation. Using the rod calibration curve from each control rod, the remaining reactivity value Δρ to the fully withdrawn rod position can be determined.

The criticality experiment were conducted based on inverse-multiplication method consisting of loading fuel elements in steps, measuring the count rates on the pulse mode detectors, after each step and plotting 1/M as a function of fuel mass or number of elements. M is the subcritical multiplication as shown in Eq. 5 and 6. Criticality is approached by adding the fuel rods in the core. Inverse multiplication rate is measured. Plot of inverse multiplication rate versus fuel mass is made after each fuel loading step, and the critical mass is estimated by extrapolating the curve to zero.

$$M = \frac{total\ thermal\ neutron\ flux\ (source + fission)}{thermal\ neutron\ flux\ due\ to\ source} \quad (5) \qquad M = \frac{1}{1 - k_{eff}} \quad (6)$$

2.3. COOLOD-N2 Computer Model

The COOLOD-N2 computer code[8] is used in conjunction with the MCNPX to perform the thermal-hydraulic analysis of the steady-state operation of the TRR-1/M1. In the COOLOD-N2 code, a Heat Transfer Package[9] is used for calculate the heat transfer coefficient, DNB heat flux. In the rod type fuel, DNB heat flux is calculated by both the Heat Transfer package and Bernath DNB heat flux correlation.[10]

The goal of the thermal and hydraulic design for TRR-1/M1 is to ensure that fuel integrity is maintained during steady-state as well as during those abnormal conditions which might be postulated for reactor operation. During the steady-state operation, fuel integrity is maintained by limiting reactor powers to the values which assure that the heat transfer rate from the fuel meat to the cladding, to the reactor coolant is fast enough such that clad rupture will not occur. In order to assure the fuel element integrity, the maximum temperature in the fuel meat is limited to 600°C which is the designed limit in the operational limits and condition (OLC) for TRR-1/M1 adopted from the fuel temp trip set point from the manufacture.[10] The thermal-hydraulic model of the TRR-1/M1 fuel rod consists of 3 layers: fuel meat (Uranium Zirconium Hydride: UZrH), gap (assumed to be He gas), and cladding (Stainless steel). Heat transfers out radially from the center of the fuel meat through the gap, cladding, and the water coolant, respectively. Figure 3 shows the calculation model used in the code for fuel temperature distribution in the rod. These parameters in Fig. 3 are used to calculate max fuel temperature using 1D heat conduction with heat generation through multiple layers of material.

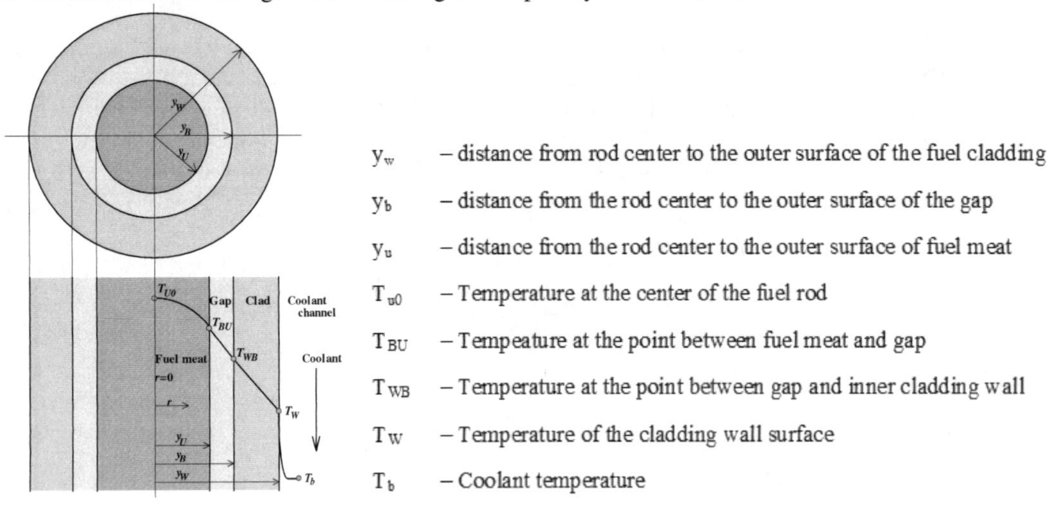

Figure 3. Fuel rod temperature calculation model

The COOLOD-N2 model for the flow channels between fuel rods of TRR-1/M1 which are not in the form of circular tube, it is necessary to compute the hydraulic diameter in order to implement equations for circular pipe flow to evaluate the thermal hydraulics of the system of interest. Figure 4 is an example of the core flow channel cross section. The cross section of an assumed flow channel is shown as a shaded area. The flow area is the differential area between the area of a hexagonal and a circle (representing a fuel rod). The length L can be calculated from the equilateral triangle. Note that the diameter of the fuel rod is approximately 3.7338 cm.

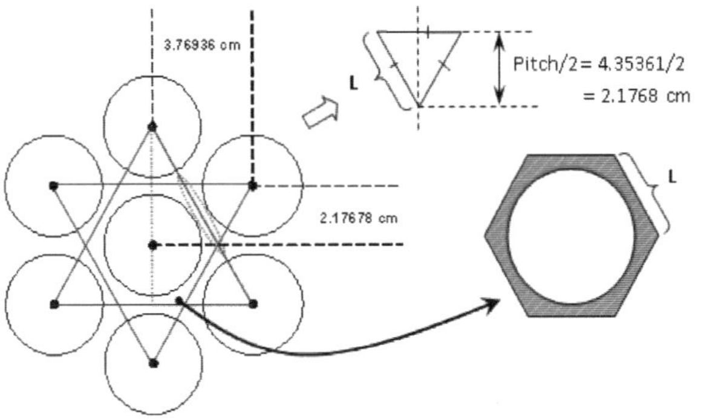

Number of fuel elements in a core	101
Fuel pallet radius, cm	1.816
Cladding thickness, cm	0.051
Fuel element heated length, cm	38.1
Unheated length of fuel rod at channel inlet, cm	14.351
Unheated length of fuel rod at channel outlet, cm	8.89
Coolant mass flow rate, kg/sec	66.47
Inlet coolant temperature, °C	40.0
Inlet coolant pressure, kg/cm2	1.677
Effective flow area for one fuel rod, cm2	5.465
Fuel rod hydraulic diameter, mm	1.864
Pitch between fuel rods, cm	4.3536
Initial gap thickness between fuel pellet and cladding, cm	0.0023
Cladding thermal conductivity, W/cm-K	0.81 (at 100°C) 1.075 (at 500°C)
Fuel meat thermal conductivity, W/cm-K	0.1758 (at 0°C) 0.1833(at 100°C) 0.1908 (at 200°C) 0.1983 (at 300°C) 0.2133 (at 500°C) 0.2508 (at 1000°C) 0.3257 (at 2000°C)

Figure 4. Cross section view of the flow channel inside the TRR-1/M1 core

3. Results and Analysis

3.1. The criticality (k_{eff}) and core excess reactivity

Approach to criticality experiment was performed on the basis of inverse multiplication method. MCNPX is used to calculate the k_{eff} using KCODE and BURN card. MCNPX approximates k_{eff} by estimating the number of fission neutrons produced per fission neutron started for a given generation. By repeating this process for more than ten thousands of generations MCNPX arrives at a good approximation of a multiplication factor.

Table 1. Results of criticality measurements and calculations for burned core configurations

Core Number	K_{eff} (measured)	K_{eff} (MCNPX calculation)
1	1.00168 ± 0.00015	$1.00196 \pm 0.0404\%$
5	1.00185 ± 0.00015	$1.00270 \pm 0.0365\%$
7	1.00148 ± 0.00015	$1.00225 \pm 0.0309\%$
9	1.00212 ± 0.00015	$1.00283 \pm 0.0335\%$

The result shows that the discrepancy is very low for Core 1 since it is a fresh core and the discrepancy increases for the burned core. The calculated values are overestimated than the measurement values. The discrepancy in k_{eff} measurement and calculation may result from the decay time for the fission products poisoning, Xe poisoning, several fission products and actinides produced during the irradiation of the fuel which related to the isotopic compositions of burned fuels.

Figure 5 shows the MCNPX calculated integral reactivity curves for safety, shim I, shim II, and regulating rod, respectively. From these curves, it can be observed that the MCNPX calculated integral reactivity worth of shim I and shim II control rods are consistent with the measurement values. The measured reactivity worth of safety and regulating control rods are slightly below to the calculated results. However the small disagreement remain acceptable related to other TRIGA reactors.[7] In all control rods, the largest differential reactivity worth occurred when the rods position are fully withdrawn. The calculated and measured reactivity worth resulting from the withdrawal of the rods from 90% to 100% are less than 0.1$ and when withdrawal from 40% to 50%, the reactivity worth are less than 0.5$.

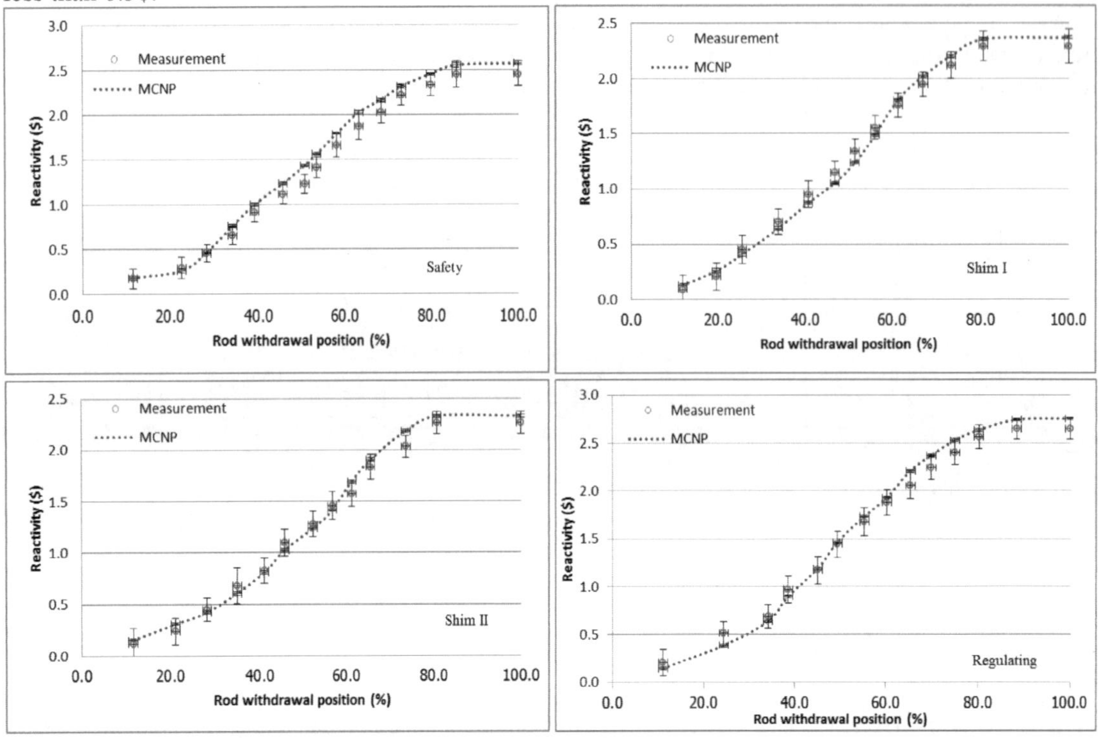

Figure 5. MCNPX calculations and experimental integral rod worth for safety, shim I, shim II and regulating.

3.2. The axial power peaking factor

The axial power distribution within the fuel meat of the hottest fuel elements calculated by MCNPX is shown in Figure 6. The figure shows the cosine shape of axial power distribution with a hot rod power factor (f_{HR}) of 1.894. The axial power peaking factor (f_Z) is 1.268. The value for f_{HR} x f_Z is 2.40. The small increments show in the left and right sides of the axial power profile are resulting from the lower and upper reflectors of TRIGA fuel. The results show that the core excess reactivity (%Δk/k) is 7.07. It is assumed that the power density is directly proportional to the fission density. The maximum power produced in the hottest fuel element is 23.4 kW.

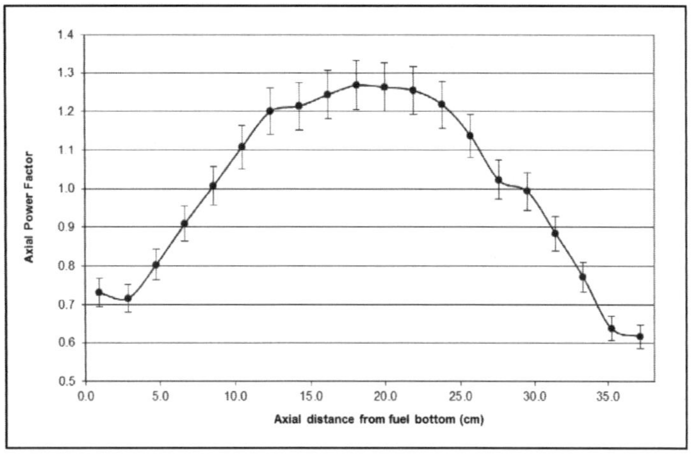

Figure 6. Hot channel fuel axial power factor profile.

3.3. The steady state thermal hydraulic analysis
Considered the case that all control rods are withdrawal at position 400, given the maximum f_{HR} x f_Z value, it is found that based on the temperature criteria, the maximum allowable power of TRR-1/M1 is 1.5 MW. Figure 7 and 8 show temperature distribution of the coolant and the fuel meat along the axial distance of the "hottest" location in the reference reactor core for the maximum allowable power of 1.5 MW (anticipated transient condition) and the normal operating power of 1.3 MW. T_{sat} is the saturated coolant temperature. Axial distance is the relative distance which the maximum is 38.1 cm.

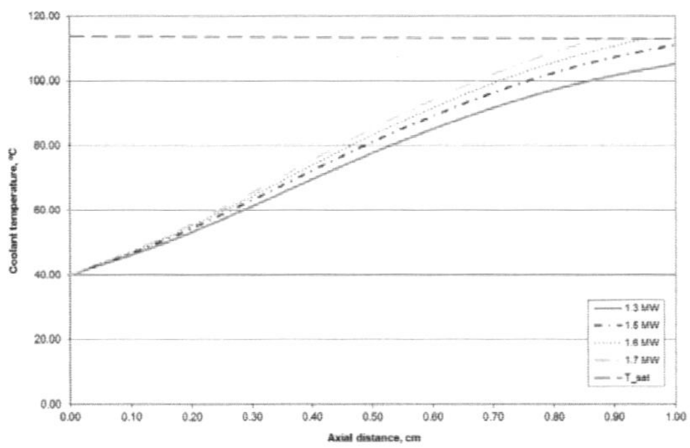

Figure 7. Coolant temperature profiles along axial direction

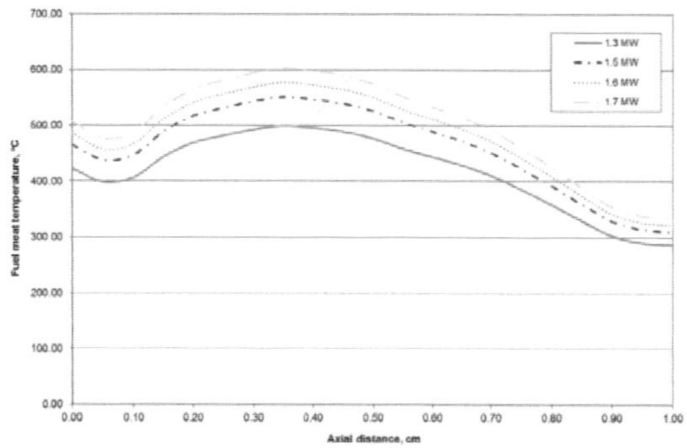

Figure 8. Fuel temperature profiles along the axial direction

From the results, at the maximum allowable power (1.5 MW), the peak of fuel meat temperature is around 552°C which is still below the designed maximum fuel temperature of 600°C. The calculated fuel temperature is far below the fuel swelling temperature limit of 950°C recommended by General Atomic.[11] The outlet coolant temperature is approximately 111°C while the saturated temperature at that point is 113°C. This result infers that there will be no bulk boiling of the water coolant in any flow channel of the reactor core.

3.3.1. DNB ratio (DNBR). The hot channel DNBR evaluated by COOLOD-N2 is depicted in Figure 9 for various operating power. If consider only DNBR, TRR-1/M1 can be operated up to 1.7 MW without breaking the DNBR design criteria which is 2.0. For the maximum allowable power of TRR-1/M1 at 1.5 MW, the maximum fuel temperature is 552°C and the minimum DNBR is 2.43 without bulk boiling in the core and allow approximately 15% thermal safety margin based on the nominal operating power of 1.3 MW. The SCRAM point for reactor power safety channels is set at 1.43 MW (110% of the nominal power) while, for the purpose of testing the reactor steady-state power level scram, the reactor power can be varied up to 1.5 MW.

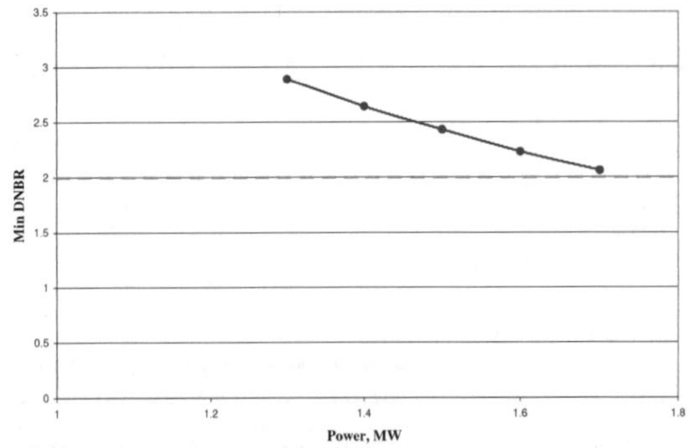

Figure 9. Minimum DNBR in at various power levels.

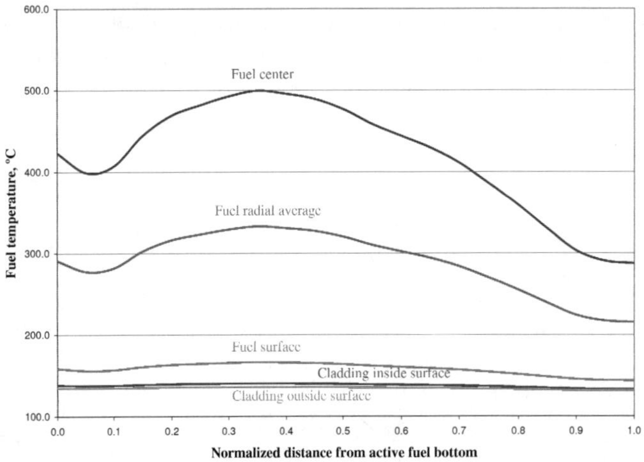

Figure 10. Temperature profiles along the axial direction of the hottest fuel element at 1.3 MW.

3.3.2. Axial temperature profiles. The comparison of axial temperature profile of TRR-1/M1 hot channel using COOLOD-N2 is presented in Fig. 10 with the highest power peaking factor (f_R x f_Z) is 2.40 and all control rods are withdrawal at 400 unit position. It can be seen from the figure that the maximum temperature of fuel centerline, fuel surface, and clad surface are observed at the center of the reactor core. The fuel surface temperature, clad surface temperature is found to be far below the fuel centerline temperature. As the fuel burns up, radiation induced swelling reduces the gap size. This phenomenon enhances the heat transfer at the fuel cladding interface, which will result in lowering the peak temperature at the centerline of the fuel rod. From the calculation, as shown in Fig. 10, the fuel element maximum temperature for steady-state operation at 1.3 MWt at BOL at the location of expected maximum power peaking factor is around 500°C, which is below the operating limit of 600°C.

3.3.3. The validation of the COOLOD-N2 with the experiments. The results of the COOLOD-N2 were validated with the experimental results measured coolant at 1200 kW of flow channel between C1, C2 and D2 position shown in Fig. 11. The measurement was conducted using K-type thermocouple and temperature processing unit. It can be found that the temperature distribution obtained from the calculation and the experiment are in good agreement with minor discrepancies. With this validation, the thermal hydraulic model using COOLOD-N2 can be used to estimate safety parameter for reactor in a steady state with confidence in some extent.

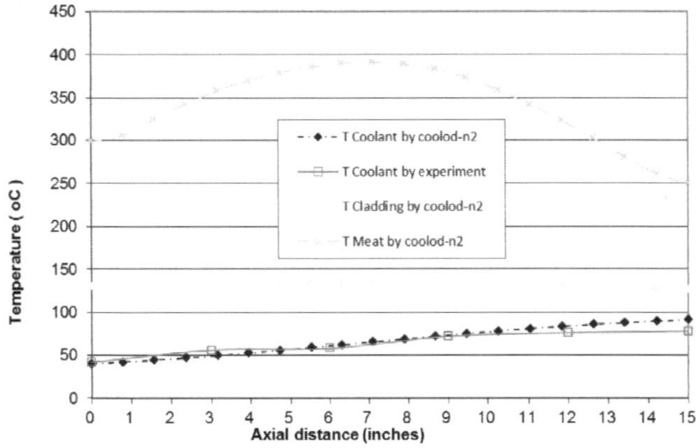

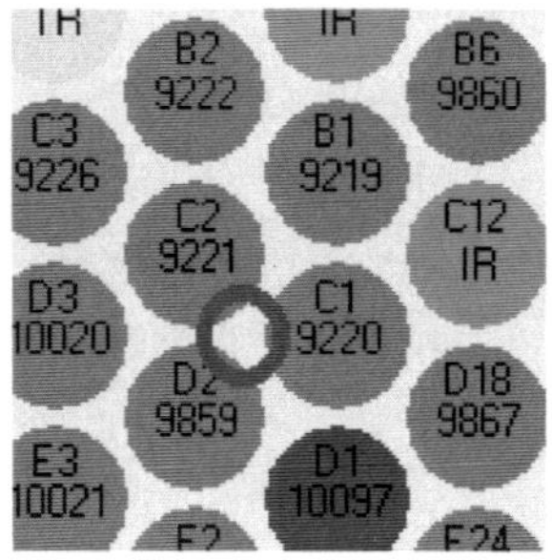

Figure 11. Comparison between the measured hot channel coolant temperature and the predicted values from the COOLOD-N2.

4. Conclusion

The TRR-1/M1 research reactor is model using MCNPX and COOLOD-N2. The practical calculation such as criticality value (k_{eff}), core excess reactivity, total and integral control rod worth and power peaking analysis were performed and compared with the experiments. Good agreement between calculations and experiments ensure the validity and reliability of the model. The validation of the COOLOD-N2 model was performed comparing their calculation results with the experiments. The thermal hydraulic analysis in steady state results shows good agreement between the calculations and the experiments. From this analysis, all safety related thermal hydraulic parameters are within the thermal design limits for steady state operating conditions. Further neutronic study should be conducted to be more consistency and improvement by reducing the discrepancies between the calculation and the measurements. For the thermal hydraulic analysis, the verification of reactivity feedback effects for several transient events should be conducted.

References

[1] Denise B. Pelowitz 2008 MCNPX User Manual Version 2.6.0, Los Alamos National Laboratory, April.
[2] Bock, H. and Khan, R. 2010 Neutronic Analysis of TRIGA Vienna Mixed Core, Proceeding of the International Conference Nuclear Energy for New Europe, 6-9 September, p. 203.1 – 203.8.
[3] Matsumoto, T. 1998 Benchmark Analysis of Criticality Experiments in the TRIGA Mark II Using a Continuous Energy Monte Carlo Code MCNP. Journal of Nuclear Science and Technology 35(9), p. 662-670.
[4] Boulaich, Y., Nacir, B., Bardouni, T. El, Zoubair, M., Bakkari, B. El., Merroun, O., Younoussi, C. El, Htet, A., Boukhal, H., and Chakir, E. 2011 Steady State Thermal Hydraulic Analysis of the Moroccan TRIGA Mark II Reactor by Using PARET/ANL and COOLOD-N2 Codes, Nuclear Engineering and Design Vol. 241, p. 270-273.
[5] Negut, Ch., 2006 Fuel Behavior Comparison for a Research Reactor, Journal of Nuclear Materials, Vol. 352, p. 157-164.
[6] Matsumoto, T., Hayakawa, Nobuhiro. 2000 Benchmark Analysis of TRIGA MARKII Reactivity Experiment Using a Continuous Energy Monte Carlo Code MCNP. Journal of Nuclear Science and Technology 37(12), p. 1082-1087.

[7] Dalle, H.M. 2002 Neutronic Calculation of the TRIGA Ipr-R1 reactor using WIMSD4 and CITATION codes. Annals of Nuclear Energy 29, p. 901-912.
[8] Kaminaga, M. 1994 COOLOD-N2: A Computer code, for the analyses of steady-state thermal-hydraulics in research reactors, JAERI-M94-052.
[9] Sudo, Y., Ikawa, H. and Kaminaga, M. 1985 Development of Heat Transfer package for Core Thermal Hydraulics Design and Analysis of Upgraded JRR-3. Proceeding of the International Meeting of Reduced Enrichment for Research and Test Reactors, Petten, 14-16 October 1985, p. 363-372.
[10] General Atomics 1979 10 MW TRIGA-LEU Fuel and Reactor Design Description, General Atomics, San Diego.
[11] U. S. Nuclear Regulatory Commission 1987 Office of Nuclear Reactor Regulation, "Safety Evaluation Report on High-Uranium Content, Low-Enriched Uranium Zirconium Hydride Fuels for TRIGA Reactor," NUREG-1282, August.

Renovation status of neutron radiography facility at TRR-1/M1 reactor

R Picha[1], J Channuie, T Liamsuwan, J Promping, W Ratanatongchai, and S Wonglee

Thailand Institute of Nuclear Technology, Ongkharak, Nakorn Nayok, Thailand

[1] Corresponding author e-mail: aeroppon@tint.or.th

Abstract. Thai Research Reactor-1/Modification 1 (TRR-1/M1) at Thailand Institute of Nuclear Technology (TINT) has a beamline facility dedicated for neutron radiography (NR) research. The renovation has been ongoing to enhance the radiation safety and to facilitate research operations of the only reactor NR station in the country. The design of beam shutter and shielding walls has been conducted using Monte Carlo simulations along with attenuation measurements in the lab to ensure the shielding capability of the components and efficient operation. The progress and future plans TINT NR will be presented.

1. Introduction

Thai Research Reactor-1/Modification 1 (TRR-1/M1) is the only research reactor in Thailand. Operated at 1.2 MW, its main applications are radioisotope production, gemstone irradiation, and neutron activation analysis. There has recently been a plan to renovate the neutron radiography (NR) facility, which at current state can be used with a limited capacity due to a low L/D ratio (currently about 50), lack of automation, and a small working area. The details of the original NR setup can be found in [1].

NR facility at TRR-1/M1 has been in used for a long time and the original wall integrity has been compromised due to block dislocation. The initial improvement effort was to construct a new special concrete with barite and copper slag as aggregate, based on past literature. However, efforts to secure domestic supplier and contract were proven to be challenging. Therefore, the research team decided to take a more practical route, by using high-density concrete blocks from a hospital radiotherapy treatment room, which was recent decommissioned, to serve as components of the new NR wall. The NR upgrade is part of the large scale improvement of the neutron beam experimental facilities at TRR-1/M1, which includes the prompt gamma neutron activation analysis (PGNAA) facility [2].

Besides the new shielding, the ongoing renovation involve replacing the beam shutter with a larger volume and automated system, installing a frame for future collimators, and the digital imaging system. The content is organized in the following manner. In section 2, the methods of study are described. Section 3 goes through the results on concrete properties. The shutter description is given in Section 4. In the last section, the summary and conclusion are provided.

2. Materials and Methods

The NR system consists of six main parts: 1) the fission neutron source from the reactor core, 2) the beam shutter, 3) the collimator, 4) the sample stage, 5) the image capture system, and 6) the shielding walls. In the renovation plan, the beam shutter is designed based on [3] and the system is to be driven by a motor and placed on a sturdy table with tracks. The shutter also has a part where a collimator can be inserted in the future. This collimator will help focusing the beam and improve the image quality. The sample stage is currently being designed so that it can be remotely controlled via a computer. For the image capture system, the facility will mainly utilize a digital camera attached to an L-shaped light-tight box. Finally, shielding walls are planned to be built with ledite concrete. To design and evaluate the composition for the new shielding walls, the blocks were tested with a gamma radioisotope source.

In summary, the renovation plan timeline is as indicated in Fig. 1.

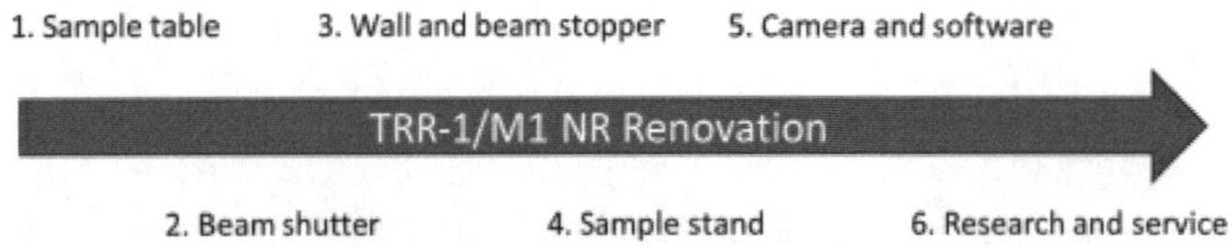

Figure 1. Radiography renovation plan.

2.1. Gamma-shielding concrete

Ledite concrete blocks were acquired from a hospital. The dimensions of a block (Fig. 2) are roughly 15 cm x 15 cm x 30 cm. It has an interlock design. The current blocks have all flat surfaces.

Figure 2. Ledite concrete block (top). Gamma shielding test of old (L) and new (R) concrete blocks.

Gamma attenuation using a cobalt-60 source was measured and calculated using the Beer-Lambert law, $I(x) = I(0)e^{-\mu x}$. The coefficients μ will be calculated and compared between different concrete samples to evaluate gamma shielding effectiveness.

Dislocation which lead to gaps between blocks is one of the major problem at the current wall. The ridge design allows for an improved shielding capability than the flat block design currently used.

2.2. Beam shutter

The new beam shutter design is shown in Figs. 3-4, and the actual pictures in Fig. 5. The outside walls are made of iron. Its size (approximately 50 x 50 x 50 cm^3) is appreciably larger than the current shutter (approximately 40 x 40 x 40 cm^3). The first layer is borated polyethylene, which will moderate and capture neutrons. The outside layers are lead, which will shield the gamma radiation.

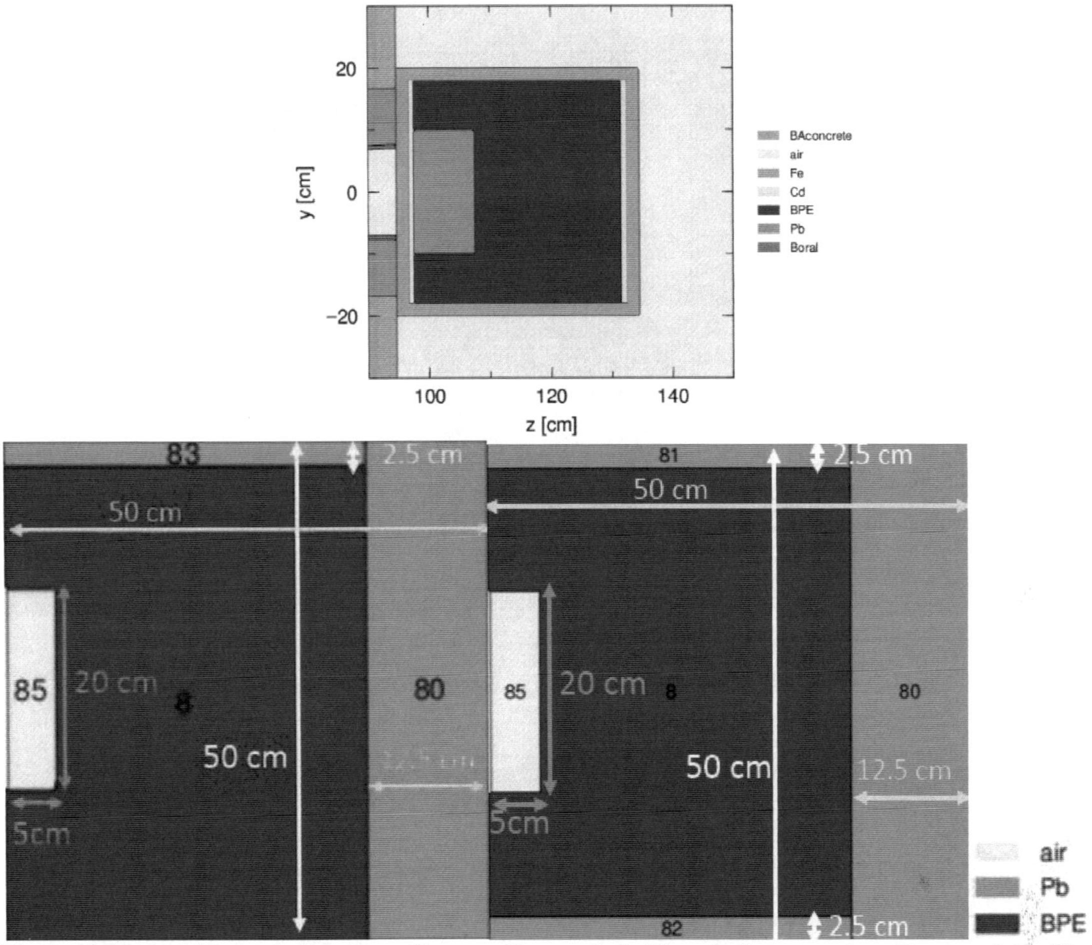

Figure 3. Beam shutter design. Top: Current. Bottom left: New design, top view. Bottom right: New design, side view. New design has lead and borated polyethylene as shielding materials

IOP Conf. Series: Journal of Physics: Conf. Series **860** (2017) 012036 doi:10.1088/1742-6596/860/1/012036

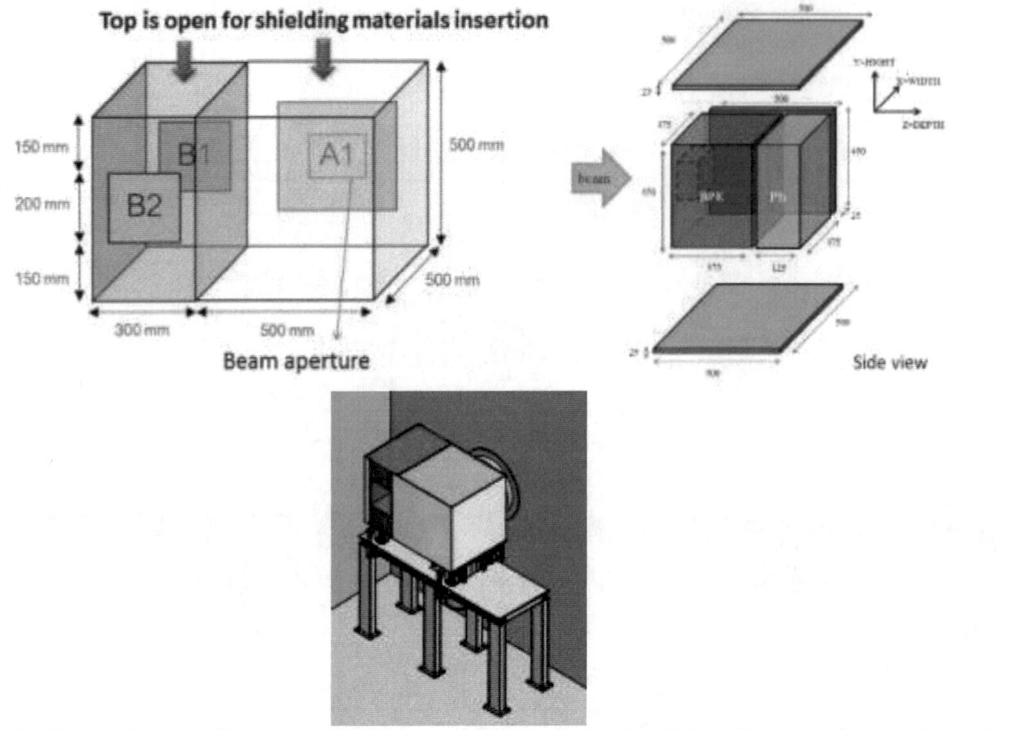

Figure 4. Shutter box configuration (left) and shielding materials (right). Bottom: Orientation with the reactor wall.

Figure 5. Shielding materials, consisting of BPE and lead, being assembled into the shutter box.

Radiation flux maps can be seen in Fig. 6. The fluxes are computed using PHITS Monte Carlo simulation software. It can be seen from the figures that the neutron dose is much reduced when using the new shutter. The gamma dose is comparably low in both cases.

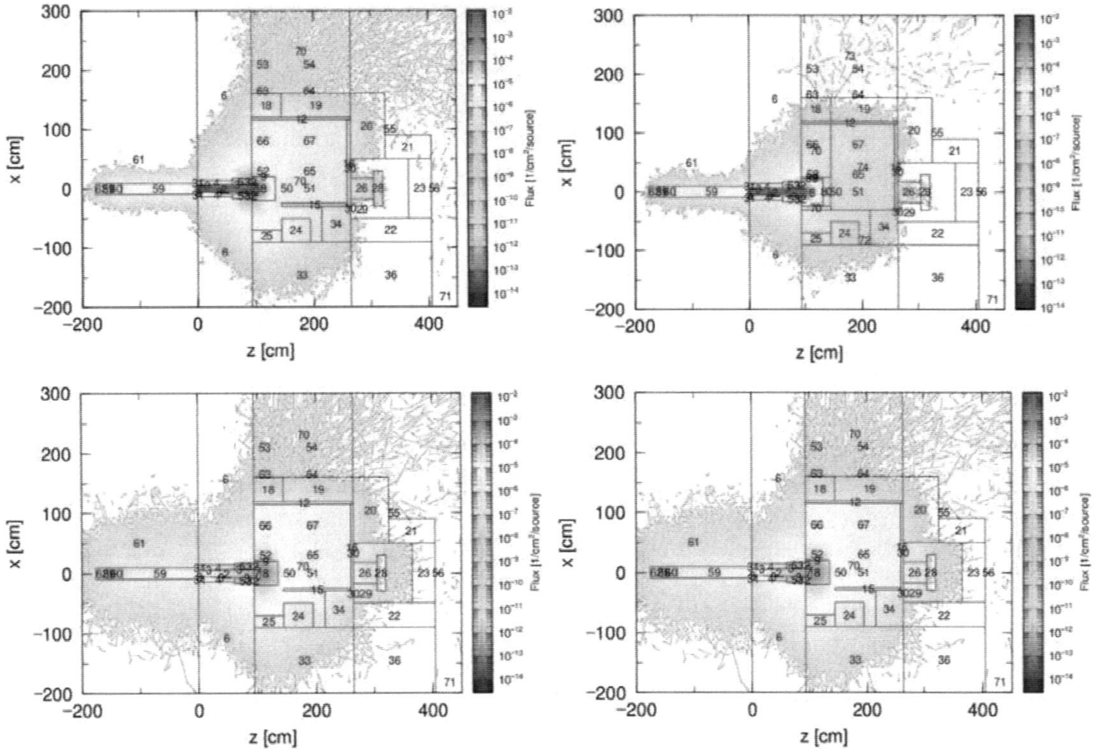

Figure 6. Radiation flux maps, comparing neutron and gamma fluxes of the old and the new shutters. The display is the top view, in which the left side of the figures is the reactor core and the right side is the NR area while the shutter is closed. The numbers inside the figures indicate different volumes. (Top-left) Neutron flux using old shutter; (Top-right) Neutron flux using new shutter; (Bottom-left) Gamma flux using old shutter; (Bottom-right) Gamma flux using new shutter.

Lanthanum bromide ($LaBr_3(Ce)$) scintillation detector (Canberra LABR-1.5x1.5) was used to measure gamma transmission. A linear setup with the Co-60 source and the detector separated by approximately 40 cm was used, along with partial collimation using lead blocks. Results are shown in Table 1.

Table 1. Gamma attenuation coefficients (μ) of different type of materials.

Sample type		1.17 MeV	1.33 MeV
	Thickness (cm)	μ (cm^{-1})	μ (cm^{-1})
Barite concrete (current)	20	0.1904 ± 0.0001	0.1762 ± 0.0004
hematite	15	0.2661 ± 0.0005	0.2407 ± 0.0017
Fe	0.95	0.3991 ± 0.0051	0.3704 ± 0.0112
Pb (1 block)	0.3	0.7024 ± 0.0060	0.6356 ± 0.0059
Pb (2 blocks)	0.6	0.7176 ± 0.0066	0.6374 ± 0.0055
Ledite (1 block)	15	0.2318 ± 0.0016	0.2151 ± 0.0013
Ledite (2 blocks)	30	0.2431 ± 0.0001	0.2370 ± 0.0008

IOP Conf. Series: Journal of Physics: Conf. Series **860** (2017) 012036 doi:10.1088/1742-6596/860/1/012036

3. Results and Discussions

In combination with a new 3D imaging software and camera system, the new concrete wall and beam shutter setups for TRR-1/M1 neutron radiography should provide a better operation in terms of safety and imaging capability. From the gamma shielding tests, the ledite concrete blocks have higher gamma ray attenuation coefficients than that of the current barite concrete blocks. The shielding capability should also be further improved with an interlock connection. The new concrete blocks also serve as an effective option in terms of the gamma shielding property, availability, and the cost. Neutron radiography provides a unique probe into the physical structures of objects, which can be valuable for sample analyses and can be used in combination with other techniques such as X-ray radiography, X-ray fluorescence, X-ray diffraction, and neutron activation analysis. The new neutron beam shutter is considerably larger than the current one and can be operated automatically. The simulations also show that the new shutter can substantially confine the neutron dose in the surrounding area.

With the new upgrade, the neutron radiography facility should be able to construct 3D images of objects such as those from archeological, industrial, and other fields in Thailand [4].

4. Conclusions

The neutron radiography area at Thai research reactor TRR-1/M1 is one of the beam experiments being renovated. Ledite concrete blocks are used to construct the new shielding wall. A beam shutter consisting of borated polyethylene and lead and driven by a motor can improve safety and ease of operation. The new facility is expected to be complete before the end of 2016 and, when used along with an appropriate imaging software, should have a higher capacity to perform 3D neutron radiography.

References

[1] Picha R, Channuie J, Khaweerat S, Liamsuwan T, Promping J, Ratanatongchai W, Silva K, and Wonglee S 2015 Gamma and neutron attenuation properties of barite-ceent mixture. *J. Phys. Conf. Ser.* **611** 012002

[2] Ratanatongchai W, Khaweerat S, Picha R, Channuie J, and Sangaroon S 2016 Measurements of TRR-1/M1 Prompt Gamma Neutron Activation Beam. INST2016 Proceedings, Bangkok, Thailand

[3] Turkoglu, D J 2012 Design, construction and characterization of an external neutron beam facility *at. Diss. The Ohio State University*

[4] Khaweerat S, Ratanatongchai W, and Wonglee S 2016 Development of a Neutron Tomography System in Thailand. INST2016 Proceedings, Bangkok, Thailand

IOP Conf. Series: Journal of Physics: Conf. Series **860** (2017) 012037 doi:10.1088/1742-6596/860/1/012037

The MCNP Simulation of a PGNAA System at TRR-1/M1

S Sangaroon[1,3], W Ratanatongchai[2], R Picha[2], S Khaweerat[2], and J Channuie[2]

[1] Department of Physics, Faculty of Science, Mahasarakham University, Mahasarakhum, 44150 Thailand
[2] Thailand Institute of Nuclear Technology (Public Organization), Ongkharak, Nakhon Nayok 26120, Thailand

[3] Corresponding author e-mail: siriyaporn.s@msu.ac.th

Abstract. The prompt-gamma neutron activation analysis system (PGNAA) has been installed at Thai Research Reactor-1/ Modified 1 (TRR-1/M1) since 1999. The purpose of the system is for elemental and isotopic analyses. The system mainly consists of a series of the moderator and collimator, neutron and gamma-ray shielding and the HPGe detector. In this work, the condition of the system is carried out based on the Monte Carlo method using Monte Carlo N-Particle transport code and the experiment. The flux ratios ($\Phi_{thermal}/\Phi_{epithermal}$ and $\Phi_{thermal}/\Phi_{fast}$) and thermal neutron flux have been obtained. The simulated prompt gamma rays of the Portland cement sample have been carried out. The simulation provides significant contribution in upgrading the PGNAA station to be available in various applications.

1. Introduction

The prompt gamma neutron activation analysis (PGNAA) is a non-destructive technique for an elemental and isotopic analytical. The PGNAA system has been widely used for analyzing the elements. The neutron interacts with the sample and was captured with the nucleus of the sample so called a compound nucleus. The compound nucleus is formed in an excited state which then de-excites to the ground state in 10^{-14} s by released the prompt characteristic gamma ray for each element. The PGNAA technique is based on the neutron inelastic scattering (n, n′γ) and thermal neutron capture (n_{th}, γ) and measuring a prompt gamma spectrum which emitted from the sample.

The PGNAA have been designed based on different neutron source such as the accelerator neutron source [1-4] and reactor thermal neutron source [5-9]. The system at Thai Research Reactor 1/Modified 1 (TRR-1/M1) was designed based on the reactor thermal neutron source. It was constructed and installed since 1999 [10]. The aim is to use this technique in Thailand to analyze elements in various samples. The TRR-1/M1 is a pool type reactor with light water moderator and cooling. It is operated with the maximum power of 1.2 MW. It provides a high thermal and fast neutron flux. There are four beam experimental ports available for the utilization. One of the four is for the PGNAA system. The facility descriptions and the simulation set up of the PGNAA system at TRR-1/M1 are described in section 2.

However, the existing PGNAA system suffers from the complicate structural, neutron source and neutron and gamma background. To expand the experimental capability of the PGNAA system, the calculations using a Monte Carlo code have been obtained using the Monte Carlo N-Particle transport code (MCNP) version MCNPX [11]. The MCNP code is the most widely used for all types of particle transport, e.g. neutron, photon, electron based on the evaluated nuclear data library (ENDF/B). It has

IOP Conf. Series: Journal of Physics: Conf. Series **860** (2017) 012037 doi:10.1088/1742-6596/860/1/012037

been widely used to validate the design calculation of the PGNAA [12, 13]. The simulation and experiment results are reported in section 3. Moreover, the yield of the prompt gamma rays from Portland cement sample have been obtained and presented in this section. The conclusion is in section 4.

2. Description of the facility and MCNP model

The main components of the PGNAA system at TRR-1/M1 include a series of annular collimators, beam shutter, shielding and a high purity germanium (HPGe) detector. The layout of the system shows in figure 1. The detector and detector shielding is not taken into account in this work.

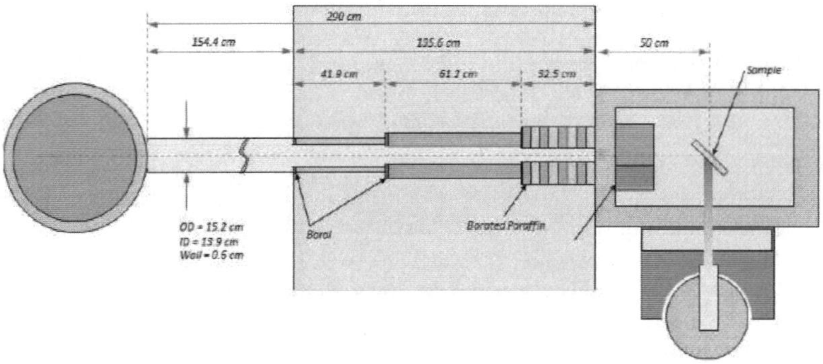

Figure 1. Horizontal cross section view of the PGNAA facility at TRR-1/M1, showing a series of collimators, beam shutter, shielding, sample area and detector position

2.1. Neutron source

The PGNAA system at TRR-1/M1 was designed for the research reactor operating at 1.2 MW with the maximum thermal neutron flux and fast neutron flux at the central of the reactor core of 3.1×10^{13} $n.cm^{-2}.sec^{-1}$ and 2.3×10^{13} $n.cm^{-2}.sec^{-1}$ respectively. The thermal and epithermal neutron fluxes at the sample position were measured by irradiating gold-foil (the cadmium was used to separate the activities of epithermal neutron) and found to be $4.63 \pm 2.15 \times 10^{6}$ $n.cm^{-2}.sec^{-1}$ and $4.43 \pm 1.13 \times 10^{5}$ $n.cm^{-2}.sec^{-1}$ respectively [14].

The calculated neutron source from MCNP at the reactor beam port which including thermal neutron epithermal neutron and fast neutron have been provided. The uniform neutron source is placed at the beginning of the collimator for the simulation as depicted in figure 2. The diameter of the disk source is 4 cm which almost matching that of the collimator tube.

2.2. Collimator

The collimator consists of a series of lead and boron rings and teakwood. The purpose is to reduce the fast neutron flux, maintaining the high thermal neutron flux and collimate the neutron beam to the sample. The collimator was installed with the configuration described in figure 1. The inner diameter of the entrance collimator is 13.9 cm. The entrance of the collimator connects to the reactor core at a southeast beam port of the reactor pool. The end of the collimator connects to the shutter with the inner diameter of 3.68 cm. The overall collimator length is 290 cm.

The MCNP model of the annular collimator was provided and depicted in figure 2. The shell of the collimator model consists of stainless steel, lead, teakwood and borated polyethylene. A point detector (F5) tally was used to obtain the radiation flux at the sample area. Additional, the surface-crossing (F1) mesh tally at the collimator mid-plane is used to study the neutron and photon distribution.

2.3. Shutter

The rotating shutter is placed at the end of collimator. The shutter is used to stop the beam from the collimator to the sample area. The thickness of the rectangular shutter box is 30 cm. It composes of

IOP Conf. Series: Journal of Physics: Conf. Series **860** (2017) 012037 doi:10.1088/1742-6596/860/1/012037

borated paraffin and lead and has a hole of about 4-cm diameter. In MCNP model, the shutter was modeled at the end of the collimator.

2.4. Biological shield

The shutter and sample are surrounded by a biological shield. It is a 30-cm thick box which is filled with steel shots and paraffin. It is mounted to the reactor wall and aimed to reduce background gamma and neutron radiation levels around the facility. Paraffin is used for the neutron moderator and steel shots are used for the thermal neutron capture and absorb prompt gamma and gamma rays from other structure. Sample was held at 45° angle to the beam and to the detector. The sample holder was held vertically from the 24 cm × 24 cm hold on the top side of the biological shield. The biological shield was modeled in this calculation.

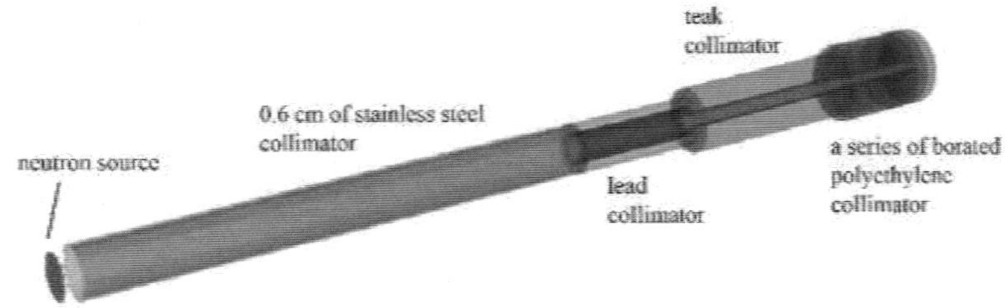

Figure 2. The collimator arrangement

3. Results and discussions

3.1. The collimator characterization

In this section, the transmission of neutron from the reactor core to the biological shielding through the collimator was calculated. The mesh tally results of neutron and photon flux through the collimator and biological shield have been obtained and plotted with the geometry as shown in figure 3(a) and figure 3(b) respectively. The results show that the neutrons have been collimated by the collimator and transferred to the biological shield. Nevertheless, the results show that the biological shielding is not sufficiently thick enough that high neutron and photon fluxes leak out through the shielding.

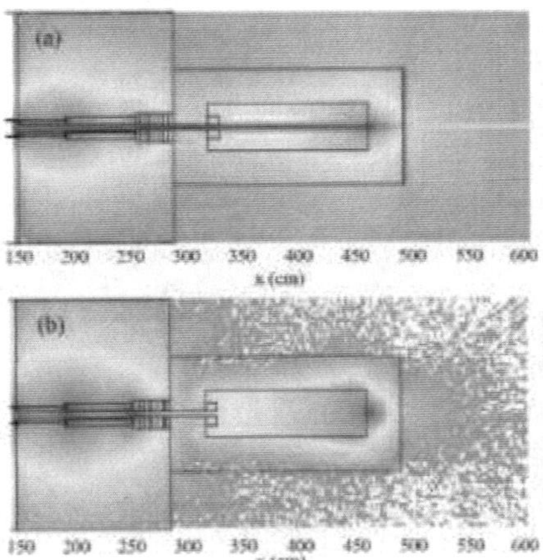

Figure 3. Mesh tally results from MCNP simulation of (a) neutron and (b) photon distribution in collimator and biological shielding

IOP Conf. Series: Journal of Physics: Conf. Series **860** (2017) 012037 doi:10.1088/1742-6596/860/1/012037

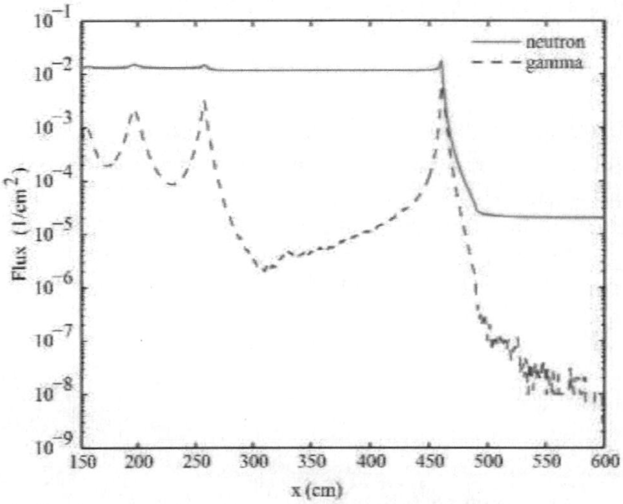

Figure 4. The mid-plane collimator neutron flux per source neutron (red line) and photon flux per source neutron (blue dash)

The neutron and photon fluxes per source neutron at the collimator mid-plane are presented in figure 4. The neutron flux is constant through the collimator and biological shielding. About three orders of magnitude of neutron flux reduce at the end of the biological shield. In addition, the gamma fluxes produced by the interaction of neutron with the collimator and surrounding structure in the system have been obtained.

The neutron spectra per source neutron at the sample position have been calculated. It consists of the thermal neutron, epithermal neutron and fast neutron fluxes. The neutron flux ratios ($\Phi_{thermal}/\Phi_{epithermal}$ and $\Phi_{thermal}/\Phi_{fast}$) have been conducted. The ratio of $\Phi_{thermal}/\Phi_{epithermal}$ is equal to 10.15 while $\Phi_{thermal}/\Phi_{fast}$ is equal to 5.11. The thermal neutron flux was 3.55×10^7 n.cm^{-2}.sec^{-1} which higher than the experimental results by a factor of seven [14]. The in-scattering neutrons in collimator at sample position have been obtained as shown in figure 5. The 1.88 percent of in-scattering neutron in the collimator is found.

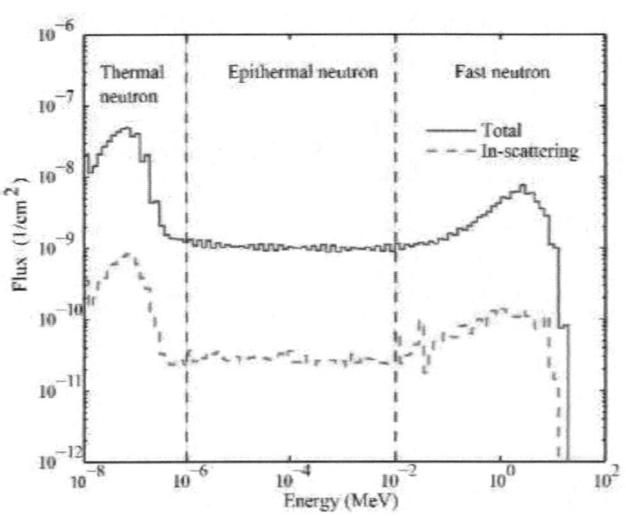

Figure 5. Total neutron spectrum per source neutron (blue line) and in-scattering neutron spectrum per source neutron (red dash) at the sample area

3.2. Beam characterization

3.2.1. Neutron. In this section the neutron beam shape was characterized by the experiment and simulation. In the experimental set up, the neutron radiograph imaging plates were used to measure the neutron while the shutter opened and closed. The imaging plates were placed perpendicular to the neutron beam from the collimator at sample position at 20 cm from the shutter. The results were used to determine the dimension and shape of neutron beam. The image and profile of measured neutron beam shape while the shutter opened show in figure 6(a). The results show that the neutrons beam relatively distribute in a Gaussian function over 2-cm diameter, approximately. The outer beam is not symmetrical, it distributes more toward to the right side. The beam size is 6-cm diameter approximately. The image and profile of neutron distribution at the sample position while the shutter was closed are presented in figure 6(b). The intensity of the neutron at the center of the beam is higher than the outside the beam by a factor of 1.2 approximately.

For the simulation using MCNP code, the results of the neutron beam distribution and profile show in figure 7. The beam distributes with a Gaussian function over 6-cm diameter approximately.

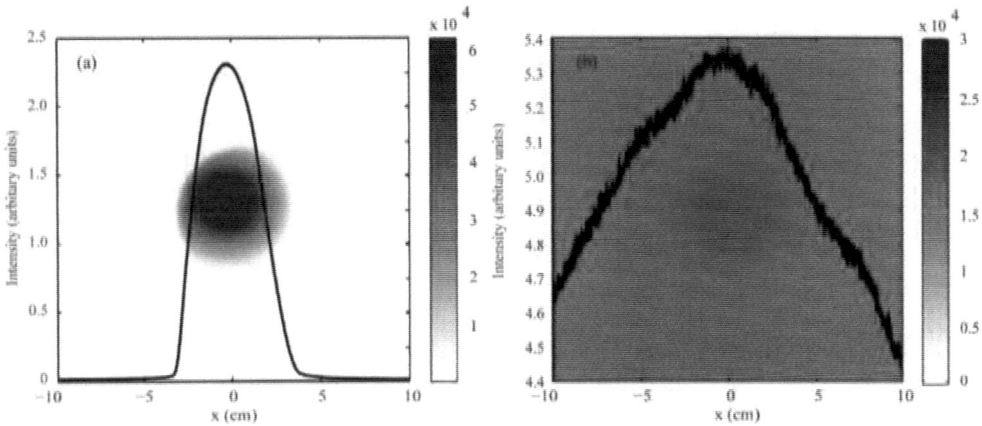

Figure 6. The image and profile of the neutron beam at the sample position: (a) opened shutter and (b) closed shutter

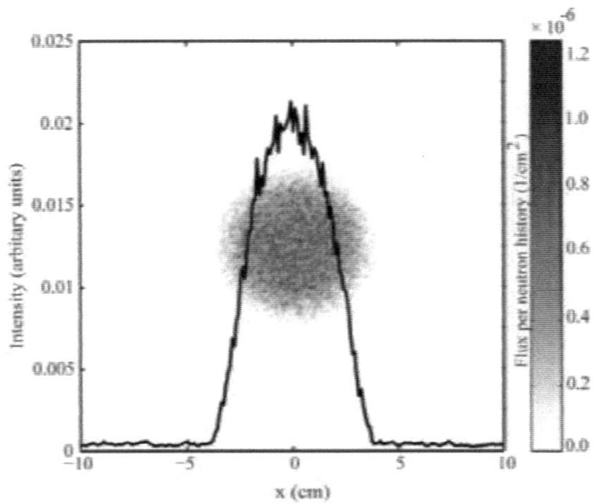

Figure 7. The distribution and profile of the neutron beam at the sample position using the MCNP simulation

IOP Conf. Series: Journal of Physics: Conf. Series **860** (2017) 012037 doi:10.1088/1742-6596/860/1/012037

3.2.2. Gamma. In the experimental set up, the gamma ray was determined using the sensitive gamma radiograph imaging plates while the beam shutter opened and closed as neutron beam. The image and profile of measured gamma ray while the shutter opened show in figure 8(a). The results show the non-symmetrical. The flux is high on the left side. This is similar to the results while the shutter closed in figure 8(b) which the distribution shows a high flux on the left side of the beam due to the non- symmetrical of the shutter.

In the simulation, the gamma ray was calculated to determine the effectiveness of the neutron to the surrounding structure. The 0.47 MeV and 7.6 MeV peak of gamma ray produced from boron and iron are found as presented in figure 9(a).

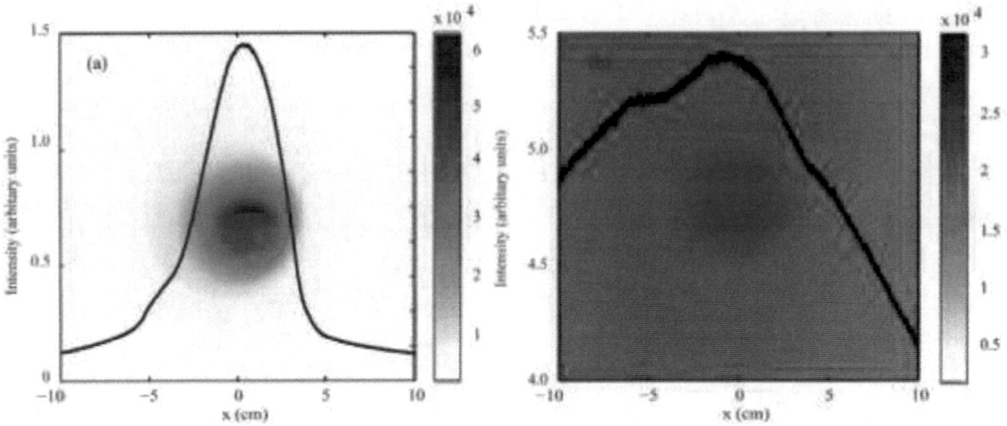

Figure 8. The image and profile of the gamma rays at the sample position: (a) opened shutter and (b) closed shutter

3.3. Simulation of the Prompt gamma ray yield of Portland cement sample

In this section, the yields of prompt gamma rays from the cylinder Portland cement sample have been calculated. The sample model held at 45° angle to the beam at distance 20 cm from the beam shutter with the thickness of 1 cm and radius of 6 cm. The chemical composition of the Portland cement is given in table 1.

Table 1. The chemical composition of Portland cement [4]

Compound	**Composition weight (%)**
SiO_2	20.52
Al_2O_3	5.64
Fe_2O_3	3.80
CaO	64.35
MgO	2.11
SO_3	2.10
K_2O	0.36
Na_2O	0.19

The energies of the prompt gamma ray have been carried out by subtracting background radiation and presented in figure 9(b). The 2.23 MeV photon peak is found due to the neutron capture reaction (n,γ) of the neutron in the polyethylene shielding. Moreover, the 3.55 MeV and 4.95 MeV of the prompt gamma rays due to capture the thermal neutron in silicon in the cement sample are observed.

The 7.63 MeV and 7.64 MeV of the prompt gamma rays due to the thermal neutron capture reactions of the iron in cement sample are found.

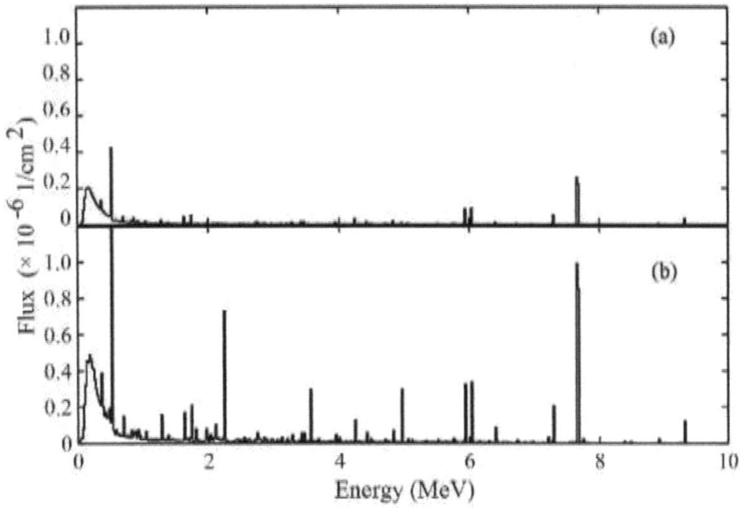

Figure 9. (a) Background gamma ray spectrum per source neutron and (b) the prompt gamma ray spectrum per source neutron of the Portland cement without the background radiation

4. Conclusion

The MCNP model of the PGNAA system at TRR-1/M1 have been set up. The simulation results show that the neutrons are well collimated through the collimator. At the sample position, the number of $\Phi_{thermal}/\Phi_{epithermal}$ and $\Phi_{thermal}/\Phi_{fast}$ have been obtained. The calculated thermal neutron flux at the sample position is higher than the measured one by a factor of seven. However, to improve the results, the detector efficiency will be considered. The neutron beam shape at the sample position is characterized by both experimental set up and calculation. The disagreement between the experiment and simulation are found. The non-symmetrical shape shows in the experimental result while the simulation result is well symmetry. The improvement of the MCNP model is needed by taken the surrounding structure in the reactor hall into account. In addition, the prompt gamma ray spectrum from a sample of Portland cement is calculated. The results provide an information in order to fabricate the experimental set up. For the future work, the detector, detector shielding and sample size will model and optimize using the MCNP simulation. Since, there is a plan to upgrade the PGNAA system in 2016-2017, thus the simulation results will provide significant contribution in upgrading the system. The improvement of the facillity will be a great asset to be available in various applications.

Acknowledgments

This work was funded by Mahasarakham University. The authors would like to express sincere thanks to Department of Physics, Faculty of Science, Mahasarakham University and Thailand Institute of Nuclear Technology (Public Organization). The authors acknowledge the High Performance Computing Server of Thailand Institute of Nuclear Technology for providing computing resources that have contributed to the research results reported within this paper.

References

[1] Naqvi AA, Nagadi MM, Al-Amoudi OSB (2005) Measurement of lime/silica ratio in concrete using PGNAA technique. Nucl. Instr. Meth. Phys. Res. A. 554:540–545
[2] Naqvi AA, Nagadi MM, Al-Amoudi OSB (2004) Elemental analysis of concrete samples using an accelerator-based PGNAA setup. Nucl., Instr. Meth. Phys. Res. B. 225:331–338
[3] Naqvi AA, Nagadi MM, Rehman K, Maslehuddin M, Kidwai S (2003) Monte Carlo simulation for design of the KFUPM PGNAA facility. Radiat. Phys. Chem. 66:89-98

[4] Naqvi AA, Garwan MA (2004) Validity test of design calculations of a PGNAA setup. Nucl. Instr. Meth. Phys. Res. B. 215:283-291

[5] Canella L, Kudejova P, Schulze R, Tuerler A, Jolie J (2011) Characterisation and optimisation of the new Prompt Gamma-ray Activation Analysis (PGAA) facility at FRM II. Nucl. Instrum. Meth. A. 636:108-113

[6] Gardner RP, Mayo CW, El-Sayyed ES, Metwally WA, Zheng Y, Poezart M (2000) A feasibility study of a coincidence counting approach for PGNAA applications. Appl. Radiat. Isot. 53:515–526

[7] Gardner RP, Sayyed E, Zheng YS, Hayden S, Mayo CW (2000) NaI detector neutron activation spectra for PGNAA applications. Appl. Radiat. Isot. 53:483–497

[8] Revay Zs, Harrison RK, Alvarez E, Biegalski SR, Landsberger S (2007) Construction and characterization of the redesigned PGAA facility at the University of Texas at Austin. Nucl. Instr. Meth. Phys. Res. A. 577:611-618

[9] Robinson A, Hartman R, Reese R (2010) Design, construction and characterization of a prompt gamma activation analysis facility at the Oregon State University TRIGA® reactor. J Radioannal Nucl. Chem. Lett. 283:359-369

[10] Asvavijnijkulchai C, Dharmavanij W, Siangsanan P, Ratanathongchai W, Chongkum S (1989) Status of prompt gamma neutron activation analysis (PGAA) at TRR-1/M1 (Thai Research Reactor-1/Modified 1). JAERI-Conf 99-006

[11] Sweezy, J. E. Los Alamos National Laboratory: MCNP-A general Monte Carlo N-Particle transport code, version 5. 2008, LA-CP-03-0245

[12] Burgio N, Rosa R (2004) Monte Carlo design for a new neutron collimator at the ENEA Casaccia TRIGA reactor. Appl. Radiat. Isot. 61:663-666

[13] Khan R, Karimzadeh S, Stummer T, Bock H (2011) Monte Carlo simulation of the thermal column and beam tube of the TRIGA Mark II research reactor. Nucl. Eng. Des. 241:2859-2863

[14] Ratanatongchai W, Khaweerat S, Picha R, Channuie J, Sangaroon S (2016) Measurements of M1/1-TRR prompt gamma neutron activation beam. International Nuclear Science and Technology Conference, 4-6 August 2016, Bangkok, Thailand

IOP Conf. Series: Journal of Physics: Conf. Series **860** (2017) 012038 doi:10.1088/1742-6596/860/1/012038

Baseline quantity of ^{131}I, ^{137}Cs, ^{134}Cs and ^{40}K in urinary excretions from Thai people and internal exposure dose

D Peekhunthod[1], J Bangvirunrak, S Sansakon, A Nukultham and T Pukkhaw

Bureau of Technical Support for Safety Regulation, Office of Atoms for Peace, Bangkok, Thailand

[1] Corresponding author e-mail: d_peekhunthod@hotmail.com

Abstract: Today, sealed and unsealed radioactive materials have been used in Thailand for various purposes such as medical, agricultural and industrial applications. There is a growing trend in the use of radioactive materials. Moreover, neighboring countries are planning to construct and operate nuclear power plants. In case of nuclear power plant accidents, radioactive releases in environment and intakes into human body by inhalation and ingestion causing long term health effects. This research aims to determine the radiation baseline quantity of interested relevant radionuclides such as ^{131}I, ^{137}Cs, ^{134}Cs as well as a natural radionuclide, ^{40}K in urine samples of Thai people by gamma spectrometry. Two types of detectors (NaI and HpGe detectors) are calibrated by mixed radionuclide standards of ^{109}Cd, ^{57}Co, ^{133}Ba, ^{54}Mn, ^{137}Cs and ^{60}Co, (energy range from 88 to 1,331 keV). 720 urine samples are collected over a 24 hour period from Thai volunteers with the age older than 18 years old, who lived in eight locations of Thailand. To reduce the effect of geometric difference, 30 ml of urine samples are prepared for counting measurement and efficiency determination. The radiation baseline quantity of ^{131}I, ^{137}Cs, ^{134}Cs and ^{40}K in 30 ml of urine samples are 0.37±0.09, 0.63±0.13, 0.39±0.08 and 7.84±1.63 Bq, respectively. Based on the assumption of intake (50% of the intake by ingestion and 50% of the intake by inhalation), internal dose for members of public are assessed. The committed dose equivalent due to an intake of ^{131}I, ^{137}Cs, ^{134}Cs and ^{40}K are 2.36E-03±1.66-03, 1.15E-01±8.61E-02, 1.16E-01±7.77E-02, 9.44E-01±3.56E-01 mSv per year, respectively.

1. Introduction

Radiological contamination has been widely found in the normal environment in Thailand, for example in the air, soil, food and drinking water. According to the wide spreading contaminants, radioactive material can enter through the body via ingestion and inhalation. Some radionuclides such as K-40 has been found as the primary source of radiation inside human body. Due to radioactive decay, the radiological contaminants may affect not only people but also plants and animals. The internal intake of radioactive material has been considered as one of the major concerns for radiation protection. To make sure that the level of the internal contamination is not higher than the accepted level; biological samples such as urine, excreted from human body should be monitored. Natural intake of radioactive material is determined and assessed.

This research mainly determined the quantity of radioactivity and assessed the internal contamination of radioactive material in urine samples, collected from volunteers of 8 locations in Thailand. The interested radionuclides include radioactive materials; leaked from nuclear accidents, nuclear activities, nuclear fuel cycles and other applications; for example ^{134}Cs, ^{137}Cs, ^{131}I, ^{14}C, ^{90}Sr, ^{3}H, ^{241}Pu, ^{232}Th and ^{238}U. The radiation baseline of interested nuclide will be applied as reference data for Thai people in case of nuclear accidents both of Thailand and neighboring countries.

2. Methods

2.1. Network Creations

Collecting urine samples is difficult to manage by only one institute. Convenience sampling method is applied by network creations in order to provide volunteers who are conveniently available to participate in this research for collecting urine samples. The appropriated universities should have a gamma ray spectrometer or locate in suitable location where dispersed in many parts of Thailand. They were invited to participate as networks. Coordinated network with 10 universities where located in 10 parts of Thailand was created. Participated universities played a part in collecting and/or measuring radioactivity in urine samples. The urine samples and radioactivity determinations were submitted to the Office of Atoms for Peace (OAP) for the measurement of internal dose assessment. Unfortunately, the raw data of 2 locations had problems and could not be conclusion. The urine samples collecting will be repeated and reanalyzed next time. Therefore, there are only the results of 8 locations in this research as show in Figure 1.

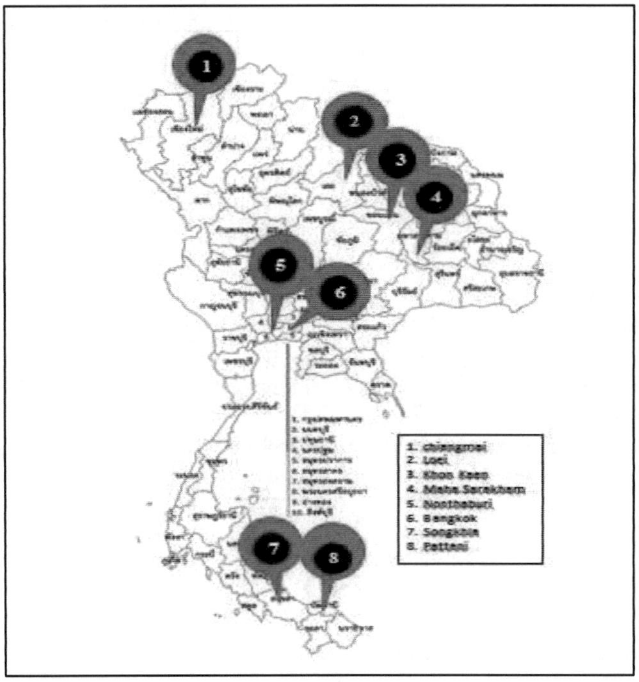

Figure 1. Urine sampling locations.

2.2 Urine sampling

Urine samples are collected over a 24 hour period from Thai volunteers with the age older than 18 years old, who lived in 8 locations of networks. The networks provided at least thirty volunteers for collecting urine samples which are kept in 500 ml of wide mouth polyethylene bottles. Urine samples are continuously collected for 3 days. There were at least 90 urine samples per each location. The total samples are 720.

2.3. Sample preparation
Urine samples are continuously collected for 3 days. Then, they were mixed together and the total volume was measured. 720 urine samples were prepared by adding 10% (v/v) of concentrated nitric acid. 30 ml of each sample was withdrawn, transferred to plastic bottle and measured by appropriated detector in order to determine the radiation baseline quantities of ^{131}I, ^{134}Cs, ^{137}Cs and ^{40}K. The rest of urine samples were kept in cool temperature for the next measurement.

2.4. Calibration
Two types of radiation detectors were used. NaI and HpGe detectors with Genie 2000 gamma spectroscopy software in this study were similarly calibrated by mixed radionuclide standards (point source) of ^{109}Cd, ^{57}Co, ^{133}Ba, ^{54}Mn, ^{137}Cs and ^{60}Co (energy range from 88 to 1,331 keV). The efficiency of each radionuclide was calculated in equation 1, when cps is counts per second, dps is activity in Bq and I is gamma ray abundance. The efficiencies from calibration were fitted a curve. Due to the difference of geometry between sample and calibration point source, the prepared solution containing ^{131}I in the same geometry type as the urine sample was used for correction counting efficiencies. The efficiency ratio(R) was determined by dividing the efficiency of ^{131}I in urine by the efficiency of point sources at the same energy. The efficiencies for urine sample counting were determined by using R value multiply by the efficiencies from calibration fitted curve of point sources.

$$\text{eff} = \frac{\text{cps}}{\text{dps x I}} \tag{1}$$

2.5. Gamma analysis
Referred to RS-G-1.2 [3], gamma emission from radionuclides in urine samples can be detected by NaI or HpGe detector with Genie 2000 spectroscopy software. To determine the quantities of radioactivity in urine samples, 30 ml of empty plastic bottle was used as a blank. Radioactivities of ^{131}I, ^{137}Cs, ^{134}Cs and ^{40}K were measured at 356, 662, 796 and 1,461 keV respectively.

$$A = \frac{\text{cps}}{\text{eff}} \tag{2}$$

When eff is efficiency for urine sample counting of radionuclide multiplied by its gamma ray abundance.

2.6. Internal dose assessment
According to ICRP 78 [1], biokinetic models which describe activity in excreta, as a function of time following intake. To calculate an estimate of intake of radionuclides by the members of public, the daily activities of ^{131}I, ^{137}Cs, ^{134}Cs and ^{40}K in urine samples, M, is divided by the predicted values of the measured quantity in urine at time t days after intake of 1 Bq or m(t) as show in equation 3. Most of m(t) parameters were taken from ICRP 78 except for m(t) parameter of ^{40}K for ingestion was taken from Turban *et al.,* [4] and the inhalation was taken from calculation on CINDY code program.

$$\text{Intake} = \frac{M}{m(t)} \tag{3}$$

The intake can be multiplied by the effective dose coefficients, e(70), to give the committed effective dose equivalent(CEDE) as show in equation 4. The effective dose coefficients, e(70), were taken from Health Physics 2012 [2].

$$\text{CEDE} = \text{Intake} \times e(70) \tag{4}$$

2.7. Minimum detectable activity
The minimum significant activity (MSA) is related to the smallest recorded signal. It is a significant excess of the background response indicating the presence of radioactivity at 95% of probability which can be calculated in equation 5.. To ensure that the radionuclide signals correspond to the MSA, the minimum detectable activity (MDA) was calculated as equation 6.

$$\mathrm{MSA} = \frac{1.64}{\mathrm{F}}\sqrt{\frac{\mathrm{n_b}}{\mathrm{t_s}}\left[1+\frac{\mathrm{t_s}}{\mathrm{t_b}}\right]} \quad (5)$$

$$\mathrm{MDA} = \frac{2.71}{\mathrm{Ft_s}} + 2\mathrm{MSA} \quad (6)$$

When n_b is the background count rate, t_s and t_b are the count times for the sample and for the background, respectively.

2.8. Uncertainty and Statistic
Generally, the measurement uncertainty is the most straightforward means for dose estimation. In the situation where the activity level is low and closed to the detection limit, the counting uncertainty can dominate the overall uncertainty. The combine uncertainty was calculated as equation 7, calculation expanded uncertainty at 95% confidence (α=0.05) from URACHEM [5], where U_{rep} is the standard uncertainty from repeatability, U_{eff} is the standard uncertainty from efficiency calibration and U_{cps} is the standard uncertainty from the count per second.

$$u_c = x_i\sqrt{\left(\frac{u_{rep}}{x_i}\right)^2+\left(\frac{u_{Eff}}{Eff}\right)^2+\left(\frac{u_{cps}}{CPS}\right)^2} \quad (7)$$

3. Results

3.1. Calibration
The efficiency calibration fitted curves of two types of radiation detectors, NaI and HpGe detectors with Genie 2000 gamma spectroscopy software shown in figure 2.

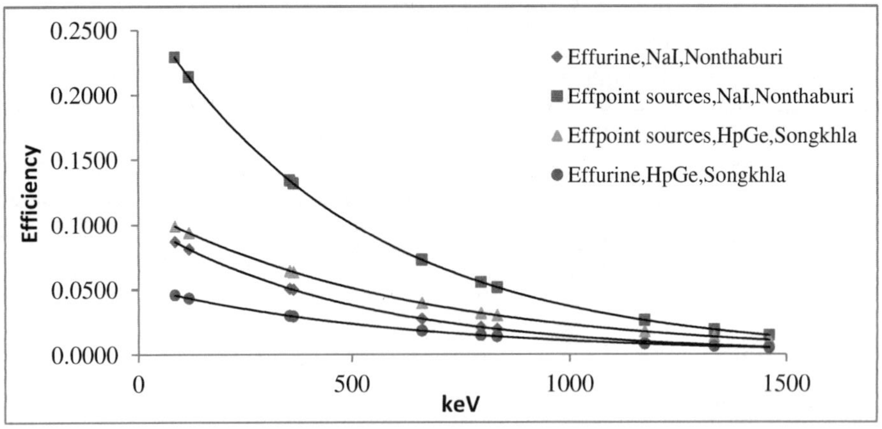

Figure 2. The efficiency calibration fitted curves for point sources and urine sample counting by NaI and HpGe detectors by assuming 100% gamma ray abundance.

3.2. Gamma analysis

The spectrum of each radionuclide was calculated as equation 2. Table 1, figure 2 and 3 show the average activity of ^{131}I, ^{137}Cs, ^{134}Cs and ^{40}K in eight locations.

Table 1. The baseline quantities of ^{131}I, ^{137}Cs, ^{134}Cs and ^{40}K in 30 ml of urinary excretions From Thai people.

location	Average activity (Bq)				detector
	^{131}I	^{137}Cs	^{134}Cs	^{40}K	
Chiangmai	0.04±0.03	0.25±0.13	0.25±0.14	7.41±6.83	NaI
Nonthaburi	0.48±0.17	0.40±0.14	0.52±0.19	9.99±6.55	NaI
Maha Sarakham	0.30±0.11	1.22±0.42	0.68±0.38	9.03±5.34	NaI
Loei	0.05±0.02	0.06±0.02	0.04±0.02	0.33±0.21	NaI
Khon Kaen	0.17±0.07	0.36±0.14	0.22±0.09	8.86±7.21	Hp(Ge)
Bangkok	0.07±0.03	0.28±0.10	0.06±0.11	3.05±2.09	Hp(Ge)
Songkhla	0.07±0.03	0.06±0.02	0.05±0.02	2.82±1.48	Hp(Ge)
Pattani	1.83±0.75	2.46±0.99	1.31±0.53	21.19±10.49	Hp(Ge)
average	0.19±0.17	0.43±0.41	0.30±0.25	6.45±3.08	

The continuous uncertainty can be calculated in equation 8.

$$S_y = \sqrt{S_a^2 + S_b^2 + S_c^2 + ...} \tag{8}$$

Where S_a, S_b, S_c, … are uncertainties from determination of average activities in each location. When using the one-way analysis of variance (ANOVA), the most of the average activities in eight locations are significant differences. Because each location is difference of diet and environment.

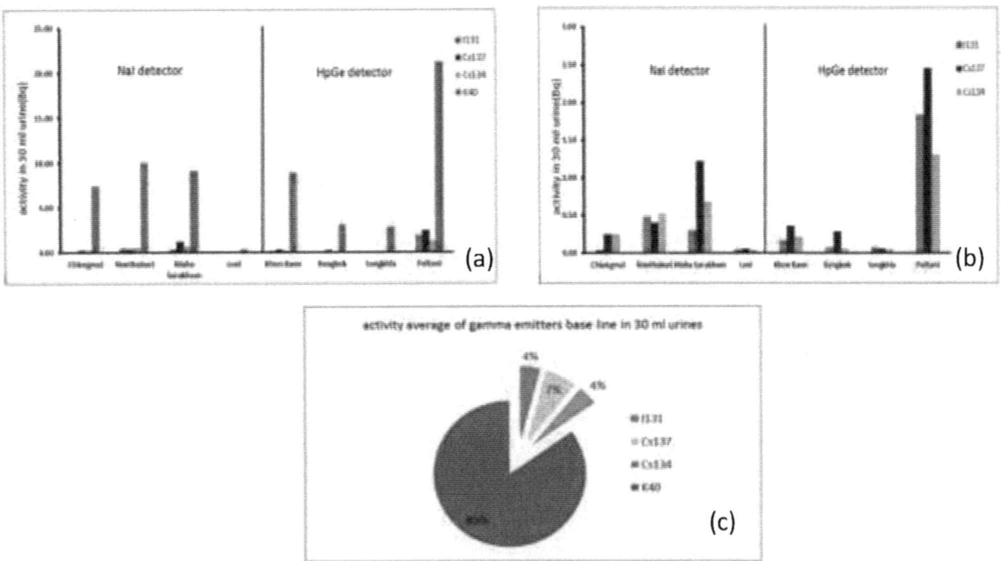

Figure 3. Radioactivity of ^{131}I, ^{137}Cs, ^{134}Cs and ^{40}K in 30 ml urine samples obtained from different locations of Thailand (a and b), the average activity is showed in (c).

3.3. Internal dose assessment

Due to the activities of ^{131}I, ^{137}Cs, ^{134}Cs and ^{40}K in urine samples are less than MDA, it means that the activities have no significantly in excess of the background level. The activities in 30 ml urine samples are not depend on the volume of urine. Therefore, the activity in daily excretion is equivalent to the activity from measurement. For ^{40}K, the predicted value of the measured quantity in urine from daily intake by ingestion equivalent to daily excretion because of its rapidly excretion after intake. The percentage of dietary potassium excreted in the urine was high. It is approximately 63 to 92% of intake from ref[4].

The estimation of intake can be calculated by assuming default parameter, absorption type F. Time after acute intake 1 day, intake by inhalation, the m(t) of ^{131}I, ^{137}Cs, ^{134}Cs and ^{40}K, are 2.80E-01, 7.90E-03, 7.90E-03 and 6.90E-03 Bq per Bq intake, respectively. The intake by ingestion, the m(t) of ^{131}I, ^{137}Cs, ^{134}Cs and ^{40}K are 2.80E-01, 7.90E-03, 7.90E-03 and 6.90E-03 Bq per Bq intake, respectively. In addition, this research assumed that members of the public intake 50% by ingestion and 50% by inhalation. The dose coefficients for AMAD 1 μm for inhalation of ^{131}I, ^{137}Cs, ^{134}Cs and ^{40}K are 7.40E-09, 4.60E-09, 6.60E-09 and 2.10E-09 SvBq^{-1}, respectively. The dose coefficients for AMAD 1 μm for ingestion are 2.20E-08, 1.30E-08, 1.90E-08 and 6.20E-09 SvBq^{-1}, respectively. The committed effective dose equivalent due to intake of these radionuclides is shown in table 2.

Table 2. The committed effective dose equivalent in members of public due to intake of ^{131}I, ^{137}Cs, ^{134}Cs and ^{40}K by assuming 50% acute intake by ingestion and 50% acute intake by Inhalation.

CEDE (mSv.y^{-1})			
^{131}I	^{137}Cs	^{134}Cs	^{40}K
2.36E-03±1.66E-03	1.15E-01±8.61E-02	1.16E-01±7.77E-02	9.44E-01±3.56E-01

3.4. Minimum detectable activity

The activities of ^{131}I, ^{137}Cs, ^{134}Cs except for ^{40}K in 30 ml of urine, obtained from 8 locations of Thailand were lower than the MDA. It means that the activities have no significant in the excess of the background level, except in Pattani, which seems higher than the other. They showed that the activities are significantly in the excess of the background. The MDAs of ^{131}I, ^{137}Cs, ^{134}Cs and ^{40}K are shown in table 3.

Table 3. The MDAs of ^{131}I, ^{137}Cs, ^{134}Cs and ^{40}K of urine sample counting in each location.

location	MDA (Bq)				detector
	^{131}I	^{137}Cs	^{134}Cs	^{40}K	
Chiangmai	0.88	1.39	1.60	50.81	NaI
Nonthaburi	0.88	1.39	1.60	50.81	NaI
Maha Sarakham	0.88	1.39	1.60	50.81	NaI
Loei	0.88	1.39	1.60	50.81	NaI
Khon Kaen	0.86	1.77	0.85	29.04	HpGe
Bangkok	0.40	0.50	0.73	9.76	HpGe
Songkhla	0.31	0.37	0.35	8.13	HpGe
Pattani	0.05	0.80	1.39	27.10	HpGe

IOP Conf. Series: Journal of Physics: Conf. Series **860** (2017) 012038 doi:10.1088/1742-6596/860/1/012038

4. Conclusion

This research shows that the average activities in 30 ml urines of ^{131}I, ^{137}Cs, ^{134}Cs and ^{40}K are 0.37 ± 0.09, 0.63 ± 0.13, 0.39 ± 0.08 and 7.84 ± 1.63 Bq, respectively. The average activities in eight locations are significant differences due to each location is difference of diet and environment when using the one-way analysis of variance (ANOVA). The intake of radionuclides and the assessed internal exposure dose in members of the public were calculated by assuming 50% intake by ingestion and 50% intake by inhalation. The committed effective dose equivalent of ^{131}I, ^{137}Cs, ^{134}Cs and ^{40}K are 2.36E-03±1.66E-03, 1.15E-01±8.61E-02, 1.16E-01±7.77E-02 and 9.44E-01±3.56E-01 mSv per year, respectively. The internal dose, CEDE values are less than 1 mSv a year. These data from this research can be used as the baseline quantities of ^{131}I, ^{137}Cs, ^{134}Cs and ^{40}K in urinary excretions from Thai people and internal exposure dose. For further study, the radiation baseline quantities from this study can be applied as a reference data in case of nuclear accidents and can be used to study the long term health effect of internal contamination for Thai people.

References

[1] Individual Monitoring for Internal Exposure of Workers due to intake of radionuclide, 1997, *ICRP Publication* **78.**

[2] Thomas E, John Son and Brain 2012, *Health Physics and Radiological health*, **4th edition**,

[3] Occupational exposure due to intakes of radionuclides. 1999b, *IAEA, Safety Guide* **RS-G-1.2**. International Atomic Energy Agency, Vienna, Austria.

[4] Sharon Turban ,et al,.Racial Differences in Urinary Potassium Excretion, 2008 Jul; **19(7), JASN**,: 1396–1402

[5] Quantifying Uncertainty in Analytical Measurement, 2000, **EURACHEM/CITAC guide, 2nd ed.** QUAM.

Acknowledgements

The successful research was done by the collaborating with the networks for collecting and measurement of urine samples. The authors would like to thank Chiang Mai University, Thammasat University, Naresuan University, Khon Kaen University, Mahasarakham University, Prince of Songkla University, Thaksin University, Pibulsongkram Rajabhat University, Loei Rajabhat University and Surat Thani Rajabhat University.

IOP Conf. Series: Journal of Physics: Conf. Series **860** (2017) 012039 doi:10.1088/1742-6596/860/1/012039

Preliminary results of consequence assessment of a hypothetical severe accident using Thai meteorological data

K Silva[1,3], S Lawawirojwong[2], J Promping[1]

[1] Thailand Institute of Nuclear Technology (Public Organization), 9/9 Moo 7, Sai Moon, Ongkharak, Nakhon Nayok 26120, Thailand
[2] Geo-Informatics and Space Technology Development Agency (Public Organization), Ratthaprasasanabhakti Building 6th and 7th Floor, Chaengwattana Road, Laksi Bangkok 10210, Thailand

[3] Corresponding author e-mail: kampanarts@tint.or.th

Abstract. Consequence assessment of a hypothetical severe accident is one of the important elements of the risk assessment of a nuclear power plant. It is widely known that the meteorological conditions can significantly influence the outcomes of such assessment, since it determines the results of the calculation of the radionuclide environmental transport. This study aims to assess the impacts of the meteorological conditions to the results of the consequence assessment. The consequence assessment code, OSCAAR, of Japan Atomic Energy Agency (JAEA) is used for the assessment. The results of the consequence assessment using Thai meteorological data are compared with those using Japanese meteorological data. The Thai case has following characteristics. Low wind speed made the radionuclides concentrate at the center comparing to the Japanese case. The squalls induced the peaks in the ground concentration distribution. The evacuated land is larger than the Japanese case though the relocated land is smaller, which is attributed to the concentration of the radionuclides near the release point.

1. Introduction

Probabilistic risk assessment (PRA) is a common way to assess the risks of accidents in a nuclear power plant. PRA is divided into three levels. In level 1 PRA, the accidents which can possibly damage the reactor core are selected, and the core damage frequencies (CDFs) are calculated. Then in level 2 PRA, the accidents which can cause release of radionuclides to the environment are selected out of the core damage accidents, and the containment failure frequencies (CFFs) and the large early release frequencies (LERF) are evaluated along with the source term. Finally, in level 3 PRA, the consequence assessment of the release of the radionuclides to the environment are performed. This is a very important step of the risk assessment, since it helps the assessor understand the extent of the consequences which may be attributed to a severe accident, and helps them plan for appropriate accident management strategies. It starts from the determination of the source term i.e. the amount, time and period of the release of radionuclides, then continues with the evaluation of the environmental transport where the atmospheric dispersion calculation is performed. After that the radiation dose is estimated, and the associated risks are finally calculated in terms of health effects, number of affected people, or size of the affected area [1].

It is widely known that the meteorological conditions can significantly influence the outcomes of such assessment, since it determines the results of the calculation of the radionuclide environmental transport [2]. Thus it is important to correctly select the representative meteorological conditions of the site of interest.

Several pieces of research have been performed in the past inside [3] and outside of Thailand [2,4–8], while the effects of different meteorological conditions in Thailand on the results of the consequence assessment of a hypothetical severe accident in a nuclear power plant have not yet been systematically performed. Therefore, the objective of this research is to study the effects of Thai meteorological conditions to the consequence assessment results. The results of the consequence assessment using Thai meteorological data are compared with those using Japanese meteorological data in order to identify the characteristics of the consequences of a hypothetical severe accident in Thailand. The influences of the meteorological conditions are indicated in terms of differences in air and ground concentrations of the released radionuclides, and differences of consequences, e.g. collective dose, number of affected people and size of affected area.

2. Methodology

2.1. Consequence assessment code, OSCAAR

The consequence assessment code, OSCAAR (Off-Site Consequence Analysis code for Atmospheric Releases in reactor accidents) of Japan Atomic Energy Agency (JAEA) is used for the assessment [9]. It can take into account the meteorological conditions hourly, and has the ability to perform the sampling of the meteorological data. The calculation scheme of OSCAAR is shown in Figure 1, and its detailed description can be found in Homma et al. [3].

2.2. Calculation conditions

The basic calculation conditions are shown in Table 1. THA and JPN stand for the cases where Thai and Japanese meteorological data sets are used. The authors assumed the long-term station blackout accident [10] as the hypothetical accident in this study, since it has a large CFF and the scenario is similar to the accident happened in the Fukushima Daiichi Nuclear Power Plant. The source term data is shown in Table 2, where the release is divided into two steps in order to accurately represent the release. Population distributions of the two cases are shown in Figure 2. The peaks represent the highly populated cities, which is Bangkok and its vicinities for the Thai case, and Mito and Tokyo for the Japanese case.

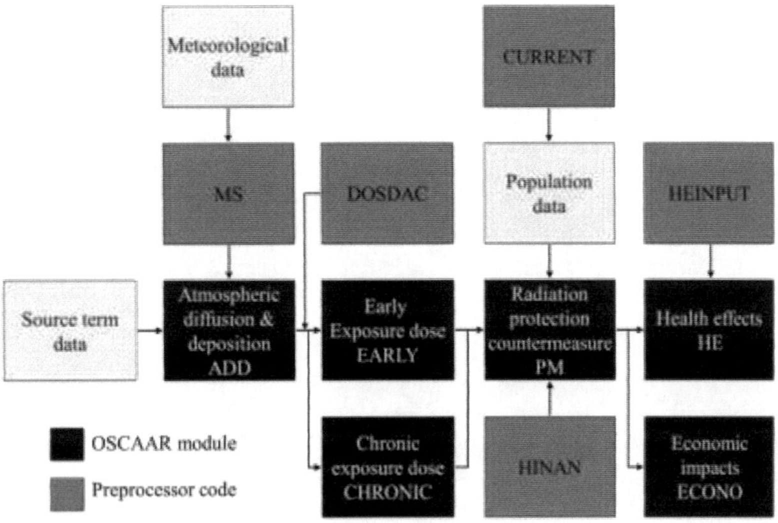

Figure 1. Calculation scheme of OSCAAR.

Table 1. Calculation conditions.

Items	Conditions
Reactor type	1,100 MWe BWR-5
Location	THA: Headquarters of Thailand Institute of Nuclear Technology JPN: Tokai Research and Development Center, Japan Atomic Energy Agency
Hypothetical accident	Long-term station blackout [10]
Coverage of meteorological data	Within the radius of 200 km from the release point
Meteorological data taken into account	Hour wind speed, wind direction, precipitation, weather stability
Year	THA: 2014 JPN: 1990
Sampling method	Random sampling (300 samples)
Calculated parameters	Air concentration, ground deposition, collective dose, evacuated and relocated people, evacuated and relocated area
Dose criteria for radiation protective actions	Sheltering: 10 mSv/1 day [11] Evacuation: 50 mSv/7 days [11] Relocation: 20 mSv/year [12]

Table 2. Source term data of the hypothetical accident.

Release time [hr]	Release duration [hr]	Release ratio to core inventory [-]							
		Noble gas 1	Organic I 2	Inorganic I 3	Cs-Rb 4	Te-Sb 5	Sr-Ba 6	Ru 7	La 8
12.7	4.0	2.9E-1	1.7E-4	3.1E-3	5.4E-3	1.1E-3	2.5E-4	4.5E-9	3.1E-7
16.7	25.0	7.1E-1	6.3E-3	1.2E-1	4.2E-2	7.4E-2	2.7E-3	3.1E-8	3.3E-6

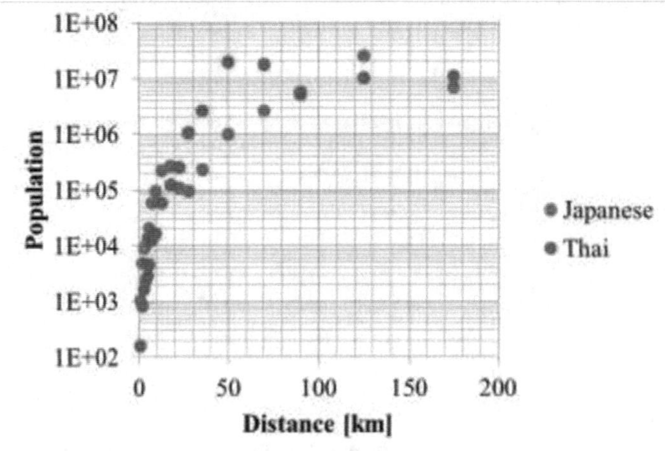

Figure 2. Population distributions of Thai and Japanese cases.

3. Results and Discussion

Figures 3–7 show (1) the average air concentration of radionuclides relative to distance from release point, (2) the average ground concentration of deposited radionuclides relative to distance from release point, (3) the total collective dose relative to distance from release point, (4) the total collective dose from all pathways and (5) the numbers of evacuated and relocated people, and the areas of abandoned land, respectively. Since a pile of meteorological data is used for the calculation, the results above are shown in expected values, i.e. the probability weighted averages.

It can be seen from Figure 3 that the air concentration when the Thai meteorological data is used (hereinafter referred to as Thai case) is larger until the radius of around 100 kilometers. However, at the distance further than 100 kilometers, the case that the Japanese meteorological data is used (hereinafter referred to as Japanese case) gives larger air concentration. Similar trends can be observed in Figure 4, though not as obvious as Figure 3. This is because the wind speed of the Thai case is basically lower than that of the Japanese case, thus the radionuclides could not migrate as far as the Japanese case.

The average ground concentrations of the deposited radionuclides of both cases gradually decrease with the distance. However, some peaks can be observed in the Thai case (notable peak can be found at 90 km from the release point). This can be attributed to the difference in the characteristics of the rainfall. Except for the case of typhoons, the rain in Japan normally falls constantly and continues for a long time (several hours or days). On the other hand, the rain in Thailand is usually in the forms of squall, which is a heavy rain but lasts only for a short period of time and falls only at specific area. If the squall passes the area where the radioactive plume stays in the atmosphere, it will bring the radionuclides down to the ground, and consequently create the peaks in the ground concentration at some specific distance.

The average air concentration of radionuclides and the average ground concentration of deposited radionuclides are used to calculate the effective doses from each pathway: cloudshine, groundshine, inhalation, resuspension and food intake. The total effective doses are used to determine sheltering, evacuation and relocation based on the criteria specified in Table 1. The sums of the effective dose after taking into account all radiation protective actions of the whole population, i.e. collective dose, at each distance band which represent the extent of the stochastic health effects from the accident is shown in Figure 5. It can be observed from the figure that the total collective doses of the Japanese case has two peaks, while there is only a single peak in the Thai case. This may be attributed to the difference in the distribution of the population of the two countries. The peaks match with the location of large cities of the two countries. A single peak in the Thai case matches with the location of Bangkok, and the two peaks in the Japanese case represent Mito and Tokyo.

The difference of the total collective doses within the radius of 200 kilometers of the two cases in Figure 6 is relatively small comparing to the uncertainties associated with the inputs of the assessment. This implies that most of the radionuclides have not travel beyond the radius of 200 kilometers. It can thus be assumed that the size of the target area is appropriate for the assessment. Note that the slight difference of the total collective doses between the two cases is probably due to the difference in the sizes of the total population.

It can be seen from Figure 7 that the numbers of evacuated people of the two cases are quite the same whereas the evacuated land of the Thai case is larger than that of the Japanese case. This is attributed to the larger air concentration and ground deposition in the area near the release point as shown in Figures 3 and 4. On the other hand, the relocated area, and the time-integrated relocated area of the Thai case is smaller, since the air concentration and ground deposition are smaller than the Japan case in the area further from the release point. These differences are also caused by the lower average wind speed of the Thai case.

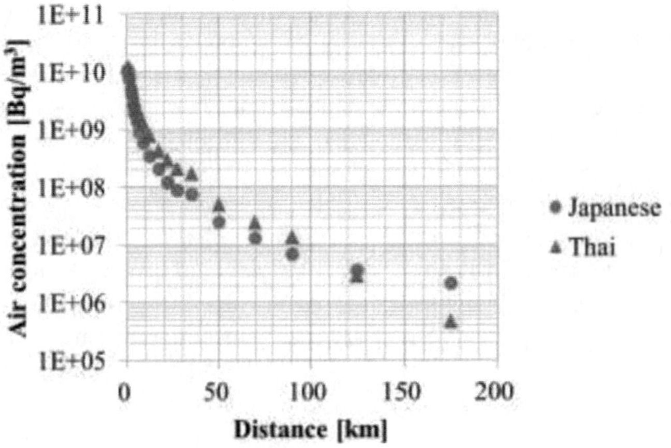

Figure 3. Average air concentration of radionuclides relative to distance from release point.

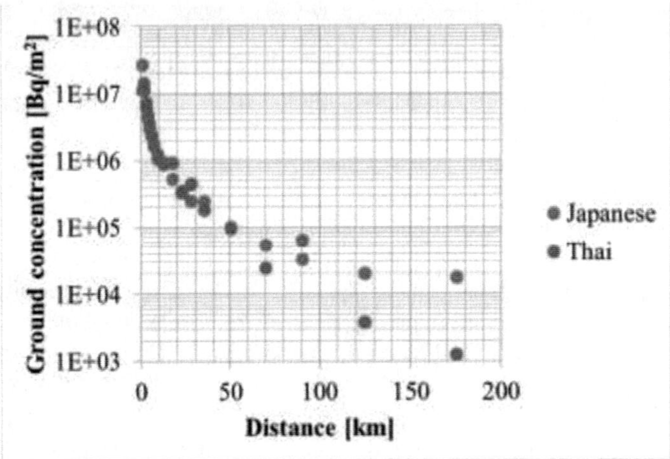

Figure 4. Average ground concentration of deposited radionuclides relative to distance from release point.

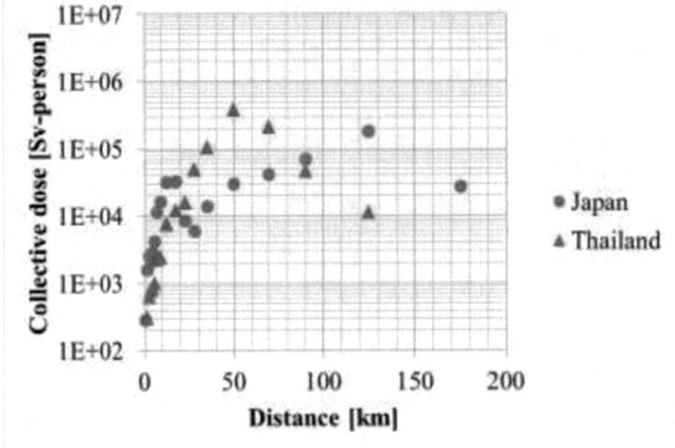

Figure 5. Total collective dose relative to distance from release point.

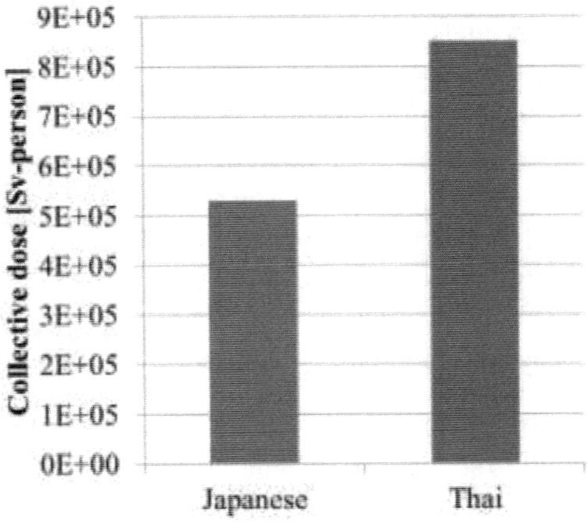

Figure 6. Total collective dose from all pathways.

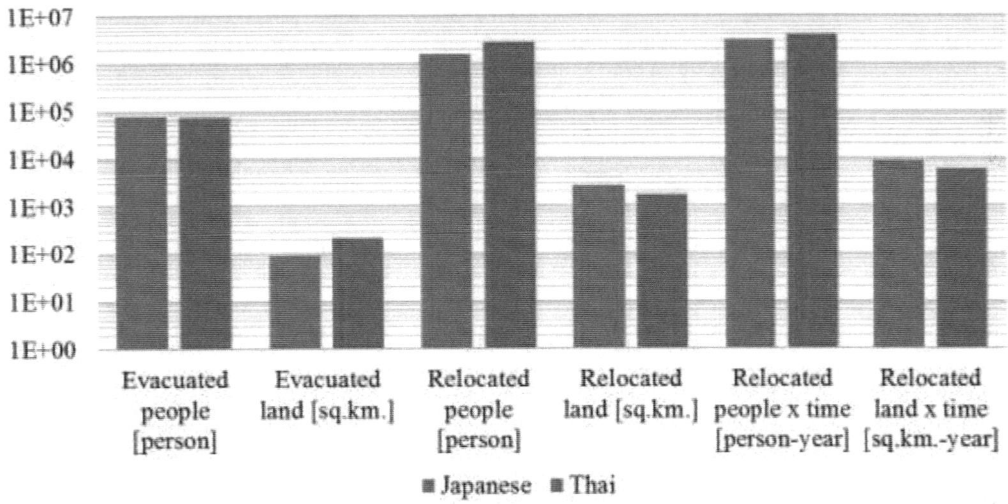

Figure 7. Number of affected people and size of affected area.

4. Conclusions

Consequence assessment using Thai meteorological data was performed in order to study the effects of meteorological conditions to the consequence assessment results. The results are compared with the case where Japanese meteorological data was used, and following characteristics are confirmed. Lower wind speed of the Thai case made the radionuclides concentrate at the center, while the radionuclides migrated and deposited further in the Japanese case. The squalls induced the peaks in the ground concentration in the Thai case. The evacuated land of the Thai case is larger than the Japanese case though the relocated land is smaller. This is because the radionuclides concentrated in area near the release point in the Thai case, which is again attributed to the low wind speed.

Acknowledgment

The authors would like to thank the Nuclear Safety Research Center, Japan Atomic Energy Agency for the OSCAAR code used in this study.

IOP Conf. Series: Journal of Physics: Conf. Series **860** (2017) 012039 doi:10.1088/1742-6596/860/1/012039

References

[1] Till J E 2008 *Radiological Risk Assessment and Environmental analysis* ed Till J E and Grogan H A (New York: Oxford University Press) chapter 1 pp 1–30
[2] Homma T, Tomita K and Hato S 2005 *Nuc. Eng. Technol.* **37(3)** 245–258
[3] Kuasakul P and Wongsawang D 2015 Proc. Conf. STOU Graduate Research (Bangkok) vol 5 (Bangkok: Sukhothai Thammathirat Open University) P-ST 012
[4] Helton J C, Johnson J D, Shiver A W and Sprung J L. 1995 *Reliab. Eng. Syst. Saf.* **48(2)** 91–127
[5] Helton J C, Johnson J D, Rollstin J A, Shiver A W and Sprung J L. 1995 *Reliab. Eng. Syst. Saf.* **50(2)** 137–177
[6] Haste T, Birchley J, Cazzoli E and Vitazkova J 2006 *Nucl. Eng. Des.* **236(10)** 1099–1112
[7] Baklanov A, Mahura A, Jaffe D, Thaning L, Bregman R and Andres R 2002 *J. Environ. Radioact.* **60(1–2)** 23–48
[8] Cao J Z, Yeung M R, Wong S K, Ehrhardt J and Yu K N 2000 *J. Environ. Radioact.* **48(3)** 265–277
[9] Homma T, Ishikawa J, Tomita K and Muramatsu K 2000 *Radiological Consequence Assessments of Degraded Core Accident Scenarios Derived from a Generic Level 2 PSA of a BWR: JAERI-Research-2000-060* (Ibaraki: Japan Atomic Energy Agency) chapter 2 pp 2–15
[10] Japan Nuclear Energy Safety Organization 2006 *Methodology of level 2 PSA (BWR): JNES/SAE06-046* (Tokyo: Japan Nuclear Energy Safety Organization)
[11] International Atomic Energy Agency 1994 *Intervention Criteria in a Nuclear or Radiation Emergency* (Vienna: International Atomic Energy Agency)
[12] International Commission on Radiological Protection 2007 *Annals of the ICRP Publication 103: The 2007 Recommendations of the International Commission on Radiological Protection* (Ontario: International Commission on Radiological Protection)

IOP Conf. Series: Journal of Physics: Conf. Series **860** (2017) 012040 doi:10.1088/1742-6596/860/1/012040

Development of remaining wall thickness measurement system for boiler wall tube using gamma scattering technique

K Durongsak, C Yenjai, and S Rassame[1]

Department of Nuclear Engineering, Faculty of Engineering, Chulalongkorn University, Bangkok, 10330 Thailand

[1] Corresponding author e-mail: somboon.ra@chula.ac.th

Abstract. In this study, the General Monte Carlo N-Particle version 5 (MCNP5) simulation of the measuring system for the remaining thickness of the wall tube using gamma scattering technique is performed to investigate the applicability of this technique and to find the optimum geometrical setup for the experimental setup. The numerical results show that the optimal geometry condition to provide the highest ratio between the gamma flux changes per thickness variation, namely, the measurement sensitivity, is the alignment of the source incidence angle of 30 degree and detector scattering angle of 30 degrees. Sequentially, the preliminary experiment of thickness measurement system for the tube using the gamma scattering technique is conducted based on the selected geometrical test setup by the simulation. It is found that the experimental results have generally a good agreement with the calculated results. Conclusively, it is suggested that the gamma scattering technique has a potential method to measure the remaining wall thickness in the boiler wall tube.

1. Introduction

The wall tubes in boiler of the fossil-fired power plant are very essential component which function in heat transferring from the boiler combustion room into the water inside the tubes. The inspection of the remaining thickness of wall tubes is required to perform annually during the maintenance period of the plant. To estimate the remaining thickness of tube, several techniques can be used as the non-destructive testing methods such as the gamma radiography, thermograph, Eddy current, infrared thermograph, and Ultrasonic Testing (UT) technique. Among those techniques, the UT technique is quite more popular in the power plant industry due to several advantages such as its good accuracy and acceptable measuring cost. However, the disadvantages of the UT technique are the long-time measuring and the need of professional skills in the UT.

The applicability of gamma scattering techniques in the tubes or pipes thickness measuring has been investigated by several researchers. For instance, Abdul-Majid and Tayyeb tried to measure the difference thickness of the carbon tubes at 100, 160, and 200 mm thick tubes with 50 mm of the covered insulation by the gamma ray scatterings using the setting the Cs-137 radioisotope source, the radiation at 10 mCi, and 2"×2" NaI(Tl) detector [1]. The test results from his study shows that the tube should be less than 10 mm in thickness in order to be measure the tube thickness accurately.

In addition, Abdul-Majid and Balamesh [2] used the gamma ray scatterings to measure the iron carbon tube 280 mm diameter, thickness with 9 mm and 40 mm asbestos covered with the insulation. They applied gamma ray from 5 mCi Cs-137 radioisotope source with 7 mm collimator. The backscattered radiation was measured by 2"×2" NaI(Tl) detector connected with multichannel

IOP Conf. Series: Journal of Physics: Conf. Series **860** (2017) 012040 doi:10.1088/1742-6596/860/1/012040

analyzer. The test results found that the system could analyze defect at the tubes about 70 mm in diameter. Also, Abdul-Majid and Balamesh [3] studied the measuring system installed into the waterproof pipe for the underwater test. The signal was transferred via 100 m cable line. The backscattered radiation was measured by 2"×2" NaI(Tl) detector connected with multichannel analyzer. The results of his study indicated that the amount of backscattered radiation depended on the tube's thickness.

Moreover, Vo Hoang Nguyen et al. [4] simulated the thickness measurement of pipe-wall by gamma scattering technique using the MCNP4C codes and Geant4. The related experiments were conducted to investigate the correction of the simulation results.

The purpose of this study is preliminary investigating the applicability of gamma scattering technique to the weared thickness of wall tube measurement system. The investigation result will also determine the optimum geometrical setup condition for the system. The MCNP5 code is applied to simulate the system. The sensitivity of the scattered gamma flux change due to the thickness change is calculated in several geometrical set-ups with the relevant angles conditions. The experiments are performed in accordance with the selected test conditions in order to verify and validate the numerical results.

2. The gamma scattering technique

Figure 1 shows the schematic diagram of the thickness measurement using the gamma scattering technique. The radiation source is placed in the collimator at a possible angle (0°-180°) to the radiation detector respected to the plane of workpiece. This technique is applicable to measure the thickness of one-side tube thickness or the tube that the radiation source or the radiation detector cannot put conveniently inside the tube.

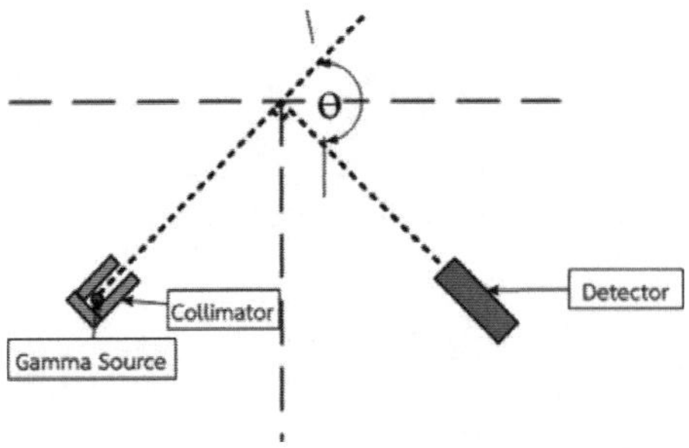

Figure 1. Set up for gamma scattering.

In gamma ray scatterings technique, the relation between the scattered gamma ray, the energy of gamma ray reflected on the workpiece, the mass energy of electron in the break condition, and the reflected angle can be formulated as [5]:

$$h\nu' = \frac{h\nu}{1 + \left[\frac{h\nu}{m_0c^2}\right][1-\cos\theta]} \quad (1)$$

$h\nu'$ is the energy of the scattered gamma ray

$h\nu$ is the energy of gamma ray reflected on the workpiece

m_0c^2 is rest mass energy that is 0.511 MeV

θ is the scattering angle

The equation (1) obviously indicates that the energy of Compton scattered gamma ray depends on the energy of gamma ray interacted to the workpiece and the scattering angle of gamma ray. In order to estimatethe workpiece's thicknesses (T), the equation (2) can be used [4]:

$$T = -\frac{1}{\mu_{eff}} \ln\left(1 - \frac{I}{I_s}\right) \quad (2)$$

I is the intensity of scattered gamma ray

I_s is the constant intensity of scattered gamma ray at saturation thickness of workpiece

μ_{eff} is the linear absorption coefficient of the workpiece

It is noted that the equation (2) is developed based on the workpiece which has a shape in planar. It should be verified further if the equation directly applicable to the tube thickness estimation.

3. Simulation and experiment

3.1. MCNP Simulation

The General Monte Carlo N-Particle code developed by Los Alamos National Laboratory is a general-purpose code that has widely used to simulate the coupled neutron, photon, and electron transport problem [6]. It has a capability to simulate a variety of nuclear particles with flexible geometry and materials used in samples or detectors. In this study, the MCNP version 5 (MCNP5) is used to simulate wall thickness measurement system of boiler tubes using gamma scattering technique.

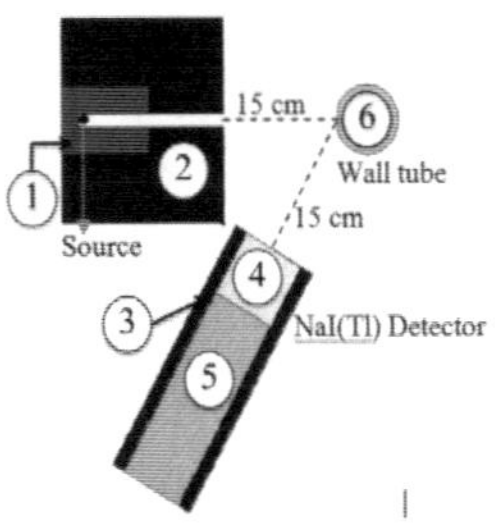

No.	Material	Geometrical	
1	Copper	Diameter Length Collimator dia.	5 cm 7 cm 1 cm
2	Lead	Diameter Length Collimator dia.	15 cm 12 cm 1 cm
3	Lead	Inner diameter Outer diameter Length	5.6 cm 7.6 cm 18.2 cm
4	NaI(Tl)	Diameter Length	5.6 cm 5.2 cm
5	PMT (Aluminium)	Diameter Length	5.6 cm 13 cm
6	Carbon Steel (Type A201)	Inner diameter Outer diameter Length	4.08 cm 5.08 cm 10 cm

Figure 2. Set-up geometry of measurement in the MCNP5 simulation.

Figure 2 displays the geometry model for the thickness measurement system in the MCNP5 simulation. In the MCNP simulation, a gamma source, a shielding cylinder, a NaI(Tl) detector, and a wall tube are key modelled components using the MCNP5 code.

The scintillation crystal of NaI(Tl) detector is a cylinder with diameter of 5.08 cm and length of 5.08 cm. The detector is shielded by the lead cylinder with length, inner diameter, and outer diameter of 20.08 cm, 5.08 cm, and 7.08 cm, respectively.

A Cs-137 radioactive source emitting 661.6 keV of gamma ray energy is used in this simulation. The gamma source is specified as the point source which contain inside the cavity of a cylindrical copper with length of 7 cm and diameter of 5 cm. Gamma ray emitted from the source are collimated by a lead cylindrical collimator with length, inner diameter, and outer diameter of 12 cm, 1 cm, and 15 cm, respectively.

The wall tubes used in the experiment are made of carbon steel with the content of each element as follows: Fe: 99.60%, Mn: 0.93% C: 0.27% Si: 0.10%, K: 0.035%, S: 0.035%, Cr: <0.01%, Mo: < 0.01%, and Ni: <0.01%. Five different thicknesses of wall tubes were prepared with remaining thicknesses of 1, 2, 3, 4, and 5 mm. All tube samples have inner diameter of tube is 4.08 mm, outer diameter is 5.08 mm, and length is 100 mm.

The distance from the surface of the source collimator to the surface of the wall tube was maintained at 15 cm and the distance from the surface of the NaI(Tl) detector to the surface of the wall tube was set at 15 cm.

In the MCNP5 simulation, the calculated conditions with the varied incidence and scattering angles are showed in the table 1. For each calculated condition, the simulation is performed with the wall tube thickness of 1, 2, 3, 4 ,and 5 mm. Energy distribution of scattering gamma created in a detector is calculated using the F8 tallies in the MCNP5. The intensity at area of the scattering peak will be subtracted with the intensity at area of the Compton scattering from the detector.

Based on the MCNP5 calculated results, the total area of single scattering peak at each thickness values is estimated and the relationship between the intensity of single scattering peak and the thickness of wall tube are gained. The sensitivity of the thickness measurement system is estimated by the slope of graph showing the relation of the intensity of single scattering and thickness of wall tube.

Table 1. Calculated conditions in the MCNP5 simulation.

Calculated condition No.	Incidence angle (degree)	Scattering angle (degree)	Calculated condition No.	Incidence angle (degree)	Scattering angle (degree)
1	0	60	15	30	50
2	0	70	16	30	60
3	0	80	17	40	20
4	0	90	18	40	30
5	10	50	19	40	40
6	10	60	20	40	50
7	10	70	21	50	10
8	10	80	22	50	20
9	20	40	23	50	30
10	20	50	24	50	40
11	20	60	25	60	0
12	20	70	26	60	10
13	30	30	27	60	20
14	30	40	28	60	30

3.2. Experiments

3.2.1. Wall tube. The wall tubes used in the experiment are made of carbon steel. The composition of this material is identical to an actual wall tube used in boiler of a coal-fired power plant. Five samples of wall tubes with the different remaining thicknesses of 1, 2, 3, 4, and 5 mm were prepared in order to find the relation between the intensity of single scattering and remaining thickness of wall tubes. Figure 3 shows the photographs of wall tube samples used in the experiment.

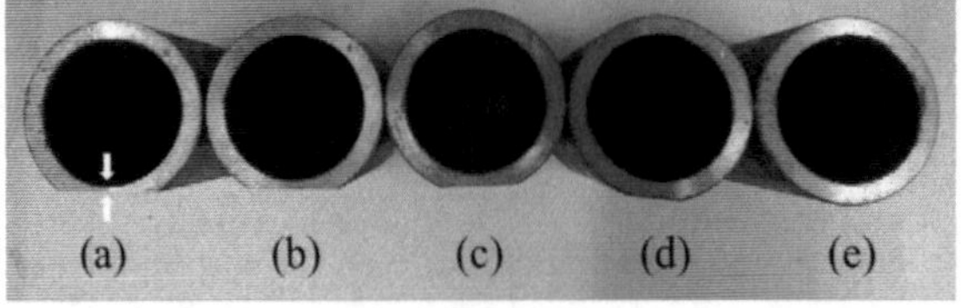

Figure 3. Remaining thickness of wall tube (a) 1 mm (b) 2 mm (c) 3 mm (d) 4 mm (e) 5 mm.

3.2.2. Experiment set-up and performing. The dimension of detector, detector shielding, radiation source and source collimator and shielding including their materials used and the equipment arrangement are identical to what are specified in the MCNP5 simulation. The experiments used the 5.08 cm × 5.08 cm NaI(Tl) detector that consists of the photo multiplier at the end and shielded by lead cylinder in one side. Another ended side of detector was connected to the tube base that serves as signal amplifier. The measured signal was transferred to the computer and display as a gamma spectrum. The experiments used 0.1 mCi of Cs-137 radiation source which contained in a lead

IOP Conf. Series: Journal of Physics: Conf. Series **860** (2017) 012040 doi:10.1088/1742-6596/860/1/012040

shielding and collimator. Experimental set-up is illustrated in the figure 4. To verity the calculated results of the MCNP5 calculation, some experiments at the selected condition are three of the best performed. The intensity of the single scattering peak at the each thickness of wall tube is estimated based on the measured results in each condition.

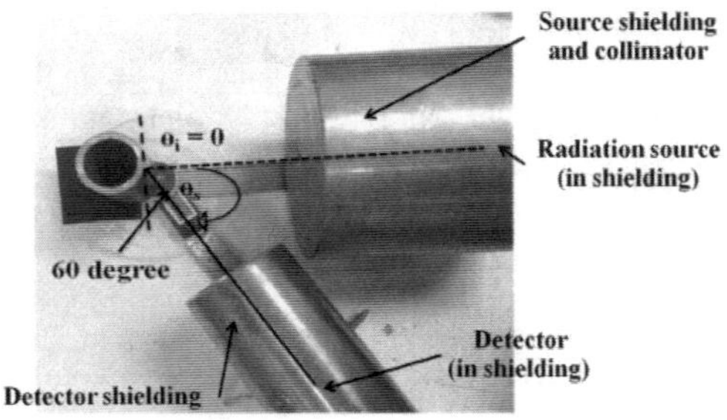

Figure 4. Experimental set-up for remaining thickness measurement of a wall tube.

4. Results and Discussion

4.1. Analysis of simulation's results

Figure 5 show the geometries modelling of experiment using in MCNP5 which demonstrates the gamma ray scattering in the simulation. It is clearly shown in the MCNP5 displaying program that some gamma rays enter directly to the detector without the scattering to the tube consequently, the Compton scattering occurred in detector without scattering play some roles to the measurement. Therefore, the intensity at area of the gamma scattering peak in the detector is necessary to subtract with the intensity at area of the Compton scattering from the gamma ray directly to the detector.

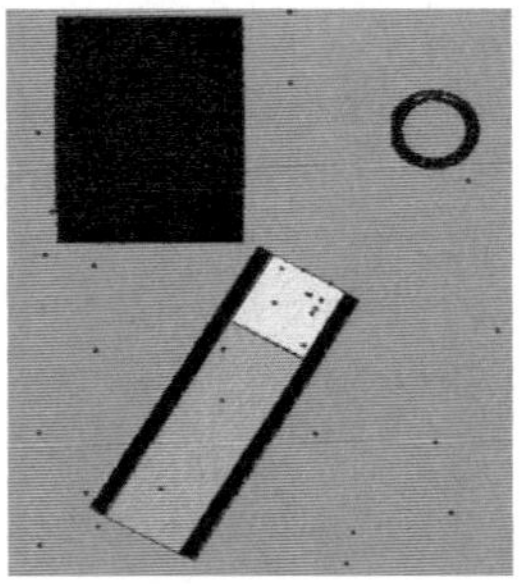

Figure 5. Example of particles displaying by MCNP5.

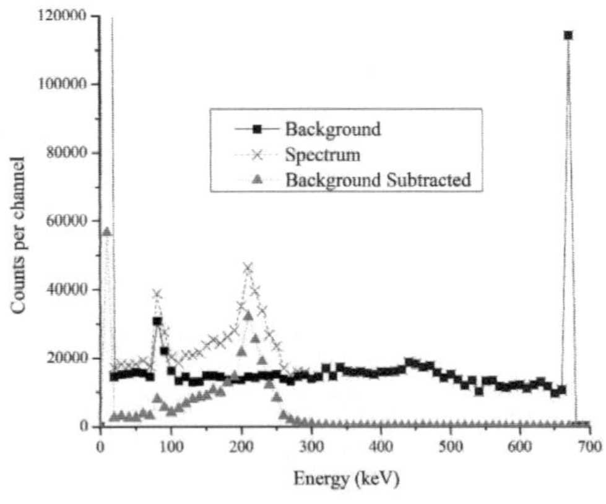

Figure 6. An example of MCNP5 simulation scattering and background spectra with a 3 mm remaining thickness of wall tube at the calculated condition no. 13.

Table 2. Sensitivity estimated by the MCNP5 calculated results.

Calculated condition No.	Incidence angle (degree)	Scattering angle (degree)	Sensitivity (counts/mm)	Calculated condition No.	Incidence angle (degree)	Scattering angle (degree)	Sensitivity (counts/mm)
1	0	60	16,752	15	30	50	16,024
2	0	70	15,024	16	30	60	15,490
3	0	80	15,242	17	40	20	16,218
4	0	90	15,218	18	40	30	15,488
5	10	50	16,444	19	40	40	14,694
6	10	60	15,024	20	40	50	13,988
7	10	70	15,126	21	50	10	16,342
8	10	80	14,694	22	50	20	15,498
9	20	40	16,544	23	50	30	15,286
10	20	50	15,798	24	50	40	14,846
11	20	60	15,548	25	60	0	17,298
12	20	70	15,940	26	60	10	16,544
13	30	30	19,434	27	60	20	15,898
14	30	40	16,668	28	60	30	16,004

The sensitivity (count/mm) is estimated by the slope of graph showing the relation of the intensity of single scattering and thickness of wall tube. Based on the calculated results shown in the table 2, it is concluded that the calculated condition no. 1, 13, and 25 are the top three conditions that provide the highest value of the sensitivity. These three conditions of incidence angles and scattering angles according to the calculated condition no. 1, 13, and 25 are selected to perform the tests to validate the calculation results by the MCNP5 simulation.

4.2. Comparison between experimental and calculated results

Table 3 shows the comparison between the intensity of the single scattering peak obtained from the experiment and estimated by the MCNP5 simulation. It can be seen that the calculated and experimental results are good agreement. Additionally, it is seen that the intensity of the single scattering peak is higher with thickness increasing in both of the experimental and calculated results.

A scattering spectrum of 662 keV gamma photon beam scattered on a 3 mm remaining thickness of wall tube obtained by experiment and simulation are shown in the figure 7. The energy position of single scattering peak in the experimental and simulated results is closed to approximately 210 keV which is the E′ the scattered gamma ray calculated by the equation (1).

Table 3. Area of single scattering peak obtained from experiment and simulation of condition no. 1, 13 and 25.

Condition No.	Incidence angle (degree)	Scattering Angle (degree)	Thickness of wall tube (mm)	Area of single scattering peak (counts)		
				Experiment	MCNP5	% Relative
			1	145,789	132,180	90.67 %
			2	165,892	153,240	92.37 %
1	0	60	3	180,546	170,100	94.21 %
			4	193,648	182,400	94.19 %
			5	202,445	191,280	94.48 %
			1	178,860	133,740	74.77 %
			2	196,542	154,680	78.70 %
13	30	30	3	215,350	174,240	80.91 %
			4	234,791	192,000	81.77 %
			5	245,132	202,680	82.68 %
			1	168,945	162,120	95.96 %
			2	184,243	182,160	98.87 %
25	60	0	3	201,544	198,720	98.60 %
			4	217,839	214,260	98.36 %
			5	220,144	219,960	99.92 %

IOP Conf. Series: Journal of Physics: Conf. Series **860** (2017) 012040 doi:10.1088/1742-6596/860/1/012040

Table 4 shows the sensitivity obtained by the experimental results and the MCNP5 calculation at the condition no. 1, 13, and 25. Among those three conditions of the related angles, the maximum sensitivity is found to be 19,434 and 18,660 counts/mm in the simulation and experiment, respectively, at the condition no. 13 of 30 degree of incidence angle and 30 degree of scattering angle.

Figure 8 shows an example of the intensity of single scattered photon and the remaining thickness of wall tube in the condition no. 13. It can be seen that the similar trend on the changing of intensity of single scattering peak area with the remaining wall tube thickness is found. It is noted again that the sensitivity of measurement is estimated by the slope of graph. For the condition no. 13, the sensitivity of 19,434 and 18,660 counts/mm are obtained in the MCNP5 calculation and experiment, respectively.

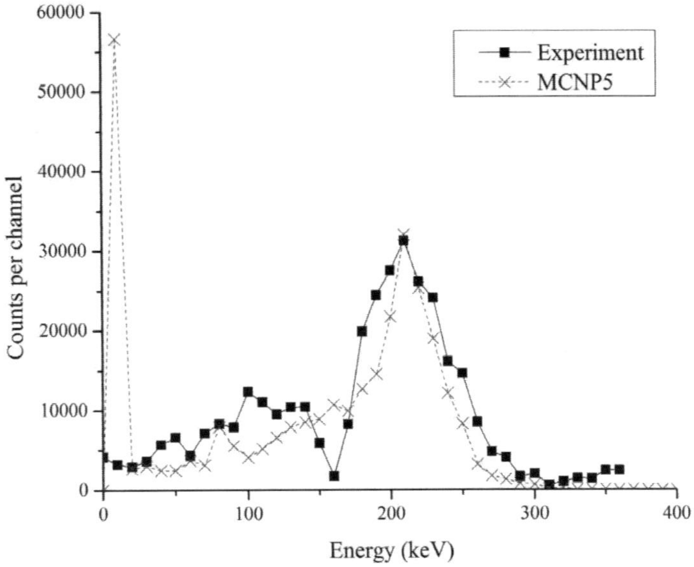

Figure 7. An example of scattering spectrum of 662 keV gamma photon beam scattered on a 3 mm remaining thickness of wall tube by experiment and simulation in the condition no. 13.

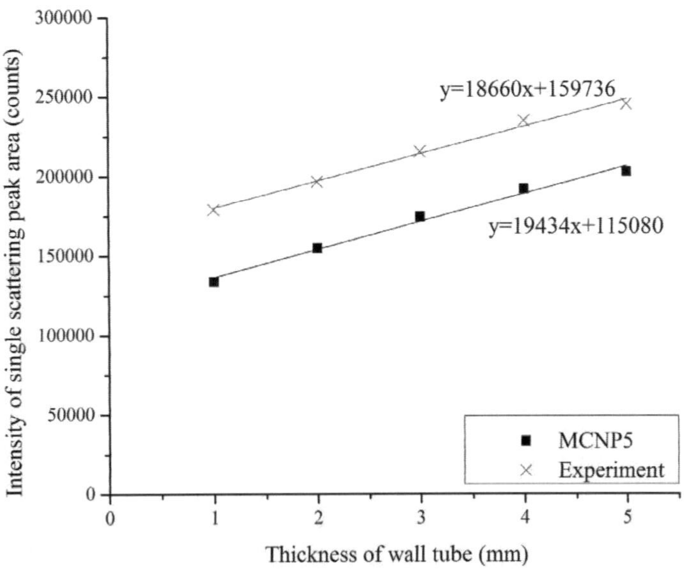

Figure 8. Intensity of single scattered photon and remaining thickness of wall tube in the condition no. 13.

Table 4. An example data of sensitivity obtained by experiments and simulations.

Condition No.	Incidence angle (degree)	Scattering angle (degree)	Sensitivity (counts/mm)		
			Experiment	MCNP5	% Relative
1	0	60	15,823	16,752	94.45 %
13	30	30	18,660	19,434	96.02 %
25	60	0	16,398	17,298	94.80 %

5. Conclusion

The MCNP5 code is applied to simulate the measuring system for the remaining thickness of the wall tube using gamma scattering technique. The alignment of radiation source and detector setup as well as the tube remaining thickness is varied as input parameters in the MCNP calculation. Three conditions of the measure equipment alignment providing the highest values of measurement sensitivity are selected to perform the experiment in order to confirm the validity of numerical results. It is found that the relation between the intensity of the single scattering peak and the tube remaining thickness obtained by the experiment and MCNP calculation has well correlated. The alignment of the source incidence angle of 30 degree and detector scattering angle of 30 degrees are considered as the best geometric set-up to provide the highest measurement sensitivity at approximately 19,434 and 18,660 counts/mm. It is suggested that the measuring system using gamma scattering technique has a potential to use as the alternative technique in wall tube remaining thickness measurement but the detailed analysis may require for further improvement the accuracy of the measurement system.

References

[1] Samir A M and Zuhair T 2005 Use of Gamma ray back scattering method for inspection of corrosion under insulation. *3rd Middle East Nondestructive Testing Conference and ExhibitionMore references*. 173

[2] Samir A M and Ahmed B 2012 Imaging corrosion under insulation by gamma ray backscattering method. *18th World Conference on Nondestructive Testing*. **17**

[3] Samir A M and Ahmed B 2012 Underwater pipe wall thickness measurements by gamma backscattering. *6th Middle East Nondestructive Testing Conference & Exhibition*. **18**

[4] Vo H M , Hoang D T, Le B T, Tran T T, Hua T L, Le D M Q, Huynh D H, Tran N T N, and Chau V T A 2016 Semi-empirical method for measuring thickness of pipe-wall using gamma scattering technique. *Journal of Radioanalytical and Nuclear Chemistry*. **308** 1011–16

[5] Knoll G F 2000 Radiation detection and measurement (John Wiley & Sons, Inc)

[6] X-5 Monte Carlo Team 2003 *MCNP A General Monte Carlo N-Particle Transport Code, Version 5* Vol 2 (User's Guide. Los Alamos National Laboratory Report LA-CP- 03-0245)

Gamma-radiation-induced corrosion of aluminum alloy: low dose effect

K Kanjana[1, 3], P Ampornrat[2] and J Channuie[1]

[1] Thailand Institute of Nuclear Technology (Public Organization), Ongkharak, Nakhon Nayok 26120, Thailand
[2] Office of Atoms for Peace, Chatuchak, Bangkok 10900, Thailand

[3] Corresponding author e-mail: kotchaphank@tint.or.th

Abstract. Gamma-radiation-induced corrosion of aluminium alloy 6061 (AA6061) immersed in demineralized water was studied at radiation dose up to 206 kGy using a Co-60 gamma radiation source. The surface morphology and chemical composition of the oxide produced on the post-irradiated samples were investigated using SEM-EDS. The electrochemical corrosion potentials (E_{corr}) of the post-irradiated samples were measured. The corrosion behavior of AA6061 appeared to be dose dependent under the experimental conditions. A dramatic change in surface morphology was observed in the samples exposed to gamma radiation at 206 kGy. At this radiation dose the aluminium oxide scale developed can be clearly seen. The results from electrochemical corrosion tests have shown that the corrosion potentials (E_{corr}) can be undoubtedly decreased by gamma irradiation, giving corrosion rate of 7×10^{-4} mm/yr.

1. Introduction

Aluminium Alloys (AAs) have been the material of choice for application in water-cooled-and-moderated Research Reactor (RR) systems due to their unique mechanical properties, high corrosion resistance in water and low neutron cross-section. Although AAs are known to develop highly stable aluminium oxides under RR conditions (pH 5.6, T<100°C and atmospheric pressure), corrosion in this group of materials has commonly been observed by the reactor personnel. Corrosion of material in RRs is believed to be mostly attributed to ionizing radiations[1-5]. The information regarding effects of ionizing radiations on nuclear materials is highly required for assessment of operation and maintenance as well as safety margins of nuclear reactors. Ionizing radiation can interact with material via both direct and indirect mechanisms (effects). Direct interactions of ionizing radiation with material can result in lattice vacancies, self-interstitial atoms and transmutation products (radiation damage microstructure), whereas indirect effect which stems from the chemical reactions between the material surface and the radiolytic primary species ($(e^-)_{aq}$, $^\bullet OH$, $H^\bullet$, H_2O_2, O_2, and $^\bullet O_2^-$) can dictate electrochemistry of the solution and may ultimately lead to corrosion of the material[6-12]. As of now, the effects of ionizing radiation on nuclear materials corrosion are still not yet fully understood. Only limited information regarding radiation-induced/accelerated corrosion of AAs is available. Many key aspects of this system have not been investigated, including corrosion kinetics, corrosion products and surface morphology of the irradiated materials. An earlier study on the effects of neutron and gamma radiation on corrosion of AA claimed that neutron and gamma radiation had no different effects on corrosion of these materials. It was also found in this study that pitting corrosion of the AA samples appeared to be in higher density at a greater distance from the radiation source[13]. This conclusion is completely inconsistent with the findings from the recent works which indicated that corrosion of metal alloys such as stainless steel can be considerably induced and accelerated by radiation. The studies also demonstrated that exposing to high dose radiation can initiate stress corrosion cracking and intergranular corrosion in AA[14-16].

To develop accurate understanding of the effects of ionizing radiation - gamma radiation on corrosion of AA, an in-depth study of this system under the RR conditions (low dose effects) has been carried out. The results obtained will not only benefit the reactor personnel in evaluating corrosion level of the AA components, but also provide fundamental understanding of surface chemistry of this chemical system.

2. Experimental Section

An AA6061 sheet (Southwest Aluminium, China) was cut into square-shaped samples with dimensions of 1 inch × 1 inch × 2 mm. The samples were later wet polished with abrasive paper up to 1200 grit using a Buehler MetaServ 250 Twin polisher. To remove polishing residues, the samples were subsequently ultrasonic cleaned in acetone, methanol and ethanol, respectively. Prior to experiment, the samples were weighed and their chemical composition was analyzed using an X-ray Fluorescence (XRF) spectrometer (Bruker, S8 TIGER).

A cobalt-60 irradiator (30366 Ci) at the Thailand Institute of Nuclear Technology was used as radiation source throughout the experiment. During gamma irradiation, the samples were completely immersed in demineralized water in a glass container (using suspension glass hooks). Irradiation time and dose were varied with the maximum dose rate of 4.30 kGy/hr. Dosimetry was performed using Harwell Amber Perspex dosimeter. The sample temperature during irradiation was around 32°C.

The post-irradiated samples were weighed, kept in a desiccator and lastly characterized using SEM-EDS (Hitachi SU8020). An Autolab potentiostat, model PGSTAT302N, with a saturated calomel electrode (reference electrode) and a platinum electrode (counter electrode) was employed to determine the corrosion rates of the irradiated surface.

3. Results and Discussion

Dose effect was investigated by irradiating the AA6061 samples for 48 hours at different dose rates (1.08, 2.14 and 4.30 kGy/hr). The SEM micrographs of the unirradiated and post-irradiated samples can be seen in Figure1.

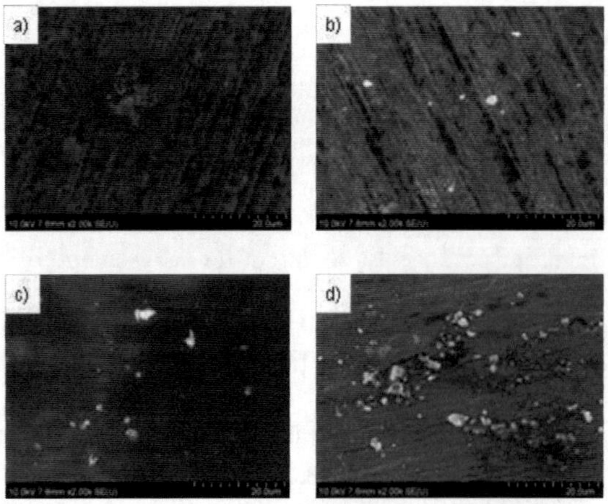

Figure 1. SEM micrographs of AA6061 samples after exposing to gamma radiation for 48 hours at different dose rates: a) unirradiated (immersed in demineralized water for 48 hours) b) 1.08 kGy/hr c) 2.14 kGy/hr and d) 4.30 kGy/hr.

It can be clearly seen in Figure 1a. that uneven white spots were developed on the surface of the blank sample which was fully immersed in demineralized water for 48 hours without exposing to radiation. Presumably, the product is alumina $Al(OH)_3$, a water insoluble product commonly found when Al is oxidized in water[17-18]. Upon irradiation, the growth of the white product, (presumably $Al(OH)_3$) was obviously noticed as illustrated in Figure 1b, 1c and 1d. Increasing the dose rate increases the density of this corrosion product. As the dose rate changes from 2.15 to 4.30 kGy/hr, the density rises considerably. There was no correlation between the product size and dose rate under the experimental conditions.

The irradiation time experiment was performed with the dose rate of 4.30 kGy/hr. The effects of radiation on the morphology of AA6061 from this experiment are shown in Figure 2. It is evident that irradiation time has great influence on the density of the corrosion product. There was also no correlation observed between the product size and time. The elemental composition analysis of the irradiated surface by Energy Dispersive X-ray Spectrometer (EDS) revealed that the white product was composed of aluminium, oxygen and carbon (Figure 3.), suggesting the formation of aluminium oxide as expected. Though it is well-known that at room temperature the corrosion product generated from oxidation of Al in water is in the form of amorphous alumina[17,19], there is no information regarding aluminium oxide structure in the system involving radiation reported. This challenging issue is left to be explored in our future work. The carbon component detected is likely to come from contamination during the sample preparation process. In addition, SEM images of the sample cross-section exposed to gamma radiation indicated that pit formation is undetectable at this range of radiation dose. The differences in weights of pre- and post-irradiated samples are not statistically significant, thus indicating no effects on weight loss at the radiation doses studied (see Table 1).

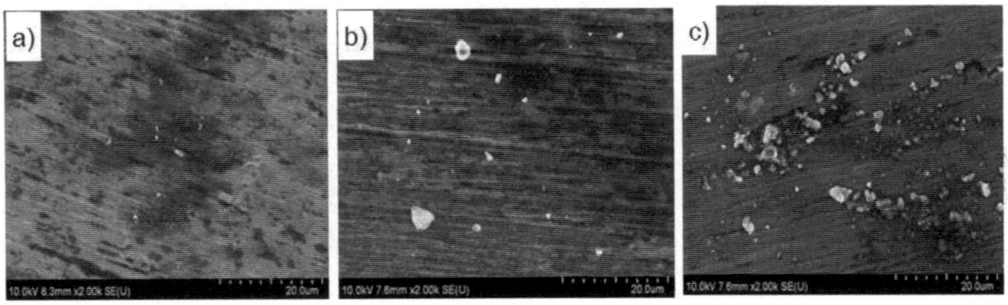

Figure 2. SEM micrographs of AA6061 samples after exposing to gamma radiation at different time (dose rate 4.30 kGy/hr): a) 8 hours b) 24 hours and c) 48 hours.

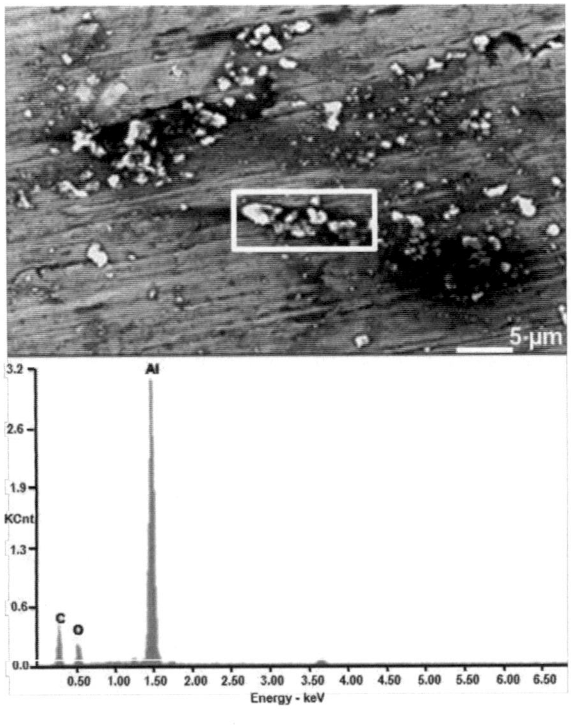

Figure. 3. a) SEM micrograph of AA6061 samples after exposing to gamma radiation for 48 hours at dose rate 4.30 kGy/hr and b) EDS spectrum of the corrosion product indicated in Figure 3a.

Table 1. Weight change of AA6061 samples after immersion/irradiation.

Immersion/Irradiation Time (hr)	Dose Rate (kGy/hr)	Weight (g)		Weight change (g)
		Pre-	Post-	
		3.0416	3.0413	0.0003
8	0	2.9925	2.992	0.0005
		3.0956	3.0949	0.0007
		2.8126	2.8125	0.0001
8	4.30	2.7192	2.7189	0.0003
		3.0332	3.0325	0.0007
		3.0608	3.0618	0.001
24	0	3.1784	3.1782	0.0002
		3.0199	3.0197	0.0002
		3.0093	3.0087	0.0006
24	4.30	3.0622	3.0623	0.0001
		2.6823	2.6825	0.0002
		3.2207	3.2203	0.0004
48	0	3.1627	3.1619	0.0008
		3.0376	3.0371	0.0005
		2.8443	2.8437	0.0006
48	1.08	2.7154	2.7146	0.0008
		2.705	2.7041	0.0009
		2.3393	2.3381	0.0012
48	2.14	2.8539	2.853	0.0009
		2.9875	2.9873	0.0002
		2.8523	2.8516	0.0007
48	4.30	3.1071	3.1067	0.0004
		3.0416	3.0413	0.0003

The results from the electrochemical measurements of the blank (unirradiated), 16-hour-irradiated and 24-hour-irradiated samples (4.30kGy/hr) in demineralized water are presented in Figure 4. A similar shape of potentiodynamic polarization curves was observed for all 3 samples. The values of corrosion potential (E_{corr}) and corrosion current (I_{corr}) density obtained, presented in (E_{corr}, I_{corr}), are (-0.181 V, 7.54×10^{-8} A/cm^2), (-0.284 V, 7.18×10^{-8} A/cm^2) and (-0.494 V, 3.68×10^{-10} A/cm^2) for the blank, 16-hour-irradiated and 24-hour-irradiated samples, respectively. Based on this study, it is clear that gamma irradiation can shift E_{corr} of the AA6061 sample to more negative values, thus decreasing corrosion resistance. The I_{corr} density values of the irradiated samples were found to be not systematic. The evidence from E_{corr} measurements confirms that gamma exposure can increase corrosion susceptibility of AAs. The corrosion rate of the irradiated samples given by this electrochemical study is *ca.* 7×10^{-4} mm/yr. The calculation of this corrosion rate was based on Faraday's Law shown in Equation (1)[20]:

$$CR = K_1 \frac{I_{corr}}{\rho} EW \quad (1)$$

Where:

CR is corrosion rate in mm/yr,

K_1 is a constant equal to 3.27×10^{-3} mm·g/μA·cm·yr,

ρ is density in g/cm^3 of the material used in corrosion testing (ρ_{Al6061}= 2.70 g/cm^3), and

EW is equivalent weight, mass of the material in g that will be oxidized by the passage of one Faraday of electric charge, (EW_{Al6061} = 9.01 g).

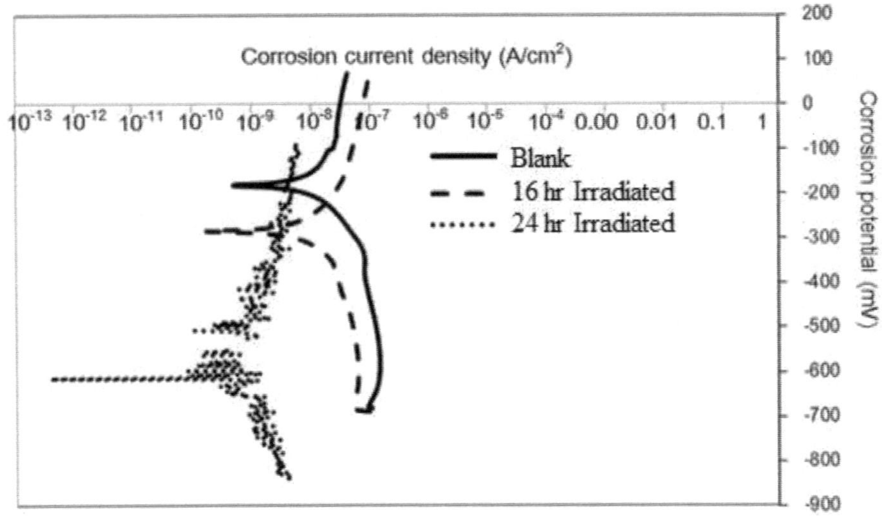

Figure 4. Polarization curves for the AA6061 samples obtained in demineralized water.

4. Conclusion

Gamma-radiation-induced corrosion of AAs can be revealed by SEM-EDS in conjunction with an electrochemical technique. Pitting corrosion was not detected under the experimental conditions. It is strongly evident that gamma irradiation can significantly accelerate aluminium oxide formation on AA6061. The density of aluminium oxide was found to increase with irradiation time and dose. The correlation between irradiation dose and E_{corr} was observed. However, I_{corr} is dose insensitive. The corrosion rate extracted from the electrochemical data is approximately 7×10^{-4} mm/yr.

Acknowledgements

The research described herein was supported by the Thailand Institute of Nuclear Technology (TINT). The authors also wish to acknowledge Mr. Piya Khamsuk and Dr. Namuratha Sathirachinda at the National Metal and Materials Technology Center (MTEC) for their kind assistance with electrochemical measurements.

References

[1] Farrell K 2012 *Performance of aluminum in research reactors. Comprehensive nuclear materials*. Vol. 5.07. Elsevier, Oxford

[2] Farrell K 1981 Microstructure and tensile properties of heavily irradiated 5052-0 aluminum alloy, *J. Nucl. Mater.* **97** 33-43

[3] Weeks J R, Czajkowski C J and Tichler P R 1990 *Effects of high thermal and high fast fluences on the mechanical properties of type 6061 aluminum in the HFBR.* 441ASTM STP 1046, Pennsylvania

[4] Munitz A Shtechman A, Cotler C, Talianker M and Dahan S 1998 Mechanical properties and microstructure of neutron irradiated cold worked A1-6063 alloy, *J. Nucl. Mater.* **252** 79-88

[5] Yahr G T 1997 Prevention of nonductile fracture in 6061-T6 aluminum nuclear pressure vessel, *J. Press. Vessel Technol.* **119** 150-56

[6] Spinks J W T and Woods R J 1990 *An introduction to radiation chemistry (3rd edition)*. John-Wiley and Sons, Toronto

[7] Mozumder A 1999 *Fundamentals of Radiation Chemistry*. Academic Press, New York

[8] Elliot A J and Bartels D M 2009 The reaction set, rate constants and g-values for the simulation

of the radiolysis of light water over the range 20° to 350°C based on information available in 2008, *AECL Report 153-1 271 60-450-001*

[9] Spotheim-Maurizot M, Mostafavi M, Douki T and Belloni J 2008 *Radiation chemistry: from basics to applications in material and life sciences*. EDP Sciences, Les Ulis Cedex A

[10] Waldemar A 2016 *Radiation effects in materials*. Intech, Editors, Monteiro

[11] Scott P A 1994 A review of irradiation assisted stress corrosion cracking, *J. Nucl. Mater.* **211** 101-122

[12] Cuba V, Mucka V and Pospisil M 2012 *Advances in nuclear fuel: radiation induced corrosion of nuclear fuel and materials*. Intech, Vienna

[13] Kumar R, Baheti G L and Chacharkar M P 1989 Corrosion behaviour of aluminium in the presence of neutrons and gamma radiations, *J. Radio. Nucl. Chem. Articles*. **132** 3-9

[14] Sindelar R L, Lam P S, Louthan Jr M R and Lyer N C 1999 Corrosion of metals and alloys in high radiation fields, *Mater. Charact.* **43** 147-157

[15] Maksimkin O P, Yarovchuk AV, Turubarova L G, Aulova D S, Karbysheva S A and Rusakova A V 2011 Influence of neutron irradiation on intergranular corrosion and corrosion cracking of low-alloyed aluminium alloy SAV-1, *Probl. Atom. Sci. Tech.* 108-115

[16] Was G S, Ashida Y and Andresen P L 2011 Irradiation-assisted stress corrosion cracking, *Corros. Rev.* **29** 7-49

[17] Vargel C 2004 Corrosion of aluminium, 1st edition. Elsevier, New York

[18] Fouquet M D, Razvi J and Whittemore LW 2003 TRIGA research reactors: a pathway to the peaceful applications of nuclear energy, *Nuclear News* 46-56

[19] IAEA 2003 *Technical reports series 418, corrosion of research reactor aluminium clad spent fuel in water*. IAEA Library Cataloguing in Publication Data, Vienna

[20] ASTM 1999 Annual Book of Standards *G 102: Standard practice for calculation of corrosion rates and related information from electrochemical measurements* (West Conshohocken: ASTM International) p 1

IOP Conf. Series: Journal of Physics: Conf. Series **860** (2017) 012042 doi:10.1088/1742-6596/860/1/012042

Preliminary study of degradation from neutron effects of core-structural materials of Thai Research Reactor TRR-1/M1

P Ampornrat[1], P Boonsuwan, S Sangkaew and T Angwongtrakool

Bureau of Nuclear Safety Regulation, Office of Atoms for Peace
16 Vibhavadi-Rangsit, Chatuchak, Bangkok 10900, Thailand

[1] Corresponding author e-mail: pantipam@gmail.com

Abstract. Thai research reactor went first critical in 1962. The reactor was converted in 1977 from an MTR-type with high-enriched uranium fuel to a TRIGA-MARK III type using low-enriched uranium fuel, called TRR-1/M1. Since the TRR-1/M1 has been operated for almost 40 years, degradation of reactor structural materials is expected. In this preliminary study, the potential degradation from neutron effects of core-structural materials, e.g., fuel clad (SS304) and core components (Al6061) were studied. Assessment included calculation of neutron energy, flux and fluence in the reactor core to evaluate displacement rate (dpa) and irradiation effects on the material properties. Results showed maximum displacement rates on SS304 was 5.24×10^{-8} per $cm^3\cdot sec$ and on Al6061 was 1.14×10^{-8} per $cm^3\cdot sec$. The corresponding maximum displacement levels were ~17 dpa for SS304, and ~4 dpa for Al6061. At these levels of displacement, it is possible for the materials to result in tensile strength increasing and ductility reduction. Further inspection on the core-structural materials needs to be conducted to validate the assessment results from this study.

1. Introduction

Thai research reactor has been in operation since 1962. The reactor was converted from a Material Testing Reactor (MTR) – type utilizing high-enriched uranium fuel to a TRIGA-MARK III type using low-enriched uranium fuel in 1977, called TRR-1/M1. Until now (2016) the TRR-1/M1 has been operated for approximately 39 years. It is possible for its structural systems and components (SSC) to undergo degradation or aging. Especially in-core structural materials (Figure 1) being exposed to cooling water and high radiation simultaneously, these materials would be damaged by radiation as well as corrosion. The International Atomic Energy Agency (IAEA) safety requirements emphasize on Safety of Research Reactor that the aging effects shall be taken into account for all operational states, including periods of maintenance and shutdown [1]. Therefore Office of Atoms for Peace (OAP), as the regulatory authority, has determined degradation of core structural materials of TRR-1/M1 in order to ensure safe operation of the research reactor and determine the SSC that are vulnerable to the aging mechanisms.

Degradation of materials used in nuclear reactors caused by several parameters, namely, temperature, pressure, radiation, mechanical stress and water chemistry. Most of available information arose from nuclear power reactors. Only little information of aging in research reactors is available because the reactors exhibit less degradation evidence than that of the power reactors. The degradation mechanisms usually found in the nuclear reactors are; changes in physical or chemical properties (e.g., swelling, chemical decomposition, phase changed, strength or ductility), irradiation and thermal embrittlement, creep, fatigue, and corrosion (e.g., galvanic corrosion, pitting corrosion, stress corrosion cracking (SCC)).

Aging management of the TRR-1/M1 should be conducted to ensure the safety after long time operation. First aging investigation program of the TRR-1/M1 was conducted with concrete pool structure by the OAP in cooperation with Kasetsart University in 2009. Results showed that there were no leakage or seep from the pool wall. The reactor pool can be continued to use for 25 – 30 years with fluence limit of 10^{17} neutrons per m^2 at the pool wall. The preliminary study reported in this paper focuses on the evaluation of degradation of in-core structural components that have been exposed to neutron radiation. Such in-core structural components consist of core shroud, control rod, fuel elements and instrumented fuel elements. Due to hazardous work conditions caused by high radiation exposure on specimen from the TRR-1/M1, the study on corrosion effects has been planned at a later date. Expected outcomes of the study are: i) to be able to evaluate neutron damage rates and effects on mechanical properties of in-core components after 39 years of service, and ii) to identify the most vulnerable components to such mechanical degradations, which may need immediate attention. The results will be used for further inspection and maintenance to ensure safe operation of the reactor.

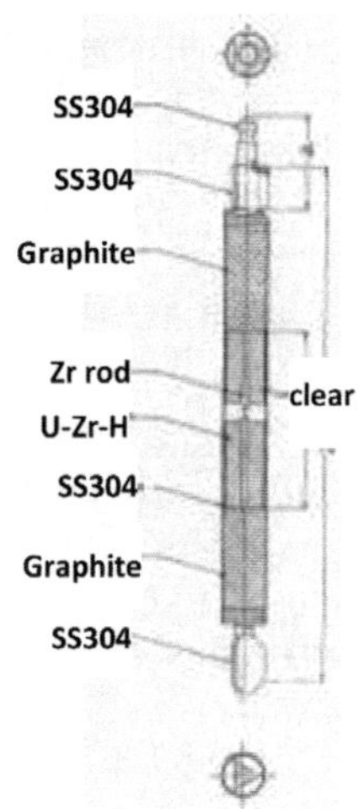

(a)

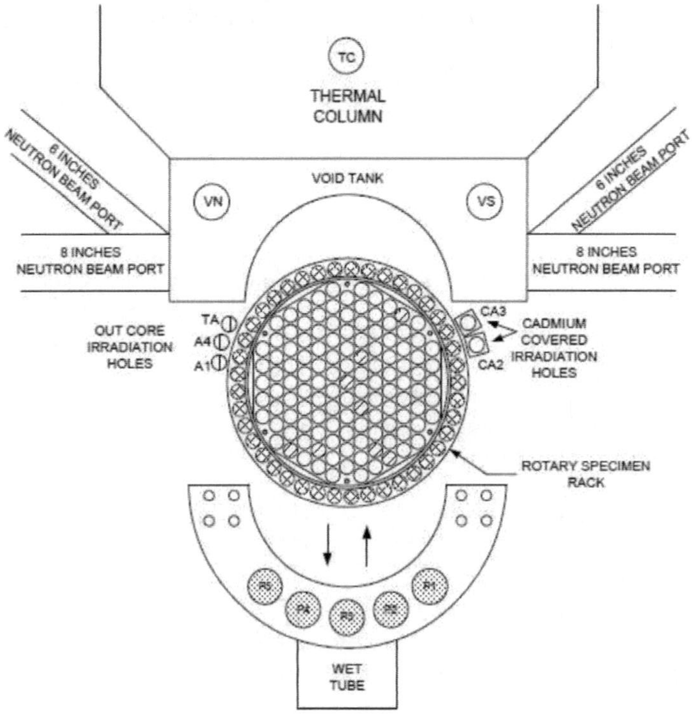

(b)

Figure 1. (a) Fuel element [4], (b) Top view of reactor core and out-core irradiation facilities [4].

2. Methodology

Methodology of aging management program (AMP) followed an IAEA safety standard series no. SSG-10 "*Aging management for research reactor*" [2]. Elements of the AMP included; screening of SSCs, identification and understanding of aging degradation, minimization of expected aging degradation, detection, monitoring and trending of aging degradation, and ongoing improvement of the AMP and accurate record keeping.

This work involved first three parts of the AMP listed above. First, the SSCs of the reactor core were classified in terms of materials, importance to safety, ease of replacement, possible aging mechanism (according to their service conditions), and effects of the mechanisms. Identification and understanding of aging degradation were conducted through determination of service conditions and compare with literature data of the structural materials. Major degradation mechanisms of radiation effects were studied in this work.

To evaluate the radiation effects, neutron fluxes of three energy ranges; thermal (0-0.025 eV), epithermal (0.025-0.42 eV) and fast neutrons (1-15 MeV) were evaluated using software tool Monte Carlo N-Particle eXtended Transport Code (MCNPX) developed by Los Alamos National Laboratory. Model was developed with core number 1 arrangement consisting of 97 fuel elements, four fuel follower control rods, a transient control rod, a central thimble irradiation tube, and three neutron detector tubes. Fuel elements were modeled as drawing in Figure 1(a). The control rods were modeled with the sections of boron carbide, fuel follower with zirconium rod through the center of active fuel section and void region. Two types of fuel contained 8.5%wt and 20%wt uranium with hydrogen to zirconium ratio of 1.6. Erbium content of 0.5%wt was added in the fuel with 20%wt uranium. Enrichment of U-235 was 20%wt for both types. More information of modelling were explained in a work of S. Sangkaew [3].

3. Classification of core-internal structural components

The TRR-1/M1 is a pool type TRIGA Mark III research reactor. The pool contains reactor core, irradiation facilities, coolant water, and used fuel storage racks. Geometry of the TRR-1/M1 reactor core and irradiation facilities inside the pool is shown in Figure 1(b). The reactor core is contained in core shroud assembly. Inside the assembly, the core consists of fuel elements, instrumented fuel elements, control rods, neutron detectors, neutron source, grid plates and safety plate.

List of the in-core structure components, material, important to safety, ease of replacement, potential aging mechanism and effects from the aging mechanisms is presented in Table 1. Most of components made of stainless steel SS304 and aluminum alloy Al6061. Clad of fuel elements and control rods is made of SS304 due to its good properties in corrosion resistance, high strength and low susceptible to radiation swelling. Other core structure components are made of Al6061, which has fair corrosion resistance, good fracture properties and ability to withstand irradiation damage. The importance to safety was classified based on their involvement in the reactor protection system. Possible aging mechanisms were listed following guidance in IAEA SSG-10 [2]. It is found that most probable aging mechanisms are corrosion and radiation damages. The neutron radiation damage was reported in this study. The corrosion effects on structural materials will be studied at a later date up on specimen availability.

Table 1. In-core structural components of TRR-1/M1 with properties and possible aging mechanism.

SSC	Materials	Important to safety	Ease of replacement	Possible aging mechanisms	Effects
Core shroud	Al6061	No	No	Corrosion, radiation effects, temperature effects, mechanical stress	Reduce strength, brittle, crack, physical deformation
Control rod (shim, safety and regulating rods)	Clad – SS304 Absorber – B_4C Fuel–UErZrH (20% U)	Yes	Yes	Corrosion, radiation effects, temperature effects, mechanical stress	Reduce strength, brittle, physical deformation
Control rod drive motor (over pool)	Drive motor assembly	Yes	Yes	Temperature effects, mechanical stress	Unable to control properly

IOP Conf. Series: Journal of Physics: Conf. Series **860** (2017) 012042 doi:10.1088/1742-6596/860/1/012042

Fuel element and instrumented fuel element	Clad – SS304 Fuel – UZrH (8.5% U) and UErZrH (20% U) Reflector – Graphite	Yes	Yes	Corrosion, radiation effects, temperature effects	Reduce strength, brittle, crack, physical deformation
Neutron detector	Detector assembly	Yes	Yes	Corrosion, radiation effects, supply voltage instability	Reduce strength, physical deformation, unable to count properly
Neutron source	Source – Am-Be Clad – double layers of SS304	Yes	Yes	Corrosion, radiation effects, temperature effects	Reduce strength, brittle, physical deformation
Irradiation tube	Al6061	No	No	Corrosion, radiation effects, temperature effects, mechanical stress	Reduce strength, brittle, crack, physical deformation
Grid plate	Anodized Al	No	No	Corrosion, radiation effects, temperature effects, mechanical stress	Reduce strength, brittle, crack, physical deformation

4. Evaluation of radiation effect

Fission reaction in the reactor core creates various types of radiation, e.g., neutrons, alpha particles, beta, and fission products. Energetic radiation, especially radiation in forms of particles or ions, can cause damage to target materials via displacement or dislocation of the target atoms. This work first focused on effects of neutrons, the dominant radiation in the reactor core.

4.1. Neutron energy and flux

Using the MCNPX software, neutron fluxes of three energy ranges were calculated at central thimble position at power 1 MW, Figure 2. This position yielded the highest neutron fluxes in the reactor core. The fluxes were calculated along axial direction with 10 centimetres increment. The maximum fluxes of three energy ranges were used to determine possible aging mechanisms by applying the conservative approach assessment.

Three ranges of neutron energies determined in this study include; thermal neutrons (0 – 0.025 eV), epithermal neutrons (0.025 – 0.42 eV) and fast neutrons (1 – 15 MeV). Note that the flux at energy 0.42 eV – 1 MeV was omitted in this study to avoid the cross section fluctuation zone.

Results showed that neutron flux of epithermal neutrons is the highest, followed by fluxes of fast and thermal neutrons, respectively. Fluxes of three energy ranges are within magnitude of 10^{13} n/cm^{2}s. Table **2** presents average and maximum fluxes of thermal, epithermal and fast neutrons along axial direction at central thimble position. A plot of neutron flux versus vertical distance from center (Figure 2) shows that fluxes of neutrons at three energy ranges are highest at the center position.

Table 2. Flux and fluence of thermal, epithermal and fast neutrons at central thimble.

Neutron energy	**Average neutron flux (n·cm^{-2}-s^{-1})**	**Maximum neutron flux (n·cm^{-2}-s^{-1})**	**Neutron fluence (39 years) (n·cm^{-2})**
Thermal neutrons (0.0-0.025 eV)	4.72×10^{12} ±0.32%	9.04×10^{12} ±0.84%	2.92×10^{21} ±0.32%
Epithermal neutrons (0.025 - 0.42 eV)	1.58×10^{13} ±0.26%	3.05×10^{13} ±0.66%	9.85×10^{21} ±0.26%
Fast neutrons (1.0-15.0 MeV)	5.39×10^{12} ±0.44%	1.10×10^{13} ±1.10%	3.55×10^{21} ±0.44%
Total (0.0-15.0 MeV)	4.01×10^{13} ±0.18%	7.93×10^{13} ±0.43%	2.56×10^{22} ±0.18%

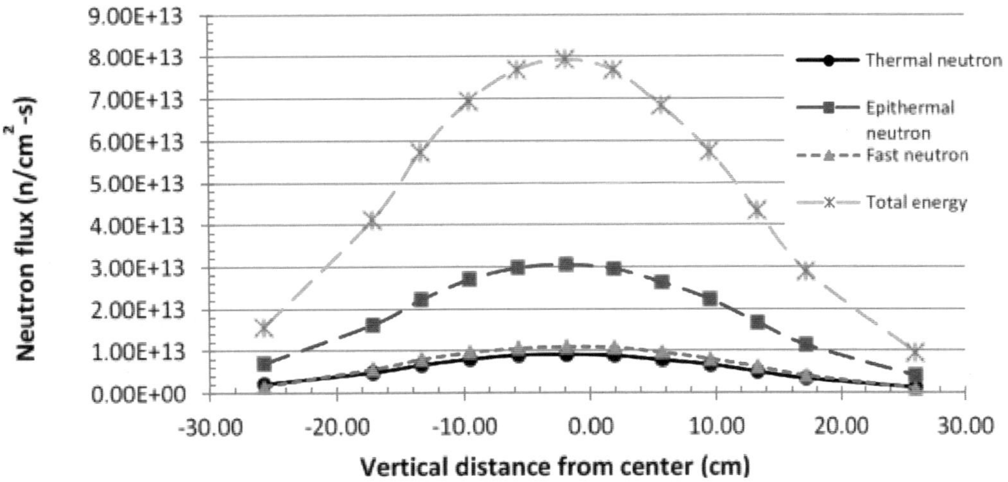

Figure 2. Plot of axial neutron flux at central thimble position at power 1 MW

Neutron fluence was evaluated using assumptions of;

- TRR-1/M1 had been operated for 39 years with 46 hours per week and 50 weeks per year, and
- the operation power is 1 MW.

The fluence was calculated using the maximum neutron fluxes to follow the conservative approach. Results of neutron fluence over 39 years are presented in Table 2. It was found that the neutron fluence is in the order of 10^{21} n·cm^{-2}. The trend of neutron fluence is closely resembling that of neutron flux, where fluence of epithermal neutrons is highest following by fast and thermal neutrons. The fluence was found to reach its maximum at the middle of fuel. Neutron flux and fluence at each energy range shown in Table 2 will be used to calculate displacement rates.

4.2. Displacement rate

Displacement rates of atoms in SS304 and Al6061 were calculated using the radiation damage rate equation;

$$R_d = N \int_{E_1}^{E_2} \phi(E_i)\sigma_D(E_i)dE_i \qquad (1)$$

where: R_d is the damage rate
N is the lattice atom density
E_i is the incident neutron energy
$\phi(E_i)$ is the energy dependent particle fluence
$\sigma_D(E_i)$ is the energy dependent cross section

The energy dependent cross section is a probability for displace lattice atoms with the incident neutrons.

$$\sigma_D(E_i) = \int_{T_1}^{T_2} \phi(E_i, T)\, \nu(T)dT \qquad (2)$$

where: $\sigma(E_i, T)$ is the energy transfer dependent cross section
T is the energy transfer to the lattice atoms
$\nu(T)$ is the number of displaced atoms

Kinchin-Pease Model determines the number of displaced atoms.

$$\nu(T) = \begin{cases} 0 & for\ T < E_D \\ 1 & for\ E_D < T < 2E_D \\ \frac{T}{2E_D} & for\ 2E_D < T < E_c \\ \frac{E_c}{2E_D} & for\ T \geq E_c \end{cases} \quad (3)$$

where: E_D is the displacement energy
E_c is the cut-off energy

The displacement rates of SS304 and Al6061 of three neutron energy ranges were calculated separately due to discrete values of the flux. The calculation was done under assumption that the dominant neutrons and target atoms interaction was isotopic scattering. Results showed that the displacement of atoms was mainly caused by fast neutrons. This is due to the fact that the incident energy of thermal and epithermal neutrons were not sufficient to displace atoms in SS304 and Al6061. The displacement rates over 39 years were also calculated using the neutron fluence. Calculation results are shown in

Table 3. The results shows that the displacement in 39 years of SS304 reached ~17 displacements per atom (dpa), and that of Al6061 reached ~4 dpa.

Table 3. Calculated displacement rate and displacement in 39 years of SS304 and Al6061.

Neutron energy	Displacement rate (displacement per second)	Displacement in 39 years (dpa)
SS304		
Fast neutrons	$5.24x10^{-8} \pm 18.44\%$	16.91 ± 4
Al6061		
Fast neutrons	$1.14x10^{-8} \pm 7.44\%$	3.67 ± 1.83

In comparison, a nuclear research reactor is typically experiencing significantly less neutron damage rates than those of the power reactors due to much lower fast neutron flux during operation. While fast neutron flux is within the order of 10^{13} n/cm^{2}s in a research reactor, fast neutron flux in a typical power reactor usually is greater than 10^{15} n/cm^{2}s resulting in displacement rate of at least 15 times greater than that of a research reactor. Nonetheless, changes in mechanical properties due to neutron effect can still be observed in a research reactor after many years of service.

4.3. Change in material property due to radiation

The results in previous section showed that fast neutrons caused displacements of atoms of the reactor components. Such displacement can cause changes in the material properties. Most of metal or alloys will undergo radiation strengthening phenomena in which the yield stress will increase and the ductility and the fracture toughness will decrease. These phenomena are caused by various mechanisms in the materials, e.g., displacement cascade and radiation-induced segregation. Due to limitation of specimen and high radiation field in TRR-1/M1, mechanical tests cannot be performed directly on such specimen. In the study, data from literature will be gathered and compared in order to evaluate changes in mechanical properties. Once specimens are available in the future, they will be tested for hardness and other predetermined mechanical properties. Data from other studies has suggested that yield strength (YS) and ultimate tensile strength (UTS) of neutron irradiated SS304 increased tremendously compared to those of neutron irradiated Al6061 [5-8]. Similar trends were observed to have total elongation. The YS and UTS of neutron irradiated SS304 increased rapidly with dpa and reached the maximum of 800 MPa in a sample irradiated to 10 dpa [5, 9]. Neutron irradiation is found to have smaller effects on Al6061, where YS and UTS slightly change after being irradiated up to 5 dpa [8]. Displacements after 39 years of irradiation in SS304 (17 dpa) and Al6061 (4 dpa) were interpolated using the aforementioned data [5-8]. Corresponding YS, UTS and total elongation of both alloys are reported in Table 4. Results indicate that after 39 year of service in the reactor core, YS, UTS and total elongation of SS304 has increased 225%, 58% and 85%, respectively. Meanwhile, YS, UTS and total elongation of Al6061 has found to only increase less than 20%. Therefore, SS304 clad would require careful attention.

Table 4. Comparison of unirradiated and irradiated (literature data) YS, UTS and elongation of SS304 and Al6061.

Property	Unirradiated	After irradiation	Change
SS304		17 dpa	
YS (MPa)	215	~700 [5, 6],	225%
UTS (MPa)	505	~800 [7]	58%
Elongation (%)	70	~10 [5], 8 [6]	85%, 88%
Al6061		4 dpa	
YS (MPa)	276	325 [8]	18%
UTS (MPa)	310	360 [8]	16%
Elongation (%)	12	11 [8]	7%

5. Summary

Analyses showed that the core-structural materials could be effected from aging mechanism after 39 year service in TRR-1/M1. Major possible degradation mechanisms are caused by radiation effects. The neutron irradiation reached fluence of 10^{21} n/cm^2. Determination of displacement rate of SS304 and AL6061 showed that only fast neutrons had influence on the radiation damage. Corresponding displacement per atom of SS304 is approximately 17 dpa, and that of Al6061 is 4 dpa. Comparison with the literature data revealed that the neutron fluence was high enough to effect material properties. Radiation affected YS, UTS and elongation of SS304 tremendously. The effects on Al6061 were found to be smaller than those on SS304. Continuous study should be conducted on corrosion of the structural materials. Moreover, further inspection should be conducted in the reactor based on these results.

References

[1] International Atomic Energy Agency 2005 *Safety of Research Reactor: safety requirements* Safety standard series, no. NS-R-4 (Vienna: International Atomic Energy Agency) p 53
[2] International Atomic Energy Agency 2010 *Aging management for research reactors: specific safety guide* Safety standard series, no. SSG-10 (Vienna: International Atomic Energy Agency) chapter 5 pp 22 – 29
[3] Sangkaew S, Angwongtrakool T and Srimok B 2016 *Proc. Int. Nuclear Science and Technology Conf. 2016* (Bangkok) *Preprint*
[4] Chue-Inta S, Klaysuban N, Tippayakul C and Konduangkaeo A 2010 *Technical meeting on commercial products and services of research reactors* (Vienna) IAEA-TM-38728
[5] Lucas GE 1993 *J. Nucl. Mater.* **206** 287
[6] Chen Y, Alexandreanu B, Soppet WK, Shack WJ, Natesan K, Rao AS 2011 *Proc. 15th Int. Conf. on Environmental Degradation of Materials in Nuclear Power System* (La Grange Park, IL: ANS)
[7] Dadfarnia M, Sofronis P and Robertson IM 2012 *Irradiation Effects on Material Properties for 304L Stainless Steel Base Metal and Welds* Final report to Canadian Nuclear Safety Commission (Urbana, Illinois)
[8] Farrell K and King RT 1979 *Effect of Radiation on Structural Materials* ASTM STP 683 pp 440-449

IOP Conf. Series: Journal of Physics: Conf. Series **860** (2017) 012042 doi:10.1088/1742-6596/860/1/012042

[9] Maloy SA, James MR, Sommer WF, Ferguson P, Willcutt G, Alexander D, Louthan MR, Hamilton ML, Snead L and Sokolov MA 1996 *The Change in the Mechanical Properties of Alloy 718, 304L and 316L Stainless Steel and Al6061 after Irradiation in a High Energy Proton Beam* UNT Digital Library: http://digital.library.unt.edu/ark:/67531/metadc722804/ (New Mexico)

Acknowledgments
Authors would like to acknowledge assistance from the colleagues from Reactor Center of Thailand Institute of Nuclear Technology for useful discussions and information used in this study.

IOP Conf. Series: Journal of Physics: Conf. Series **860** (2017) 012043 doi:10.1088/1742-6596/860/1/012043

Monte carlo simulation of innovative neutron and photon shielding material composing of high density concrete, waste rubber, lead and boron carbide

P Aim-O[1,3], D Wongsawaeng[1,4], P Phruksarojanakun[2] and S Tancharakorn[3]

[1] Department of Nuclear Engineering, Faculty of Engineering, Chulalongkorn University, Bangkok, Thailand, 10330
[2] Office of Atoms for Peace, 16 Vibhavadi Rangsit Road, Bangkok, Thailand, 10900
[3] Synchrotron Light Research Institute, Nakhon Ratchasima, Thailand, 30000

[4] Corresponding author e-mail: doonyapong.w@chula.ac.th

Abstract. High-density concrete exhibits high strength and can perform an important role of gamma ray attenuation. In order to upgrade this material's radiation-shielding performance, hydrogen-rich material can be incorporated. Waste rubber from vehicles has high hydrogen content which is the prominent characteristic to attenuate neutron. The objective of this work was to evaluate the radiation-shielding properties of this composite material against neutron and photon radiations. Monte Carlo transport simulation was conducted to simulate radiation through the composite material. Am-241/Be was utilized for neutron source and Co-60 for photon source. Parameters of the study included volume percentages of waste rubber, lead and boron carbide and thickness of the shielding material. These designs were also fabricated and the radiation shielding properties were experimentally evaluated. The best neutron and gamma ray shielding material was determined to be high-density concrete mixed with 5 vol% crumb rubber and 5 vol% lead powder. This shielding material increased the neutron attenuation by 64% and photon attenuation by 68% compared to ordinary concrete. Also, increasing the waste rubber content to greater than 5% resulted in a decrease in the radiation attenuation. This innovative composite radiation shielding material not only benefits nuclear science and engineering applications, but also helps solve the environmental issue of waste rubber.

1. Introduction

Crumb rubber is a material that is easy to find and inexpensive. It is made by recycling used automobile tires into powder. The recycling process helps reduce waste and the environmental problem. Used rubber has been utilized in many engineering works. For example, researchers in ref. [1] added used automobile tire to concrete and studied the strength of the compound. It was found that as the rubber aggregate percentage increased, the compressive strength decreased. Researchers in ref. [2] found that by adding a small amount of crumb rubber to concrete, the tensile splitting strength increased by 7% because the powder bonded with cement better than sand.

Today, there are many utilizations of radiation in medical, research and other fields. Safety in working with radiation must be emphasized, especially with neutron and gamma radiation which have high penetrating power. Thus, a shielding material capable of shielding both neutron and gamma ray should be designed. An example of a radiation source emitting both neutron and gamma ray is the high-energy gamma ray emitted from a synchrotron light source with the superconducting wiggler installed in the storage ring. Neutron radiation can be emitted when high-energy gamma ray interacted with the beryllium target, a process called photonuclear reaction [3].

IOP Conf. Series: Journal of Physics: Conf. Series **860** (2017) 012043 doi:10.1088/1742-6596/860/1/012043

Concrete has been widely used as gamma ray and, in certain applications, as neutron shielding material. Rubber is a hydrocarbon material composing mainly of carbon and hydrogen, which are effective neutron moderators. Moderating fast neutron into thermal one increases the chance of being absorbed [4], thus enhancing the neutron shielding efficiency of the shielding material. The objective of the present research is to design and study the neutron and gamma ray shielding efficiency of the shielding material composing of high-density concrete and waste rubber. Lead and boron carbide are also added to increase the shielding efficiency. The MCNP simulation as well as radiation attenuation experiments were performed to determine efficiency of the designed shielding materials.

2. Materials and Methods

2.1. Material Selection and Analysis

Concrete has been widely used as gamma ray shielding material because it is easy to find, inexpensive and easy to fabricate. High-density concrete has also been used to shield gamma radiation because it exhibits a higher density and, thus, can shield gamma ray more effectively, resulting in thickness reduction [5]. Addition of different quantities of materials into concrete affects neutron and gamma ray shielding. Important factors to consider when determining neutron attenuation are elastic scattering reaction, inelastic scattering reaction and neutron capture interaction process [6]. Incorporating material composing of heavy elements helps increase the inelastic scattering interaction, while incorporating material composing of light elements such as hydrogen enhances the elastic scattering. Both interactions decrease neutron energy. Neutron-capturing elements help reduce the generation of secondary gamma and thermal neutron [7].

Materials used in the research are Portland cement type I, crumb rubber from waste rubber, barite powder, lead powder and boron carbide powder. Elemental analysis of the materials were performed at Scientific and Technological Research Equipment Centre Chulalongkorn University using scanning electron microscope model JSM6610LV (JEOL) with Energy Dispersive Spectroscopy (EDS). Figures 1 and 2 show results of crumb rubber analysis.

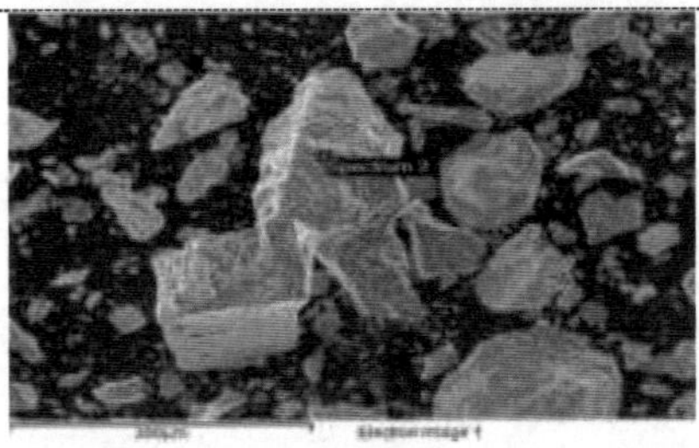

Figure 1. SEM micrograph of crumb rubber

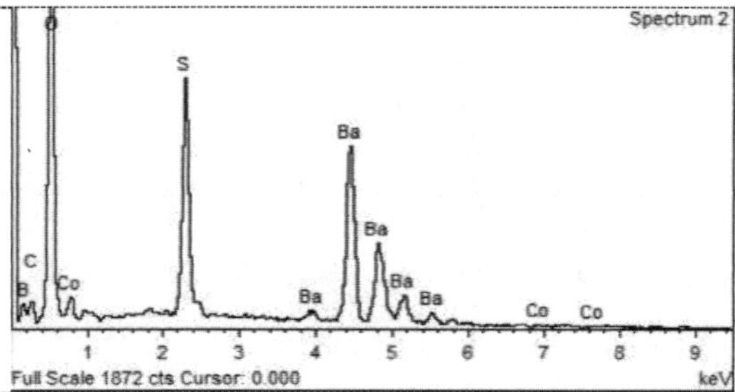

Figure 2. Elemental analysis spectrum of crumb rubber

Carbon/Hydrogen/Nitrogen Determinator was used to determine the elemental composition of waste rubber. Results are illustrated in Table 1.

Table 1. Chemical composition analysis of waste rubber

	% C	% H	% N	Mass (gram)
Waste rubber #1	80.91	9.52	0	0.023
Waste rubber #2	80.68	8.54	0	0.036
Average	80.79	9.03	0	0.029

IOP Conf. Series: Journal of Physics: Conf. Series **860** (2017) 012043 doi:10.1088/1742-6596/860/1/012043

2.2. Monte Carlo Simulation

The present study used MCNP version 2.7.0 to simulate the radiation shielding efficiency of each shielding design to determine the most appropriate composition. Am-241/Be was used as a neutron source and Co-60 was used as a gamma ray source. An isotropic point source was assumed with the neutron energy range 5.0×10^{-7} - 1.0×10^{2} MeV. The source was placed at the center of one side of the shielding material. The shielding was simulated as a cube with the dimension on each side of 12.7 cm according to British Standard. Surface flux tally was used to determine the amount of radiation passing through the material in the unit of particles/cm^2. The standard deviation error was calculated from the simulation and the Dose Energy (DE) and Dose Function (DF) were used to convert flux into dose equivalent rate with the unit of (rem/hr)/particle. The simulated 3D geometries using Vised to study the attenuation and shielding thickness are shown in Figures 3 and 4, respectively.

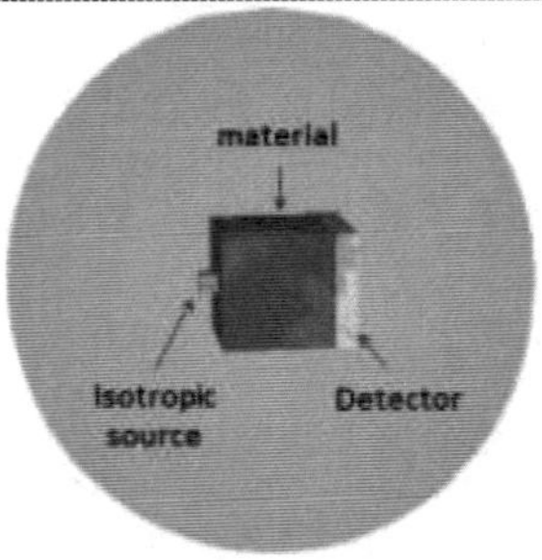

Figure 3. Simulated 3D geometry using Vised to study attenuation in MCNP

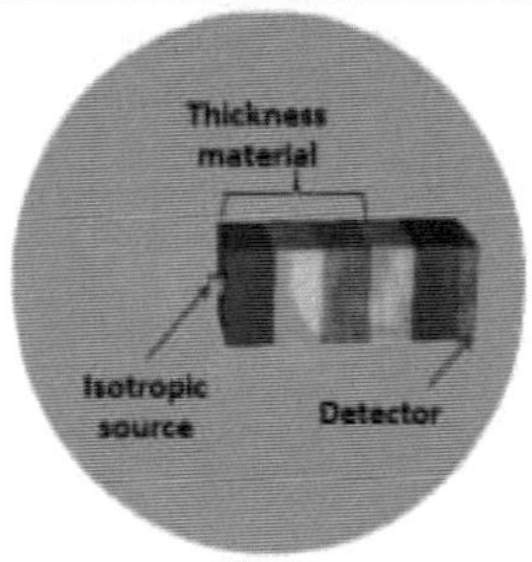

Figure 4. Simulated 3D geometry using Vised to study shielding thickness in MCNP

Three different compositions of shielding materials were designed and investigated in this research. They were: high-density concrete mixed with crumb rubber (HDCR), high-density concrete mixed with crumb rubber and lead (HDCRL) and high-density concrete mixed with crumb rubber and boron carbide (HDCRB). Table 2 shows the composition of each shielding material. The number after a letter indicates the volume percentage of the component. For example, HDCR5 represents high-density concrete mixed with 5 vol% rubber, HDCR5L2 represents high-density concrete mixed with 5 vol% rubber and 2 vol% lead and HDCR5B2 represents high-density concrete mixed with 5 vol% rubber and 2 vol% boron carbide.

Table 2. Material composition of shielding materials

Specimen assignment	Waste rubber (gram)	Cement (gram)	Barite (gram)	Pb/B_4C (gram)	Water (gram)	Density (gram/cm^3)
HDC	-	1,000	5,669	-	450	3.35
HDCR5	94.22	1,000	5,669	-	450	2.831
HDCR10	188.25	1,000	5,370	-	450	2.782
HDCR15	282.67	1,000	5,070	-	450	2.733
HDCR20	376.91	1,000	4,771	-	450	2.636
HDCR25	471.00	1,000	4,472	-	450	2.587
HDCR5L2	94.22	1,000	5,669	464.98	450	2.957
HDCR5L4	94.22	1,000	5,669	929.96	450	3.080
HDCR5L5	94.22	1,000	5,669	1,162.45	450	3.154
HDCR5L6	94.22	1,000	5,669	1,394.94	450	3.219
HDCR5L8	94.22	1,000	5,669	1,859.92	450	3.350
HDCR5B2	94.22	1,000	5,669	103.23	450	2.812
HDCR5B4	94.22	1,000	5,669	206.47	450	2.798
HDCR5B6	94.22	1,000	5,669	309.71	450	2.785
HDCR5B8	94.22	1,000	5,669	412.95	450	2.771

2.3. Specimen Fabrication and Attenuation Experiments

2.3.1. Preparation of specimens. Portland cement, barite, water, crumb rubber and/or lead and/or boron carbide with specific portion of each ingredient were mixed thoroughly in a mortar mixer. Waste rubber and fine aggregates were bonded together by Portland cement through the hydration process in the mold. After the curing process in water for 7 days, the shielding materials were tested for radiation attenuation.

2.3.2. *Radiation shielding tests.*

2.3.2.1. Neutron Attenuation Experiment. The neutron attenuation coefficients were determined by performing a transmission experiment. All measurements were performed under the same geometry and experimental setup. The detection system consisted of Am-241/Be neutron source with strength 3.08 x 10^3 n-cm^{-2} s^{-1}, acrylic plates and boron trifluoride (BF_3) detector. The distance from source to neutron detector was 51.5 cm. The BF_3 detector was operated with the high voltage of 1,075 V. The counting time was 10 minutes. The experiment setup is shown graphically in Figure 5.

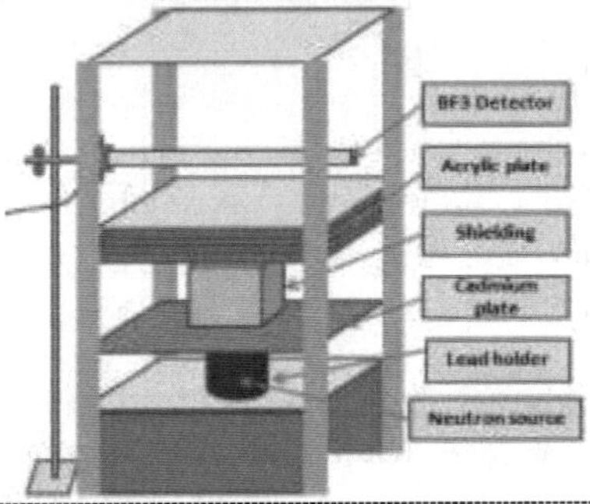

Figure 5. Graphical representation of fast neutron attenuation measurement

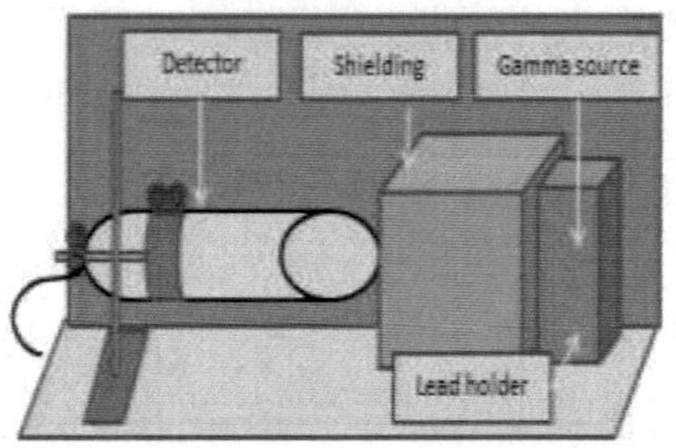

Figure 6. Graphical representation of photon attenuation measurement

2.3.2.2. Photon Attenuation Experiment. The attenuation coefficients of gamma ray of shielding materials were measured at energies 1173 and 1332 keV using Co-60 source. The NaI(Tl) detector was CANBERRA Model 802-2x2, crystal size 2 x 2 inch with the source-to-detector distance of 17 cm. The height of the detector from floor was 9.8 cm. The photon measurement set up is shown graphically in Figure 6.

3. Result and Discussions

3.1. Simulation results
The MCNP simulation was first performed to determine the most appropriate portion of waste rubber in the high-density concrete. The studied volume percentages were 5, 10, 15, 20 and 25. Results revealed that neutron attenuation increased with increasing waste rubber percentage as shown in Figure 7a. This is because light elements in rubber are effective in attenuating fast neutrons. However, for the case of gamma ray attenuation, high-density concrete alone exhibited the best performance as shown in Figure 7b. This is because heavy elements as well as the high-density characteristic of the high-density concrete are effective in gamma ray shielding. Because the aim of this research was to design the shielding material for both neutron and gamma ray and because the mechanical strength of the shielding material needs to be maintained as well, the waste rubber percentage of 5% was decided to be most suitable and this composition was used throughout the remaining simulations when adding lead or boron carbide to the shielding material.

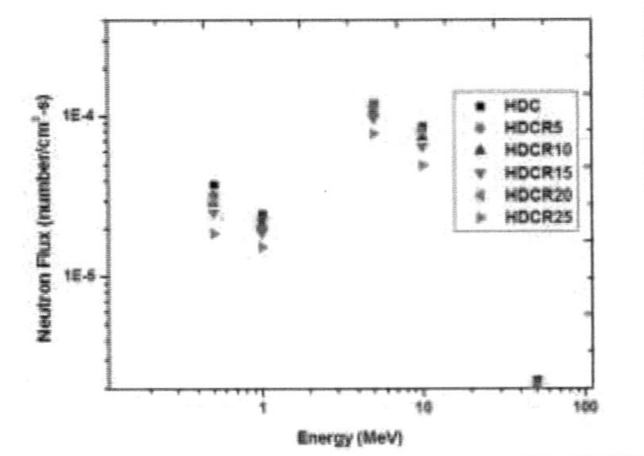

Figure 7a. Simulation result of neutron attenuation of high-density concrete mixed with various volume percentages of waste rubber

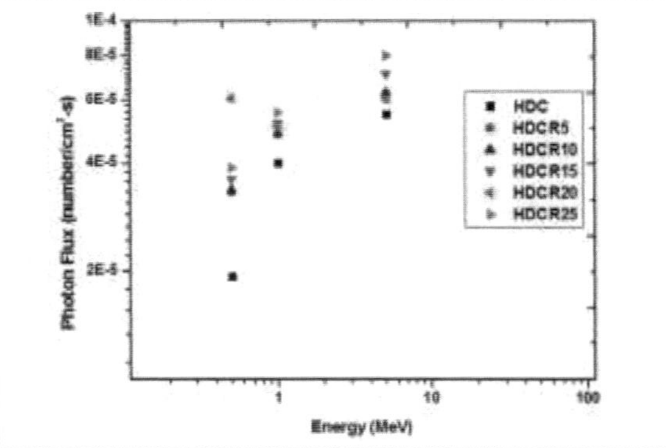

Figure 7b. Simulation result of photon attenuation of high-density concrete mixed with various volume percentages of waste rubber

Simulation results of gamma ray and neutron transmissions are illustrated in Figures 8 and 9 for different shielding thickness. For the case of gamma ray transmission, lead was used as additive to HDCR5, and for the case of neutron transmission, boron carbide was used as additive to HDCR5. Simulated shielding thicknesses were 6.38, 12.7, 19.08 and 25.46 cm. It was found that the $\ln(I/I_0)$ value, where I and I_0 represent transmitted radiation in the presence of the shielding and without the shielding, respectively, decreased with increasing shielding thickness according to the Beer Lambert's law, as the curve of $\ln(I/I_0)$ vs. shielding thickness exhibited a linear form. Moreover, for the case of gamma ray transmission, the more volume percentage of lead added, the better the shielding efficiency. This behavior is expected since lead is a very effective gamma ray shielding material.

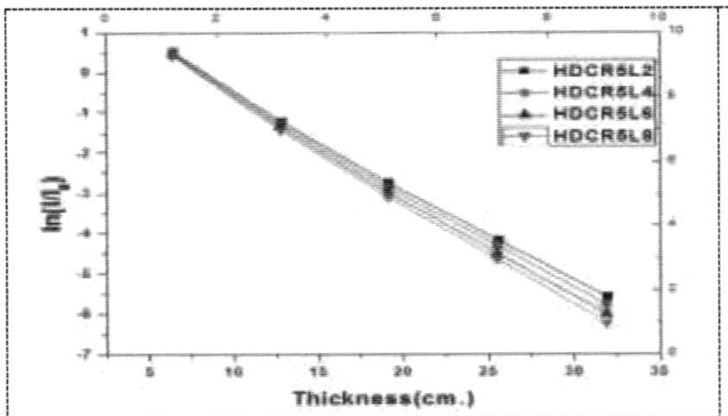

Figure 8. Simulation results of gamma ray transmission of HDCR5 with different vol% of lead addition

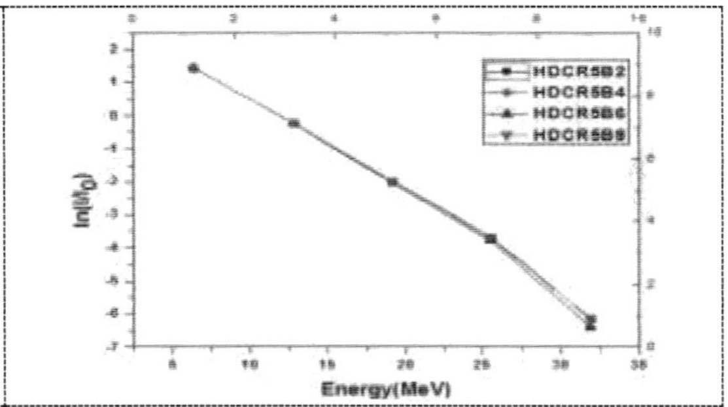

Figure 9. Simulation results of neutron transmission of HDCR5 with different vol% of boron carbide addition

3.2. Experimental results

Results from the attenuation experiments indicate that high-density concrete mixed with 5, 10, 15, 20 and 25 vol% of waste rubber can shield fast neutron better than high-density concrete alone, and that addition of waste rubber more than 5 vol% resulted in the detrimental effect as the fast neutron transmission increased, as shown in Figure 10. The neutron transmission was lowest at 43.2% for 5 vol% waste rubber addition, and this optimum content was chosen to perform all remaining experiments. Waste rubber enhanced fast neutron attenuation because of the hydrocarbon content of waste rubber. Atomic hydrogen and carbon undergo elastic scattering with fast and intermediate energy neutrons to reduce their energy. However, addition of waste rubber more than 5 vol%

effectively reduced the amount of high-density concrete and, subsequencly, reduced the fast neutron attenuation. This is because barite in high-density concrete (presence as $BaSO_4$) contributed 45.45 % by weight. Reduction of Ba quantity in the shielding material (as a result of increasing the waste rubber content) reduced the probability of fast neutron undergoing inelastic scatterring interaction with Ba.

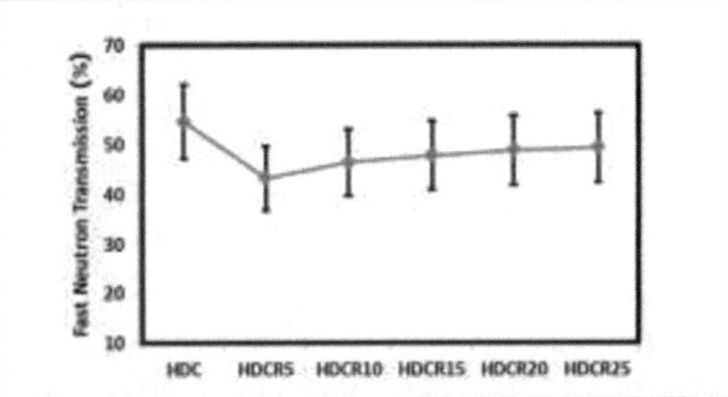

Figure 10. Experimental results of fast neutron transmission for HDC with various vol% of waste rubber

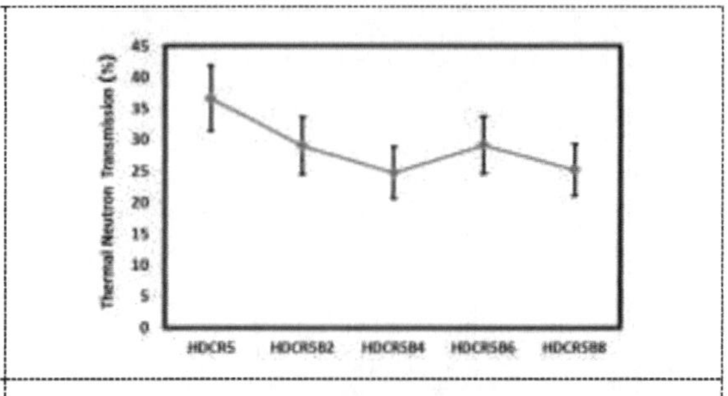

Figure 11. Experimental results of thermal neutron transmission for HDCR5 with various vol% of boron carbide

With the waste rubber content held constant at 5 vol%, addition of 2, 4, 6 and 8 vol% of boron carbide provided better thermal neutron shielding than HDCR5 alone, as can be seen in Figure 11. As boron is a very effective thermal neutron absorber, addition of boron carbide into concrete shielding significantly increased thermal neutron shielding efficiency. Moreover, as the hydration process of cement in concrete increased, the neutron attenuation was enhanced [8]. Experimental results in Figure 11 may be compared to simulation results in Figure 9. From Figure 9, although it appears that HDCR5 with 6 vol% boron carbide exhibited a slightly better performance than the rest, the difference is small and it can be said that addition of boron carbide anywhere from 2 – 8 vol% (the range used in the present study) provided the same benefit. In Figure 11, although 4 and 8 vol% boron carbide addition offerred a slightly better performance than the rest, the difference is not large and there can always be small fluctuations in experimental outcomes resulting from many factors inherent to radiation counting experiments. Thus, the same conclusion can be made with experimental results that addition of boron carbide anywhere from 2 – 8 vol% provided roughly the same benefit.

According to Figure 12, addition of waste rubber of 5, 10, 15, 20 and 25 vol% resulted in a slightly lower gamma ray attenuation at 1332 keV compared to 1173 keV. However, when HDCR5 was added with 2, 4, 6 and 8 vol% lead content, the attenuation of 1332 keV gamma ray became better than that of 1173 keV gamma ray, as can be observed from the higher linear attenuation coefficient value in Figure 13. Moreover, the more lead percentage, the better the gamma ray shielding capability because lead is a very effective gamma ray shielding material. This experimental result is in good agreement with the simulated result in Figure 8. Another added bebefit of lead is the increased inelastic scattering with fast neutron resulting in the enhanced neutron attenuation ability.

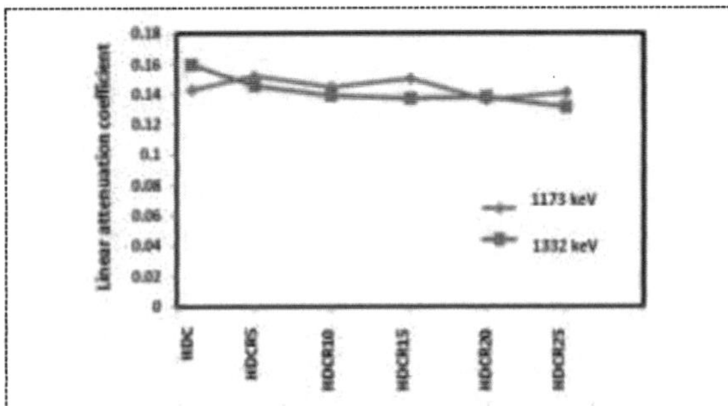

Figure 12. Gamma ray attenuation coefficient of HDC mixed with various waste rubber content

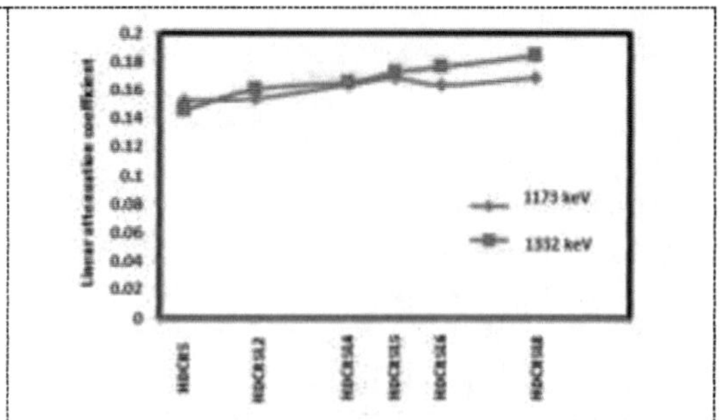

Figure 13. Gamma ray attenuation coefficient of HDCR5 mixed with various lead content

IOP Conf. Series: Journal of Physics: Conf. Series **860** (2017) 012043 doi:10.1088/1742-6596/860/1/012043

Figure 14 shows results of fast neutron transmission experiment when lead powder was added to HDCR5. With 5 vol% of added lead powder, the fast neutron transmission was lowest. Lead, a heavy element, undergoes inelastic scattering interaction with fast neutron and reduces the transmitted fast neutron flux. Thus, 5 vol% of added lead powder was the optimal composition.

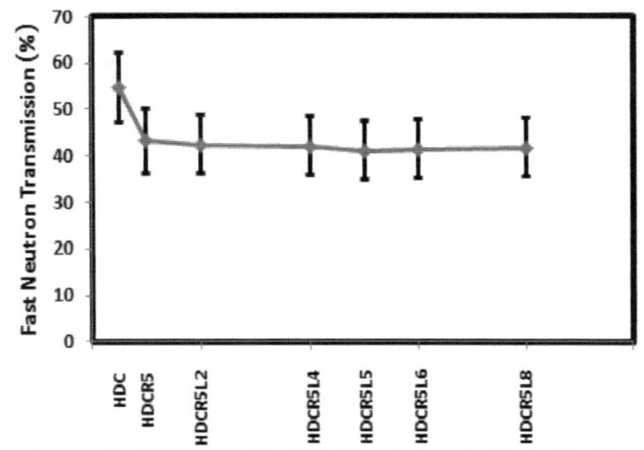

Figure 14. Experimental results of fast neutron transmission for HDCR5 with various lead powder content

The last step of this work is to compare the effectiveness of lead addition to boron carbide addition on neutron and photon transmissions. With crumb rubber content being 5 vol%, 5 vol% lead content was compared to 8 vol% boron carbide content. Simulated results of ratio of neutron and photon transmission percentages of various shielding material designs to those of ordinary concrete are shown in Figure 15. In the Figure, HDCR5L5 performed the best for photon attenuation and almost the best for neutron attenuation. When considering both photon and neutron attenuations together, HDCR5L5 was the most suitable shielding material. It is interesting to observe that HDCR5L5 shielded against neutron even more effectively than HDCR5B8. The latter contained boron, a strong thermal neutron absorber, so it should absorb neutrons much more effectively. The reason is that HDCR5B8 contained no lead to undergo inelastic scattering interaction with fast neutron, so it was not an effective fast neutron moderator compared to HDCR5L5. With less thermal neutron generated, the presence of boron provided little benefit. For the case of HDCR5L5, barium present in barite and hydrogen present in high-density concrete as well as in waste rubber help capture thermal neutron.

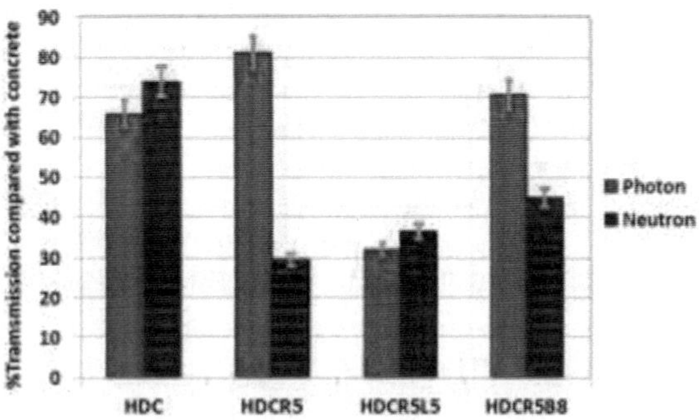

Figure 15. Ratio of neutron and photon transmission percentages of various shielding material designs to those of ordinary concrete

IOP Conf. Series: Journal of Physics: Conf. Series **860** (2017) 012043 doi:10.1088/1742-6596/860/1/012043

4. Conclusions

The radiation shielding materials fabricated from high-density concrete mixed with crumb rubber capable of shielding both neutron and gamma ray were designed using the MCNP computer code and verified by radiation attenuation experiments. From experimental results, it was found that addition of crumb rubber increased the neutron shielding efficiency, and that addition of more than 5 vol% actually lowered the neutron shielding efficiency because the barite content of high-density concrete was reduced. Addition of lead to high-density concrete mixed with 5 vol% crumb rubber increased both neutron and gamma ray attenuation. Moreover, addition of boron carbide, a strong thermal neutron absorber, to high-density concrete mixed with 5 vol% crumb rubber increased thermal neutron attenuation. However, the best neutron and gamma ray shielding material was determined to be high-density concrete mixed with 5 vol% crumb rubber and 5 vol% lead powder. The precence of lead effectively moderates fast neutron through inelastic scattering. This shielding material increased the neutron attenuation by 64% and photon attenuation by 68% compared to ordinary concrete. This research offerred a new method to recycle used rubber for radiation shielding purposes. This not only helps reduce environmental problems, but also enhances radiation safety in nuclear and other related works.

Acknowledgement

This study was supported by Synchrotron Light Research Institute (Public Organization) and Ministry of Science and Technology of Thailand. The authors would like to thank Dr. Nawin Juntong for providing the MCNP program.

References

[1] Kotresh K.M, M.G.B., *Study On Waste Tyre Rubber As Concrete Aggregates.* International Journal of Scientific Engineering and Technology, 2014. Vol. 3(4): p. 433-436.

[2] Audrius Grinys, H.S., *Tyre Rubber Additive Effect on Concrete Mixture Strength.* Journal of Civil Engineering and Management, 2012. Vol. 18(3): p. 393-401.

[3] Y. Kawashima,. *Proposal Of A Synchrotron Radiation Facility To Supply Ultraviolet Light, X-Ray, MeV-Photon, GeV-Photon And Neutron.* Proceedings of FLS 2006, 2006: p. 24-26.

[4] Jumpee, C., Wongsawaeng, D., *Fabrication Of Neutron Shielding Material Comprising Of Natural Rubber/Synthetic SBR Blends And Neutron Absorber.* Journal of the Nuclear Society of Thailand, 2016. Vol. 1(No.1): p. 40-54.

[5] Tonnayopas, D., *Radiation Shielding of Barite Aggregate Concrete Containing Smectite.* The 7th Engineering Conference Prince of Songkla University, 2009: p. 27-32.

[6] E. Yilmaz, H.B., E. Kiris, I. Ustabas, U. Cevik, A.M. El-Khayatt, 2011, *Gamma ray and neutron shielding properties of some concrete materials*. Annals of Nuclear Energy, (Amsterdam: Elsevier), p. 2204-2212.

[7] Hausi Hu, Q.W., Juan Qin, Yuelei Wu, Tiankui Zhang, Zhongsheng Xie, Xinbiao Jiang, Guoguang Zhang, Hu Xu, Xiangyang Zheng, Jing Zhang, Wenhao Liu, Zhenghong Li, Boping Zhang, Linbo Li, Zhaohui Song, Xiaoping Ouyang, Jun Zhu, Yaolin Zhao, Xiaoqin Mi, Zhengping Dong, Cheng Li, Zhenyu Jiang, and Yuanping Zhan, *Study on Composite Material for Shielding Mixed Neutron and Gamma-Rays.* IEEE Transaction On Nuclear Science, 2008. Vol. **55**: p. 2376-2384.

[8] Yosof Abdullah, M.R.Y., Rusnah Mustaffa, Abdul Aziz Mohamed, Fatin Nabilah Tajul Ariffin and Nurhaslinda Ee Abdullah, *Properties of concrete/boron carbide as neutron shielding materials.* Journal of nuclear and related technologies, 2011. **8**(2): p. 15-25.

IOP Conf. Series: Journal of Physics: Conf. Series **860** (2017) 012044 doi:10.1088/1742-6596/860/1/012044

Study on Developing Safety Infrastructure for Mineral Processing Waste (NORM Waste) and Contamination Monitoring at the TINT Rare Earth Research & Development Center, Khlong 5, Pathumthani, Thailand

N Yaanant[1], V Kasemtanasak, A Pattanasub, A O-manee, S Khaweerat, P Pruantonsai, T Akharawutchayanon, P Nuanjan, S Punbut, P Srimork, and N Prasertchiewchan

Thailand Institute of Nuclear Technology (Public Organization), Ongkharak, Nakhon Nayok 26120, Thailand

[1]Corresponding author e-mail: nanthavan@tint.or.th

Abstract. The objective of this study is to prepare all technical and administrative actions leading to the release of the disused radiological facilities from regulatory control and safe management of radioactive waste. In the first phase, the study covers: the preliminary site characterization, waste characterization, the preparation for authorization of the foreseen strategy and activities. The first phase study shows that several areas of the TINT's Rare Earth Research & Development Center are contaminated, such as the U/Th extraction, monazite processing and NORM residues storage.

1. Introduction

What is NORM? NORM stands for "naturally occurring radioactive material" or in other words, a substance that naturally contains one or more radioactive isotopes, also called radionuclides. NORM waste typically is produced by an industrial, mining, or manufacturing process that uses NORM.

In in this paper, the NORM wastes were generated from the monazite processing and the U-Th extraction pilot plans from the Rare-Earth Research and Development Center (RE R&D Center), Thailand Institute of Nuclear Technology (TINT). In Thailand, Monazite (Ce, La, Nd, Y, Th)PO4, occurs in heavy-mineral sand deposits, vein type deposits in low grade tin ores from the south of Thailand, with a rare earth content of about 60.20 % REO [1]. Rare earth mineral concentrates are chemically processed to extract intermediate groups of mixed rare earth compounds. In general, chemical treatment of mineral concentrates derived from hard rock deposits may start with roasting in air (calcining) to drive off carbon dioxide and oxidize cerium to the tetravalent state. This is, in many steps, grinding, caustic leach then followed by treatment with hydrochloric acid to dissolve non-cerium rare earths, yielding a marketable cerium concentrate which can be used directly as a low value product or further separated into high purity individual rare earths. Figure 1 shows the monazite processing chart from IAEA Safety Series [1].

IOP Conf. Series: Journal of Physics: Conf. Series **860** (2017) 012044 doi:10.1088/1742-6596/860/1/012044

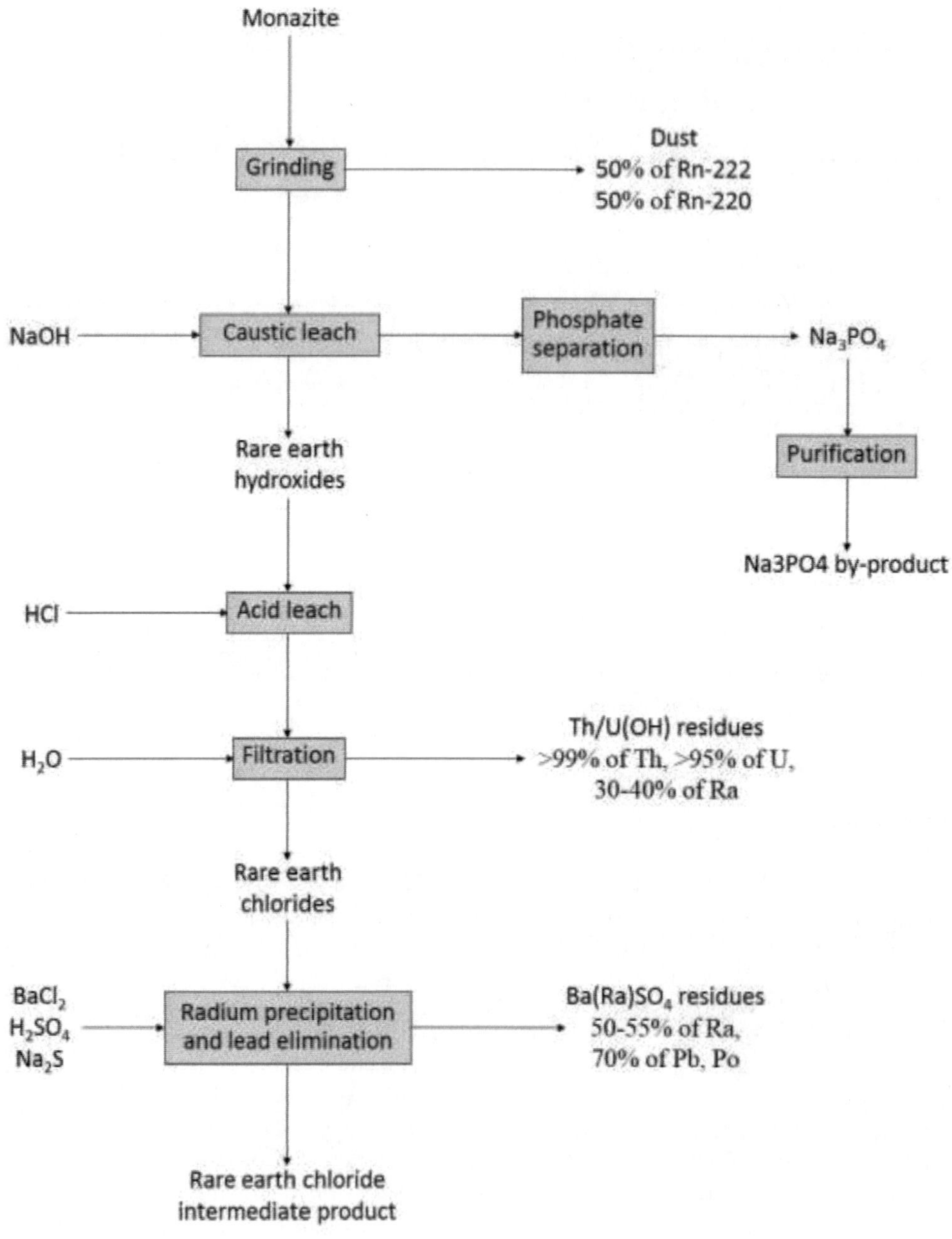

Figure 1. Monazite processing chart [1]

IOP Conf. Series: Journal of Physics: Conf. Series **860** (2017) 012044 doi:10.1088/1742-6596/860/1/012044

2. Historical Information and Infrastructure of the RE R&D Center

The RE R&D Center is located at Khlong Luang district, Pathumthani province, Thailand, about 40 km away from the TINT, Bangkok. This center was operated by Office of Atom for Peace during 1995 – 2005 for extraction of rare earth elements from monazite sand and also the U-Th extraction activities. After the re-organization on December 2006, the RE R&D Center has been transferred to the TINT. However, the rare-earth and monazite processing and the U/Th extraction plants have been stopped in operation since 2005. Later, in 2011, the heavy flood occurred in Thailand, the pilot plants and area around were unfortunately flooded. It was possible for the contamination with naturally occurring radioactive materials (NORM).

There were about 18 buildings for several activities. The building No. 8 consisted of the monazite processing and uranium and thorium processing plants which were the main building for radiological activities. The building No. 9 was the rare earth processing plant for extraction of rare earth elements. The building No. 14 was the storage facility for uranium and thorium cake. The building No. 18 was monazite storage which contains a huge quantity of monazite sand. The figure 2 shows the map of Rare Earth Research Development Center. Up to now, those building filled with the same instruments and materials as before, safety measures are not in place. There is no classified waste. There are mixed-items, such as instruments, chemicals, NORM residues and other infected waste from birds accumulated in those facilities.

Currently, there are a lot of NORM residues in drums onsite. The radiation safety infrastructure for NORM residue management is recommended to be seriously concerned. So far, the radioactive waste, both solid and liquid wastes are awaiting to be managed in proper ways.

In addition, the historical activity concentration of radon in each building was collected during this study as shown in Table 1 and Table 2.

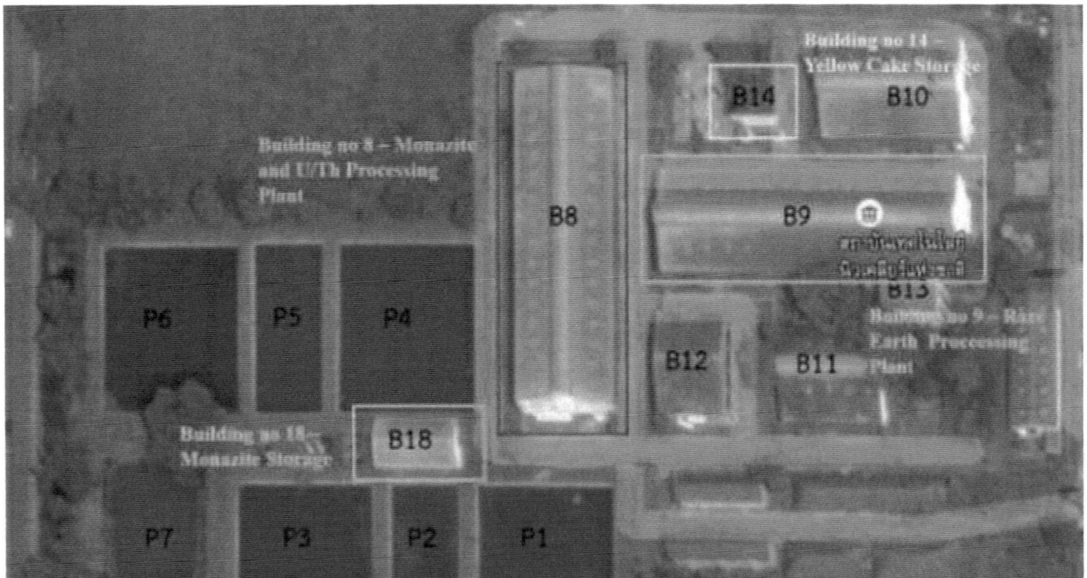

Figure 2. Map of Rare Earth Research Development Center

Table 1. Indoor Rn-222 and Rn-220 concentration at the Rare-Earth Research and Development Center (measured date: Feb 2010) [2]

Location	Range of concentration (Bq.m^{-3})	
	Rn-222	Rn-220
Rare Earth Processing Building 9	3 – 20	10 – 30
Monazite/ U-Th Processing Building 8	20 – 70	490 – 2200
U-Th Cake storage Building 14	160 – 442	870 – 5150

Remark: US EPA action level of radon in building should not exceed 4pCi/l, 148 Bq.m^{-3} [3]

Table 2. Concentration of radon in workplaces at the Rare-Earth Research and Development Center (measure dated: March 2013) [4]

Location	Rn concentration in the air (Bq.m^{-3})	Temp. (^{O}C) (RAD7)	RH (%) (RAD7)
1. Building No. 8	180 ± 10	31.6	12
2. Outside of Building No.14	52 ± 7	35.6	11
3. inside of Building No. 10	16 ± 4	35.3	11

Remark: Build No.10 is located in the opposite side of No.14 (this building has no activity)

3. Scope of Work in Phase 1

The work in phase 1 covers on study the TINT Rare Earth Research & Development Center's current safety situation. The scope of work in phase 1 is as followings;

- Onsite Survey: dose rate, and dose mapping
- Checking contamination at the monazite/ U-Th processing building (Building 8)
- Searching for NORM residue onsite
- Characterization of NORM residue

4. Method

4.1. Survey on Dose rate and Contamination Monitoring at the facilities

- The dose rate survey meter and the contamination monitor (which was able to detect alpha radiation) were used for survey the area in building 8, 9 and 18 (monazite storage building).
- The dose rate survey meter, Ludlum model 14C was used for measuring the exposure dose
- The contamination monitoring at the building No. 8, was conducted by direct method. The contamination survey meter, Ludlum model 12 with probe model 44-9, were used for measuring contamination.

Direct method is the measurement of radiological contamination by using the detector measure on the area directly. The distance from the detector from the target area is about 0.5 cm for alpha and 2.5 cm. for beta/gamma.

The area in the building No.8 was divided into 2 sections, *i.e.* the uranium and thorium processing and monazite processing facilities. The dose and contamination mapping method was performed for both facilities. The grid-method was applied coordinate to measure the highest contaminations in each grid. In the same time, the number of drums which contain liquid, sludge and residues were also counted.

4.2. Monitoring environmental water sample onsite

In order to monitor environmental water sample onsite, six water samples were collected randomly from different ponds as shown in figure 3. Water was collected from ponds and kept into 5 liters plastic bottle using manual procedure. The samples were transferred to the Radioactive Waste Management (RWM) Center's laboratory for analysis.

Pond no.1

Environmental monitoring point 1

Pong for sediment suspensions

Pond no.7

Channel beside the pond 7

Figure 3. Sampling water samples from ponds onsite

4.2.1. Gross Alpha Beta Analysis. The gross alpha and gross beta activity concentration measurement was carried out by gas flow proportional counter, using Low-level Gas Proportional Counter Berthold LB770. All samples were prepared by EPA method (EPA method 900.0). The counting time was 40 minutes for each counting period.

4.2.2. Gamma Analysis. 1 liter of each sample was poured into a Marinelli beaker in preparation of gamma spectrometry analysis. After properly tightening the threatened lid, a Marinelli beaker were sealed with adhesive tape and left for at least 4 weeks (>7 half-lives of ^{222}Rn and ^{224}Rn) before counting by gamma spectrometry in order to ensure that the daughter products of ^{226}Ra up to ^{210}Pb and of ^{228}Th up to ^{208}Pb achieve equilibrium with their respective parent radionuclides. The mass activity concentration of ^{40}K, ^{226}Ra and ^{232}Th determined gamma-ray spectrometry, using an ORTEC Hyper Purity Ge (HPGe) detector of 60% relative efficiency, coupled to a PC based digital analyser system employing ORTEC Gamma Vision software. Count time was in excess of 80,000 sec per sample, providing results with analytical precision of around 1% at the 95% level of confidence.

4.3. Characterization of solid waste (NORM residue)

During the survey of the site, NORM residue samples were collects from several places (7 points around treatment ponds and monazite storage). Solid samples were analyzed by using XRF technique provided by Physics and Engineering Group, Nuclear R&D Division, TINT. The XRF technique was applied for 60 seconds in TestAll Geo mode for each sample to find the elemental composition.

4.4. Characterization of liquid waste in drums stored at building No. 8

- Collecting samples of liquid waste from drums at the building No 8
- Analysis by Gamma Spectrometry (the same methods as in 3.3)

5. Results and Discussions

5.1. Survey on Dose rate and Contamination Monitoring at facilities

IOP Conf. Series: Journal of Physics: Conf. Series **860** (2017) 012044 doi:10.1088/1742-6596/860/1/012044

5.1.1. Dose rate and contamination map at the building No. 9. The dose rate of the over-all area in building 9 where the rare earth processing plant used to be operated, was found in the level of background (1 μSv/hr), and no contamination.

5.1.2. Dose rate and contamination map at the building No. 18. The maximum dose rate of the wall of building 18 was found to be very high dose rate, 62.2 μSv/hr. The average of dose rate around the wall was about 30-60 μSv/hr. It is about 30-60 times of background level (1 μSv/hr) at the RE R&D Center. Figure 3 shows that the building no.18 is full with monazite sand, and our radiation safety officer did not allow us to get inside due to the hazard of radon gas because the room was closed for a long time.

Figure 4. Dose rate of the wall of building 18

5.1.3. Dose rate and contamination map at the building No. 8. Building No. 8 was separated to 2 sections: the uranium and thorium processing section and the monazite processing section. The grid is applied to each section to estimate dose rate and contamination, including the volume of materials (in 80 Liter drum) in each square. The results of dose rate and contamination are shown in Figure 5 and Figure 6, respectively. Wide areas of contamination cause high dose rates in the building No. 8, mainly in the uranium and thorium processing section. The highest contamination value is at the A1 (100,000cpm in the map).

In addition, there are about 185 containers which contain NORM residues and chemical solvents in the U/Th processing section.

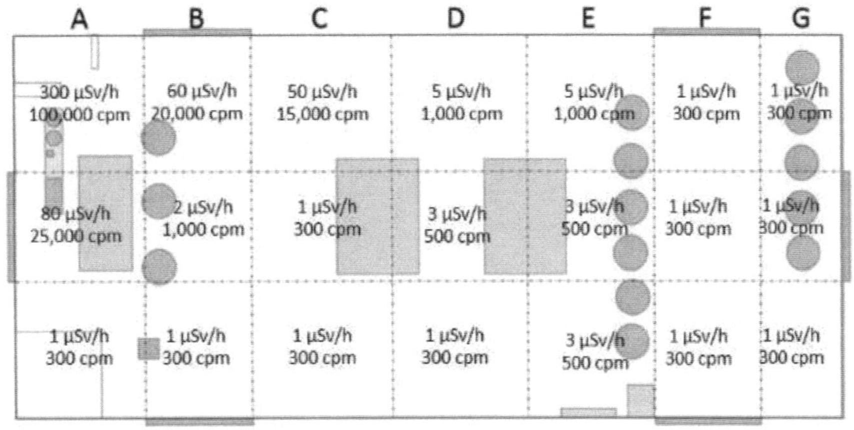

Figure 5. Dose rate and contamination map of uranium processing

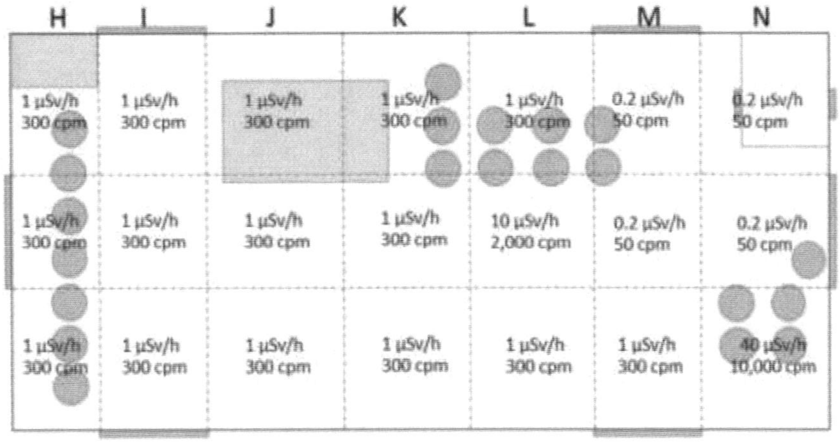

Figure 6. Dose rate and contamination map of monazite processing

5.2. Monitoring environmental sample onsite

Gross alpha- gross beta of water samples from each pond were analyzed, the results were shown in Table 3.

Table 3. Gross alpha-gross beta of water samples from ponds onsite

No	Sample	Activity Concentration (Bq/L)	
		Gross Alpha	Gross Beta
E1	Pond 7	< 0.018	0.715
E2	Channel beside the pond 7	< 0.018	0.878
E3	Pond (for sediment suspensions and filtration)	< 0.018	0.841
E4	Environmental monitoring point 1	< 0.018	0.426
E5	Pond 1	< 0.018	0.922
E6	Environmental monitoring point 2	< 0.018	0.508
Baseline Surface water before operation (1991) [5]		0.037	0.274
Detection limit, DL (Bq/L) (as July 2016)		0.018	0.012

Also the samples had been left for 4 weeks and were analyzed by using HPGe spectrometer for 24 hours to find the concentration of Ra-226, Th-232 and K-40 as shown in Table 4.

Table 4. Radionuclides and activity concentration in surface water from the ponds onsite

No	Sample	Activity Concentration (Bq/L)		
		Ra-226	Th-232	K-40
E1	Pond 7	10.41 ± 2.84	1.66 ± 0.64	65.19 ± 5.31
E2	Channel beside the pond 7	4.15 ± 0.33	1.72 ± 1.22	66.69 ± 3.12
E3	Pond filter sediment suspensions	15.49 ± 2.95	1.15 ± 0.46	66.11 ± 3.23
E4	Environmental monitoring point 1	8.84 ± 2.39	1.63 ± 0.64	66.92 ± 3.12
E5	Pond 1	7.87 ± 2.34	2.36 ± 0.67	63.91 ± 3.12
E6	Environmental monitoring point 2	8.94 ± 2.34	1.28 ± 0.67	63.92 ± 3.12
Baseline Surface water before operation (1991) [5]		0.00126		

5.3. Characterization of solid waste (NORM residue)

The NORM residue samples were collected from several points in the area of the RE R&D center, mainly around the monazite storage and the treatment ponds. The XRF technique was applied. The results are shown in Table 5. The results show that the quantities of uranium and thorium were found in high value in the sample no 5, the sludge from P5 which was the evaporation pond for treatment of waste water from uranium and thorium processing section.

Table 5. Uranium and thorium containing in NORM residue samples

No	Origin of Sample	ppm	
		Th	U
1	Sludge from Pond 7	ND	18.92
2	Sludge from the side channel of Pond 7	11.39	14.21
3	Sand in the drum at the monazite storage	ND	ND
5	Sample from other R&E laboratories	51.27	ND
6	Sludge from Pond 5	1524.12	1981.78
7	Sample from other R&E laboratories	59.92	ND
8	CeO_2 residues	ND	ND

IOP Conf. Series: Journal of Physics: Conf. Series **860** (2017) 012044 doi:10.1088/1742-6596/860/1/012044

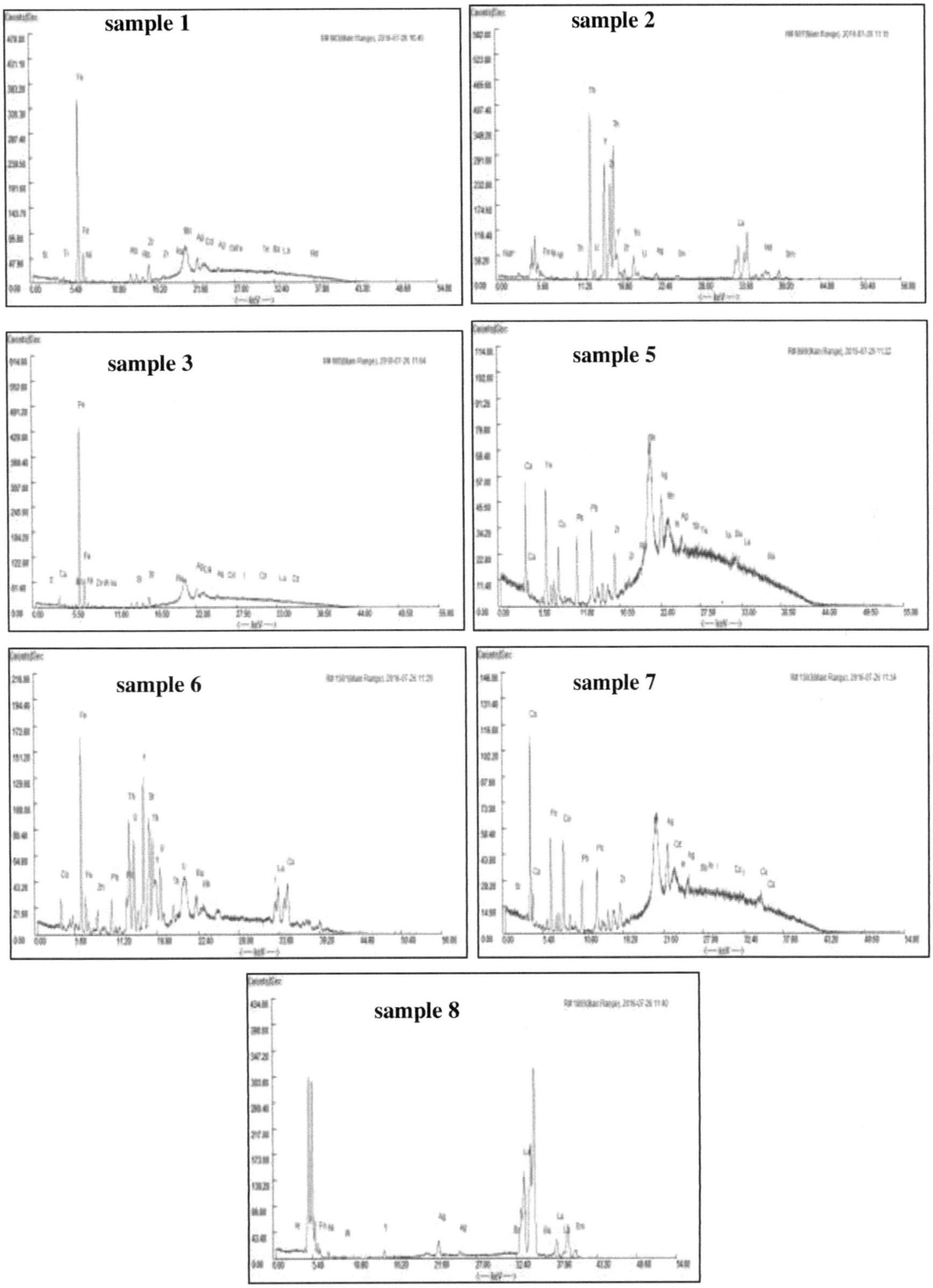

Figure 7. The XRF results from NORM residue samples

Note: The Y-axis is cps and the X-axis is energy of X-rays in keV

5.4. Characterization of liquid waste in drums stored at Building 8

Activity concentrations of liquid waste samples are as shown in Table 6. These results show that the high concentration of Ra-226, Th-232 and K-40 in liquid waste samples were found in most samples, they could be the contaminated solvents from the monazite processing. The different composition of these elements in contaminated waste samples was depend on monazite sand U-Th processing procedures.

Table 6. Activity concentration of liquid waste in drums stored at Building 8

Container no.	pH	Activity Concentration (Bq/l)		
		Ra-226	Th-232	K-40
1	2	1441.11 ± 114.72	19612.12 ± 42.5	1600.9 ± 49.5
3	5	8.87 ± 0.17	31.21 ± 1.84	66.81 ± 5.88
23	13	7.77 ± 3.35	4.15 ± 0.92	66.64 ± 3.81
24	1	1668 ± 104.51	16502.54 ± 47.82	1339.56 ± 54
25	1	683.76 ± 80.14	11373.05 ± 36.89	980.62 ± 41.19
48	1	-	32646.78 ± 51.91	-
51	1	653.91 ± 13.81	19584.13 ± 42.29	1558.32 ± 59.88
52	5	16 ± 8.53	65.87 ± 1.27	79.96 ± 5.54

6. Advices on safety infrastructures

Worker awareness and training are particularly important for supporting the introduction of radiation safety rules and for creating an understanding of the precautions embodied in such rules. Individual employee work practices may exacerbate dust generation**.** The general standard of housekeeping and spillage control also needs to be kept under regular review. Even when low activity concentration materials are handled, a reasonable standard of housekeeping may be necessary to ensure that dust and dirty from bird sham resuspension are adequately controlled. Very high standards would generally be required in the processing areas and storage areas where highly active materials, such as monazite sand and U cake/Th cake are stored.

7. Conclusion

This study is aimed to survey the storage area of NORM residue/ waste. We found contamination in some water samples from ponds and high contamination in liquid waste stored at Building8. The regulatory controls of NORM should be strengthened, e.g. management of NORM residues, residues disposal, environment monitoring, and effluent (processing water, gas) discharge to environment. There should be the clear declaration and segregation of radioactive and non-radioactive waste. The comprehensive plan should be draw up to balance of prevent contamination with workers. The survey of the whole site should be conducted. The contamination zone found should be clean up or remediated.

References

[1] International Atomic Energy Agency, *Radiation Protection and NORM Residue in the Production of Rare Earths from Thorium Containing Materials* (Vienna) p 213

[2] Kranrod C, Chanyotha S, Chankow N,Tokonali S and Ishikawa T 2013 Meansurement of radon and thoron progeny size distributions at the mineral treatment industry in Thailand. *Journal of Radioanalytical and Nuclear Chemistry*. Vol 296. Issue 2. p 625-630.

[3] US Environmental Protection Agency 2013 *Home Buyer's and Seller's Guide to Radon* (USA) p 19

[4] Sola P, Sola 2013 *Concentration of radon in workplaces at the Rare-Earth Research and Development Center* Report presented for the executives of TINT during 27 – 29 March 2013 (Bangkok)

[5] Srisuksawat, K, et al 1996 *Pre-operational Survey of a Rare-earth Research and Development Center*, Proceeding of Nuclear Science and Technology Conference 1996 (Bangkok) p 367 (http://www.iaea.org/inis/collection/NCLCollectionStore/_Public/29/044/29044568.pdf)

Acknowledgments

Authors would like to thank all radioactive waste management colleagues for their hard working during this study. Special thanks to the Physic-Engineering group from the R&D division of TINT, especially to Dumrongkiat Srilachai for his kind technical support on solid sample analysis by XRF technique. Authors would like to thank to Komon Pangsub (OAP), Chutima Kranrod, and Rawiwan Kritsananuwat (Chulalongkorn University) and Phachirarat Sola (TINT), for their kind supports for historical information and data of radon respectively.

Comparing $H_p(3)$ evaluated from the conversion coefficients from air kerma to personal dose equivalent for eye lens dosimetry calibrated on a new cylindrical PMMA phantom

J Esor[1,3], W Sudchai[1], S Monthonwattana[1], V Pungkun[2] and A Intang[2]

[1] Thailand Institute of Nuclear Technology (Public Organization), Nakhon Nayok, Thailand
[2] Office of Atoms for Peace, Bangkok, Thailand

[3] Corresponding author e-mail: jeelawat@tint.or.th

Abstract. Based on a new occupational dose limit recommended by ICRP (2011), the annual dose limit for the lens of the eye for workers should be reduced from 150 mSv/y to 20 mSv/y averaged over 5 consecutive years in which no single year exceeding 50 mSv. This new dose limit directly affects radiologists and cardiologists whose work involves high radiation exposure over 20 mSv/y. Eye lens dosimetry ($H_p(3)$) has become increasingly important and should be evaluated directly based on dosimeters that are worn closely to the eye. Normally, $H_p(3)$ dose algorithm was carried out by the combination of $H_p(0.07)$ and $H_p(10)$ values while dosimeters were calibrated on slab PMMA phantom. Recently, there were three reports from European Union that have shown the conversion coefficients from air kerma to $H_p(3)$. These conversion coefficients carried out by ORAMED, PTB and CEA Saclay projects were performed by using a new cylindrical head phantom. In this study, various delivered doses were calculated using those three conversion coefficients while nanoDot, small OSL dosimeters, were used for $H_p(3)$ measurement. These calibrations were performed with a standard X-ray generator at Secondary Standard Dosimetry Laboratory (SSDL). Delivered doses ($H_p(3)$) using those three conversion coefficients were compared with $H_p(3)$ from nanoDot measurements. The results showed that percentage differences between delivered doses evaluated from the conversion coefficient of each project and $H_p(3)$ doses evaluated from the nanoDots were found to be not exceeding -11.48 %, -8.85 % and -8.85 % for ORAMED, PTB and CEA Saclay project, respectively.

1. Introduction

The International Organization for Standardization (ISO) has issued a standard series on photon reference radiation qualities for calibrating dosimeters (ISO 4037) which included conversion coefficients of $H_p(0.07)/K_a$ and $H_p(10)/K_a$ in different types of phantom (slab, pillar, rod) but no conversion coefficients for $H_p(3)$ [1]. Personal dose equivalents are evaluated in terms of $H_p(10)$ and $H_p(0.07)$ by using optically stimulated luminescence (OSL) dosimeters (InLight) placed on a slab phantom (dimension: 30 x 30 x15 cm^3). The eye lens dose, $H_p(3)$, can be calculated by a combination of the $H_p(0.07)$ and $H_p(10)$ values and dosimeters worn at the neck. Eye lens dosimetry which can directly evaluated $H_p(3)$ by using dosimeters worn closely to the eye has become increasingly important. In 2011, International Commission on Radiological Protection (ICRP) recommended that the occupational equivalent dose limit for the lens of eye should be reduced from 150 mSv per year to 20 mSv per year, averaged over defined 5 years periods, with no single year exceeding 50 mSv. The dose limit was reduced by more than a factor of 7 due to new findings indicating that the threshold dose for the lens of the eye to be at risk from radiation is lower than previously concerned. The threshold in absorbed dose was considered to be 0.5 Gy compared with the earlier figure of 5 Gy for chronic and 0.5-2.0 Gy for acute exposures [2]. The new eye lens dose limit will impact interventional radiologists and cardiologists who often receive an annual eye dose of more than 20 mSv due to interventional procedures involving long fluoroscopic time from cine acquisitions to operation of fluoroscopic equipment in high-dose fluoroscopic modes, leading to high worker doses.

In 2009, the Optimization of RAdiation protection for MEDical staff (ORAMED) presented a project which aimed to introduce new elements in the discussion on the quantity $H_p(3)$ and to propose a more suitable theoretical cylindrical phantom to better approximate the head in which the eyes are placed. A new Monte Carlo approach helped to define the operational quantity of $H_p(3)$ [3]. Under this project, the Monte Carlo code MCNP-4C was proposed by Radiation Protection Institute Bologna team to simulate the photon interactions within an ICRU tissue equivalent cylindrical phantom, 20 cm in diameter and 20 cm in height, which is representative of the mass and shape for a human head [4]. The monoenergetic photons from 10 keV to 10 MeV were simulated for an aligned and expanded field with an angle of 0^0 and 40^0. In the same ORAMED project, the MonteCarlo code PENELOPE was proposed by the French Atomic Energy Commission (CEA Saclay Nuclear Research Centre) to simulate a set of energy and angular dependent also using ICRU cylindrical phantom. The $H_p(3)$ values have been determined in terms of dose equivalent, according to the definition of this quantity, and also with the kerma approximation as formerly reported in ICRU reports [5]. Conversion coefficients ($H_p(3)/K_{air}$) for monoenergetic photons from both projects were also presented. In 2012, Behrens (Physikalisch-Technische Bundesanstalt (PTB)) presented air kerma to $H_p(3)$ conversion coefficients ($(h_{pK}(3;R,\alpha)_{cyl}$) for a new cylinder phantom for X and gamma rays radiation qualities defined in ISO 4037. These conversion coefficients were valid for the total air kerma which calculated using the mass energy transfer coefficient $(\mu_{tr}/\rho)_{E,air}$ [6]. The purpose of this presentation is to compare the delivered doses $H_p(3)$ using three different of conversion coefficients to the $H_p(3)$ from nanoDot dosimeter measurements.

2. Materials and Methods

2.1. Materials

The cylindrical phantom used in this study composes of 40 polymethyl methacrylate (PMMA) slice rings (20 cm in diameter and 0.5 cm in slice thickness) which was designed at Meikai University. There are 10 holes at the depth of 3 mm from the surface of the slice rings. Small dosimeters can be inserted in those holes at position of left and right eyes. The nanoDot dosimeters (Landauer Inc., USA), designed for measuring a small, single point radiation exposure normally worn on a wrist or a finger, were selected in this study. Each dosimeter, 10x10 mm^2 in size and 2 mm in thickness, composes of aluminum oxide (Al_2O_3 :C) crystal, whose thickness is 0.3 mm and diameter is 7 mm, sealed in a thin polyester sheet. Measurements were read out by a microStar mobile reader (Landauer Inc., USA). In the luminescence process which is the same as optically stimulated luminescence (OSL) or InLight dosimeters, the irradiated dosimeters were stimulated by a quantum of visible green light from the light emitting diode (LED). The luminescence time, at the order of 10 – several hundred milliseconds, depends on light intensity. During the stimulation, only some fractions of the populated electrons were depopulated. The amount of luminescence was proportional to the absorbed radiation and remained significantly unchanged after the reading. This characteristic enables nanoDots to be read out many times [7].

2.2. Methods

Three nanoDot dosimeters were inserted into the cylindrical phantom holes representing left and right eyes. These dosimeters were irradiated using X-ray beam of qualities N-40, N-60, N80, N-100, N120 and N-150 at Secondary Standard Dosimetry Laboratory (SSDL), Office of Atoms for Peace, Thailand. The delivered air kerma values were 100, 1,000 and 2,000 Gy at 0 degree angle of incidence from the X-ray generator. The air kerma values were traceable to Physikalisch-Technische Bundesanstalt (PTB), Germany. Delivered ($H_p(3)$) doses using three conversion coefficients from

IOP Conf. Series: Journal of Physics: Conf. Series **860** (2017) 012045 doi:10.1088/1742-6596/860/1/012045

ORAMED, PTB and CEA Saclay were calculated. The irradiated nanoDot dosimeters were read out by a microStar reader at Thailand Institute of Nuclear Technology (TINT), Thailand (Figure 1). The microStar reader used at TINT was calibrated with an X-ray generator at 80 kVp which was traceable to National Institute of Standard and Technology (NIST), USA. After reading, $H_p(3)$ were calculated from the average counts readings multiplied by the calibration factor. The delivered doses ($H_p(3)$) based on ORAMED, PTB and CEA Saclay conversion coefficients were compared to the nanoDot measurements.

2.3. Conversion coefficients

The conversion coefficients used in this stydy are given in Table 1.

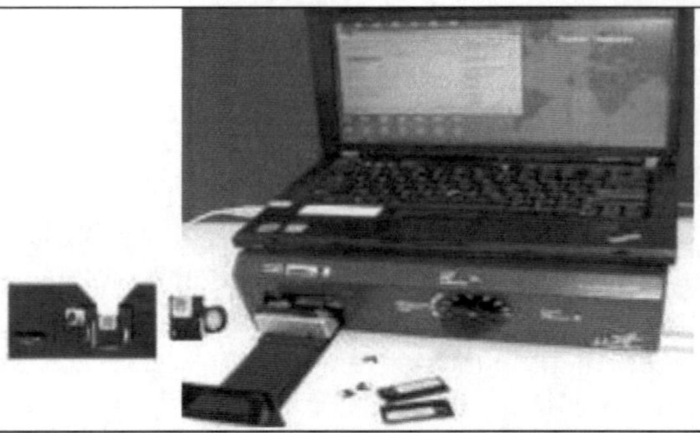

Figure 1. "NanoDot" dosimeters and a mobile reader "microStar"

Figure 2. The nanoDot dosimeters were placed in the cylindrical phantom at the depth of 3 mm.

Table 1. Conversion coefficients $H_p(3)cyl/K_a$ in Sv Gy^{-1} for different radiation qualities based on ORAMED, PTB and CEA Saclay projects.

Radiation quality	$H_p(3)cyl/K_a$		
	ORAMED Sv Gy^{-1}	PTB Sv Gy^{-1}	CEA Saclay Sv Gy^{-1}
N-40	1.207	1.280	1.289
N-60	1.469	1.540	1.553
N-80	1.628	1.660	1.670
N-100	1.606	1.630	1.638
N-120	1.567	1.580	1.580
N-150	1.532	1.520	1.534

IOP Conf. Series: Journal of Physics: Conf. Series **860** (2017) 012045 doi:10.1088/1742-6596/860/1/012045

Table 2. Delivered $H_p(3)$ doses evaluated from the conversion coefficients of ORAMED, PTB and CEA Saclay compared with $H_p(3)$ evaluated from nanoDot measurements.

	ORAMED			PTB			CEA Saclay		
Radiation quality	Delivered μSv	nanoDot μSv	% difference	Delivered μSv	nanoDot μSv	% difference	Delivered μSv	nanoDot μSv	% difference
N-40	120.66	126.12	4.53	128.00	126.12	-1.47	128.94	126.12	-2.18
	1207.98	1272.99	5.38	1281.46	1272.99	-0.66	1290.87	1272.99	-1.39
	2415.96	2542.99	5.26	2562.92	2542.99	-0.78	2581.74	2542.99	-1.50
N-60	146.91	141.57	-3.63	154.00	141.57	-8.07	155.26	141.57	-8.82
	1467.88	1565.51	6.65	1538.73	1565.51	1.74	1551.32	1565.51	0.91
	2935.77	3155.46	7.48	3077.45	3155.46	2.53	3102.63	3155.46	1.70
N-80	162.80	173.90	6.82	166.00	173.90	4.76	166.95	173.90	4.17
	1628.08	1660.74	2.01	1660.08	1660.74	0.04	1669.58	1660.74	-0.53
	3256.16	3272.38	0.50	3320.17	3272.38	-1.44	3339.17	3272.38	-2.00
N-100	161.30	152.33	-5.56	163.75	152.33	-6.97	164.50	152.33	-7.40
	1604.89	1673.13	4.25	1629.28	1673.13	2.69	1636.77	1673.13	2.22
	3209.77	3255.55	1.43	3258.55	3255.55	-0.09	3273.55	3255.55	-0.55
N-120	156.70	138.69	-11.49	158.00	144.01	-8.85	158.00	144.01	-8.85
	1569.07	1468.87	-6.39	1582.09	1468.87	-7.16	1582.09	1468.87	-7.16
	3134.73	3004.37	-4.16	3160.74	3004.37	-4.95	3160.74	3004.37	-4.95
N-150	153.20	141.82	-7.43	152.00	141.82	-6.70	153.40	141.82	-7.55
	1529.14	1495.11	-2.23	1517.16	1495.11	-1.45	1531.11	1495.11	-2.35
	3063.24	2995.71	-2.20	3039.25	2995.71	-1.43	3067.19	2995.71	-2.33

3. Results

Table 2 gives an overview in order to compare the conversion coefficients of ORAMED, PTB and CEA Saclay. The delivered doses which were corrected using three conversion coefficients were compared with evaluated $H_p(3)$ doses from nanoDot dosimeters at the depth of 3mm.

The comparison results showed that the percentage differences between $H_p(3)$ delivered dose calculated using the conversion coefficients of each project and $H_p(3)$ evaluated from the nanoDot did not exceed -11.48 %, -8.85 % and -8.85 % deviation from ORAMED, PTB and CEA Saclay project, respectively. The expanded uncertainty of the reader measurements was 13.10 % (k = 2).

4. Discussion and Conclusion

From this study, $H_p(3)$ evaluation by nanoDots of TINT's Personal Radiation Monitoring Services laboratory appeared slightly different from $H_p(3)$ calculated by using the conversion coefficients from the three pilot projects. NanoDot dosimeters are small which can be attached on lead glasses and can be calibrated directly at the depth of 3 mm in the cylindrical phantom. This new method should be better than the old method which uses only one personal dosimeter attached at the chest position or uses an additional dosimeter attached on the lead collar. The old method has to use the correction factor, which cannot be calibrated directly due to the size of the dosimeter. For the next study, $H_p(3)$ at an angle of 40^0 on the cylindrical axis will be evaluated. Measurements at this angle is better for interventional radiology application because it is in the direction of scattered radiation from the patient.

IOP Conf. Series: Journal of Physics: Conf. Series **860** (2017) 012045 doi:10.1088/1742-6596/860/1/012045

Acknowledgments
The cylindrical phantom provided by Dr. I. Kobayashi and Mr. A. Matoba from Nagase Landauer, Japan, was highly appreciated.

References

[1] ISO 4037-3, 1999 *Calibration of area and personal dosimeters and the measurement of their response as a function of energy and angle of incidence*
[2] International Commission on Radiological Protection (ICRP), 2011. *Statement on Tissue Reactions.* (International Commission on Radiological Protection (ICRP))
[3] Optimization of Radiation Protection of Medical Staff (ORAMED) *Project within the 7th framework programme of the European Union.*
[4] Marriotti F and Gualdrini G 2009 *ORAMED project Eye-lens Dosimetry A new Monte Carlo approach to define the operational quantity $H_p(3)$* (RT/2009/BAS/ENEA)
[5] Daures J, Gouriou J and Bordy J 2009 *Conversion coefficients from air kerma to personal dose equivalent Hp(3) for eye-lens dosimetry* (CEA-R-6235)
[6] Behrens R 2012 *Air kerma to Hp(3) conversion coefficients for a new cylinder phantom for photon reference radiation qualities* Radiat. Prot. Dosim. Vol 151 No3 pp 450-455
[7] Sudchai W, Thierrattanakul S, Jianklang S and Termsuk C 2015 Nanodot Dosimeter for Extremity Monitoring Service in Thailand (Proceeding of AOCRP-4)